The rich and often bizarre variety of form exhibited by orchids has long provided a fascination for amateur and professional botanists alike. Once seen as a hobby exclusively for the rich, the cultivation of orchids is now widespread and the need for an accurate and simple guide to the identification of species in cultivation has become apparent. This book aims to fulfil that need by providing botanically correct, yet easily accessible information about this unique and exceptionally diverse group of plants.

The orchid book

A guide to the identification of cultivated orchid species

The orchid book

A guide to the identification of cultivated orchid species

Edited by

J. CULLEN

Director, Stanley Smith Horticultural Trust, Cambridge

Published by the Press Syndicate of the University of Cambridge
The Pitt Building, Trumpington Street, Cambridge CB2 1RP
40 West 20th Street, New York, NY 10011–4211, USA
10 Stamford Road, Oakleigh, Victoria 3166, Australia

© Cambridge University Press 1992

First published 1992

Printed in Great Britain at the University Press, Cambridge

A catalogue record of this book is available fron the British Library

Library of Congress cataloguing in publication data

The Orchid book : a guide to the identification of cultivated orchid species / edited by J. Cullen.
p. cm.
Includes bibliographical references and index.
ISBN 0-521-41856-9 (hardback)
1. Orchids – Identification. 2. Orchids – Europe – Identification. 3. Orchid culture. 4. Orchid culture – Europe. I. Cullen, J. (James)
SB409.065 1992
635.9′3415 – dc20 91-39794 CIP

ISBN 0 521 41856 9 hardback

UP

CONTENTS

FOREWORD

Now that *The European Garden Flora* is half-completed, and the fourth of the six planned volumes is in the press, it is most appropriate that the first of the 'specialist group' series of works should be published. This volume, devoted entirely to orchids in cultivation in Europe, is based on the account of the Orchidaceae published in 1984 in volume II of the main *Flora*, but the text has been considerably updated, new taxa have been added, and in particular the illustrations have been greatly increased. This new book is timed to be available for the World Orchid Congress to be held in Glasgow in 1993, and will surely be welcomed by the many orchid growers, professional and amateur, throughout the world. It is often said that 'orchids are the rich man's hobby' and, whilst that may indeed be true, is is also the case that an increasing number of amateur gardeners with quite limited means find in the rich and often bizarre viariety of this quite exceptionally diverse and world-wide family of plants an inestimable attraction.

As former Chairman of the Editorial Committee of the *Flora* I am delighted that my successor, James Cullen, has found time from his many interests to devote to the preparation of this special volume, which we have good reason to hope will be the first of several such family or group treatments based upon *The European Garden Flora*.

S.M. Walters
Founding Chairman, European Garden Flora
Cambridge
26 August 1991

ACKNOWLEDGEMENTS

The text of this present book is based almost entirely on that published in *The European Garden Flora* volume II, pp. 137–290 (1984), and it is a great pleasure to acknowledge the work of my co-authors (all, at the time, at the Royal Botanic Garden, Edinburgh) in the production of that account:

Dr J.C.M. Alexander: *Arpophyllum*, *Arundina*, *Bletilla*, *Bletia*, *Bothriochilus*, *Brassavola*, *Broughtonia*, *Bulbophyllum*, *Cattleya*, *Chysis*, *Cirrhopetalum*, *Coelogyne*, *Cryptochilus*, *Cryptophoranthus*, *Cypripedium*, *Dactylorhiza*, *Dendrochilum*, *Dracula*, *Encyclia*, *Eria*, *Hexisea*, *Isochilus*, *Laelia*, *Laeliopsis*, *Leptotes*, *Masdevallia*, *Meiracyllium*, *Nageliella*, *Octomeria*, *Paphiopedilum*, *Phaius*, *Pholidota*, *Phragmipedium*, *Physosiphon*, *Pleione*, *Pleurothallis*, *Porroglossum*, *Restrepia*, *Schomburgkia*, *Sobralia*, *Sophronitella*, *Sophronitis*, *Spathoglottis*, *Thunia*, *Trisetella*, *Vanilla*.

Ms J. Lamond: *Liparis*, *Ophrys*, *Pterostylis*, *Serapias*, *Spiranthes*.

Ms V. Matthews: *Acacallis*, *Aerides*, *Anguloa*, *Aplectrum*, *Bifrenaria*, *Bollea*, *Calypso*, *Chiloschista*, *Chondrorrhyncha*, *Cochleanthes*, *Dichaea*, *Eriopsis*, *Galeottia*, *Gastrochilus*, *Huntleya*, *Lycaste*, *Maxillaria*, *Mormolyca*, *Ornithocephalus*, *Pabstia*, *Paraphalaenopsis*, *Pescatoria*, *Phalaenopsis*, *Promenaea*, *Pteroceras*, *Rhynchostylis*, *Sarcochilus*, *Schoenorchis*, *Scuticaria*, *Sedirea*, *Trigonidium*, *Warrea*, *Xylobium*, *Zygopetalum*.

Mr P.J.B. Woods: *Aceras*, *Anacamptis*, *Anoectochilus*, *Barlia*, *Bulbophyllum*, *Cephalanthera*, *Cirrhopetalum*, *Coeloglossum*, *Corybas*, *Dendrobium*, *Disa*, *Epipactis*, *Goodyera*, *Gymnadenia*, *Habenaria*, *Himantoglossum*, *Liparis*, *Listera*, *Macodes*, *Malaxis*, *Nigritella*, *Ophrys*, *Orchis*, *Platanthera*, *Pterostylis*, *Satyrium*, *Serapias*, *Spiranthes*, *Stenoglottis*.

The illustrations in this volume are partly taken from the account in *The European Garden Flora*, and it is a pleasure to acknowledge the work of the artists used: Ms E.J.F. Campbell, Ms S.J. Mackay and Ms V.A. Matthews. The numerous additional illustrations for this volume have been prepared by Ms Rodella Purves, and I am extremely grateful to her for her great care and accuracy. I am also grateful to the Royal Botanic Garden, Edinburgh for providing the slides which have been reproduced in the colour plate section.

The staff of Cambridge University Press offered valuable advice and much helpful comment during the preparation of the text and I am especially grateful to Dr A. Crowden, Mr C. McKeown, Dr Maria Murphy and Ms S. Thelwell.

Finally, I owe a great debt to Mr P.J.B. Woods of the Royal Botanic Garden, Edinburgh, who has provided me with corrections, much new information and many references to literature which I otherwise might well have missed. The errors and omissions which undoubtedly have occurred are, however, all mine.

James Cullen
Cambridge, August 1991

INTRODUCTION

The purpose of this book is to provide the means for the accurate identification of the genera and species of orchids in general cultivation. The main part of the text, which is based on the account of the family published in *The European Garden Flora*, volume II, pp. 137–290, Cambridge University Press, 1984, consists of keys to, and descriptions of, the various genera and species. This Introduction provides the basic necessary explanation of the descriptive text so that this can be easily and properly used; it does not provide a complete account of the structure and biology of the family as a whole. For a more complete account of these aspects, reference should be made to R.L. Dressler's *The orchids: natural history and classification* (1981), and other books and papers cited in the Bibliography (pp. xxv–xxvi)

The family Orchidaceae

The orchids comprise the botanical family Orchidaceae. This family, together with grasses, palms, lilies and others, is grouped into the smaller of the two major units into which the flowering plants (Angiospermae, informally Angiosperms) are divided. This group, the Monocotyledoneae (informally Monocotyledons) is characterised by a number of features, notably the presence of only a single seedling-leaf (cotyledon) in the embryo and young plant, the parallel venation of the leaves, and the production of new roots from the stems or other underground organs (tubers, corms, bulbs, etc.); these roots, known as adventitious roots, rapidly come to make up the whole of the plant's root-system, the original root of the seedling rapidly dying off. The Orchids show all these features, and are, in many respects, representative Monocotyledons.

The family is probably the largest of all the families of the Angiosperms. Dressler estimates that there are some 700 genera and 22,000–25,000 species; representatives of the family are found in all parts of the world that can sustain plant growth, though they are particularly abundant in the tropics and subtropics, where most of them grow epiphytically, that is, on the branches of other plants.

Among the Monocotyledons, the Orchidaceae forms a clearly distinguished group, which has been recognised as a unit for hundreds of years. In spite of the immense variability of the family, as implied by the large number of genera and species, the decision as to whether or not a particular plant is an orchid is usually easily made.

The features which allow for this easy recognition are the following (though each of them individually may have exceptions, the combination of characters is generally decisive):

(a) the flowers are usually inverted by means of a twist of 180° in the flower-stalk and ovary;

(b) the perianth of the flower is composed of 6 segments in 2 whorls of 3; of these segments, that usually appearing lowermost (because of the inversion of the flower, this is, in fact, morphologically the uppermost) is generally different from the other 5 (it may be larger, smaller, differently coloured or patterned, or may bear warts or calluses of various kinds), and is known as the lip or labellum;

(c) there is generally only 1 stamen (rarely 2 stamens plus a staminode); where the stamen is single it is on the radius of the flower opposite to that on which the lip stands;

(d) the pollen is compacted into 2, 4, 6, 8 or more structures known as pollinia; these are released from the anthers during pollination and are transferred whole to the stigmas; various other structures may be associated with the pollinia (these are generally formed from stigmatic areas of the column);

(e) the ovary is inferior, and is almost always single-celled, with 3 parietal placentas bearing very numerous ovules;

(f) the stamens, style and stigma are united into a combined structure known as the column; this is very varied in form, but is generally easily recognisable;

(g) the seeds are very numerous and very small.

In addition to these general characteristics, many of the cultivated species, which are epiphytic in the wild, have characteristically swollen stems, known as pseudobulbs, and frequently produce long, aerial roots, whose function is apparently to absorb moisture from the air.

The immense variability presented by the flowers of orchids is due to the fact that the different species have become closely adapted to their pollinators (generally insects of various kinds). The various modifications of the flower have come about as a response to closer and closer integration of the flower and its specific pollinator, and, in the wild, the species are prevented from hybridisation by the specificity of their pollinators. As a result of this close adaptation, the species have no strongly developed internal mechanisms for the prevention of fertilisation by 'foreign' pollen, such as are found in most other families of Angiosperms. Hybridisation between the species is therefore generally possible if the floral modifications can be by-passed; this sometimes happens in the wild, when alteration or disturbance of the habitat disrupts the normal patterns (for example, the hybrid swarms between different species of *Dactylorhiza* which have developed on old, dried-out salt pans in Cheshire – see Newton, A., *Flora of Cheshire*, Cheshire Community Council, Chester, 1971, pp. 33–35), and can easily be done by human intervention. Many thousands of orchid hybrids, some between species of the same genus, others between species from different genera (multiple hybrids, involving genetic material from species of 8 different genera have been produced), have been developed since the first half of the 19th century, and very many of them are grown by gardeners and enthusiasts. The identification of these is a very difficult matter, and the present book does not attempt to provide the means for this. The identification of hybrids can be based only on a secure knowledge of the species from which they were produced. More information on hybrids can be found in *Sander's list of orchid hybrids* – see the Bibliography, p. xxv.

Orchids in cultivation

The origins of orchid cultivation are lost in the mists of time, but intensive cultivation in Europe began towards the end of the 18th century. The species then grown were the native, hardy orchids, which were generally transferred from the wild to gardens by interested enthusiasts. With the opening up of the tropics by botanical exploration in the late 18th and early 19th centuries, the existence of the enormous range of floral form became better known, and wealthy enthusiasts, who could provide the necessary glasshouses and staff, became obsessed with the forming of living collections of the rare, the beautiful and the bizarre.

The rise of this class of enthusiasts was rapidly followed by the development of specialist nurseries aiming to fulfil the rising demand. These nurseries began the practice of sending out specialist orchid collectors to various areas of the tropics, to collect plants and send them back to Europe, where they could be either sold on, or used as nursery stock. Towards the end of the 19th century, the mania for new orchids reached its peak, when some growers would pay prices of 1000 guineas or more for a single pseudobulb of something new and interesting. The high economic returns from the trade meant that the collectors went through certain areas, completely denuding them of their 'interesting' species. Vast numbers of plants (in the hundreds of thousands) were collected and shipped back to Europe, where, if they arrived at all, they were sold on. The atmosphere of the times is well caught in Arthur Swinson's *Frederick Sander, the orchid king: the record of a passion*, Hodder & Stoughton, London (1970).

Fortunately, during the 20th century, the mania died away, and now, in perhaps more enlightened times, trade in wild orchids is strictly banned under the terms of CITES, the convention which regulates international trade in organisms threatened with extinction in the wild.

The orchid collectors employed by the nurseries were involved in an intensely competitive business, and hence were very secretive about their activities. Not only did they bring several species close to extinction, but they (and their employers) deliberately set out to mislead others about the origins of their various pet species. This had an unfortunate effect on the botanical taxonomists working to try to sort out the species and their distinctions and, in many of the original descriptions of the species produced at the time, the geographical information given must be viewed with considerable scepticism.

Hybridisation

As the interest in the orchids developed, nurseries began the process of making hybrids between the species which they already had, and the new ones that were regularly coming in. The first artificial hybrid was made in Veitch's nursery in 1852; this was between two species of *Calanthe*. Other successful hybridisations soon followed, involving genera such as *Cattleya* and *Paphiopedilum*, and very rapidly the production of new hybrids became an important activity in the nursery trade. This activity still continues today. New hybrids must be registered with the International Registration Authority for Orchid Hybrids, which is the Royal Horticultural Society of London. The Society maintains a database of these hybrids, and lists of them are produced regularly.

The naming of orchid hybrids is a very complex matter, and is regulated by a set of rules and guidelines (Cribb, P.J., Greatwood, J. & Hunt, P.F., *Handbook on orchid nomenclature and registration, edn* 3, 1985).

Selection of species for this volume

It is difficult to estimate the number of orchid species that have, at one time or another, been in cultivation, but it is clear that several thousand have been introduced. Of these, many have not survived to become standard members of collections, or have been lost because of difficulties with their cultivation or propagation, or by accident. The selection of just over 900 species belonging to 166 genera included in this volume has been made using information from several sources: (a) nursery catalogues listing species available for sale; (b) a survey of current orchid literature; and, (c) consultation with the curators of notable orchid collections. These 900 species represent a core of widely cultivated species likely to be found in general collections in Europe (and indeed, in the USA, South Africa, Australia and Japan). Of course, individual collections, especially those of the major botanic gardens, are likely to contain species and genera that are not included here, but these are special cases which cannot be covered by a work of this scale. The identification of such species is likely to be difficult, and may require the services of an orchid specialist.

Cultivation techniques

The cultivation of orchids is a subject which has generated much discussion, controversy and publication. It is not possible to go into detail here and the reader is referred to the relevant section of the Bibliography (pp. xxv–xxvi) for further information. Most of the species included in this book are of tropical origin, and require greenhouse protection throughout the whole of Europe; the hardy species, of which there is a small number (mainly in genera 1–28) can be grown out of doors.

Two different composts are used for the tropical species. The first is for those which are terrestrial in their native habitats (terrestrial compost), the second for those which are epiphytic (epiphytic compost).

Terrestrial compost is made up of 2 parts (by bulk) of fibrous loam, 2 parts of finely crushed bark or medium-grade peat and 1 part of grit or sand; 50 grams of a slow-release general fertiliser should be added to each cubic metre of the mixture. Species with particularly vigorous root systems will benefit from an increase in the amount of bark or peat in the mixture.

Epiphytic compost consists of 3 parts of pine bark mixed with 1 part of rough peat or sphagnum moss; 50–70 grams of slow-release general fertiliser is added to each cubic metre of this mixture. While most of the species do well in pots of suitable size, some epiphytic species grow best when attached to a piece of cork-oak bark, using the smallest possible amount of compost.

Control of pests and diseases in orchids is a rather complex and specialised matter, and the relevant literature should be consulted for precise information. Good hygiene and cultivation are, of course, essential.

Under each generic description in the main text there is a short paragraph on methods of cultivation. This is necessarily brief, but may contain references to further useful information.

Structure and terminology used in identification

The orchids are very diverse in their structure, as is to be expected of a large and highly varied family. Many of their structures are highly specialised, and the terminology that has been developed to deal with this is also extensive. This section deals only with those features of importance in the identification of the genera and species; further information should be sought in books such as that by Dressler (1981).

Growth habit

Most orchids have rhizomes which creep near or below the substrate surface; the rhizomes may creep extensively, bearing spaced stems or pseudobulbs, or may form a mat, in which case the stems or pseudobulbs are clumped. The rhizomes bear adventitious roots and stems or pseudobulbs; occasionally they also bear individual leaves. In some of the ground orchids 1, or

a. A cane-like stem. b. A simple, single-leaved pseudobulb. c. A compound pseudobulb. d. A club-shaped compound pseudobulb. e. A compound pseudobulb swollen only at the base.

more commonly 2, 'tubers' are present at the bases of the stems. The word 'tuber' is placed in inverted commas here, as these structures are morphologically composed of both root and stem tissue and are thus somewhat different from normal plant tubers (e.g. the potato), which are made up entirely of stem tissue.

Most of the species included here are evergreen, with stems or pseudobulbs and leaves persisting throughout the year; many of the hardy orchids, however, are seasonal, with all or most of the aerial growth dying down in winter, the tubers persisting below the soil surface.

The hardy orchids and some tropical species grow in the soil, as normal plants. Most of the tropical species, however, are epiphytic, growing on the branches of other plants (mainly trees), extracting food material from the detritus and moss that builds up on tree-branches. Such epiphytic plants grow under conditions of water-stress (even in areas where the general precipitation and humidity are high), and show numerous adaptations to such conditions. Two of these adaptations are important: the frequent occurrence of swollen, water-storing stems (pseudobulbs), and the presence of aerial roots arising from the stems. These roots do not usually penetrate the soil or detritus, but absorb water from the surrounding damp air.

Pseudobulbs (see figure opposite)

A *pseudobulb* is a swollen, fleshy stem. Whether or not a particular plant has pseudobulbs or more normal stems is a very important characteristic in its generic identification. Generally there is no difficulty in deciding whether or not a particular plant has pseudobulbs, but in a few species and genera there is no sharp dividing line between a 'normal' stem and a pseudobulb; the keys to the genera have been constructed, as far as possible, to allow for these to be identified whether or not they have been considered as having pseudobulbs.

Pseudobulbs, if present, are borne on the rhizomes, and may consist of several internodes, when they will bear leaves or leaf-scars along their lengths; such pseudobulbs are described here as '*compound*'. Alternatively, they may consist of a single internode, when they bear 1–several leaves or leaf-scars at the apex only; such pseudobulbs are described here as '*simple*'.

Pseudobulbs may be surrounded by leaf-sheaths at their bases; such sheaths may be persistent or may wither early, and not be present when the pseudobulb is mature. In a few genera, pseudobulbs are borne one on top of the other, forming chains.

Leaves (see figure on p. xix)

The shape of the leaves, often an important characteristic in plant identification, is of little significance in orchid identification, and is scarcely mentioned in the descriptive text which follows.

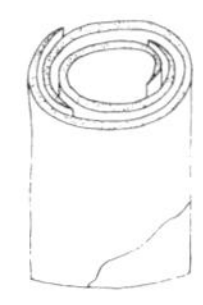

Cross-section of the bases of 2 rolled leaves.

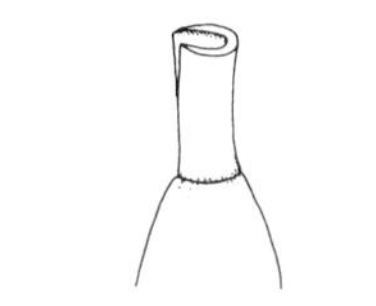

A pseudobulb with the base of a rolled leaf.

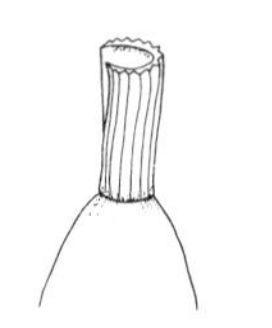

A pseudobulb with the base of a rolled and pleated leaf.

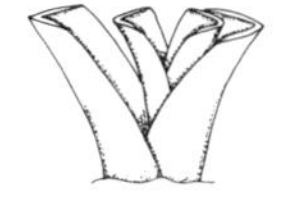

Cross-section of the bases of folded leaves.

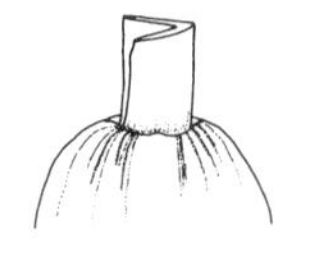

A pseudobulb with the base of a folded leaf.

The manner in which the young leaves are packed when young is, however, a very important character in generic identification. The young leaves may be rolled, so that one margin overlaps the other and the back of the young leaf is rounded (technically known as *supervolute* or, somewhat incorrectly as *convolute*, or they may be folded once longitudinally (technically, *conduplicate*. This character is generally easy to see in plants with young leaves, and it is also generally easy to decide even when the leaves are mature, as the rolling or folding persists at the base. In some genera in which the plants bear few leaves it may be difficult to decide on this character when the leaves are mature; the following features should help in making a decision. Folded leaves generally retain a longitudinal line or small fold down the middle over most of their length, and usually have a definite keel on the outside towards the base; very hard, leathery leaves are almost all folded. Rolled leaves are sometimes pleated or have prominent veins, do not usually have a single line or fold down the middle, and are rounded on their backs towards the base.

In a few genera and species the leaves are terete or cylindric with a groove or grooves; such leaves cannot be said to be either rolled or folded when young, but they are keyed here with the genera with folded leaves.

Inflorescences

The inflorescence in most orchids is a raceme. In some genera this is reduced to a single terminal flower, and in others is more highly branched, producing a racemose panicle.

The inflorescence may be terminal on the stem or pseudobulb, or lateral; in the latter case, the inflorescence can arise in a leaf-axil or a leaf-scar axil on the pseudobulb or stem, or from the rhizome at the base of the pseudobulb. In a few genera in which the stems bear only a single leaf (e.g. *Pleurothallis*), the racemes are considered here to be terminal even though they might be intepreted as axillary to the leaf.

Flowers

The flowers of orchids, however highly modified, are all built on a common plan, with 3 sepals, 3 petals, one of which is highly modified as the lip, 1 or 2 stamens borne on a complex structure, the column (see below) which consists of the stamens united to the styles and stigmas of the inferior, 3-carpellary ovary. Though this plan is generally easy to see, the variations on it are very numerous, and include such features as fusion of various parts (especially of the sepals), the occurrence of spurs (sometimes formed by the uppermost sepal, on other occasions by the base of the lip), and great variation in the relative sizes of the various parts.

In all the keys and descriptions the flowers are considered to

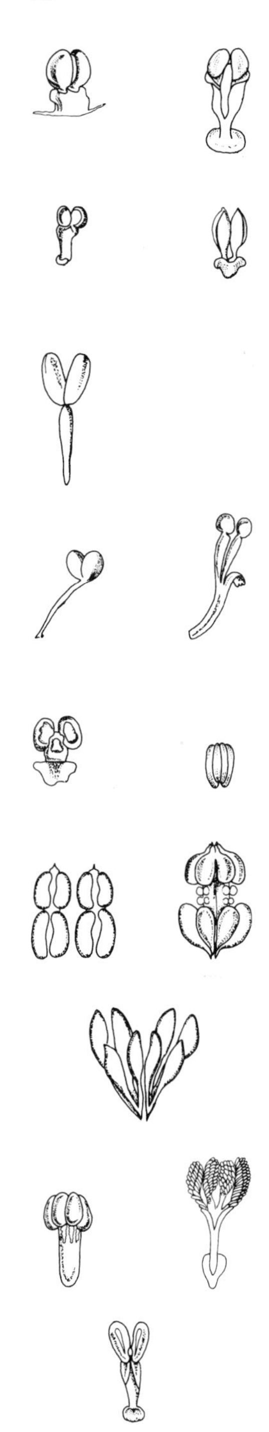
Various kinds of pollinia.

be resupinate (i.e. with their stalks and ovaries twisted through an angle of 180° so that the lip is lowermost with the column above it), unless the contrary is stated. In species with hanging or arching inflorescences the degree of resupination of the flower may vary, those flowers towards the hanging apex being non-resupinate (they are effectively turned upside-down by the hanging over of the inflorescence), while those at the more erect base may be properly resupinate. However, in some genera with hanging inflorescences (e.g. *Paphinia*), the flowers are all strictly resupinate.

Pollinia (see figure on this page)

Much of the classification of the family is based on the structure of the pollinia and their associated organs. In this book, characters from these organs are used as little as possible, but they cannot be completely ignored, particularly their number. They can be extracted from the flower by sliding a needle or match slowly upwards along the inner side of the column. They can then be examined with a hand-lens (a magnification of × 10 or × 15 is usually adequate). The number of pollinia can usually be easily seen, though care must be taken as the pollinia are sometimes not all extracted at once.

A few species have 2 deeply bilobed pollinia, which can look like 4; again, other species may have extra, infertile pollinia which are shrunken and colourless. In most of the tropical species in cultivation the pollinia are waxy to hard and of a well-defined shape. In some hardy species they are more powdery, or are made up of several to many individual packets which separate easily from each other.

The pollinia generally have a small, sticky pad (the *viscidium*, derived from stigmatic material) at the base. There is sometimes a stalk (*stipe*) between the viscidium and the individual pollinia, or each pollinium may have its own individual stipe.

Again, in some genera, the median lobe of the 3-lobed stigma is modified to play a part in the pollination process. The structure so formed is known as the *rostellum*, and can be quite conspicuous in some groups.

The Column

The column is a very variable structure in shape and size; details are given, where important, in the generic and specific descriptions. In many genera the base of the column is prolonged at an angle to the rest, forming a *column-foot*. In all genera with a foot, the lip is attached at or near the end of it. In many of these genera, the lateral sepals are also borne on the sides of the foot, which results in a humped, angular, chin-like projection being visible on the outside of the back of the flower. This projection is known as a *mentum* and is of some importance in identification.

How to use the text

The descriptive text which follows is organised as it would be in a Flora (an account of all the genera and species to be found in a certain area). Each genus is described, using as little technical language as possible. Following the description, there are observations about the genus – its size and distribution, any interesting features it may have, and some details about its cultivation. The section ends with the citation of any particularly useful references on the genus, including sound taxonomic revisions or monographs.

Then, if the genus contains more than one cultivated species, there will follow either 1 or 2 keys to the species (see below for details of how these are to be used). Following this, each species is described, and references are provided to useful published illustrations of it; there may be up to 8 of these, depending on how frequently the species in question has been illustrated. An attempt has been made to cite some up-to-date, widely available illustrations as well as more classic illustrations (such as those published in *The Botanical Magazine*, *Edwards's Botanical Register*, etc.).

The genera and species are arranged in systematic order, rather than alphabetically by name. The advantages of this arrangement are many, but the main one is that similar genera or species appear in the text close together, so that it is possible to make easy comparisons between descriptions; this makes accurate identification more likely. An alphabetical listing of the generic and specific names is provided in the Index (pp. 513–29).

In attempting to identify an unknown orchid, it is necessary to look at the plant very carefully, taking note of its important features; these may be large-scale, such as the size of the flower and the disposition of its sepals and petals, or may be very small-scale, such as the presence or absence of hairs on the lip, the number and shape of the pollinia, etc. For these latter features, the use of a hand-lens with a magnification of between ×10 and ×20 is very helpful. Other available information about the specimen should also be borne in mind, particularly its country of origin (if known), as this will be very helpful in confirming an identification.

After such consideration, the specimen should be run through the key to the genera (pp. 2–13) to determine the genus to which it belongs. Precise information on how to use the keys is given below (pp. xvii–xxiv). All the information presented in the key should be used, including the heading paragraphs to the various groups (these provide a useful check on accurate progress through the key). Having reached the name of a genus, the plant should be compared carefully with the description of that genus in the descriptive text. If the specimen matches the description in all particulars, then it can be assumed that the generic determination is correct. If it does not match, then an error in using the generic key has probably been made, and the generic key should be tried again, with extremely close observation and great care. If, finally, after such checking, no good match between specimen and description can be obtained, it must be assumed that the specimen belongs to a genus not covered in the text of this book, and help should be sought from an orchid expert.

Having accurately identified the genus, the process is repeated with the key(s) to the species under that genus, so that the species can be determined. The same procedures apply, but, additionally, use can be made of the references to illustrations to confirm the identification that has been made.

By using this procedure carefully and precisely, accurate identifications should be possible for all the widely cultivated orchid genera and species.

Keys

A key is an abstract of a set of descriptions in which the differences between the individual descriptions are set out typographically in a way that allows the user to find the one appropriate to his specimen easily. In this book there is a single key to the 166 genera included, and for each genus with more than a single cultivated species there may be one or two keys to the species within it. If there are 2 keys under a genus, one of them, in the same form as the key to the genera, will always be present, headed 'Key to species'. The other, if present, is an informal guide to the distribution of characters among the various species and is headed 'Synopsis of characters'. Instructions on using both these types of key are given below.

The 'formal' keys (the key to the genera, and those to the species headed 'Key to species') are designed to lead to the correct names by a sequential process of elimination. Their form may appear somewhat unusual to the professional botanical taxonomist, as it is something of a compromise between the standard bracketed and indented keys used generally in taxonomic works. However, this type of key (which is similar to that used in John Hayward's *A new key to British wild flowers*, 1987) should be easier for the non-specialist to use. In what follows, the key to the genera is treated as an example, but the method applies equally to the keys to the species.

The key to the genera consists of a number of separated sections: a short key to the groups into which the key is divided, then keys to the genera or subdivisions (subkeys) of each group. In determining the genus one proceeds sequentially through this structure, eliminating those groups and subgroups which do not apply to the plant in question, until a generic name is reached.

The first part of the key is that to the five groups (**A, B, C, D & E**) into which the whole key is divided. Like all the rest, this consists of short, numbered (or numbered and lettered) paragraphs (known as 'leads') arranged in a precise sequence.

One first compares the plant to the lead numbered **1**; this asks whether or not the aerial growth of the plant dies down each year. If the specimen agrees completely with this, then the plant belongs to **Group A**, as indicated in the right-hand margin at the end of the lead. If, on the other hand, the specimen does not agree, then it must be compared with the next numbered lead (that numbered **2**); this asks if the leaves are rolled when young. If they are, as there is no Group at the right-hand side of this paragraph, the specimen must be compared with the next lead below that numbered **2**, identified as **2.A.**, which asks whether the plant has pseudobulbs. If it has, then it belongs to **Group B**, as shown by this appearing at the right-hand end of the lead. If it has not, then the next lead (**2.B.**) should be considered, which indicates that the specimen belongs to **Group C**. Note that if the specimen agrees with lead **2.**, it must also agree with one or other of the leads **2.A.** or **2.B.**

If, on the other hand, lead **2.** has been rejected, then one goes on to lead **3.**, and follows the same procedure.

Having arrived at a particular group, the short paragraph at the head of its key should be considered. This paragraph, in italics, lists the characters of plants in the group and acts as a check that the correct group has been arrived at. Then, the next key is followed in exactly the same way, until a generic name, or an instruction to go to a subkey is arrived at. The various subkeys also each have a heading paragraph (in italics) which can be used as a check on progress. Finally, having arrived at a generic name, one can turn to its description in the main text; the genera are numbered, to make finding the relevant one simple.

This procedure can be summed up as follows. In any key of this type, start with the lead numbered **1.** If the specimen agrees with this, continue to the next lead under it (identified as **1.A.**), or to the next stage (group, subkey, generic name or specific name) indicated at the right-hand side of the lead. Proceed similarly through the key until a name is reached. There are never more than 3 subsidiary leads under any particular number (identified by capital letter, lower-case letter, roman numeral, e.g. **1.A.a.i.**, **1.A.a.ii.**, **1.A.b.i.**, **1.A.b.ii.**, etc.), though there may often be several lettered leads under an individual number, e.g. **1.A.**, **1.B.**, **1.C.**, **1.D.**, **1.E.**, etc.

The 'informal' keys to species are included for most genera with between 5 and 20 species covered by the text. They are headed 'Synopsis of characters'. because they list, for each character, the species (indicated by their number) which show it. Thus, in the informal key to the species of *Cypripedium* (p. 14), the first characters apply to the leaves. The whole section is as follows:

LEAVES. Paired: **14–19**; spirally arranged: **1–13**. Basal: **18**. Fan-shaped: **14**. 3-veined: **15**. Spotted: **16**. Membranous: **15**.

This means that if the plant to be identified is a *Cypripedium* with paired leaves, it must be one of the six species numbered **14–19** in the account; if the paired leaves are also basal, then the plant must belong to the species numbered **18**; if the paired leaves are both 3-veined and membranous, then it must belong to the species numbered **15**. Similar treatments follow for other characteristics (*lower stem*, *floral bract*, *flowers*, etc.). By using a combination of these characters, a species identification can usually be made, or, in difficult cases, the range of possible species to which the plant may belong can be narrowed down to a small number. This kind of key is not sequential, and may be used in any order, depending on the most striking characteristics of the plant to be identified.

By using a combination of the formal and informal keys (when the latter are present), accurate identifications can usually be made easily, as long as care is taken in both the observation of th specimen and the reading of the keys.

Alongside the formal keys, in suitable places, are small illustrations of individual organs which are difficult or complex to describe in words. These illustrations relate directly to the printed key on the page in question, and should provide help with interpreting the key as written.

Descriptions

The descriptions of the genera and species follow the normal form used by taxonomists, starting from the base of the plant and proceeding to the flowers and their parts, ending with ovary and fruit. Technical language is used as little as possible; such technical terms as are used are explained in the section on Structure and Terminology (p. xvii). The description of each organ begins with the name of that organ in small capitals, e.g. Leaves, so that comparisons between the various descriptions are easy to make. Each species description ends with a group of three items of information: the known distribution of the species (printed in italics); a code for its hardiness (H1–5, G1, G2) and an indication of when the species is likely to be found in flower.

The hardiness codes range from H1 (hardy out of doors all over Europe) to H5 (hardy only in areas relatively free of frost, e.g the Mediterranean area and the extreme western fringes of Europe) to G1 (growable only in a cool glasshouse and G2 (growable only in a warm glasshouse or stove). The code H5–G1 is occasionally used for species which are very marginally hardy in a few sheltered places in Europe, but which have to have cool glasshouse conditions over most of the continent.

Flowering time is given only a general indication ('spring', 'summer'), sometimes qualified by late or early.

Bibliography

As mentioned above, the botanical and horticultural literature on orchids is very extensive, and only a small sample of the most useful works can be cited here. Many other books cited as sources for illustrations under the various species contain valuable information, and should be used as appropriate.

This Bibliography is classified into various sections for ease of use.

1. *General taxonomic accounts*

Bechtel, H., Cribb, P. & Launert, E., *Orchideenatlas*, Stuttgart (1980); english edn, *Manual of cultivated orchid species*, Poole (1981).

Dietrich, H., *Bibliographia Orchidacearum*, Jena (1980).

Dressler, R.L., *The orchids: natural history and classification*, Harvard University Press, Cambridge, Mass. and London (1981).

Hawkes, A.D., *Encyclopaedia of cultivated orchids*, London (1965).

Holttum, R.E., *A revised Flora of Malaya, I, Orchids of Malaya*, Singapore Botanic Gardens, Singapore (1953).

Schlechter, R., *Die Orchideen*, Paul Parey, Berlin (1914); edn 2, ed. by E. Miethe (1927); edn 3, ed. by Brieger, F.G., Maatsch, R. & Senghas, K., Eugen Ulmer, Stuttgart, appearing in parts and continuing (1972–).

Sheehan, T. & Sheehan, M., *Orchid genera illustrated*, New York (1979).

2. *Horticultural guides*

Black, P.M., *The complete book of orchid growing*, London (1980).

Hartmann, W.I., *Introduction to the cultivation of orchids*, London (1965).

Northen, R.T., *Home orchid growing*, London (1962).

Richter, W., *Orchideen*, Neudamm (1969).

Sander, D., *Orchids and their cultivation*, edn. 9, Poole (1979).

Thompson, P.A., *Orchids from seed*, Kew (1979)

Williams, B.A. & Kramer, J., *Orchids for everyone*, New York (1980).

Williams, B.S. and others, *The orchid-grower's manual*, edn. 7, London (1897, reprinted 1982).

Hybrids and nomenclature

Cribb, P.J., Greatwood, J. & Hunt, P.F., *Handbook on orchid nomenclature and registration*, edn. 3, International Orchid Commission, Florida (1985).

Sander's list of orchid hybrids; first published in 1905, supplements are published as necessary by the Royal Horticultural Society (the registration authority for the names of hybrid orchids).

Journals devoted to orchids (a selection)

The Orchid Review, 1893 and continuing (British).

Orchis, 1906–20 (German).

The American Orchid Society Bulletin, 1932 and continuing (American).

Orchideen, 1934 and continuing (Dutch).

Australian Orchid Review, 1936 and continuing (Australian).

Orchid Digest, 1937 and continuing (American).

Die Orchidee, 1949 and continuing (German).

Orchid Journal, 1952–55 (American).

The Orchadian, 1963 and continuing (Australian).

Orchidata, 1965 and continuing (American).

South African Orchid Journal, 1970 and continuing (South African).

Orquidea (Mexico), 1971 and continuing (Mexican).

DESCRIPTION OF ORCHIDACEAE

HABIT: plants usually with rhizomes, sometimes with root-tubers, growing in soil, on rocks or as epiphytes some (not ours) without chlorophyll; aerial roots frequent.

STEMS: usually present, often fleshy and swollen, when referred to as pseudobulbs.

LEAVES: deciduous or persistent, borne on the stems (more rarely directly on the rhizomes) in 2 ranks or spirally arranged, rolled or folded when young, often hard and leathery.

FLOWERS: bilaterally symmetric or rarely asymmetric, solitary or in racemes or panicles borne terminally or laterally, usually on scapes which bear bracts; usually the flowers are turned upside down by a twist of 180° in the flower-stalk and ovary (flowers resupinate).

SEPALS: 3, free or united.

PETALS: 3, 2 of them usually similar to each other and to the sepals, the other (the lowermost in resupinate flowers) usually different in shape, size and colour, and referred to as the lip.

STAMEN(S), STYLE AND STIGMAS: united into a solid structure known as the column.

ANTHERS: 2 or more frequently 1, borne at or near the apex of the column.

POLLEN: aggregated into powdery or waxy masses known as pollinia, of which there may be 2, 4, 6, 8 or more; pollinia usually borne on a sticky pad (the viscidium), or more rarely on 2 viscidia, to which they may be attached by pollinial tissue, or by a structure of different hardness and texture known as the stipe.

STIGMAS: 3, the lateral 2 receptive and usually joined, forming a hollow on the inner surface of the column below the anther, or rarely borne on stalks; the median stigma not receptive, variously modified, borne just above the fertile 2 and beneath the anther, forming a structure known as the rostellum which is sometimes conspicuous, and contributes material to the viscidium.

OVARY: inferior, usually 1-celled, more rarely 3-celled, placentation parietal.

OVULES: numerous.

FRUIT: a capsule.

SEEDS: very numerous, very small and dust-like.

KEY TO GENERA

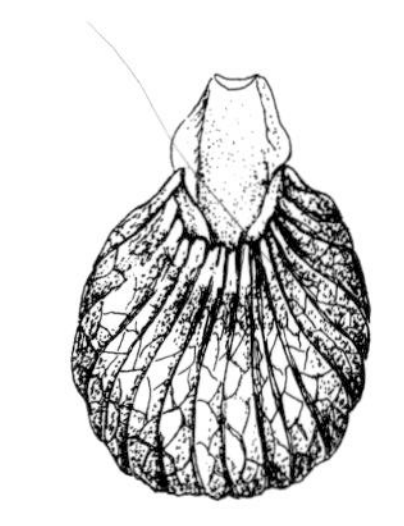

Cypripedium macranthon – lip.

Calypso bulbosa – lip.

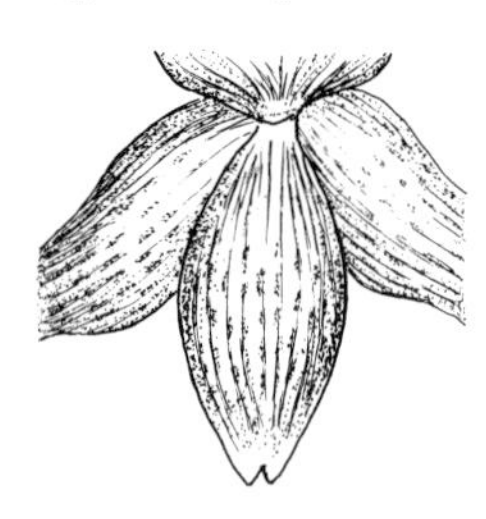

Cypripedium macranthon – lower sepals.

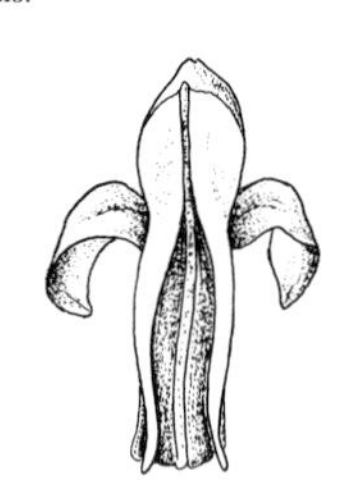

Satyrium nepalensis – spurs.

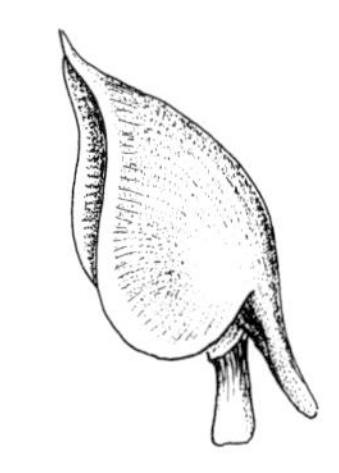

Disa uniflora – spur.

1. Annual aerial growth dying off completely each autumn or winter, leaving no part above the ground except, occasionally, a rosette of small leaves or a solitary leaf, formed after the current year's growth has died down **Group A**
2. Leaves rolled when young
2.A. Plant with distinct pseudobulbs or stems swollen above or below **Group B**
2.B. Plant without pseudobulbs or swollen stems **Group C**
3. Leaves folded when young or leaves terete or cylindric with a narrow groove
3.C. Plant with distinct pseudobulbs or stems swollen above or below **Group D**
3.D. Plant without pseudobulbs or swollen stems **Group E**

GROUP A

Annual aerial growth dying off completely each autumn or winter, leaving no part above the ground except, occasionally, a rosette of small leaves or a solitary leaf, formed after the current year's growth has died down.

1. Leaves absent at flowering; lip with 3 ridges at the base which become fleshy plates on the central lobe **80. Aplectrum**
2. Lip slipper like
2.A. Stem with 2 or more separated nodes and 2 or more leaves; lateral sepals pointing downwards, usually fused **1. Cypripedium**
2.B. Stem with 1 node and a single leaf; lateral sepals pointing upwards, free **79. Calypso**
3. One or two spurs present
3.C. Spurs 2
3.C.a. Plant with a single leaf held flat against the ground; flower 1 (rarely 2) **7. Corybas**
3.C.b. Plant with several leaves on erect stems; flowers numerous **28. Satyrium**
3.D. Spur 1 *Go to Subkey A1*
4. No spurs present *Go to Subkey A2*

Subkey A1 *Leaves present at flowering; lip not slipper-like; each flower with a single spur.*

1. Spur arising from the base of the concave or hooded central sepal **26. Disa**
2. Flower not resupinate, i.e. the lip, which is similar to the other petals, uppermost **20. Nigritella**

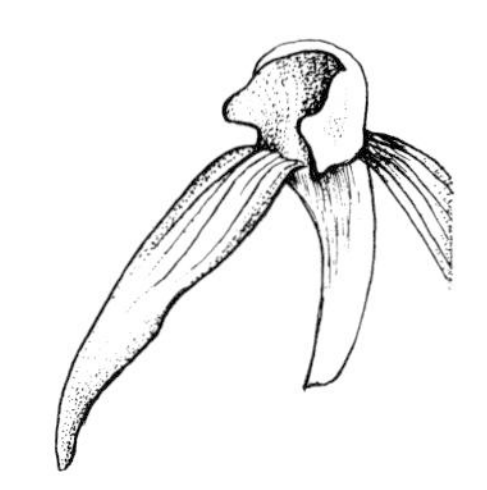
Platanthera chlorantha – lip.

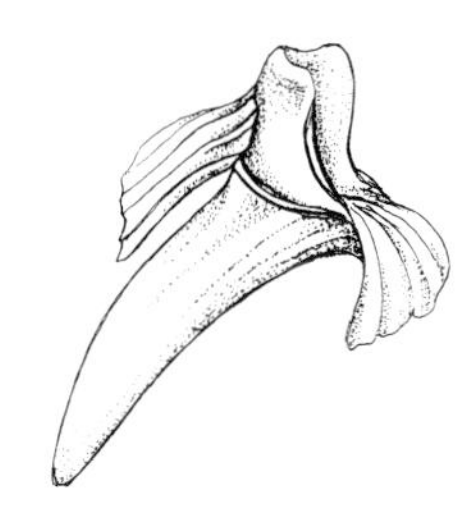
Dactylorhiza purpurella – spur.

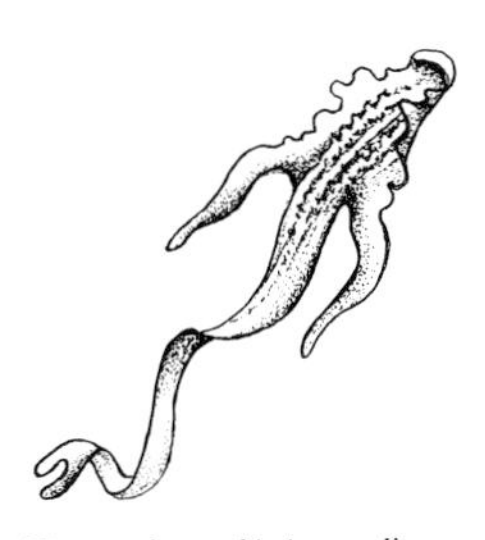
Himantoglossum hircinum – lip.

Barlia robertiana – lip.

Coeloglossum viride – lip.

3. Spur 2.5 cm or more
3.A. Lip strap-shaped, unlobed **23. Platanthera**
3.B. Uppermost leaf tightly sheathing the stem; column with 1 stigmatic area **18. Orchis**
3.C. Uppermost leaf not tightly sheathing stem; column with 2 stigmatic areas **24. Habenaria**
4. Sepals united at their bases to the column and lip **25. Stenoglottis**
5. Spur 1 cm or more, slender
5.D. Spur cylindric, obtuse; flowers usually yellow **19. Dactylorhiza**
5.E. Spur thread-like, acute; flowers not yellow
5.E.a. Raceme conical, flowers with a foxy smell; lip deeply 3-lobed with 2 basal ridges **17. Anacamptis**
5.E.b. Raceme cylindric; flowers fragrant; lip shallowly 3-lobed, without basal ridges **21. Gymnadenia**
6. Lip spirally twisted or its lateral lobes with wavy margins
6.F. Central lobe of lip 2 or more times longer than lateral lobes, strap-shaped, spirally twisted; bracts shorter than flowers **15. Himantoglossum**
6.G. Central lobe of lip less than 2 times longer than the lateral lobes, oblong, not spirally twisted; bracts longer than the flowers **16. Barlia**
7. Lip strap-shaped, terminating in 3 short teeth; flowers greenish **22. Coeloglossum**
8. Root tubers ovoid; leaves mostly in a basal rosette, the uppermost usually completely sheathing the stem; all bracts membranous, shorter than flowers **18. Orchis**
9. Root tubers lobed; leaves not in a basal rosette; upper leaf sheathing the stem at its base only, often transitional to the leaf-like bracts; bracts usually as long as or longer than the flowers **19. Dactylorhiza**

Subkey A2 Leaves present at flowering; lip not slipper-like; flowers without spurs.

1. Stem with 2 more or less opposite leaves; lip strap-shaped, 2-lobed at apex **6. Listera**
2. Flowers arranged in a tight spiral **8. Spiranthes**
3. Lip divided into 2 parts by a constriction at the middle
3.A. Roots tuberous; leaves not pleated; bracts large, coloured like the sepals, sheathing the basal parts of the flowers **13. Serapias**
3.B. Flowers shortly stalked, borne mostly on one side of the axis and horizontal or pointing downwards; sepals and petals spreading **4. Epipactis**
3.C. Flowers not stalked, borne all round the axis, pointing upwards; sepals and petals not widely spreading **5. Cephalanthera**

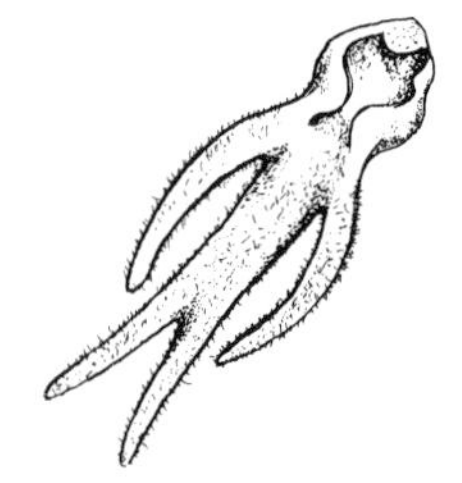
Aceras anthropophorum – lip.

4. Lip small, usually sensitive to touch, hidden within a hood formed by the sepals and petals **27. Pterostylis**
5. Lip velvety or covered in short hairs or with a geometric pattern or with a shining area or with any combination of these features; if lobed, lobes not long and slender **12. Ophrys**
6. Lip not as under 5, lateral and central lobes long and slender, like the arms and legs of a man **14. Aceras**

GROUP B

Aerial shoots persisting throughout the winter; leaves rolled when young; plant with distinct pseudobulbs, or stems swollen above or below.

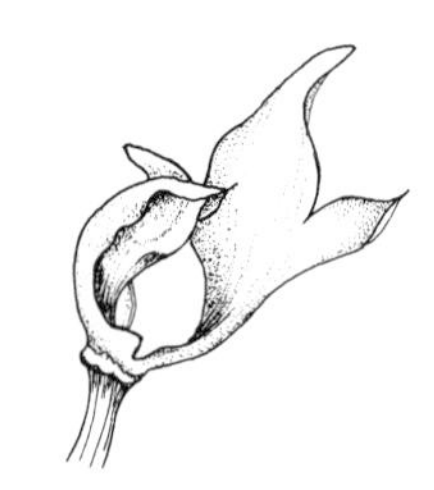
Mormodes maculata – lip and column.

1. Pollinia 2
1.A. Pseudobulbs compound *Go to Subkey B1*
1.B. Pseudobulbs simple
1.B.a. Sepals and petals directed forwards and incurved, the flower bell-shaped *Go to Subkey B2*
1.B.b. Sepals and petals spreading or reflexed, the flowers not bell-shaped *Go to Subkey B3*
2. Pollinia 8 *Go to Subkey B4*
3. Pollinia 4
3.C. Column with a distinct foot *Go to Subkey B5*
3.D. Column without a foot *Go to Subkey B6*

Cycnoches egertonianum – column.

Subkey B1 Pollinia 2; pseudobulbs compound.
1. Flowers asymmetric due to twisting of the column to one side **137. Mormodes**
2. Flowers not resupinate (i.e. lip uppermost)
2.A. Column thin but expanded conspicuously upwards towards the apex, arching, without 2 backwardly directed bristle-like outgrowths **139. Cycnoches**
2.B. Column equally thick throughout, straight, usually with 2 backwardly directed bristle-like outgrowths **138. Catasetum**
3. Raceme terminal; lip rolled around and partially concealing the column **128. Galeandra**
4. Lip with a distinct though short spur **129. Eulophia**
5. Lip with a distinct and conspicuous claw **132. Cyrtopodium**
6. Lip without a claw **131. Eulophiella**

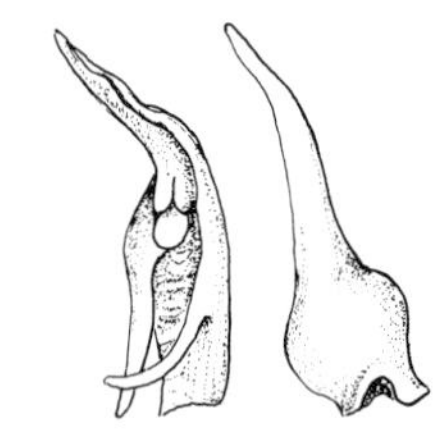
Catasetum sp. – column.

Subkey B2 Pollinia 2; pseudobulbs simple; sepals and petals directed forwards and incurved, giving the flower a bell shape.
1. Lip with a distinct claw which bears the lateral lobes near its middle **141. Acineta**
2. Anterior part of the lip freely movable on the rest **142. Peristeria**
3. Pseudobulbs bearing 1–2 leaves; central lobe of the lip not downy **143. Lacaena**

Acineta chrysantha – lip.

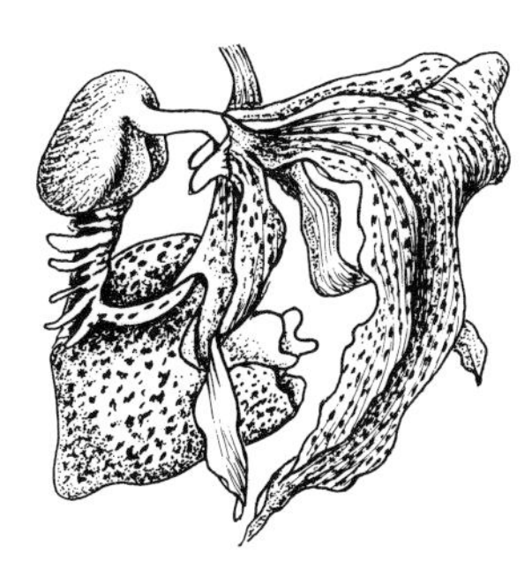
Coryanthes macrantha – whole flower.

4. Pseudobulbs bearing 3–5 leaves; central lobe of the lip downy **144. Lueddemannia**

Subkey B3 Pollinia 2; pseudobulbs simple; sepals and petals spreading, the flower not bell-shaped.

1. Column with a reflexed apex; lip very complex, bearing 2 peg-like nectar-secreting glands near the base which drip nectar into the bucket-like anterior part **151. Coryanthes**

2. Flowers in pendulous racemes, the column (as viewed in the growing position) below the lip (i.e. flowers resupinate)

2.A. Apex of lip bearing white, club-shaped, hair-like processes **147. Paphinia**

2.B. Upper sepal fused to the column for about half the length of the latter; anther at the apex of the column **149. Gongora**

2.C. Upper sepal free from column; anther apparently on the back of the column **150. Cirrhaea**

3. Basal part of the lip, or the whole lip hollowed out and sac-like **148. Stanhopea**

Stanhopea jenishiana – lip.

4. Pseudobulbs each with 2–4 leaves **95. Eriopsis**

5. Lateral lobes of lip narrow, curving backwards and upwards and then (towards the apex) forwards; axis of raceme without hairs **145. Houlletia**

6. Lateral lobes of the lip oblong-triangular, curving upwards; axis of raceme downy **146. Polycycnis**

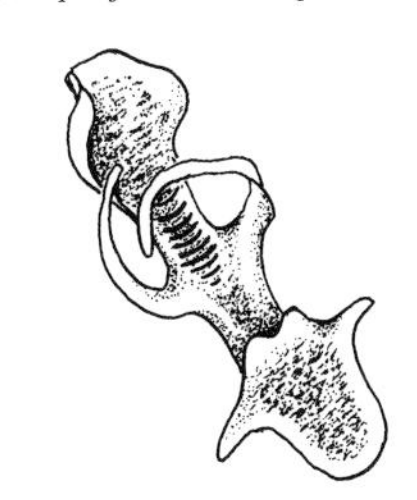
Houlletia lansbergii – lip.

Subkey B4 Pollinia 8.

1. Racemes apparently terminal

1.A. Column with a distinct foot to which the lateral sepals are attached, forming a mentum **61. Eria**

1.B. Column without a foot, mentum absent **30. Bletilla**

2. Column with a conspicuous foot to which the lateral sepals are attached, forming a mentum

2.A. Pseudobulbs simple **31. Bothriochilus**

2.B. Pseudobulbs compound **36. Chysis**

3. Lip with a projecting spur

3.A. Leaves absent at flowering; axis of raceme, flower-stalks and backs of the flowers hairy **32. Calanthe**

3.B. Leaves present at flowering; axis of raceme, flower-stalks and backs of the flowers not hairy **34. Phaius**

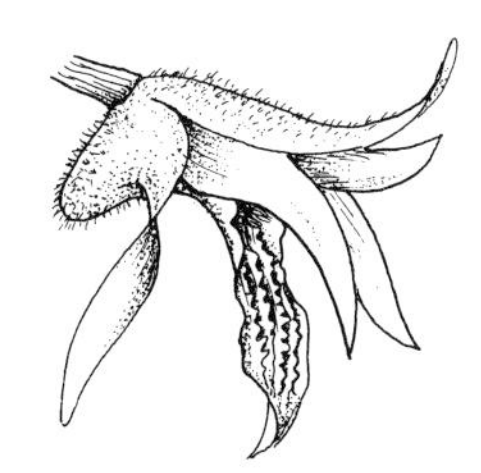
Eria javanica – whole flower.

4. Central lobe of the lip with a long, narrow claw **35. Spathoglottis**

5. Bracts broad, almost as long as the flower-stalks, often falling early **34. Phaius**

6. Bracts narrow, much longer than the flower-stalks **33. Bletia**

Bletilla striata – whole flower.

Subkey B5 Pollinia 4; column with a distinct foot.

1. Flowers solitary

Anguloa clowesii – flower.

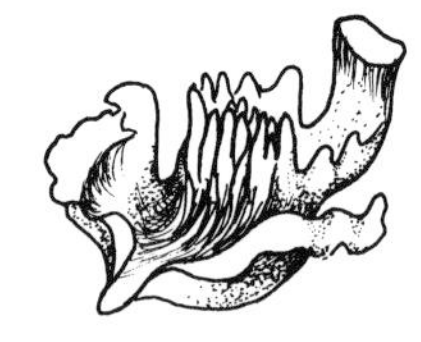

Acacallis cyanea – lip.

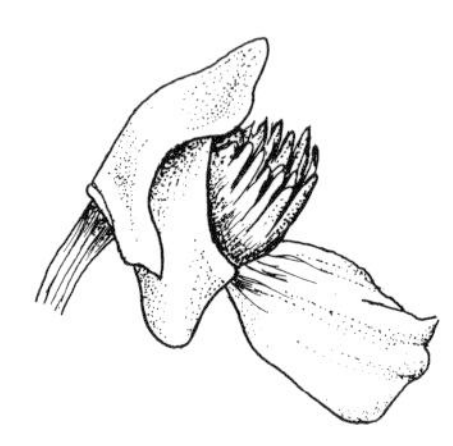

Zygopetalum mackayi – lip.

Pleione humilis – lip.

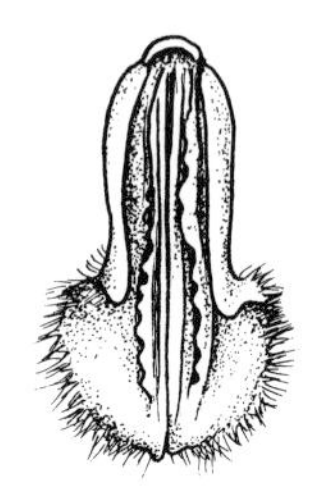

Coelogyne fuliginosa – lip.

1.A. Sepals and petals incurved, flower almost spherical **85. Anguloa**

1.B. Sepals and petals spreading, flower widely open **84. Lycaste**

2. Lip freely movable on the column; flowers many, in 2 ranks in slender racemes **44. Dendrochilum**

3. Callus of lip divided into finger-like projections **88. Acacallis**

4. Callus of lip transverse, forming a curved frill across the base of the lip **87. Zygopetalum**

5. Pollinia in 2 pairs, each pair on its own stipe; column winged above

5.C. Each pseudobulb with 1 leaf **83. Bifrenaria**

5.D. Each pseudobulb with 2 or 3 leaves **96. Galeottia**

6. Petals more or less equal in size to the upper sepal **95. Eriopsis**

7. Leaves in 2 ranks; column somewhat swollen above **81. Warrea**

8. Leaves not in 2 ranks; column cylindric **82. Xylobium**

Subkey B6 Pollinia 4; *column without a foot.*

1. Lip uppermost (flowers not resupinate), with auricles which extend downwards below the column **77. Malaxis**

2. Lip pouched or sac-like towards the base; racemes dense with more than 4 flowers **45. Pholidota**

3. Lip bent sharply backwards; flower at most 2.5 cm in diameter **78. Liparis**

4. Leaves membranous, deciduous; lip unlobed or very shallowly 3-lobed, base rolled around and concealing the column **43. Pleione**

5. Central lobe of the lip hairy, lateral lobes spreading **86. Pabstia**

6. Central lobe of the lip hairless, lateral lobes erect **42. Coelogyne**

GROUP C

Aerial growth persisting through the winter; *leaves rolled when young*; *plant without pseudobulbs or swollen stems.*

1. Leaves with veins differently coloured from the rest of the surface; pollinia made up of easily separable masses

1.A. Lip uppermost (i.e. flower not resupinate); flower asymmetric due to the twisting of the lip and column in different directions **10. Macodes**

1.B. Lip not widening towards the apex; claw not fringed or toothed **9. Goodyera**

1.C. Lip abruptly widened to a transverse blade; claw fringed or toothed **11. Anoectochilus**

2. Pollinia 2, borne on a triangular stipe **132. Cyrtopodium**

Phaius tankervilleae – lip.

3. Axis of raceme, flower-stalks and backs of flowers hairy **32. Calanthe**

4. Lip with an obtuse spur

4.A. Raceme axillary, erect **34. Phaius**

4.B. Raceme terminal, pendulous **37. Thunia**

5. Leaves many in 2 ranks; column without a foot, mentum absent **39. Sobralia**

6. Leaves 2 or 4, in opposite pairs; column with a distinct foot to which the lateral sepals are attached, forming a mentum **61. Eria**

GROUP D

Aerial growth persisting through the winter; leaves folded when young, or terete, or cylindric with a narrow groove; plant with distinct pseudobulbs or stems conspicuously swollen above or below

1. Column with a distinct foot, mentum often conspicuous

1.A. Pseudobulbs compound *Go to Subkey D1*

1.B. Pollinia 8 *Go to Subkey D2*

1.C. Pollinia 2 or 4 *Go to Subkey D3*

2. Column without a foot, mentum absent

2.D. Flowers spurred *Go to Subkey D4*

2.E. Sides of the lip attached to the column *Go to Subkey D5*

2.F. Pollinia 2 *Go to Subkey D6*

2.G. Pollinia 4 or 8 *Go to Subkey D7*

Grammangis ellisii – whole flower.

Subkey D1 Column with a distinct foot, mentum often conspicuous; pseudobulbs compound.

1. Panicle large, terminal **133. Ansellia**

2. Petals less than one-third as long as the upper sepal; raceme arising near base of pseudobulb **135. Grammangis**

3. Petals at least two-thirds as long as upper sepal; raceme borne on the pseudobulb

3.A. Pollinia 4 **73. Dendrobium**

3.B. Pollinia 8 **61. Eria**

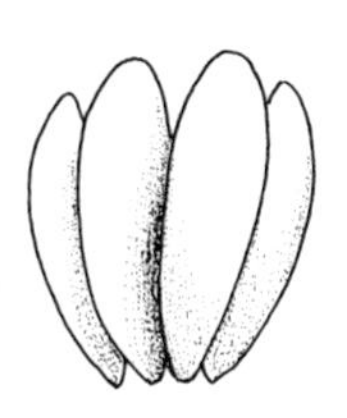

Dendrobium aduncum – pollinia.

Subkey D2 Column with a distinct foot, mentum often conspicuous; pseudobulbs simple; pollinia 8

1. Flower solitary **63. Meiracyllium**

2. Sepals fused at the base into an urn-shaped cup which conceals the petals **62. Cryptochilus**

3. Sepals free from each other, but fused to the column-foot; petals exposed **61. Eria**

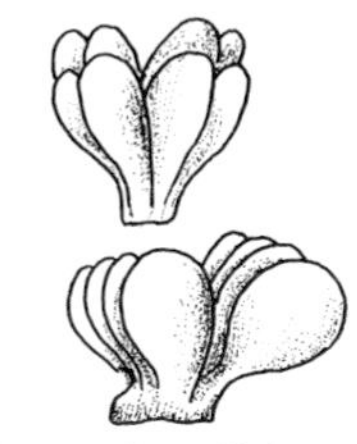

Eria marginalis – pollinia.

Subkey D3 Column with a distinct foot, mentum often conspicuous; pseudobulbs simple; pollina 2 or 4

1. Pseudobulbs borne one on top of each other, forming chains

1.A. Flowers white to purplish pink **59. Scaphyglottis**

1.B. Flowers bright reddish orange **46. Hexisea**

Trichocentrum albo-coccineum – lip.

Broughtonia coccinea – flower.

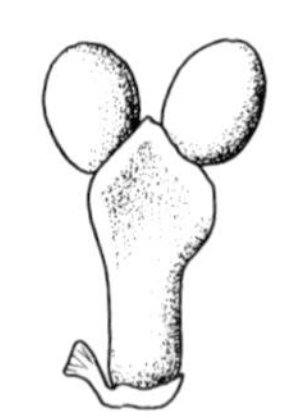

Cochlioda noezliana – pollinia.

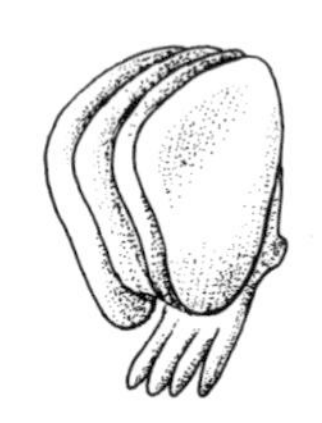

Epidendrum parkinsonianum – pollinia.

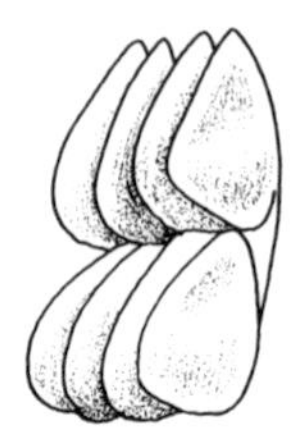

Sophronitis coccinea – pollinia.

2. Flowers or racemes terminal on the pseudobulbs

2.C. Axis of raceme downy; flower with lip uppermost (i.e. flower not resupinate) **127. Polystachya**

2.D. Axis of raceme hairless; flower with lip lowermost (i.e. flower resupinate)

2.D.a. Flowers more than 4.5 cm in diameter; lip 3-lobed, not pouched at apex **74. Epigeneium**

2.D.b. Flowers less than 2 cm in diameter; lip not 3-lobed, pouched at apex **60. Nageliella**

2.E. Lateral sepals much larger than upper sepal **76. Cirrhopetalum**

2.F. Bracts on the scape clasping, overlapping or almost so, rarely their apices spreading **98. Maxillaria**

2.G. Lip not 3-lobed (sometimes fringed) **75. Bulbophyllum**

2.H. Leaves terete **97. Scuticaria**

Subkey D4 Column without a foot, mentum absent; flower with a spur.

1. Spur formed by the base of the lip

1.A. Flowers 1 or 2, 2.5 cm or more in diameter **152. Trichocentrum**

1.B. Flowers 5 or more, at most 2.5 cm in diameter **130. Oeceoclades**

2. Spur attached to the ovary **49. Broughtonia**

3. Lip with 1 or 2 appendages which project back into the sepal-spur

3.C. Lip with 1 appendage **154. Rodriguezia**

3.D. Lip with 2 appendages **155. Comparettia**

4. Lip without appendages projecting back into the sepal-spur **153. Ionopsis**

Subkey D5 Column without a foot, mentum absent; flower without a spur; sides of the lip attached to the column.

1. Pollinia 2

1.A. Pseudobulbs compound **134. Cymbidium**

1.B. Stigmatic surface divided into 2; sepals and petals free from column **157. Cochlioda**

1.C. Stigmatic surface not divided; upper sepal and petals fused to the back of the column **161. Aspasia**

2. Pollinia 4

2.D. Sides of the lip attached to the column for almost the whole length of the latter **47. Epidendrum**

2.E. Sides of the lip attached to the column for only the lower half of the latter **48. Encyclia**

3. Pollinia 8

3.F. Leaves ovate, elliptic or oblong, leathery **56. Sophronitis**

3.G. Leaves linear, fleshy **57. Sophronitella**

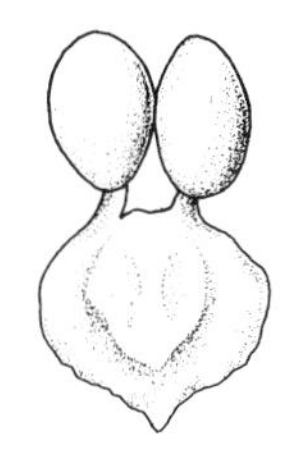

Grammatophyllum scriptum – pollinia.

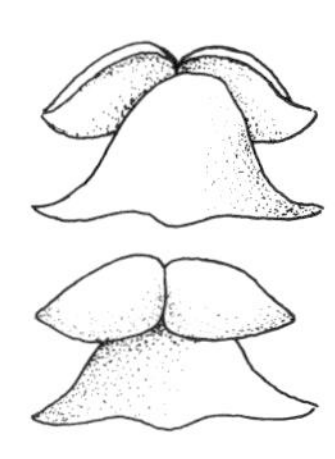

Cymbidium tigrinum – pollinia.

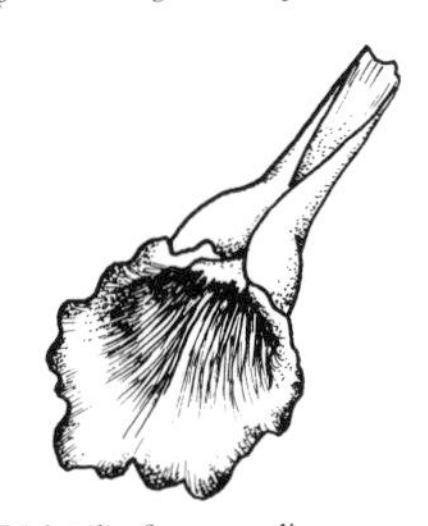

Trichopilia fragrans – lip.

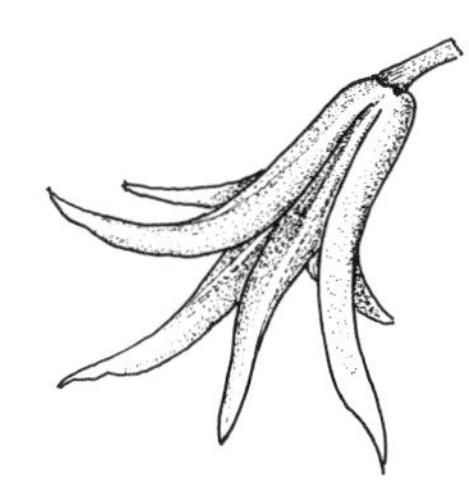

Ada aurantiaca – flower.

Helcia sanguinolenta – lip.

Subkey D6 Column without a foot, mentum absent; flowers not spurred; sides of lip free from column; pollinia 2

1. Pseudobulbs compound
1.A. Stipe of pollinia divided; plants usually very large (raceme to 2 m) **136. Grammatophyllum**
1.B. Stipe of pollinia not divided; plant much smaller **134. Cymbidium**
2. Sides of the lip rolled over the column **156. Trichopilia**
3. Sepals and petals directed forwards, incurved **162. Ada**
4. Sepals and petals at least 6 times as long as broad **163. Brassia**
5. Base of lip ascending more or less parallel with the column
5.C. Lip with 2 basal auricles which are fused to the base of the column **159. Helcia**
5.D. Ridges on the base of the lip reaching up to and slightly clasping the column; upper sepal and petals directed upwards, lateral sepals directed downwards; flowers basically yellow **158. Gomesa**
5.E. Ridges on the base of the lip not reaching the column; sepals and petals usually evenly spreading; flowers usually not basically yellow **160. Odontoglossum**
6. Lip with a distinct, parallel-sided claw **166. Sigmatostalix**
7. Lip without a claw, or, if a claw is present, then not parallel-sided
7.F. Lip usually simple, with 2 (rarely more) longitudinal ridges or lines of hairs near the base; flowers more or less flattened, not basically yellow; column not thickened towards the base **164. Miltonia**
7.G. Lip usually lobed and bearing a complex callus of tubercles or ridges and tubercles; flowers not flattened, basically yellow; column thickened with a fleshy plate below the stigma **165. Oncidium**

GROUP E

Aerial growth persisting through the winter; leaves folded when young, or tertete or cylindric with a narrow groove; plant without pseudobulbs or swollen stems

1. A broad, flat staminode present on the column and extending beyond it; fertile anthers 2
1.A. Ovary 3-celled; sepals edge-to-edge in bud **2. Phragmipedium**
1.B. Ovary 1-celled; sepals overlapping in bud **3. Paphiopedilum**
2. Inflorescence lateral, usually axillary
2.C. Lip consisting mainly of a vertical pouch or sac, its blade relatively small, consisting of 3 lobes borne on the margin of the pouch *Go to Subkey E1*
2.D. Flower with a distinct spur projecting from the back
2.D.a. Lip distinctly and obviously 3-lobed *Go to Subkey E2*

2.D.b. Lip entire or very inconspicuously 3-lobed *Go to Subkey E3*
2.E. Column without a foot *Go to Subkey E4*
2.F. Column with a distinct foot *Go to Subkey E5*
3. Inflorescence terminal or borne directly on the rhizome
3.G. Fertile pollinia 2, 4 or 6, sometimes with additional, sterile pollinia
3.G.a. Lateral sepals conspicuously fused at the base (upper sepal may also be fused, or free) *Go to Subkey E6*
3.G.b. Lateral sepals free or only very shortly and inconspicuously united at the base *Go to Subkey E7*
3.H. Fertile pollinia 8 *Go to Subkey E8*

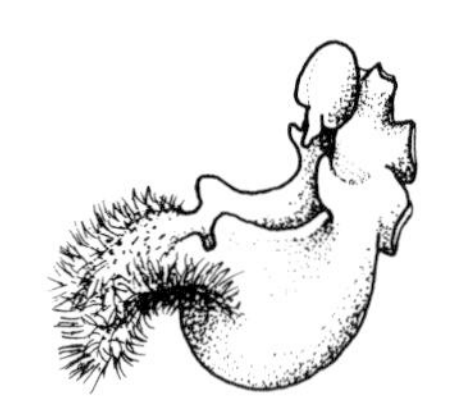

Gastrochilus acutifolius – lip.

Subkey E1 Staminode absent; fertile anther 1; inflorescences lateral, usually axillary; lip consisting mainly of a vertical pouch or sac, its blade relatively small, consisting of 3 lobes borne on the margin of the pouch

1. Column without a foot; pollinia 2
1.A. Lateral lobes of lip fused to the column; central lobe hairy or fringed **118. Gastrochilus**
1.B. Lateral lobes of lip free from column; central lobe neither hairy nor fringed **117. Ascocentrum**
2. Column with a foot; pollinia 4, often in 2 pairs
2.C. Plant leafless when flowering (though with small green bracts on the scape) **104. Chiloschista**
2.D. Spur or pouch of the lip without any calluses inside it **105. Pteroceras**
2.E. Leaves in 2 ranks; central lobe of lip smaller than the lateral lobes **103. Sarcochilus**
2.F. Leaves borne all round the stem or 1-sidedly; central lobe of lip larger than lateral lobes **120. Cleisostoma**

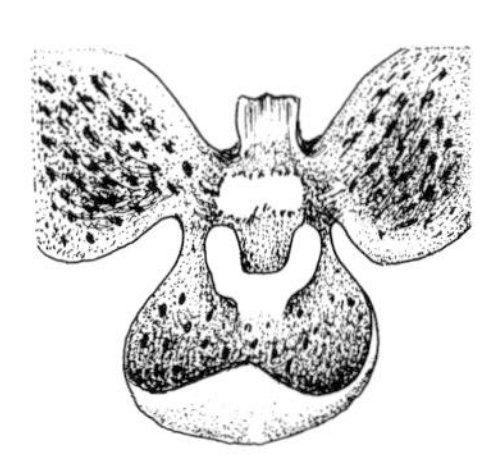

Sarcochilus moorei – lip.

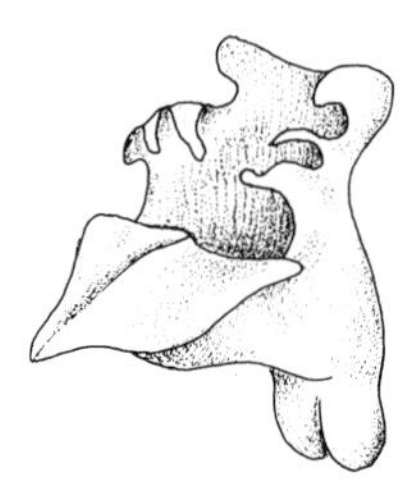

Cleisostoma filiforme – lip.

Subkey E2 Staminode absent; inflorescences lateral, usually axillary; lip not in the form of a pouch or sac, with a distinct spur projecting from the back; blade of lip distinctly and obviously 3-lobed

1. Column without a foot
1.A. Sepals and petals 2–5 mm; rostellum linear, bifid, very conspicuous **119. Schoenorchis**
1.B. Pollinia 4; flowers in panicles **116. Renanthera**
1.C. Flowers at most 3 cm in diameter, entirely white **123. Neofinetia**

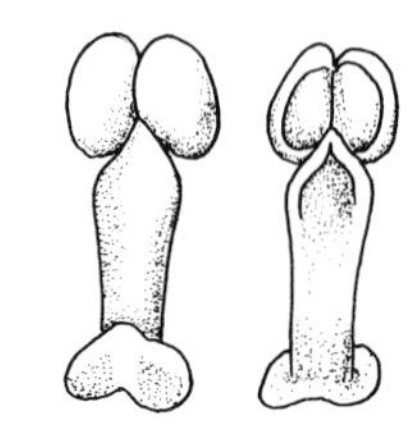

Renanthera storiei – pollinia.

2. Column with a foot
2.D. Lip freely movable on its attachment to the column-foot; spur pointing forwards **107. Sedirea**
2.E. Pollinia borne on a short, broad stipe; flowers 8–10 cm in diameter **112. Vanda**
2.F. Pollinia borne on a long, slender stipe; flowers to 5 cm in diameter **106. Aerides**

Subkey E3 Staminode absent; inflorescences lateral, usually axillary; lip not in the form of a pouch or sac, with a distinct spur projecting from the back; blade of lip entire or very inconspicuously 3-lobed

1. Column with a long foot **106. Aerides**
2. Rostellum small, inconspicuous, bifid **121. Angraecum**
3. Sepals conspicuously broader than sepals; flower usually white, spotted or blotched with red or pinkish purple **108. Rhynchostylis**
4. Pollinia 2, each borne on a separate stipe

4.A. Lip about the same size as the petals and sepals, without a tooth-like callus in front of the mouth of the spur **124. Cyrtorchis**

4.B. Lip much broader than sepals and petals, with a tooth-like callus in front of the mouth of the spur **125. Diaphananthe**

5. Pollinia 2, both borne on the one common stipe

5.C. Spur with a distinct, narrow mouth; lip equalling sepals and petals in breadth, or very slightly broader **122. Aerangis**

5.D. Spur with a wide, indistinct mouth; lip considerably broader than the sepals and petals **126. Eurychone**

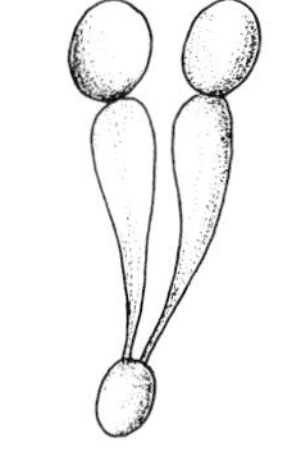

Diaphananthe bidens – pollinia.

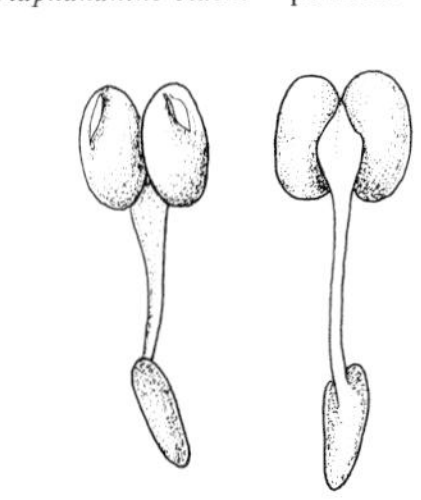

Aerangis kotschyana – pollinia.

Subkey E4 Staminode absent; inflorescences lateral, usually axillary; lip not in the form of a pouch or sac, without a spur; column without a foot

1. Pollinia 2

1.A. Sepals and petals 7–15 mm; leaves strongly keeled, with overlapping bases **140. Lockhartia**

1.B. Claw of lip fused to the column; leaves very fleshy or leathery; plant a scrambler **29. Vanilla**

1.C. Leaves conspicuously veined, apex acute or 2-lobed; lateral lobes of lip erect and partially encircling the column **134. Cymbidium**

1.D. Leaves not conspicuously veined, apex irregularly toothed; lateral lobes of lip not as under C above **112. Vanda**

2. Pollinia 4

2.E. Flowers of 2 kinds in each raceme; raceme-axis and the backs of the flowers woolly **115. Dimorphorchis**

2.F. Lip above the column (i.e. flower not resupinate); rostellum long-beaked, anther with a long terminal appendage **102. Ornithocephalus**

2.G. Flowers solitary; lip usually without a callus or longitudinal ridges, not hollowed out near the base **101. Dichaea**

2.H. Lip freely movable on the column, with erect, oblong lateral lobes which are free from each other **114. Esmeralda**

2.I. Lip not freely movable on the column, the lateral lobes joined together by a fleshy plate above the ridge of the central lobe **113. Vandopsis**

Subkey E5 Staminode absent; inflorescences lateral, usually axillary; lip not consisting mainly of a vertical pouch or sac; flower without a spur; column with a distinct foot.

1. Pollinia 2
1.A. Leaves flat **110. Phalaenopsis**
1.B. Sepals and petals greenish yellow blotched with reddish brown **97. Scuticaria**
1.C. Sepals and petals yellow or greenish yellow flushed with pink or violet **111. Paraphalaenopsis**
2. Pollinia 4
2.D. Flowers few to many, in racemes
2.D.a. Lip 3-lobed with 2 linear appendages on the claw; leaves distinctly 2-ranked **109. Doritis**
2.D.b. Lip 3-lobed, without appendages on the claw; leaves usually not distinctly 2-ranked **73. Dendrobium**
2.E. Lip at most 1 cm **98. Maxillaria**
2.F. Callus of lip fringed **94. Huntleya**
2.G. Petals yellow, with fringed or toothed margins **90. Chondrorhyncha**
2.H. Lip distinctly 3-lobed
2.H.a. Lateral lobes of lip rolled over the column and overlapping; lip with no claw, or claw very short **91. Cochleanthes**
2.H.b. Lateral lobes of lip spreading; lip with a conspicuous claw **92. Pescatoria**
2.I. Lip not 3-lobed
2.I.a. Column hooded, protruding over the lip-callus **93. Bollea**
2.I.b. Column neither hooded nor protruding over the lip-callus **91. Cochleanthes**
3. Pollinia 8 **61. Eria**

Huntleya burtii – lip.

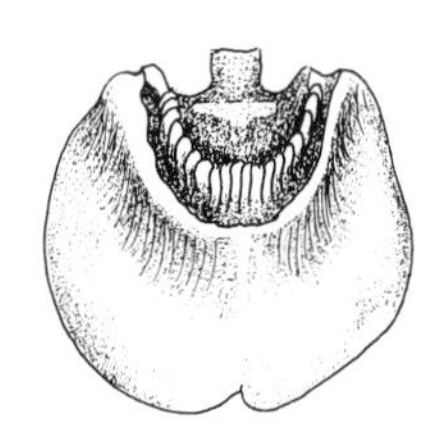

Pescatoria dayana – lip.

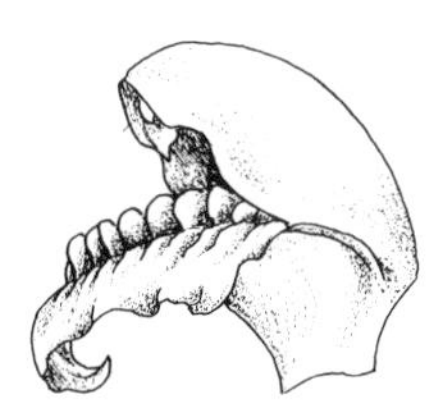

Bollea coelestis – lip and column.

Subkey E6 Staminode absent; inflorescences terminal or borne directly on the rhizome; fertile pollinia 2, 4 *or* 6, *sometimes with additional, sterile pollinia; lateral sepals conspicuously fused at the base, the upper sepal free or also fused to the others*

1. Pollinia 4
1.A. Stems with several leaves which are up to 3 mm broad; upper sepal 8–9 mm **40. Isochilus**
1.B. Stems each with a single leaf which is at least 2.5 cm broad; upper sepal at least 2 cm **70. Restrepia**
2. Upper sepal free (or almost so) from the fused lateral sepals **69. Pleurothallis**
3. Sepals fused at the base and towards the apex, leaving 3 'windows' to the interior of the flower **71. Cryptophoranthus**
4. Sepal-tube forming a 3-angled cup slightly constricted at the mouth; column without a foot **64. Physosiphon**
5. Lip with a distinct claw

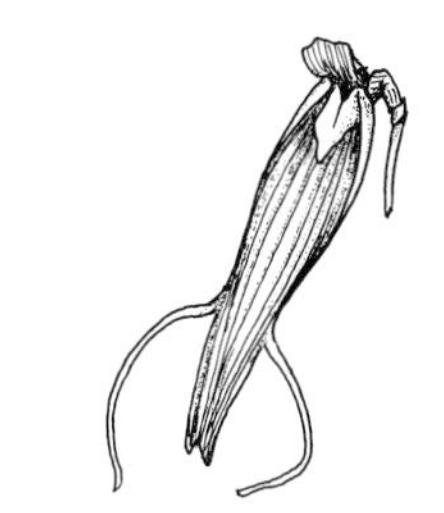

Trisetella huebneri – lip.

5.A. Leaves papillose on the upper surface; lip sensitive to touch; flowering stem densely hairy **67. Porroglossum**
5.B. Leaves not papillose; lip not sensitive; flowering stem not densely hairy **66. Dracula**
6. Fused lateral sepals forming an oblong, acute blade which bears 2 lateral tails **68. Trisetella**
7. Fused lateral sepals not as under 6., with a sinus or notch between the lateral lobes, which may be fused or free **65. Masdevallia**

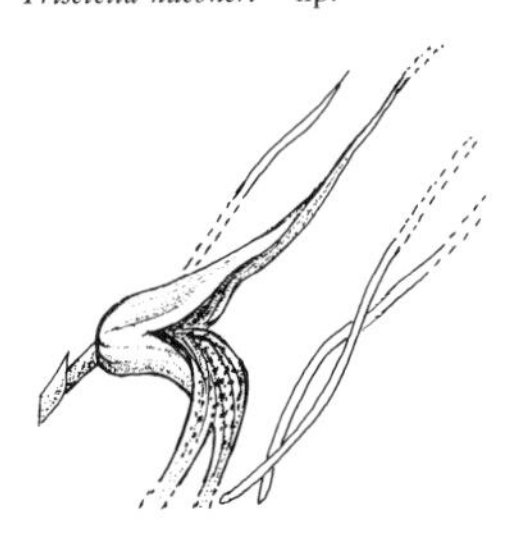

Masdevallia macrura – sepals.

Subkey E7 Staminode absent; inflorescences terminal or borne directly on the rhizome; fertile pollinia 2, 4 *or* 6, *sometimes with additional, sterile pollinia; lateral sepals free from each other or very shortly united at the extreme base*

1. Lip fused to at least the lower half of the column
1.A. Lip fused to the column for most of the length of the latter **47. Epidendrum**
1.B. Upper half of the column free from the lip **51. Barkeria**
2. Pollinia 2; leaves and racemes or panicles borne directly on the rhizome (stem absent) **165. Oncidium**
3. Pollinia 4, all equal; stem present, more than 4 cm (often much more) **52. Cattleya**
4. Pollinia 6, 4 large and 2 smaller; stem present but not more than 4 cm **58. Leptotes**

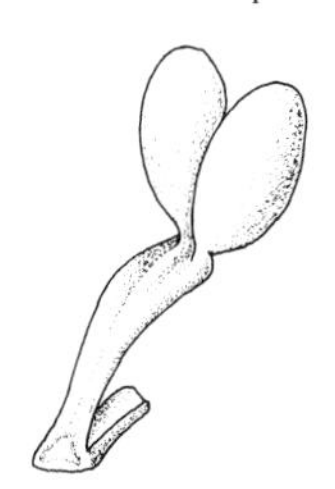

Oncidium concolor – pollinia.

Subkey E8 Staminode absent; inflorescences terminal or borne directly on the rhizome; fertile pollinia 8; *lateral sepals free from each other or very shortly united at the extreme base*

1. Sepals to 2 cm; lip not rolled around column
1.A. Lip deeply concave or pouched; flowers pinkish purple, at least in part **41. Arpophyllum**
1.B. Lip flat; flower white, yellow or greenish yellow **72. Octomeria**
2. Lip very conspicuously 3-lobed, without a claw, its lateral lobes rolled around the column **53. Laelia**
3. Lateral sepals spreading **55. Brassavola**
4. Lateral sepals not spreading, lying close together under the lip **38. Arundina**

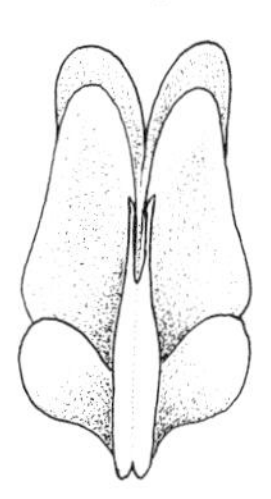

Leptotes bicolor – pollinia.

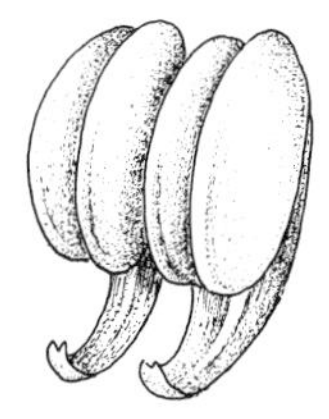

Cattleya schilleriana – pollinia.

DESCRIPTIONS OF GENERA AND SPECIES

1. CYPRIPEDIUM Linnaeus

A genus of about 35 species, widely distributed throughout the northern hemisphere, mostly north of 35° N, though extending further south in Asia and Mexico. They are known as 'Slipper Orchids', because of the shape of the lip. Their aerial parts die down completely in winter.

They can be cultivated out-of-doors on well-drained, humus-rich soils, performing best in partial shade. Though mostly hardy, they may need protection from severe frost, if not covered by snow. The name *Cypripedium* is still sometimes applied to species of *Paphiopedilum* (p. 00) and *Phragmipedium* (p. 00), though both have long been considered distinct.

Literature: Franchet, A., Les Cypripedium de l'Asie Centrale et de l'Asie Orientale, *Journal de Botanique* **8**: 225–33, 249–56, 265–71 (1894); Luer, C.A., *Native orchids of the United States and Canada* (1975).

HABIT: terrestrial, aerial parts dying away in winter.
LEAVES: deciduous, pleated (except No. **9**), rolled when young, spirally arranged or in a single, almost opposite pair.
FLOWERS: persistent, solitary or up to 12 in a raceme.
BRACTS: usually leaf-like.
SEPALS: edge-to-edge in bud; upper sepal free, lateral sepals fused into a single lower sepal below the lip (free in No. **6**).
PETALS: free.
LIP: extended into a sac-like pouch with an inturned rim (except No. 13), the opening partially blocked by a staminode.
FERTILE ANTHERS: 2; pollen in exposed granular masses.
STAMINODE: 1, borne on the column.
OVARY: single-celled.

Synopsis of characters.

Leaves. Paired: **14–19**; spirally arranged: **1–13**. Basal: **18**. Fan-shaped: **14**. 3-veined: **15**. Spotted: **16**. Membranous: **15**.
Lower stem. Hairless: **15,16**; sparsely to moderately hairy: **1–13,18,19**; shaggy: **14,17**.
Floral bract. Linear: **15**. Absent: **16**.
Flowers. Solitary: **4,5–8,11,14–19**; 1–3 per raceme: **1–3,12,13,17**; more than 4 per raceme: **9,10,17**. Clustered: **17**.
Lateral sepals. Free: **11**; fused into a rounded lower sepal: **12,14,15**; fused into a pointed or bifid lower sepal: **1–10,14,16–19**.

Petals. Longer than lip: **1–8,11,14–17**; equal to or shorter than lip: **9,10,12,13,18,19**. Linear or narrowly lanceolate: **1–4,11**; spathulate: **19**; acute: **1–9,11,14,16–18**; rounded: **10,12,13,15,19**. Twisted or curled: **1–3**.
Lip. Furrowed in front: **14–18**. With upstanding or turned-out rim: **19**. Spurred: **11**.

Key to species

1. Leaves 3 or more, spirally arranged *Go to Subkey 1*
2. Leaves 2, almost opposite, basal or borne on the stem *Go to Subkey 2*

Subkey 1. Leaves 3 or more, spirally arranged on the flowering stem.

1. Lateral sepals free from each other; lip with a blunt, conical spur **11. arietinum**
2. Petals linear to narrowly lanceolate, more than 2 cm, more than 3 times as longer than broad
2.A. Petals not twisted or curled; lower sepal entire or bifid to less than 5 mm **4. cordigerum**
2.B. Lip dull to bright yellow **1. calceolus**
2.C. Sepals and petals greenish with purple marks **2. candidum**
2.D. Sepals and petals uniformly brownish purple **3. montanum**
3. Petals acute at apex
3.A. Leaves 5–16; flowers 1–8; lip yellow **9. irapeanum**
3.B. Lip almost spherical **8. fasciolatum**
3.C. Sepals, petals and lip uniformly violet to purplish pink or greenish brown with darker veins **5. macranthon**
3.D. Sepals and petals pinkish yellow marked with a reddish purple network; lip very dark to almost black **6. tibeticum**
3.E. Sepals and petals greenish yellow; lip purple, paler around the mouth **7. himalaicum**
4. Flowers 5–10; petals less than 2 cm; upper sepal less than 2.5 cm **10. californicum**
5. Plant to 1 m, densely hairy; sepals and petals white, lip pink or white **12. reginae**
6. Plant to 60 cm, less densely hairy; sepals, petals and lip yellow faintly spotted with purplish brown **13. flavum**

Subkey 2. Leaves 2, almost opposite, basal or on the stem.

1. Sepals and petals equal to or shorter than lip
1.A. Petals rounded, less than 2.5 cm; flowers blotched; lip with out-turned rim **19. guttatum**
1.B. Petals acute, more than 3.5 cm; flowers neither spotted nor blotched; lip with inturned rim **18. acaule**
2. Petals less than 3.5 cm; upper sepal less than 3 cm

2.A. Lower part of stem hairless; leaves heart-shaped, membranous; petals rounded **15. debile**

2.B. Lower stem shaggy; leaves elliptic, pleated; petals acute **17. fasciculatum**

3. Leaves spotted, broadly elliptic, obtuse; stem hairless below; floral bract absent **16. margaritaceum**

4. Leaves not spotted, fan-shaped; stem shaggy below; floral bract present **14. japonicum**

1. C. calceolus Linnaeus

HABIT: to 80 cm, sparsely to moderately hairy.

LEAVES: 3–5, 5–20 × 2–10 cm, ovate to elliptic, acute, spirally arranged.

FLOWERS: 1 or 2.

SEPALS: upper and lower sepals ovate, acuminate.

PETALS: 3–7 cm, linear to narrowly lanceolate, longer than the lip, twisted or curled, acute.

SEPALS AND PETALS: greenish yellow with purple markings or evenly purplish or reddish brown.

LIP: dull to bright yellow, spotted inside with red or purple.

Var. **calceolus**

SYNONYM: *C. microsaccus* Kränzlin.

ILLUSTRATIONS: Sundermann, Europäische und mediterrane Orchideen, 216 (1975); Williams et al., Orchids of Britain and Europe, 25 (1978).

HABIT: to 30 cm, sparsely hairy.

LEAVES: 10–18 × 5–6 cm.

SEPALS AND PETALS: purplish brown.

DISTRIBUTION: from N & C Europe across Siberia to Korea.

HARDINESS: H2.

FLOWERING: spring–summer.

Var. **pubescens** (Willdenow) Correll

SYNONYMS: *C. pubescens* Willdenow; *C. veganum* Cockerell Barker; *C. luteum* Aiton.

ILLUSTRATIONS: Botanical Magazine, 911 (1806); American Orchid Society Bulletin 38: 905 (1969); Luer, Native orchids of the United States and Canada, 45, 47 (1975); American Orchid Society Bulletin 53: 470 (1984).

HABIT: taller and hairier than var. *calceolus* and larger in its floral parts.

SEPALS AND PETALS: greenish yellow with purple markings.

DISTRIBUTION: N America.

HARDINESS: H2.

FLOWERING: spring–summer.

Var. **parviflorum** (Salisbury) Fernald

SYNONYM: *C. parviflorum* Salisbury.

ILLUSTRATIONS: Botanical Magazine, 3024 (1830); American

Orchid Society Bulletin 38: 904 (1969); Luer, Native orchids of the United States and Canada, 51 (1975).

HABIT: very similar to var. *calceolus* though smaller in its floral parts, sparsely downy.

PETALS: reddish brown, very twisted.

DISTRIBUTION: northeast N America.

HARDINESS: H2.

FLOWERING: summer.

2. C. **candidum** Willdenow

ILLUSTRATIONS: Botanical Magazine, 5855 (1870); American Orchid Society Bulletin 38: 906 (1969); Luer, Native orchids of the United States and Canada, 53 (1975).

Like *C. calceolus* except as below.

HABIT: stem with up to 5 leaves held very upright.

FLOWERS: similar in size to those of var. *parviflorum*.

SEPALS AND PETALS: green marked with purple, or almost completely purple.

LIP: white, sometimes veined with purple, with red or purple spots inside.

DISTRIBUTION: E & C USA, adjacent Canada.

HARDINESS: H2.

FLOWERING: spring–summer.

3. C. **montanum** Lindley

ILLUSTRATIONS: Botanical Magazine, 7319 (1893); American Orchid Society Bulletin 38: 904 (1969); Luer, Native orchids of the United States and Canada, 55 (1975).

Like *C. calceolus* except as below.

HABIT: with up to 6 leaves on each stem.

FLOWERS: similar in size to those of var. *pubescens*.

SEPALS AND PETALS: uniformly brownish purple.

LIP: compressed, white flushed with purple near the base.

DISTRIBUTION: western N America east to Wyoming.

HARDINESS: H2.

FLOWERING: summer.

4. C. **cordigerum** Don

ILLUSTRATIONS: Botanical Magazine, 9364 (1934); Pradhan, Indian Orchids I: 35 (1976); Senghas (editor), Proceedings of the 8th world orchid conference, t. 4 (1976).

Like *C. calceolus* and its allies except as below.

LOWER SEPAL: bifid to 1.8 cm.

PETALS: shorter and broader.

SEPALS AND PETALS: pale yellow, white or pale green.

LIP: white with a few purple spots inside.

DISTRIBUTION: Himalaya.

HARDINESS: H2.

FLOWERING: spring.

5. C. macranthon Swartz

SYNONYMS: *C. speciosum* Rolfe; *C. franchetii* Rolfe.

ILLUSTRATIONS: Botanical Magazine, 2938 (1829); American Orchid Society Bulletin **44**: 769 (1975); Williams et al., Orchids of Britain and Europe, 27 (1978); Orchid Review **92**: 211, 212 (1984).

HABIT: to 50 cm, sparsely to moderately hairy.

LEAVES: 3 or 4, 8–16 × 4–7 cm, ovate to elliptic, acute, spirally arranged.

FLOWERS: usually solitary.

UPPER AND LOWER SEPALS: broadly ovate, acuminate, violet to purplish pink or greenish brown, with darker veins.

PETALS: 4–6.5 × 1.5–3 cm, broadly ovate, acute, a little longer than lip.

LIP: similar in colour to petals, paler round the rim.

DISTRIBUTION: USSR south of *c.* 60° N, Mongolia, N & W China, Korea, Japan.

HARDINESS: H2.

FLOWERING: spring–summer.

C. × ventricosum Swartz

ILLUSTRATIONS: Williams et al., Orchids of Britain and Europe, 26, 1978. The presumed hybrid between *C. macranthon* and *C. calceolus.* It occurs in the wild and in cultivation and may be merely a colour variant of *C. macranthon.*

6. C. tibeticum Hemsley

ILLUSTRATIONS: Botanical Magazine, 8070 (1906); Pradhan, Indian Orchids **1**: 35 (1976); Senghas (editor), Proceedings of the 8th world orchid conference, t. 8 (1976).

Similar to *C. macranthon* except as below.

FLOWERS: larger.

SEPALS AND PETALS: pinkish yellow marked with a reddish purple network.

LIP: uniformly dark (sometimes almost black).

DISTRIBUTION: SW China.

HARDINESS: H2.

7. C. himalaicum Hemsley

ILLUSTRATIONS: Botanical Magazine, 8965 (1922); Morley & Everard, Wild flowers of the world, t. 109 (1970).

Like *C. macranthon* except as below.

FLOWERS: slightly smaller.

SEPALS AND PETALS: narrower, greenish yellow, strongly marked with dark red lines.

LIP: purple, paler behind and around the mouth.

DISTRIBUTION: Himalaya & W China.

HARDINESS: H2.

8. C. fasciolatum Franchet

SYNONYM: *C. wilsonii* Rolfe.

Like *C. macranthon* except as below.

SEPALS AND PETALS: spotted and striped with violet.

LIP: almost spherical, spotted with violet below and broadly striped above.

DISTRIBUTION: W. China.

HARDINESS: H2.

9. C. irapeanum Llave & Lexarza

ILLUSTRATIONS: Luer, Native orchids of the United States and Canada, 59 (1975); Senghas (editor), Proceedings of the 8th world orchid conference, t. 4 (1976); Orchid Digest **41**: 209 (1977).

HABIT: to 1.5 m, moderately hairy, especially below.

LEAVES: 5–16, 5–15 × 3–9 cm, ovate to elliptic, acute, spirally arranged.

FLOWERS: 1–8.

UPPER SEPAL: lanceolate, acuminate.

LOWER SEPAL: ovate, acuminate, entire.

PETALS: 3–6.5 cm, oblong, acuminate, equal to or shorter than lip.

SEPALS AND PETALS: all bright yellow.

LIP: bright yellow, flushed with red inside.

DISTRIBUTION: C & S Mexico, Guatemala.

HARDINESS: G1.

FLOWERING: summer.

10. C. californicum Gray

ILLUSTRATIONS: Botanical Magazine, 7188 (1891); Luer, Native orchids of the United States and Canada, 63 (1975); Orchid Digest **43**: 112 (1979).

HABIT: to 1.2 m, sparsely hairy.

LEAVES: 5–10, 5–15 × 2–6 cm, elliptic to lanceolate, acute, spirally arranged.

FLOWERS: 5–10.

UPPER SEPAL: oblong to elliptic, acuminate.

LOWER SEPAL: elliptic, shortly bifid.

PETALS: 1.4–2 cm, oblong-lanceolate, blunt, equal to or shorter than lip.

SEPALS AND PETALS: all yellowish green.

LIP: white.

DISTRIBUTION: USA (S Oregon, N California).

HARDINESS: H2.

FLOWERING: spring–summer.

11. C. arietinum Brown

SYNONYM: *C. plectrochilum* Franchet.

ILLUSTRATIONS: Botanical Magazine, 1569 (1813); Luer,

Native orchids of the United States and Canada, 43 (1975); American Orchid Society Bulletin **44**: 384 (1977), **55**: 696 (1986).

HABIT: to 30 cm, sparsely hairy.

LEAVES: 3–4, 1.5–3.5 cm, elliptic to lanceolate, acute, spirally arranged.

FLOWER: solitary.

UPPER SEPAL: ovate-lanceolate, acuminate, green with purple veins.

LATERAL SEPALS: free, linear-lanceolate, greenish purple.

PETALS: 1.3–2.2 cm, linear, acute, greenish purple, a little longer than lip.

LIP: with a blunt conical spur, white above with purplish veins coalescing below.

DISTRIBUTION: east & central N America, from 42° to 52° N; C & SW China, N Burma.

HARDINESS: H2.

FLOWERING: spring–summer.

12. C. reginae Walter

SYNONYMS: *C. album* Aiton; *C. spectabile* Salisbury; *C. canadensis* Michaux.

ILLUSTRATIONS: Botanical Magazine, 216 (1793); Shuttleworth et al., Orchids, 17 (1970); Luer, Native orchids of the United States and Canada, 57 (1975); American Orchid Society Bulletin **54**: 1194 (1985).

HABIT: to 1 m, densely hairy.

LEAVES: 3–7, 10–25 × 6–16 cm, ovate to lanceolate, spirally arranged.

FLOWERS: 1–3.

UPPER SEPAL: broadly elliptic to circular, acuminate.

LOWER SEPAL: similar to upper, but blunt.

PETALS: 2.5–5 cm, oblong-elliptic, rounded.

SEPALS AND PETALS: white; petals shorter than lip.

LIP: pink streaked with white, or pure white.

DISTRIBUTION: eastern N America.

HARDINESS: H2.

FLOWERING: spring–summer.

The name *C. reginae* has apparently been misused for the hybrid *Paphiopedilum leeanum* × *fairrieanum* (see p. 00).

13. C. flavum Hunt & Summerhayes

SYNONYM: *C. luteum* Franchet, not Aiton, not Rafinesque.

ILLUSTRATION: Gardeners' Chronicle **57**: 257 (1915).

Like *C. reginae* except as below.

HABIT: slightly smaller (to 60 cm), with less dense, brownish hair.

FLOWERS: bright yellow, faintly spotted or striped with purplish brown, particularly on the lip.

DISTRIBUTION: W China.
HARDINESS: H2.

14. C. japonicum Thunberg
SYNONYM: *C. formosanum* Hayata.
ILLUSTRATIONS: Botanical Magazine, 9520 (1938); Shuttleworth et al., Orchids, 19 (1970); American Orchid Society Bulletin **44**: 768 (1975); Kew Magazine **4**: pl. 74 (1987).
HABIT: to 50 cm, shaggy below, sparsely hairy above.
LEAVES: 2, in the upper half of the stem, 13–20 × 13–20 cm, broadly fan-shaped.
FLOWER: solitary.
SEPALS AND PETALS: green with purple spots at base; the acute lower sepal sometimes slightly bifid.
PETALS: pointed, 4–6 cm, just longer than lip.
LIP: pale pink with purple spots, deeply furrowed in front.
DISTRIBUTION: Japan, Taiwan, S & C China.
HARDINESS: H2.

15. C. debile Reichenbach
ILLUSTRATIONS: Botanical Magazine, 8183 (1908); Maekawa, The wild orchids of Japan, 85 (1971); Kitamura et al., Herbaceous plants of Japan (Monocotyledons), t. 1 (1978).
HABIT: to 25 cm, hairless.
LEAVES: 2, about halfway up stem, 3–6 cm, heart-shaped, membranous, with 3–7 prominent veins.
BRACTS: linear.
FLOWER: solitary, pendent.
SEPALS AND PETALS: pale green; petals 1.3–2 cm, lanceolate, rounded, just longer than lip.
LIP: spherical, white marked with purple around the mouth.
DISTRIBUTION: Japan, W China.
HARDINESS: H2.

16. C. margaritaceum Franchet
SYNONYM: *C. ebracteatum* Rolfe.
ILLUSTRATION: Gardeners' Chronicle **46**: 419 (1909).
HABIT: to 15 cm, hairless below.
LEAVES: 2, about halfway up stem, 10–18 × 10–18 cm, broadly elliptic, spotted with purple.
BRACT: absent.
FLOWER: solitary.
SEPALS AND PETALS: broadly ovate, acute, yellowish green spotted with purple; petals 4–5 cm, longer than lip.
LIP: fleshy, angular, glandular-hairy, yellow with purple spots.
DISTRIBUTION: W China (Yunnan).
HARDINESS: H2.
FLOWERING: summer.

17. C. fasciculatum Watson

SYNONYM: *C. knightae* Nelson.

ILLUSTRATIONS: Botanical Magazine, 7275 (1893); American Orchid Society Bulletin **38**: 906 (1969); Luer, Native orchids of the United States and Canada, 65 (1975).

HABIT: to 30 cm, shaggy below.

LEAVES: 2, in middle or upper half of stem, 5–12 × 3–8 cm, elliptic, obtuse.

FLOWERS: in clusters of up to 4, purplish brown to green.

SEPALS AND PETALS: acute, about twice as long as lip; petals 1.5–2.5 cm.

DISTRIBUTION: Western N America.

HARDINESS: H2.

FLOWERING: spring–summer.

C. elegans Reichenbach from the Himalaya is similar and should perhaps be regarded as belonging to the same species.

18. C. acaule Aiton

ILLUSTRATIONS: Botanical Magazine, 192 (1792); Shuttleworth et al., Orchids, 19 (1970); Luer, Native orchids of the United States and Canada, 41 (1975); American Orchid Society Bulletin **53**: 475 (1984); Orchid Review **97**: 114 (1989).

HABIT: to 40 cm, sparsely hairy.

LEAVES: 2, 10–28 × 5–15 cm, basal, elliptic, obtuse.

FLOWER: solitary.

SEPALS AND PETALS: shorter than lip, lanceolate, acute, yellowish green to purple; petals 4–6 cm.

LIP: pink or white, deeply furrowed in front.

DISTRIBUTION: northeastern N America.

HARDINESS: H2.

FLOWERING: spring–summer.

19. C. guttatum Swartz

SYNONYM: *C. yatabeanum* Makino.

ILLUSTRATIONS: Botanical Magazine, 7746 (1900); Luer, Native orchids of the United States and Canada, 67 (1975); Williams et al., Orchids of Britain and Europe, 29 (1978).

HABIT: to 30 cm, sparsely hairy.

LEAVES: 2, 5–12 × 2.5–7 cm, bluish green drying black, elliptic, acute.

FLOWER: solitary, white blotched with purple or brown.

UPPER SEPAL: elliptic, acuminate.

LOWER SEPAL: lanceolate, shortly bifid.

PETALS: 1.3–2 cm, spathulate or with oblong, rounded tips, shorter than lip.

LIP: a deep open pouch with upstanding or out-turned rim

DISTRIBUTION: from European Russia across temperate Asia to Alaska & NW Canada.

HARDINESS: H2.
FLOWERING: summer.

2. PHRAGMIPEDIUM (Pfitzer) Rolfe

A genus of 15–20 species from C & S America, most of which were originally placed in *Cypripedium* or *Selenipedium*; the latter name is frequently used for them in the horticultural trade. No true *Selenipedium* species are cultivated. *Phragmipedium* species require cultivation conditions similar to those for the unmottled species of *Paphiopedilum* (pp. 33–7).

Literature: Garay, L.A., The genus Phragmipedium, *Orchid Digest* **43**: 133–48 (1979).

HABIT: terrestrial, rarely epiphytic or on rocks.
LEAVES: persistent, strap-shaped, leathery, 2-ranked, folded when young.
FLOWERS: in terminal racemes or panicles, soon falling.
SEPALS: edge-to-edge in bud.
UPPER SEPAL AND PETALS: free.
LATERAL SEPALS: fused into a single lower sepal.
LIP: a sac-like pouch (rarely petal-like), sometimes with 2 triangular appendages on the rim; opening of lip partially blocked by a staminode.
FERTILE ANTHERS AND STAMINODE: borne on the column; anthers 2; pollen in exposed granular masses.
OVARY: 3-celled.

Synopsis of characters.

Leaves. 30 cm or less: **1**; more than 30 cm: **2–8**. Less than 2 cm wide: **6**; 2–5 cm wide: **1–3,5,7,8**; more than 5 cm wide: **4**. With yellow margins: **2,3**.
Flowers. 2–4 per inflorescence: **2–5**; 6 or more per inflorescence: **1–3,6,7**. 8 cm wide or less: **1–6**; more than 8 cm wide **2–8**.
Petals. 20 cm or more, more than 3 times as long as sepals: **4,5**; 15 cm or less, less than 3 times as long as sepals: **1–3,6–8**. Similarly shaped to sepals: **1**. Wavy: **2–7,8**; crisped: **7,8**; not wavy or crisped: **1**.
Staminode. Elliptic to circular: **1**; diamond-shaped to 3-lobed: **2–5**; triangular to semicircular: **6–8**.

Key to species

1. Lip similar to sepals and petals, not pouched — **5. lindenii**
2. Petals more than 3 times as long as sepals — **4. caudatum**
3. Petals similar in shape to sepals (though larger) — **1. schlimii**
4. Opening of lip with 2 vertical triangular lobes

4.A. Petals drooping, with crisped margins — **8. boissierianum**
4.B. Petals not drooping, twisted — **7. longifolium**

5. Petals not tapering; leaves more than 3 cm wide, with yellow margins

5.A. Raceme to 40 cm, with 2–4 flowers, each to 10 cm wide **3. sargentianum**

5.B. Raceme or panicle to 80 cm, with 3–7 flowers, each 7–8 cm wide **2. lindleyanum**

6. Petals narrow and tapering; leaves less than 2 cm wide, without yellow margins **6. caricinum**

1. **P. schlimii** (Linden & Reichenbach)Rolfe

ILLUSTRATIONS: Botanical Magazine, 5614 (1866); Orchid Digest **43**: 134 (1979); Bechtel et al., Manual of cultivated orchid species, 262 (1981); American Orchid Society Bulletin **55**: 364 (1986).

LEAVES: 4–8, 20–30 × 2.5–3 cm, strap-shaped, acute.

RACEMES OR PANICLES: to 50 cm, bearing 6–10 flowers; stems hairy.

FLOWERS: 5–6 cm wide.

SEPALS: *c.* 2 cm, ovate, obtuse, white flushed with pink.

PETALS: similar to sepals but larger, speckled with pink at base.

LIP: an ellipsoid pouch, mottled dark on pale pink.

STAMINODE: elliptic to circular, bright yellow.

DISTRIBUTION: Colombia.

HARDINESS: G2.

FLOWERING: summer.

2. **P. lindleyanum** (Lindley) Rolfe

ILLUSTRATIONS: American Orchid Society Bulletin **44**: 235 (1975); Orchid Digest **43**: 134 (1979).

LEAVES: 4–7, 30–50 × 4–6.5 cm, lanceolate, acute.

RACEMES OR PANICLES: to 80 cm, bearing 3–7 flowers; stems reddish green, downy.

FLOWERS: 7–8 cm wide.

SEPALS: elliptic, slightly hooded, yellowish green.

UPPER SEPAL: 3 × 2 cm.

LOWER SEPAL: a little broader.

PETALS: *c.* 5 cm, oblong, rounded, with wavy margins, yellowish green at the base fading to greenish white at the tips; tips strongly purple-veined.

LIP: a broad shallow pouch with incurved margins, yellow with brown veins.

STAMINODE: diamond-shaped to weakly 3-lobed, yellow.

DISTRIBUTION: Venezuela & Guyana.

HARDINESS: G2.

3. **P. sargentianum** (Rolfe) Rolfe

ILLUSTRATIONS: Botanical Magazine, 7446 (1895); Orchid Digest **43**: 134 (1979).

Like *P. lindleyanum* except as below.

RACEME: shorter, *c.* 40 cm, with 2–4 flowers.

FLOWERS: to 10 cm wide.
LIP: narrower.
DISTRIBUTION: Brazil.
HARDINESS: G2.

4. P. caudatum (Lindley) Rolfe
SYNONYM: *P. warscewiczianum* (Reichenbach) Schlechter.
ILLUSTRATIONS: Orchid Digest **43**: 138 (1979); Bechtel et al., Manual of cultivated orchid species, 262 (1981); Shuttleworth et al., Orchids, 20 (1970); American Orchid Society Bulletin **55**: 364 (1986); Orchid Review **94**: 183 (1986).
HABIT: epiphytic or on rocks.
LEAVES: 5–9, 40–60 × 6 cm, strap-shaped, light green; apex irregularly toothed.
RACEMES: to 1 m, bearing 2–4 flowers; stems downy.
SEPALS: lanceolate, tapering, off-white to yellowish green with darker green or orange veins; upper sepal 10–15 × 2–3 cm, curled forward over flower; lower sepal 7–10 × 3–4 cm.
PETALS: to 60 × 1 cm, trailing, dark reddish to greenish brown, paler at the base.
LIP: a deep pinkish white pouch with pink or brown net-venation; rim turned out, yellow.
STAMINODE: diamond-shaped, off-white with dark purple corners.
DISTRIBUTION: Mexico through C America to Peru, Ecuador, Colombia & Venezuela.
HARDINESS: G2.
FLOWERING: spring–summer.

5. P. lindenii (Lindley) Dressler & Williams
ILLUSTRATIONS: Orchid Digest **43**: 139 (1979); Bechtel et al., Manual of cultivated orchid species, 262 (1981).
Like *P. caudatum* except as below.
LIP: simple, unpouched, similar in shape and colour to the petals, though a little broader.
DISTRIBUTION: Colombia, Ecuador & Peru.
FLOWERING: summer.

6. P. caricinum (Lindley) Rolfe
ILLUSTRATIONS: Botanical Magazine, 5466 (1864); American Orchid Society Bulletin **44**: 236 (1955), **55**: 364 (1986); Orchid Digest **43**: 143 (1979).
LEAVES: 3–7, 35–50 cm × 5–15 mm, linear, stiff and sedge-like.
RACEME OR PANICLE: 30–40 cm, bearing several flowers in succession; stems downy.
FLOWERS: 5–8 cm long.
SEPALS: 2–3 × 1–1.5 cm, ovate to lanceolate with wavy margins, yellowish or purplish green with a green base.

PETALS: 5–10 cm × 5 mm, tapering, twisted, similar to sepals in colour.
LIP: a rounded greenish yellow pouch with brown veins.
STAMINODE: triangular, green.
DISTRIBUTION: Peru, Bolivia & Brazil.
HARDINESS: G2.
FLOWERING: spring–autumn.

7. **P. longifolium** (Warscewicz & Reichenbach) Rolfe
ILLUSTRATIONS: Botanical Magazine, 5970 (1872); American Orchid Society Bulletin **44**: 225 (1975); Orchid Digest **43**: 146 (1979).
LEAVES: 5–8, 40–60 × 2–3 cm, strap-shaped, tapering to an acute tip.
RACEME: to 60 cm, bearing several flowers in succession; stems reddish purple, sparsely downy.
FLOWERS: to 18 cm wide.
SEPALS: green or greenish yellow with dark green or greenish red veins; upper sepal ovate-lanceolate, 5 × 1.5 cm; lower sepal broadly ovate.
PETALS: to 12 × 1 cm, tapering and twisted with wavy margins, greenish yellow with purple margins, solid purple at tips.
LIP: a deep pouch with 2 vertical triangular lobes on the rim, greenish yellow flushed with purple.
STAMINODE: broadly triangular to semicircular, greenish yellow; upper edge deep purple.
DISTRIBUTION: Costa Rica to Colombia.
HARDINESS: G2.
FLOWERING: winter.

8. **P. boissierianum** (Reichenbach) Rolfe
ILLUSTRATION: Orchid Digest **43**: 146 (1979).
Like *P. longifolium* except as below.
LEAVES: shorter.
PETALS: drooping, with crisped margins.
DISTRIBUTION: Peru.
HARDINESS: G2.
FLOWERING: summer–autumn.

3. PAPHIOPEDILUM Pfitzer

A genus of about 60 species from the Himalaya and S India through SE Asia to the Philippines, New Guinea and the Solomon Islands. Most of the species were originally placed in *Cypripedium* which now contains only temperate species with rolled leaves. Many artificial hybrids have been produced which exploit the showiness of the upper sepal seen in the wild species. The petals are often strikingly marked with spots or warts and frequently bear marginal hairs. The flowers persist for several weeks if not pollinated. All the species make striking ornamental plants, but several are bizarre rather than beautiful.

Plants can be grown on bark or in well-drained pots in epiphytic compost. Lacking pseudobulbs, they must not be allowed to dry out, though less water should be given in winter. They need shading from direct summer sunlight. Species with mottled leaves do best in a warm, moist house; those with plain leaves generally prefer cooler conditions. Propagation is usually effected by division of large plants when repotting in early spring; some species have been successfully grown from seed, e.g. *P. delenatii*.

Literature: Asher, J.R., A checklist for the genus Paphiopedilum, *Orchid Digest* **44**: 175–84, 213–28 (1980), **45**: 15–26, 57–65 (1981); Graham, R. & Roy, R., *Slipper orchids* (1982); Cribb, P., *The genus Paphiopedilum* (1987); Braem, G., *Paphiopedilum* (1988).

HABIT: epiphytic, terrestrial or on rocks.
STEMS: very short, enclosed in leaf-bases.
LEAVES: 4–several, ovate to strap-shaped, persistent, leathery, often mottled, in 2 opposite ranks, folded when young.
FLOWERS: solitary, or in terminal racemes bearing 2–8 flowers all open together, or with many flowers in succession over several months.
SEPALS: overlapping in bud.
UPPER SEPAL: free.
LATERAL SEPALS: united into a single lower sepal under the lip.
PETALS: free.
LIP: extended into a sac-like pouch usually with side lobes on the rim; opening of pouch partially blocked by a broad staminode borne on the column.
POLLEN: in exposed granular masses.
OVARY: 1-celled.

Key to species

1. Lip without 2 side-lobes on the rim *Go to Subkey* 1
2. Flowers 2 or more, arising from different points on the flowering stem
2.A. Petals spathulate, flat or slightly twisted
2.A.a. Upper sepal pale, marked with large, irregular, brown blotches **41. haynaldianum**
2.A.b. Upper sepal yellowish green flushed with purple and streaked with brown at the base **40. lowii**
2.B. Flowers open 1 at a time; upper sepal almost circular; petals horizontal, hairy on margins
2.B.a. Leaves glaucous **38. glaucophyllum**

2.B.b. Leaves green **37. victoria-regina**

2.C. Flowers open all together; upper sepal elliptic; petals drooping, not hairy on margin **39. parishii**

3. Leaves mottled on the upper surface *Go to Subkey 2*

4. Leaves not mottled on the upper surface *Go to Subkey 3*

Subkey 1. Lip without 2 side-lobes on the rim.

1. Petals similar to sepals in shape and size; flowers solitary or rarely 2 or 3 together arising from the same point on the stem

1.A. Lip cup-shaped to spherical; petals to 4 cm; upper sepal unspotted

1.A.a. Lip of the same colour as sepals and petals; staminode white and yellow **4. niveum**

1.A.b. Lip pink; sepals and petals pink and white; staminode pink and yellow **5. delenatii**

1.B. Flowers pale or deep yellow, finely flecked **1. concolor**

1.C. Flowering stem 3–10 cm; flower marked with coalescing blotches **3. godefroyae**

1.D. Flowering stem very short; flower marked with spots which coalesce only at the bases of the sepals and petals **2. bellatulum**

2. Upper sepal unstriped or with 1–5 bold stripes **9. stonei**

3. Basal parts of the petals evenly spotted with purple **6. rothschildianum**

4. Basal parts of the petals unspotted or with a few uneven spots

4.A. Upper sepal to 3.5 cm; petals to 15 cm **7. philippinense**

4.B. Upper sepal to 5 cm; petals to 12 cm **8. glanduliferum**

Subkey 2. Leaves mottled on the upper surface.

1. Lip with a prominent tooth on the outer edge of the rim **28. acmodontum**

2. Petals without spots or warts

2.A. Upper sepal to 6 cm; petals narrowly elliptic to spathulate, to 10 cm, greenish brown at the base, pink at the tip **18. dayanum**

2.B. Upper sepal at most 3 cm; petals sickle-shaped, to 4.5 cm, purple **19. violascens**

3. Each petal with 50 or more spots or warts evenly distributed across at least the basal part of the blade

3.A. Spots or warts extending well into the apical part of the petals though not necessarily to the extreme tip

3.A.a. Petals broadest in the upper half **29. javanicum**

3.A.b. Petals acute, gradually tapering, almost flat, not curled, twisted or curved downwards

3.A.b.i. Petals yellowish green spotted with purplish black, almost horizontal **35. sukhakulii**

3.A.b.ii. Petals brown, pointing obliquely downwards **36. wardii**

3.A.c. Petals evenly marked with small spots or warts
3.A.c.i. Petals curled **32. superbiens**
3.A.c.ii. Petals not curled **33. ciliolare**
3.A.d. Petals irregularly marked with large spots, some with pale centres **34. argus**
3.B. Petals elliptic, mostly purplish brown with dark spots or warts; upper sepal with *c.* 12 distinct purple and green veins **30. purpuratum**
3.C. Petals spathulate, usually pinkish purple with faint pinkish purple spots; upper sepal with 15–20 indistinct veins **25. hookerae**
4. Spots or warts all borne on or near the petal-margins
4.A. Upper sepal green to greenish yellow, often brown at the base, indistinctly veined; lip with a bright green rim **23. bullenianum**
4.B. Upper sepal ovate, gradually tapering, white or off-white with green veins
4.B.a. Petals 4–5 cm, spathulate **27. venustum**
4.B.b. Petals to 10 cm, narrowly elliptic **18. dayanum**
4.C. Upper and lower petal-margins with *c.* 10 warts **22. lawrenceanum**
4.D. Petals almost horizontal, with dark purple veins; warts *c.* 1 mm in diameter **20. barbatum**
4.E. Petals curved downwards with indistinct green veins; warts *c.* 3 mm in diameter **21. callosum**
5. Petals with large, black spots which often coalesce; lip dull yellow to reddish brown, strongly veined with greenish brown **27. venustum**
6. Petals bright green and bright pink or pinkish purple
6.A. Spots or warts all in the basal half of the petal **24. appletonianum**
6.B. Petals descending, broadly spathulate, each with a prominent white patch **29. javanicum**
7. Petals pale green with *c.* 10 large black spots on upper margin and midline; upper sepal with broad white margin **26. tonsum**
8. Petals reddish brown and dull green with many small spots in the basal half, often concentrated in an upper corner **31. mastersianum**

Subkey 3. *Leaves not mottled above.*
1. Petal-tips strongly recurved; upper sepal with bold, interconnecting veins **17. fairreanum**
2. Upper sepal unspotted, finely or indistinctly speckled or with several indistinct stripes
2.A. Petals oblong, to 4 cm; upper sepal *c.* 6 cm wide **10. charlesworthii**
2.B. Bracts much shorter than ovary; petals with black or purple speckles at base **11. hirsutissimum**

2.C. Bracts almost equalling ovary; petals without speckling **12. villosum**

3. Upper sepal with a dark central stripe
3.A. Leaves uniformly dark green above **13. spicerianum**
3.B. Leaves pale green with darker veins **14. druryi**
4. Flower-stalk and ovary shaggy **12. villosum**
5. Flowers to 8 cm wide **15. exul**
6. Flowers more than 10 cm wide **16. insigne**

1. **P. concolor** (Lindley) Pfitzer

ILLUSTRATIONS: Botanical Magazine, 5513 (1865); Orchid Digest **42**: 72, 74 & 75 (1978); Bechtel et al., Manual of cultivated orchid species, 253 (1981); Cribb, The genus Paphiopedilum, pl. 3 (1987); Braem, Paphiopedilum, 48, 49 (1988).

LEAVES: 4–5, 10–15 × *c.* 6 cm, elliptic to oblong, obtuse, mottled greyish green above, reddish purple beneath.

FLOWER-STALK: to 50 cm, purple, hairy, shorter than leaves, 1- or 2-flowered.

FLOWER: *c.* 5 cm wide, pale to deep yellow, finely flecked with reddish brown.

UPPER SEPAL: circular, hooded.

LOWER SEPAL: ovate, acute.

PETALS: elliptic to oblong, slightly curved downwards.

LIP: blunt, conical, equal to or shorter than the petals, without side lobes.

DISTRIBUTION: Burma, Thailand, Cambodia, S Laos & S China.

HARDINESS: G2.

FLOWERING: winter–spring.

2. **P. bellatulum** (Reichenbach) Stein

ILLUSTRATIONS: Orchid Digest **41**: 36 (1977); Graf, Tropica, 749 (1978); Bechtel et al., Manual of cultivated orchid species, 252 (1981); Cribb, The genus Paphiopedilum, pl. 2 (1987); Braem, Paphiopedilum, 44, 45 (1988).

LEAVES: 4–6, *c.* 25 × 8 cm, elliptic to oblong mottled pale and dark green above, reddish beneath.

FLOWER-STALK: very short, 1- or 2-flowered.

FLOWER: to 8 cm wide, white or very pale yellow with bold reddish brown spots which coalesce only at the bases of the petals and sepals.

UPPER SEPAL: almost circular, hooded.

PETALS: almost circular, sometimes shallowly notched.

LIP: blunt, conical, without side lobes.

DISTRIBUTION: Burma (Maymo plateau & Shan Highlands), adjacent Thailand.

HARDINESS: G2.

FLOWERING: spring.

3. P. godefroyae (Godefroy-Lebeuf) Stein

SYNONYM: *P. leuchochilum* (Rolfe) Fowlie.

ILLUSTRATION: Botanical Magazine, 6876 (1886); Orchid Digest **41**: 33 (1977), **44**: 178 (1980); Cribb, The genus Paphiopedilum, pl. 4 (1987); Braem, Paphiopedilum, 55 (1988).

Like *P. bellatulum* except as below.

LEAVES: *c.* 4 cm wide, oblong.

FLOWER-STALK: 3–10 cm.

FLOWER: irregularly marked with coalescing spots and blotches.

DISTRIBUTION: Burma, Thailand (Kra isthmus & Birds nest Islands) & S Vietnam.

Plants with unspotted, pure white lips have been called *P. leucochilum*: see Rittershausen & Rittershausen, Orchids in colour, 94 (1979). The name *P. ang-thong* Fowlie has been given to plants with fewer and finer spots on the flowers; these may be hybrids between *P. godefroyae* and *P. niveum*. *P. godefroyae* itself is considered to be a hybrid (*P. bellatulum* × *P. concolor*) by Braem.

4. P. niveum (Reichenbach) Stein

ILLUSTRATIONS: Botanical Magazine, 5922 (1871); Orchid Digest **41**: 33 (1977); Bechtel et al., Manual of cultivated orchid species, 255 (1981); Cribb, The genus Paphiopedilum, pl. 5 (1978); Braem, Paphiopedilum, 52 (1988).

LEAVES: 6–8, 10–15 cm, elliptic to oblong, mottled grey and green above, deep purple beneath.

FLOWER-STALK: 12–15 cm, 1- or 2-flowered, deep purple with short hairs.

FLOWER: 6–8 cm wide, white with a fine stippling of reddish brown near the base of the petals.

UPPER SEPAL: circular, hooded; lower sepal smaller, ovate, acute, bifid; sepals streaked with purple near base.

PETALS: broadly ovate to circular, scarcely pointed, slightly curled at the margins.

LIP: cup-shaped, with an incurved margin, without side lobes.

STAMINODE: yellow and white.

DISTRIBUTION: Thailand, Malay peninsula & offshore islands.

HARDINESS: G2.

FLOWERING: spring–summer.

5. P. delenatii Guillaumin

ILLUSTRATIONS: Botanical Magazine n.s., 89 (1950); Orchid Digest **41**: 35 (1977); Bechtel et al., Manual of cultivated orchid species, 253 (1981); Cribb, The genus Paphiopedilum, pl. 7 (1987); Braem, Paphiopedilum, 31 (1988).

Like *P. niveum* except as below.

PETALS AND SEPALS: pure white.

LIP: pink.

STAMINODE: pink and yellow.
DISTRIBUTION: C Vietnam.

6. P. rothschildianum (Reichenbach) Stein
SYNONYM: *P. elliotianum* misapplied, not *P. elliotianum* (O'Brien) Pfitzer.
ILLUSTRATIONS: Botanical Magazine, 7102 (1890); Orchid Digest **40**: 80 (1980); Bechtel et al., Manual of cultivated orchid species, 256 (1981); American Orchid Society Bulletin **55**: 580, 581 (1986); Cribb, The genus Paphiopedilum, pl. 16 (1987); Braem, Paphiopedilum, 82 (1988).
LEAVES: to 50 × 7 cm, elliptic to oblong, unmottled, shiny.
FLOWERING STEM: longer than leaves, reddish, downy, with 2–6 flowers.
FLOWERS: *c.* 13 cm from top to bottom.
UPPER SEPAL: *c.* 8 cm, ovate, acute, cream-coloured with *c.* 15 black stripes, margins white.
PETALS: to 15 cm, narrow, tapering, drooping, wavy with fine blunt tips and marginal hairs, cream with purple spots at base, yellowish green apically, veined with purple.
LIP: pointed, without side lobes, purplish brown, yellow around rim; flattened sideways.
DISTRIBUTION: N Borneo (false records from New Guinea).
HARDINESS: G2.
FLOWERING: summer.

7. P. philippinense (Reichenbach) Stein
SYNONYM: *Cypripedium laevigatum* Bateman.
ILLUSTRATIONS: Botanical Magazine, 5508 (1865); Orchid Digest **40**: 183 (1980); Bechtel et al., Manual of cultivated orchid species, 255 (1981); Cribb, The genus Paphiopedilum, pl. 9 (1987); Braem, Paphiopedilum, 74 (1988).
LEAVES: to 30 × 4 cm, strap-shaped, unmottled, shiny.
FLOWERING STEM: to 50 cm, hairy, bearing 2–6 flowers.
FLOWERS: *c.* 8 cm from top to bottom.
UPPER SEPAL: *c.* 3.5 cm, ovate, acute, pale yellow with *c.* 12 brownish purple stripes.
PETALS: to 15 cm, narrow, tapering, drooping, twisted and wavy, with marginal hairs, yellow at base with a few uneven spots, reddish purple to green at tips.
LIP: rounded, without side lobes, yellow with green veins.
DISTRIBUTION: Philippines.
HARDINESS: G2.
FLOWERING: summer.

8. P. glanduliferum (Blume) Pfitzer
SYNONYM: *P. praestans* (Reichenbach) Pfitzer.
ILLUSTRATIONS: Gardeners' Chronicle **26**: 776 (1886); Graf, Tropica, 746 (1978); Orchid Digest **40**: 182 (1980); Cribb,

The genus Paphiopedilum, pl. 15 (1987); Braem, Paphiopedilum, 67 (1988).

Like *P. philippinense* except as below.

UPPER SEPAL: 5 cm.

PETALS: to 12 cm, less drooping.

DISTRIBUTION: New Guinea.

9. P. stonei (Hooker) Stein

ILLUSTRATIONS: Botanical Magazine, 5349 (1862); Orchid Digest **44**: 184 (1980); Bechtel et al., Manual of cultivated orchid species, 256 (1981); Cribb, The genus Paphiopedilum, pl. 13 (1987); Braem, Paphiopedilum, 88, 89 (1988).

LEAVES: 30–40 × 4 cm, strap-shaped, rounded, unmottled.

FLOWERING STEM: to 60 cm, greenish purple, bearing 3–5 flowers.

FLOWERS: 14 cm from top to bottom.

UPPER SEPAL: 6 cm, ovate, acuminate, white, unstriped or with 1–5 thick dark stripes.

PETALS: 12–15 cm, narrow, tapering, drooping, not twisted, dull yellow to green, with purple blotches, lacking hairs or warts.

LIP: rounded, without side lobes, dull pink, yellowish green below, yellow around rim.

DISTRIBUTION: NW Borneo.

HARDINESS: G2.

FLOWERING: summer–autumn.

10. P. charlesworthii (Masters) Pfitzer

ILLUSTRATIONS: Botanical Magazine, 7416 (1895); Orchid Digest **40**: 152 (1976); Bechtel et al., Manual of cultivated orchid species, 252 (1981); Cribb, The genus Paphiopedilum, pl. 25 (1987); Braem, Paphiopedilum, 108 (1988).

LEAVES: 12.5–20 cm, linear to oblong, acute, hairless, not mottled above, spotted with brownish purple beneath.

FLOWER-STALK: to 20 cm, shortly hairy, red-spotted.

FLOWER: 8–9 cm wide.

UPPER SEPAL: *c.* 6 cm wide, almost circular, pink with a fine network of darker veins, white below, slightly reflexed above.

LOWER SEPAL: much smaller, yellowish green.

PETALS: *c.* 4 cm, oblong, obtuse-tipped, yellowish green with a fine network of brown veins.

LIP: greenish yellow to reddish brown.

DISTRIBUTION: Burma & NE India.

HARDINESS: G2.

FLOWERING: autumn.

11. P. hirsutissimum (Hooker) Stein

ILLUSTRATIONS: Botanical Magazine, 4990 (1857); Orchid Digest **44**: 218 (1980); Bechtel et al., Manual of cultivated orchid species, 254 (1981); Cribb, The genus Paphiopedilum, pl. 24 (1987); Braem, Paphiopedilum, 125, 126 (1988).

LEAVES: 20–35 cm, oblong, not mottled above.

FLOWER-STALK: 15–30 cm, green with purple hairs.

BRACT: shorter than ovary.

FLOWER: 10–12 cm wide, all parts shaggy on the back.

UPPER SEPAL: *c.* 4 × 3 cm, circular to ovate, wavy, green or greenish yellow with a central purplish brown patch, indistinctly speckled and streaked.

LOWER SEPAL: smaller, pale green.

PETALS: 6–7 cm, spathulate with hairy margins, base green to brown, finely spotted with black or purple, margins wavy, tip bright purple.

LIP: green to greenish brown.

STAMINODE: green with 2 basal warts.

DISTRIBUTION: NE India, Indo-Burmese border, Thailand & S China.

HARDINESS: G1–2.

FLOWERING: spring.

12. P. villosum (Lindley) Stein

ILLUSTRATIONS: Orchid Review **77**: 273 (1969); Orchid Digest **36**: 148 (1972); Bechtel et al., Manual of cultivated orchid species, 257 (1981); Cribb, The genus Paphiopedilum, pl. 28 (1987); Braem, Paphiopedilum, 116 (1988).

LEAVES: 25–10 × 2–3 cm, oblong, uniformly dull green, purple-spotted underneath at the base.

FLOWER-STALK: 20–35 cm, green spotted with brown, shaggy.

BRACT: almost equal to ovary.

FLOWER: glossy, 12–15 cm from top to bottom.

UPPER SEPAL: to 5 cm, circular to ovate, reflexed at the sides, yellow or greenish yellow with a brownish purple base diffusing into indistinct stripes above.

PETALS: 6–7 cm, spathulate, obtuse, sometimes notched, with a dark brownish purple mid-line separating an upper yellowish brown half from a lower greenish yellow half.

LIP: conical, yellow to green, flushed with brown.

STAMINODE: obovate, yellowish brown.

DISTRIBUTION: NE India, Burma, N Thailand & Laos.

HARDINESS: G1–2.

FLOWERING: winter–spring.

Variants from Burma with boldly mottled upper sepals and pink net-venation on the paler lip have been called var. **boxallii** (Reichenbach) Pfitzer (SYNONYM: *P. boxallii*. See Orchid Digest **36**: 148, 1972).

13. P. spicerianum (Masters & Moore) Pfitzer

ILLUSTRATIONS: Botanical Magazine, 6490 (1880); Orchid Digest **36**: 28 (1972); Bechtel et al., Manual of cultivated orchid species, 256 (1981); Cribb, The genus Paphiopedilum, pl. 31 (1987); Braem, Paphiopedilum, 132 (1988).

LEAVES: 12.5–25 × 3 cm, oblong, uniformly dark green above, mottled beneath at the base.

FLOWER-STALK: to 30 cm, purple, slightly hairy.

BRACT: much shorter than ovary.

FLOWER: 7–8 cm wide.

UPPER SEPAL: broadly ovate, to 5 cm across, reflexed at the sides, bent forward, keeled above, white with a green base and a central dark purple stripe.

LOWER SEPAL: narrower, greenish yellow with a purple stripe.

PETALS: to 5 cm, oblong, obtuse, wavy, yellowish green with a dark purple stripe, flecked with purple at the base; margins wavy.

LIP: greyish brown.

STAMINODE: purple and white, circular with recurved margins.

DISTRIBUTION: NE India.

HARDINESS: G2.

FLOWERING: winter–spring.

14. **P. druryi** (Beddome) Stein

ILLUSTRATIONS: Orchid Digest **38**: 35 (1974); Botanical Magazine n.s., 764 (1978); Bechtel et al., Manual of cultivated orchid species, 253 (1981); Cribb, The genus Paphiopedilum, pl. 30 (1987); Braem, Paphiopedilum, 129, 130 (1988).

Like *P. spicerianum* except as below.

LEAVES: light green with dark veins.

UPPER SEPAL: yellow or yellowish green with a broad, purplish brown stripe.

PETALS: elliptic to oblong, not wavy, hairy, yellow with dark central stripe.

LIP: pale yellow.

DISTRIBUTION: S India; possibly extinct in the wild.

FLOWERING: spring–summer.

15. **P. exul** (Ridley) Kerchove

ILLUSTRATIONS: Botanical Magazine, 7510 (1896); Orchid Digest **40**: 149 (1976); Bechtel et al. Manual of cultivated orchid species, 253 (1981); Cribb, The genus Paphiopedilum, pl. 26 (1987).

LEAVES: 20–30 × 2–2.5 cm, oblong, obtuse, uniformly bright green.

FLOWER-STALK: 15–30 cm, with sparse red hairs.

BRACT: nearly as long as ovary.

FLOWERS: to 8 cm wide.

UPPER SEPAL: 3–1 × 2–3 cm, broadly ovate, greenish yellow, heavily spotted with brownish purple, with a broad white wavy margin.

PETALS: *c.* 5 × 2 cm, oblong to elliptic with hairy margins, yellow with a faint central band of purplish brown.
LIP: yellow, suffused with brown.
STAMINODE: yellow, spotted with brown.
DISTRIBUTION: S Thailand.
HARDINESS: G2.
FLOWERING: spring–summer.
Very similar to *P. insigne* and considered to be a synonym of it by Braem.

16. **P. insigne** (Lindley) Pfitzer
ILLUSTRATIONS: Botanical Magazine, 3412 (1835); American Orchid Society Bulletin **42**: 809 (1973); Bechtel et al., Manual of cultivated orchid species, 254 (1981); Cribb, The genus Paphiopedilum, pl. 27 (1987); Braem, Paphiopedilum, 110 (1988).
LEAVES: 20–30 cm, linear to lanceolate, uniformly yellowish green.
FLOWER-STALK: to 30 cm, purple, felted.
FLOWER: *c.* 10 cm wide.
UPPER SEPAL: *c.* 5 × 4 cm, ovate, obtuse, wavy, mostly green or greenish yellow with brown spots; tip and some of margin white, unspotted.
LOWER SEPAL: smaller, green with brown streaks.
PETALS: 6–7 cm, wavy, oblong to narrowly spathulate, greenish yellow with a network of brown veins.
LIP: greenish or reddish brown, with a yellow rim.
STAMINODE: heart-shaped.
DISTRIBUTION: E. Himalaya.
HARDINESS: G1–2.
FLOWERING: autumn–spring.
Over 100 varieties have been described of which the following are among the best known:

Var. **sanderae** Reichenbach.
ILLUSTRATIONS: Graf, Tropica, 749 (1978); Bechtel et al., Manual of cultivated orchid species, 254 (1981).
UPPER SEPAL: primrose yellow with reddish brown spots.
PETALS AND LIP: yellow.

Var. **sanderianum** Sander is similar to var. **sanderae** but lacks spots.

P. × **leeanum** (Reichenbach) Kerchove.
ILLUSTRATION: Cogniaux & Goossens, Dictionnaire Iconographique des Orchidées, Cypripedium Hybrides, t. 3 (1903).

The hybrid between *P. insigne* and *P. spicerianum*.
UPPER SEPAL: white with lines of purple spots.
PETALS: slightly drooping.
LIP: greenish brown.

DISTRIBUTION: Himalaya & garden origin.
HARDINESS: G2.

17. P. fairrieanum (Lindley) Stein

ILLUSTRATIONS: Botanical Magazine, 5024 (1857); Orchid Digest **40**: 155 (1976); Bechtel et al., Manual of cultivated orchid species, 253 (1981); Kew Magazine **2**: pl. 47 (1985); Cribb, The genus Paphiopedilum, pl. 32 (1987); Braem, Paphiopedilum, 121 (1988).
LEAVES: 9–15 cm, oblong, unmottled.
FLOWER-STALK: 9–15 cm, pale purple, sparsely hairy.
FLOWER: 6–8 cm from top to bottom.
UPPER SEPAL: circular with a broad acuminate tip, pale green with *c.* 15 finely branched dark purple veins interconnecting near the wavy margin.
LOWER SEPAL: smaller with unconnected veins.
PETALS: 4–5 cm, oblong to lanceolate, curved downwards and strongly recurved at the tips, hairy and wavy on margins, green and white with green and purple veins.
LIP: inflated, greenish yellow, with a network of purplish brown veins.
DISTRIBUTION: Bhutan & NE India.
HARDINESS: G2.
FLOWERING: autumn.
Lindley in his original description misspelt the name 'fairieanum'; the man after whom he named this species was a Mr Fairrie.

18. P. dayanum (Lindley) Stein

SYNONYM: *Cypripedium 'spectabile'* Reichenbach var. *dayanum* Lindley.
ILLUSTRATIONS: Botanical Magazine n.s., 594 (1971); Williams et al., Orchids for everyone, 152 (1980); Orchid Digest **39**: 157 (1975) & **45**: 24 (1981); Kew Magazine **3**: 5 (1986); Cribb, The genus Paphiopedilum, pl. 47 (1987); Braem, Paphiopedilum, 207 (1988).
LEAVES: 12–20 × 5 cm, oblong to lanceolate, strongly mottled pale and bluish green above.
FLOWER-STALK: 20–30 cm, downy, purple.
FLOWER: 10–12 cm wide.
UPPER SEPAL: *c.* 6 × 3 cm, narrowly ovate, acuminate, white with *c.* 20 narrow green veins.
LOWER SEPAL: similar but smaller and more pointed.
PETALS: to 10 cm, narrowly elliptic to spathulate, with long marginal hairs, greenish brown at the base to pink at the tip, veined with green and purple, sometimes with small marginal warts.
LIP: purplish brown, conical.
DISTRIBUTION: N Borneo.

HARDINESS: G2.
FLOWERING: summer (may also flower in winter).

19. P. violascens Schlechter
SYNONYM: *P. bougainvilleanum* Schoser.
ILLUSTRATIONS: Orchid Digest **42**: 100 (1978); Millar, Orchids of Papua New Guinea, 86 (1978); Bechtel et al., Manual of cultivated orchid species, 257 (1981); Kew Magazine **3**: pl. 70 (1986); Cribb, The genus Paphiopedilum, pl. 39 (1987); Braem, Paphiopedilum, 170 (1988).

Like *P. dayanum* except as below.

UPPER SEPAL: *c.* 3 × 3 cm.
PETALS: 4–4.5 cm, sickle-shaped, purple.
LIP: greenish brown.
DISTRIBUTION: Indonesia (West Irian).
FLOWERING: autumn.

20. P. barbatum (Lindley) Stein
ILLUSTRATIONS: Botanical Magazine, 4234 (1846); Orchid Digest **41**: 61 (1977); Bechtel et al., Manual of cultivated orchid species, 251 (1981); Cribb, The genus Paphiopedilum, pl. 43 (1987); Braem, Paphiopedilum, 193 (1988).
LEAVES: 10–15 × 2–3 cm, oblong to lanceolate, mottled.
FLOWER-STALK: 20–30 cm.
BRACT: ovate-lanceolate, a quarter of the ovary length.
FLOWER: 8–10 cm wide.
UPPER SEPAL: almost circular with a short abrupt tip, green at the base, white or pale pink above with *c.* 20 dark purple veins.
LOWER SEPAL: shorter and narrower with greenish purple veins.
PETALS: to 5 cm, oblong, slightly wider at the tip, hairy on margins, held just below horizontal, reddish brown at the base, shading to purplish red at the tips, dark-veined, with 4–8 dark hairy warts about 1 mm in diameter on the upper margin.
DISTRIBUTION: S Thailand & Malay peninsula.
HARDINESS: G2.
FLOWERING: spring–summer.

21. P. callosum (Reichenbach) Stein
ILLUSTRATIONS: Botanical Magazine, 9671 (1946); Orchid Digest **36**: 9 (1972); Bechtel et al., Manual of cultivated orchid species, 252 (1981); Cribb, The genus Paphiopedilum, pl. 44 (1987); Braem, Paphiopedilum, 196 (1988).
LEAVES: 10–25 × 4–6 cm, oblong to elliptic, greyish green with dark mottling.
FLOWER-STALK: to 30 cm, brownish purple, densely downy.

BRACT: 2–3 cm, ovate, acute.

FLOWER: 7–9 cm wide.

UPPER SEPAL: 6 cm, almost circular, with a short tip, white with *c.* 25 dark purple veins which are green below.

LOWER SEPAL: shorter and narrower.

PETALS: 5–7 cm, strongly curved downwards, indistinctly green-veined, pinkish yellow at the tips, with 5–8 black hairy warts *c.* 3 mm in diameter on the upper margin (occasionally 1 or 2 on the blade).

LIP: reddish brown tinged with green.

DISTRIBUTION: E Thailand, Cambodia & S Vietnam.

HARDINESS: G2.

FLOWERING: spring.

Var. **sanderae** Anon. has a flower which lacks red or yellow pigment and is white with brilliant green veins.

P. × **maudiae** Anon.

ILLUSTRATIONS: American Orchid Society Bulletin **46**: 1089 (1977); Graf, Tropica, 746 (1978).

PETALS: less deflexed than in *P. callosum*, with 1 or 2 warts on the lower margin as well as those on the upper. This cross was originally made between 'albino' varieties of *P. callosum* and *P. lawrenceanum* (lacking red or yellow pigment) resulting in an 'albino' hybrid. The name, however, strictly applies to all hybrids between these parents.

P. callosum and *P. barbatum* probably represent the extreme forms of a single species which runs from pure *P. barbatum* at the southern end of the Malay peninsula to *P. callosum* in N Thailand. Intermediates from the Kra isthmus have been referred to as *P.* × *siamense* Rolfe (*P. sublaeve* (Reichenbach) Fowlie).

22. **P. lawrenceanum** (Reichenbach) Pfitzer

ILLUSTRATIONS: Botanical Magazine, 6432 (1879); Orchid Digest **41**: 61 (1977); Bechtel et al., Manual of cultivated orchid species, 254 (1981); Cribb, The genus Paphiopedilum, pl. 46 (1987); Braem, Paphiopedilum, 200, 201 (1987).

LEAVES: 15–30 × 5–7 cm, oblong to ovate with long sheathing bases, mottled dark and yellowish green.

FLOWER-STALK: 30–40 cm, purple, hairy.

BRACT: ovate-oblong, much shorter than ovary.

FLOWER: 10–12 cm wide.

UPPER SEPAL: 5 cm, circular, white flushed with pink; veins *c.* 20, purple above, green below.

PETALS: 5–6 × 1.5 cm, horizontal, oblong to narrowly elliptic, obtuse, dull green with purplish veins; tips tinged with reddish purple; each margin has 6–10 blackish purple hairy warts.

LIP: brownish purple.

DISTRIBUTION: N Borneo.

HARDINESS: G2.

FLOWERING: summer.

Var. **hyeanum** (Reichenbach) Pfitzer lacks red pigment and is green, white and yellow.

23. P. bullenianum (Reichenbach) Pfitzer

SYNONYM: *P. linii* Schoser.

ILLUSTRATIONS: Orchid Digest **42**: 29 (1978); Bechtel et al., Manual of cultivated orchid species, 252 (1981); Cribb, The genus Paphiopedilum, pl. 34 (1987); Braem, Paphiopedilum, 154, 155 (1988).

LEAVES: 10–17.5 cm, faintly mottled with bluish green, often red beneath.

FLOWER-STALK: to 30 cm, hairy.

FLOWER: 7–8 cm wide.

UPPER SEPAL: 3 cm, acute, reflexed above, green to greenish yellow, often brown at the base with *c.* 20 indistinct darker green veins.

PETALS: spathulate, obtuse, more or less horizontal, apically pinkish purple with green margins, shading through brown to green at the base where there are a few dark shiny warts on marginal undulations.

LIP: slightly inflated, buff below, darker brown above with a green rim.

STAMINODE: with a narrow parallel-sided sinus.

DISTRIBUTION: Tioman Isles off Malay peninsula, Borneo & Indonesia (Sulawesi & Ambon).

FLOWERING: summer–autumn.

24. P. appletonianum (Gower) Rolfe

SYNONYM: *P. wolterianum* (Kraenzlin) Pfitzer.

ILLUSTRATIONS: Rittershausen & Rittershausen, Orchids in colour, 92 (1979); Orchid Digest **44**: 227 (1980); Bechtel et al., Manual of cultivated orchid species, 251 (1981); Cribb, The genus Paphiopedilum, pl. 33 (1987); Braem, Paphiopedilum, 149 (1988).

LEAVES: *c.* 20 × 3–5 cm, elliptic, mottled.

FLOWER-STALK: to 50 cm, hairy.

FLOWER: *c.* 8 cm from top to bottom.

UPPER SEPAL: *c.* 3.5 × 2.5 cm, ovate, green, paler above with darker green veins.

LOWER SEPAL: smaller and narrower, acute.

PETALS: 5–6 × 1–2 cm, elliptic to spathulate, obtuse to rounded, slightly wavy, drooping and half twisted; basal half bright green with brown spots concentrated near margins and mid-line; apical half pinkish purple.

LIP: 4–5 cm, pale greenish brown below, dark brown near the rim.

STAMINODE: with broad, diverging sinus.

DISTRIBUTION: Laos, Thailand & Cambodia.

HARDINESS: G2.

FLOWERING: winter–spring.

Taller plants with broader, more drooping petals have often been treated separately as *P. wolterianum*. However, there are no consistent differences between the 2 species.

25. P. hookerae (Reichenbach) Stein

ILLUSTRATIONS: Botanical Magazine, 5362 (1863); Orchid Digest **40**: 155 (1976), **45**: 17 (1981); Cribb, The genus Paphiopedilum, pl. 36 (1987); Braem, Paphiopedilum, 145, 146 (1988).

LEAVES: 10–15 × 3–4 cm, ovate to oblong, mottled greyish on dark green.

FLOWER-STALK: to 20 cm, greyish purple, hairy.

FLOWER: to 10 cm wide.

UPPER SEPAL: ovate, acute, yellowish with a green centre, indistinctly veined with darker green.

PETALS: spathulate, acute, 4–5 cm, held below horizontal, bright pinkish purple at the tips shading to dull greenish yellow with many small pinkish spots in basal half.

LIP: yellowish green, tinged with reddish brown above.

STAMINODE: with sinus closed by overlapping lobes.

DISTRIBUTION: N Borneo.

HARDINESS: G2.

FLOWERING: spring–summer.

Var. **volonteanum** (Rofle) Kerchove.

SYNONYM: *P. volonteanum* (Masters) Pfitzer.

ILLUSTRATIONS: Orchid Digest **39**: 166 (1975), **45**: 18 (1981).

PETALS: with spots purplish black, extending into outer half of petal but concentrated near margins and mid-line; green and purple areas separated by a patch of white or very pale green.

DISTRIBUTION: N Borneo.

HARDINESS: G2.

26. P. tonsum (Reichenbach) Stein

ILLUSTRATIONS: Orchid Digest **40**: 85 (1976); Bechtel et al., Manual of cultivated orchid species, 257 (1981); Botanical Magazine n.s., 838 (1982); Cribb, The genus Paphiopedilum, pl. 41 (1987); Braem, Paphiopedilum, 174, 175 (1988).

LEAVES: 12–20 cm, oblong, ovate, mottled dark and light green.

FLOWER-STALK: 20–45 cm, purplish green.

FLOWER: 12 cm wide.

UPPER SEPAL: 5–6 cm, broadly ovate, pale green with 25–30 green and purple veins.

PETALS: 6–7 cm, slightly below horizontal, narrowly spathulate, similar in colour and venation to the upper sepal but with a few large dark spots on the upper margin and in the mid-line; margins hairy near the tip.

LIP: greenish yellow flushed with pink.
DISTRIBUTION: Sumatra.
HARDINESS: G2.
FLOWERING: summer–autumn.

27. P. venustum (Sims) Pfitzer
ILLUSTRATIONS: Botanical Magazine, 2129 (1820); Orchid Digest **36**: 177 (1972); Bechtel et al., Manual of cultivated orchid species, 257 (1981); Cribb, The genus Paphiopedilum, pl. 56 (1987); Braem, Paphiopedilum, 139 (1988).
LEAVES: 15–25 cm, ovate-lanceolate, mottled dark on light green above, streaked and spotted with red beneath.
FLOWER-STALK: 20–30 cm, purple.
FLOWER: 8–10 cm wide.
UPPER SEPAL: ovate, acute, white or off-white with *c.* 20 green veins.
PETALS: 4–5 cm, horizontal or slightly descending, spathulate, hairy on margins, yellowish green or green at the base, reddish brown at the tip, with several irregular dark hairy spots on blade and margins.
LIP: dull yellow to reddish brown, strongly veined with greenish brown.
DISTRIBUTION: Nepal, Bhutan, Bangladesh & NE India.
HARDINESS: G1–2.
FLOWERING: winter–spring.

28. P. acmodontum Wood
ILLUSTRATIONS: Orchid Review **84**: 350 (1976); Orchid Digest **41**: 60 (1977); Bechtel et al., Manual of cultivated orchid species, 251 (1981); Cribb, The genus Paphiopedilum, pl. 50 (1987); Braem, Paphiopedilum, 217 (1987).
LEAVES: 15–20 × 4–5 cm, oblong to elliptic, mottled.
FLOWER-STALK: to 25 cm, sparsely hairy.
FLOWERS: 7–8 cm wide.
UPPER SEPAL: *c.* 4 × 3 cm, broadly ovate, acuminate, off-white with *c.* 15 purple veins, flushed with pink at base.
PETALS: 4–4.5 × 1.5–2 cm, obovate to spathulate, acute to acuminate, greenish yellow at base, reddish brown in outer half; inner half irregularly spotted with purplish black.
LIP: *c.* 4 cm, yellowish green below, brown towards rim, with a prominent acute tooth on outer edge of rim.
DISTRIBUTION: Philippines.
HARDINESS: G2.
FLOWERING: spring.

29. P. javanicum (Lindley & Paxton) Pfitzer
SYNONYMS: *P. virens* misapplied; *P. purpurascens* Fowlie.
ILLUSTRATIONS: Orchid Review **85**: 159 (1977); Orchid Digest **45**: 24 (1981); Bechtel et al., Manual of cultivated

orchid species, 251 (1981); Braem, Paphiopedilum, 219 (1988).

LEAVES: to 20 × 5 cm, oblong to elliptic, mottled.

FLOWER-STALK: to 25 cm.

FLOWER: *c.* 10 cm from top to bottom.

UPPER SEPAL: *c.* 4.5 × 2.5 cm, ovate, acuminate, greenish yellow with 20–25 dark green veins.

LOWER SEPAL: similar but smaller.

PETALS: *c.* 5 × 1.5 cm, narrowly obovate, broader in apical half, drooping, dull green, pink at tips, most of blade covered in small purplish black warts.

LIP: *c.* 4.5 cm, dull green to brownish green with darker veins.

DISTRIBUTION: Java & Borneo.

HARDINESS: G2.

FLOWERING: intermittently.

Var. **virens** (Reichenbach) Stein.

SYNONYM: *P. virens* (Reichenbach) Pfitzer.

ILLUSTRATIONS: Reichenbach, Xenia Orchidacea, t. 162 (1870); Orchid Digest **40**: 237 (1976), **45**: 22 (1981); Orchid Review, **92**: 316 (1984); Cribb, The genus Paphiopedilum, pl. 51 (1987); Braem, Paphiopedilum, 221 (1988).

PETALS: narrowly spathulate, horizontal; basal two-thirds bright green, apical third pinkish purple; warts concentrated on margins and mid-line.

LIP: pale brown, darker near rim.

DISTRIBUTION: N Borneo.

HARDINESS: G2.

Frequently confused with *P. javanicum* and thus mistakenly said to be native to Indonesia.

30. P. purpuratum (Lindley) Stein

ILLUSTRATIONS: Botanical Magazine, 4901 (1856); Orchid Review **82**: 39 (1974); Bechtel et al., Manual of cultivated orchid species, 255 (1981); Cribb, The genus Paphiopedilum, pl. 53 (1987); Braem, Paphiopedilum, 179, 180 (1988).

LEAVES: 8–12 cm, oblong-elliptic, mottled dark and pale green.

FLOWER-STALK: 20–25 cm, deep purple, hairy.

FLOWER: 8–10 cm wide.

UPPER SEPAL: circular, acuminate, white with about 12 alternating purple and green veins.

PETALS: 4–5 cm, elliptic, hairy on margins, purplish brown with many small dark warts in basal half, paler at the tip.

LIP: purplish brown.

DISTRIBUTION: Hong Kong & adjacent China.

HARDINESS: G2.

FLOWERING: summer–winter.

31. P. mastersianum (Reichenbach) Stein

ILLUSTRATIONS: Botanical magazine, 7629 (1898); Orchid

Digest **45**: 19 & 20 (1981); Bechtel et al., Manual of cultivated orchid species, 255 (1981); Cribb, The genus Paphiopedilum, pl. 37 (1987); Braem, Paphiopedilum, 161, 162 (1988).

LEAVES: 20–24 × 3–4 cm, oblong-elliptic, mottled dark and yellowish green.

FLOWER-STALK: 30–45 cm, deep purple, hairy.

FLOWER: to 10 cm wide, fleshy.

UPPER SEPAL: circular, 5 cm broad, greenish yellow with about 20 dark green veins and a broad white or off-white margin.

PETALS: 5–6 × 2 cm, horizontal, oblong to slightly spathulate, round-tipped, reddish brown in apical half, remainder dull green with many fine purplish warts above the mid-line.

LIP: dull reddish purple, yellow round the rim.

DISTRIBUTION: Indonesia (Borneo & Moluccas).

HARDINESS: G2.

FLOWERING: spring–summer.

32. **P. superbiens** (Reichenbach) Stein

SYNONYM: *P. curtisii* (Reichenbach) Stein.

ILLUSTRATIONS: Orchid Review **83**: 394 (1975); Orchid Digest **42**: 177 (1978), **45**: 60 (1981); Williams et al., Orchids for everyone, 152 (1980); Cribb, The genus Paphiopedilum, pl. 49 (1987); Braem, Paphiopedilum, 211, 213 (1988).

LEAVES: 15–20 cm, oblong-lanceolate, strongly mottled.

FLOWER-STALK: to 30 cm, purple, downy.

BRACT: ovate, a third of the ovary length.

FLOWER: to 8 cm wide.

UPPER SEPAL: broadly ovate, acuminate, pale green, purple at the base with about 25 dark green to purple veins.

LOWER SEPAL: small, ovate-acuminate.

PETALS: oblong, drooping, curled, hairy on margins, purple at the base shading to pale green, evenly marked with numerous small purplish black warts.

LIP: brownish purple.

DISTRIBUTION: Sumatra.

HARDINESS: G2.

FLOWERING: summer.

33. **P. ciliolare** (Reichenbach) Stein

SYNONYM: *P. superbiens* (Reichenbach) Stein subsp. *ciliolare* (Reichenbach) Wood.

ILLUSTRATIONS: Orchid Digest **41**: 47 (1977); Rittershausen & Rittershausen, Orchids in colour, 90 (1979); Bechtel et al., Manual of cultivated orchid species, 252 (1981); Cribb, The genus Paphiopedilum, pl. 48 (1987); Braem, Paphiopedilum, 205 (1988).

Like *P. superbiens* except as below.

PETALS: not curled, with dense marginal hairs; warts absent from outer third.

DISTRIBUTION: Philippines.

FLOWERING: spring–summer.

34. **P. argus** (Reichenbach) Stein

ILLUSTRATIONS: Botanical Magazine, 6175 (1875); Orchid Digest **41**: 61 (1977), **45**: 57 (1981); Rittershausen & Rittershausen, Orchids in colour, 90 (1979); Cribb, The genus Paphiopedilum, pl. 42 (1987); Braem, Paphiopedilum, 191 (1988).

LEAVES: 12–20 cm, oblong-lanceolate, pale green with darker mottling.

FLOWER-STALK: to 30 cm, purplish, glandular hairy.

BRACT: pale green, half the ovary length.

FLOWER: 10 cm wide.

UPPER SEPAL: broad, acuminate, off-white with about 20 dark green or purplish veins.

PETALS: 7–8 cm, oblong, slightly tapering, hairy on margin, descending, off-white at the base, shading to pink at the tip, boldly marked all over with dark purple spots, some having pale centres.

LIP: yellow with dark green net-venation, red above, flushed with brown.

DISTRIBUTION: Philippines.

HARDINESS: G2.

FLOWERING: spring.

35. **P. sukhakulii** Schoser & Senghas

ILLUSTRATIONS: Orchid Review **77**: 145 (1969); Orchid Digest **39**: 208 (1975); Bechtel et al., Manual of cultivated orchid species, 256 (1981); Cribb, The genus Paphiopedilum, pl. 54 (1987); Braem, Paphiopedilum, 182, 183 (1988).

LEAVES: 12–15 × *c.* 4.5 cm, narrowly elliptic, mottled.

FLOWER-STALK: to 25 cm.

FLOWER: *c.* 10 cm wide.

UPPER SEPAL: 3.5–4 × 3 cm, broadly ovate, acuminate, yellowish green with 15–20 green veins.

PETALS: 5–6 × 1.3–1.7 cm, oblong to narrowly elliptic, tapering to an acute point, almost horizontal, yellowish green, closely spotted with purplish black except at extreme tip.

LIP: *c.* 4.5 cm, conical, greenish brown.

DISTRIBUTION: Thailand.

HARDINESS: G2.

FLOWERING: autumn.

36. **P. wardii** Summerhayes

ILLUSTRATIONS: Botanical Magazine, 9481 (1937); Orchid

Digest **36**: 85 (1972), **45**: 15 (1981); Pradhan, Indian orchids, t. 19 (1976); Cribb, The genus Paphiopedilum, pl. 55 (1987); Braem, Paphiopedilum, 185 (1988).

Like *P. sukhakulii* except as below.

FLOWERS: usually browner.

PETALS: descending.

DISTRIBUTION: N Indo-Burmese border.

FLOWERING: winter.

Very similar to *P. sukhakulii*. They should perhaps be considered as one species under *P. wardii*, which is the older name. Braem considers that the species is of hybrid origin (*P. venustum* × *P. sukhakulii*).

37. **P. victoria-regina** (Sander) Wood

SYNONYM: *P. victoria-mariae* invalid.

ILLUSTRATIONS: Botanical Magazine, 7573 (1898); Orchid Digest **44**: 217 (1980); Cribb, The genus Paphiopedilum, pl. 23 (1987); Braem, Paphiopedilum, 235 (1988).

LEAVES: oblong, to 40 × 6 cm, often faintly mottled.

FLOWERING STEM: 30–60 cm or more, reddish brown, bearing several flowers in succession over several months.

FLOWERS: 7–8 cm wide.

UPPER SEPAL: 3–4 cm wide, circular to broadly ovate, greenish yellow with 5–8 purple veins and a white to creamy yellow border.

PETALS: 5 cm, narrowly oblong, horizontal, hairy on margins, twisted and wavy, greenish yellow, veined with reddish purple and with an unbroken border of reddish purple.

LIP: slightly compressed, purple, unspotted, with a green and yellow rim.

DISTRIBUTION: Sumatra.

HARDINESS: G2.

FLOWERING: intermittent.

The subspecies described above is endemic to a small area in central Sumatra and is uncommon in cultivation. The following subspecies is more commonly grown:

subsp. **chamberlainianum** (Sander) Wood

SYNONYM: *P. chamberlainianum* (Sander) Stein.

ILLUSTRATIONS: Botanical Magazine, 7578 (1898); Orchid Review **84**: 140 (1976); Bechtel et al., Manual of cultivated orchid species, 252 (1981).

LEAVES: to 10 cm broad.

UPPER SEPAL: yellow and purple.

LIP: inflated, evenly spotted with purple.

DISTRIBUTION: Sumatra.

HARDINESS: G2.

FLOWERING: intermittent.

38. P. glaucophyllum J.J. Smith

SYNONYM: *P. victoria-regina* subsp. *glaucophyllum* (J.J. Smith) Wood.

ILLUSTRATIONS: Orchid Review **84**: 140 (1976); Orchid Digest **44**: 216 (1980); Bechtel et al., Manual of cultivated orchid species, 253 (1981); Cribb, The genus Paphiopedilum, pl. 20 (1987); Braem, Paphiopedilum, 231, 232 (1988).

Like *P. victoria-regina* except as below.

LEAVES: glaucous.

PETALS: slightly deflexed, marked with discontinuous purple spots on a white background.

DISTRIBUTION: E & C Java.

39. P. parishii (Reichenbach) Stein

ILLUSTRATIONS: Botanical Magazine, 5791 (1869); Orchid Digest **40**: 213 (1980); Bechtel et al., Manual of cultivated orchid species, 255 (1981); Cribb, The genus Paphiopedilum, pl. 19 (1987); Braem, Paphiopedilum, 102, 103 (1988).

HABIT: epiphytic.

LEAVES: to 30 cm, strap-shaped, unmottled, bright green.

FLOWERING STEM: to 60 cm, pale green, bearing 4–7 flowers.

FLOWERS: 8 cm wide.

UPPER SEPAL: elliptic, pale yellowish green with faint veins.

PETALS: to 15 cm, linear, gently tapering, drooping and twisted, green at base with marginal purple spots, purple at tips.

LIP: green to greenish purple.

DISTRIBUTION: Burma, NW Thailand & S China.

HARDINESS: G2.

FLOWERING: spring–summer.

40. P. lowii (Lindley) Stein

ILLUSTRATIONS: Gardeners' Chronicle, 765 (1847); Orchid Digest **44**: 214 (1980); Williams et al., Orchids for everyone, 154 (1980); Cribb, The genus Paphiopedilum, pl. 18 (1987); Braem, Paphiopedilum, 98 (1988).

HABIT: epiphytic or on cliffs.

LEAVES: oblong, to 40 × 4 cm, unmottled, light green.

FLOWERING STEM: to 1 m, arched, brownish purple, bearing 2–6 flowers.

FLOWERS: 15 cm across.

UPPER SEPAL: 5 cm, broadly elliptic, short-pointed, concave above, with reflexed margins below, yellowish green flushed with purple, streaked with brown at base.

PETALS: 8 cm, spathulate, spreading, slightly drooping at tips, slightly twisted, yellowish green with brown spots at base; tips bright purple.

LIP: blunt, oblong, pale brownish green with brown markings, flushed with purple near rim.

DISTRIBUTION: Malaysia & Indonesia.
HARDINESS: G2.
FLOWERING: spring–summer.

41. P. haynaldianum (Reichenbach) Stein
ILLUSTRATIONS: Botanical Magazine, 6296 (1877); Orchid Digest 44: 214 (1980); Bechtel et al., Manual of cultivated orchid species, 254 (1981); Cribb, The genus Paphiopedilum, pl. 17 (1987); Braem, Paphiopedilum, 95–97 (1988).
Like *P. lowii* except as below.
UPPER SEPAL: pale, marked with large irregular brown blotches.
DISTRIBUTION: Philippines.
FLOWERING: spring.
Possibly better considered as a subspecies of *P.lowii*.

4. EPIPACTIS Zinn

A genus of about 20 species from north temperate regions. Cultivation is as for *Dactylorhiza* (p. 80), though some species, e.g. *E. gigantea* and *E. palustris*, prefer moister conditions.

HABIT: perennial herbs with creeping or vertical rhizomes and numerous fleshy but not tuberous roots; aerial growth dying off in winter.
LEAVES: spaced along the stem, spirally arranged or in 2 opposite rows, strongly veined, pleated.
FLOWERS: held horizontally or hanging, often only on 1 side of the axis.
BRACTS: green and leaf-like, not sheathing the basal parts of the flowers.
SEPALS AND PETALS: spreading or sometimes scarcely opening.
LIP: in 2 distinct parts, the basal cup-like, the apical flat and variously shaped; spur absent.
POLLINIA: rapidly breaking up.

Synopsis of characters

Rhizome. Creeping: 4,5; not creeping: 1–3.
Leaves. Spirally arranged: 1,4,5; in opposite rows: 2,3.
Lip. Lateral lobes present: 4,5; lateral lobes absent: 1–3.

Key to species

1. Lip with lateral lobes; rhizome creeping
1.A. Basal part of the lip narrower than the ovate, basal part **4. palustris**
1.B. Basal part of lip broader than the oblong-lanceolate apical part **5. gigantea**
2. Leaves spirally arranged **1. helleborine**

3. Raceme-axis, flower-stalks and ovary downy; flowers purple **2. atrorubens**

4. Raceme-axis, flower-stalks and ovary hairless; flowers yellowish green. sometimes tinged with red or reddish violet **3. phyllanthes**

1. E. helleborine (Linneaus) Crantz

SYNONYM: *E. latifolia* (Linnaeus) Allioni.

ILLUSTRATIONS: Danesch & Danesch, Orchideen Europas, Mitteleuropa, edn 3, 190, 191 (1972); Sundermann, Europäische und mediterrane Orchideen, 206 (1975); Williams et al., Orchids of Britain and Europe, 15 (1978); Mossberg & Nilsson, Orchids of northern Europe, 35 (1979); American Orchid Society Bulletin **54**: 558 (1985).

RHIZOME: short.

STEMS: 1–3, arising close together, 35–90 cm, sparsely hairy.

LEAVES: 3–10, 5–17 × 2.5–10 cm, spirally arranged, ovate or broadly elliptic, the upper lanceolate, acute or acuminate.

RACEME: 10–40 cm with 15–50 flowers, borne on 1 side of the axis.

BRACTS: lowest longer than flowers, the upper shorter.

SEPALS: 1–1.3 cm, ovate, greenish.

PETALS: slightly smaller, pinkish or purplish.

LIP: apical part ovate, acute, recurved, with 2 basal bosses.

DISTRIBUTION: Europe, N Africa, SW Asia to Himalaya.

HARDINESS: H2.

FLOWERING: summer.

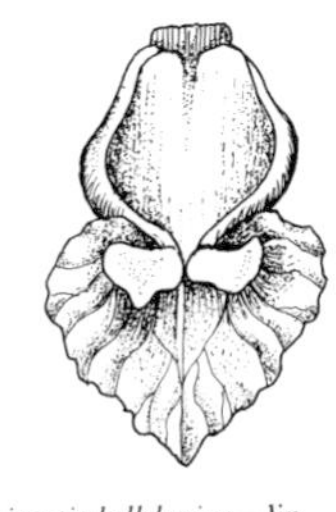

Epipactis helleborine – lip.

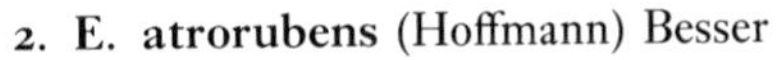

2. E. atrorubens (Hoffmann) Besser

SYNONYM: *E. atropurpurea* Rafinesque.

ILLUSTRATIONS: Danesch & Danesch, Orchideen Europas, Mitteleuropa, edn 3, 186, 187, 194 (1972); Sundermann, Europäische und mediterrane Orchideen, 212 (1975); Williams et al., Orchids of Britain and Europe, 158 (1978); Mossberg & Nilsson, Orchids of northern Europe, 33 (1979).

RHIZOME: short.

STEMS: 15–60 cm, downy, purplish-tinged.

LEAVES: 5–10, 4–10 × 1.5–4.5 cm, arranged in opposite ranks along the stem, ovate to ovate-lanceolate, acute.

RACEME: to 25 cm with 8–18 flowers.

BRACTS: the lower as long as flowers.

FLOWERS: wine-red, faintly fragrant.

SEPALS: 6–7 mm, ovate, acuminate.

PETALS: slightly narrower than sepals.

LIP: 5.5–6.5 mm, basal part cup-shaped, apical part wider than long, with 3 rough basal bosses, apex acute or blunt, recurved.

DISTRIBUTION: Europe to C Asia.

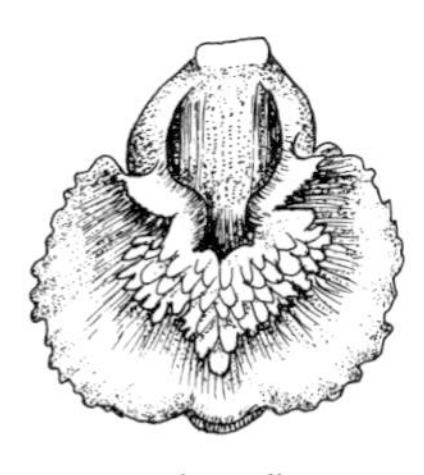

Epipactis atrorubens – lip.

HARDINESS: H2.
FLOWERING: summer.

3. E. phyllanthes G.E. Smith
ILLUSTRATION: Landwehr, Wilde Orchideeën van Europa **2**: 493, 495 (1977); Williams et al., Orchids of Britain and Europe, 155 (1978); Mossberg & Nilsson, Orchids of northern Europe, 41 (1979).
RHIZOME: short.
STEMS: 1 or occasionally 3, 20–45 cm, hairless or slightly hairy.
LEAVES: 3–16, arranged in opposite rows on the stem, 3.5–7 × 3–5 cm, ovate or lanceolate, acute.
RACEME: to 15 cm with 15–35 flowers.
BRACTS: longer than flowers.
FLOWERS: greenish, hanging, opening only a little.
SEPALS: 8–10 mm.
PETALS: sometimes slightly purplish-tinged.
LIP: basal and apical parts not always clearly distinct.
DISTRIBUTION: NW & C Europe.
HARDINESS: H2.
FLOWERING: summer.

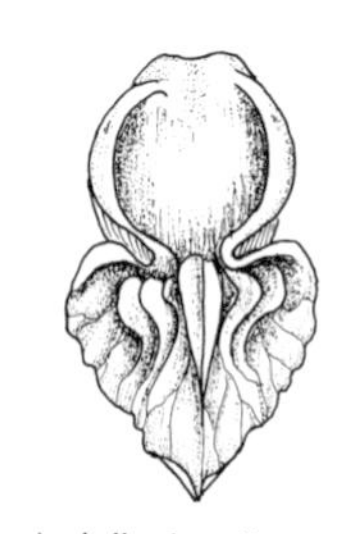

Epipactis phyllanthes – lip.

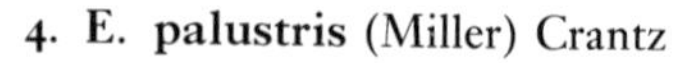

4. E. palustris (Miller) Crantz
ILLUSTRATION: Danesch & Danesch, Orchideen Europas, Mitteleuropa, edn 3, 183–5 (1972); Sundermann, Europäische und mediterrane Orchideen, 214 (1975); Williams et al., Orchids of Britain and Europe, 149 (1978); Mossberg & Nilsson, Orchids of northern Europe, 43 (1979); American Orchid Society Bulletin **54**: 561, 562 (1985).
RHIZOME: creeping.
STEMS: to 50 cm, downy.
LEAVES: 4–8, 5–15 × 2–4 cm, spirally arranged and often crowded, oblong to lanceolate, acuminate.
RACEME: to 15 cm, with up to 15 flowers.
BRACTS: lower bracts equalling the flowers.
SEPALS: 8–12 mm, ovate-lanceolate, greenish lined with red.
PETALS: whitish flushed with pink.
LIP: 1–1.2 cm, basal part concave with erect lateral lobes, pinkish white, purple-lined and orange-dotted, narrower than the apical part which is ovate, as broad as long, wavy-margined, white with a yellow bar across the base.
DISTRIBUTION: Eurasia, N Africa.
HARDINESS: H2.
FLOWERING: summer.

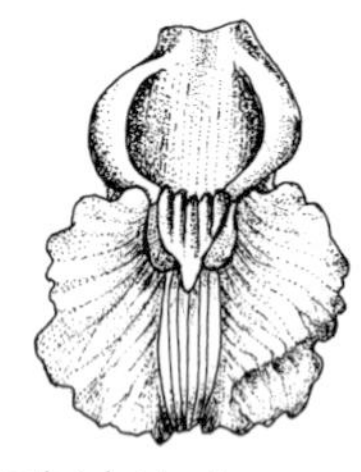

Epipactis palustris – lip.

5. E. gigantea Hooker
SYNONYM: *E. royleana* Lindley.
ILLUSTRATIONS: Jahresberichte des Naturwissenschaftlichen Vereins in Wuppertal **23**: t. 5 (1970); American Orchid

Society Bulletin 71: 240 (1971); Luer, Native orchids of the United States and Canada, 79 (1975); Flora Iranica **126**: t. 15, 16 (1978).

RHIZOME: creeping.

STEMS: to 1 m, hairless.

LEAVES: 4–12, 5–20 × 2–7 cm, spirally arranged, ovate to lanceolate, acuminate.

RACEME: to 30 cm with up to 15 flowers.

BRACTS: longer than flowers.

SEPALS: 1.5–2 cm, ovate-lanceolate, greenish yellow, purple-veined.

PETALS: ovate, 1.3–1.5 cm, flushed with pink.

LIP: 1.4–1.5 cm, basal part concave with red warts, lateral lobes bluntly triangular, yellowish with reddish nerves, separated by a rigid fold from the apical part which is oblong-lanceolate with 2 basal ridges, orange or yellowish, the apex suffused with pink.

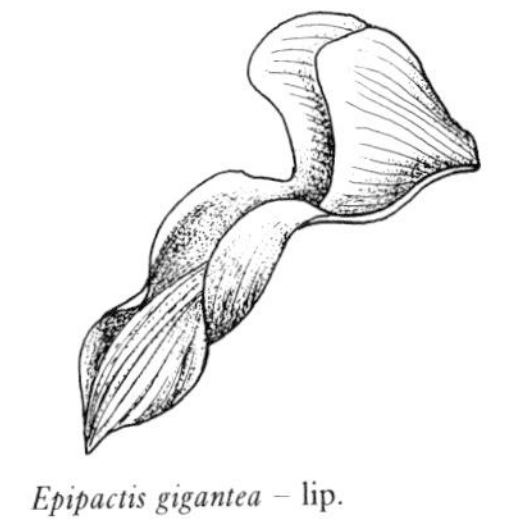

Epipactis gigantea – lip.

DISTRIBUTION: NW America, Himalaya.

HARDINESS: H2.

FLOWERING: summer.

5. CEPHALANTHERA Richard

A genus of about 14 species from north temperate regions. Cultivation as for *Dactylorhiza* (p. 80).

HABIT: perennial herbs with erect or short, creeping rhizomes and fibrous roots; aerial growth dying down in winter.

LEAVES: evenly spaced along stem, pleated.

FLOWERS: stalkless or very shortly stalked in a loose terminal spike, pointing upwards.

BRACTS: green and leaf-like, not sheathing the basal parts of the flowers.

SEPALS AND PETALS: hooded, scarcely opening.

LIP: constricted at the middle into 2 distinct parts, the base concave and clasping the base of the column, the apex with several ridges; spur absent.

POLLINIA: 2, club-shaped, each longitudinally divided, powdery.

Key to species

1. Flowers pinkish red; ovary hairy — **3. rubra**
2. Bracts as long as or longer than the ovary; lip with 3–5 ridges — **1. damasonium**
3. Bracts (except for the lowermost) shorter than the ovary; lip with 4–6 ridges — **2. longifolia**

1. C. damasonium (Miller) Druce

ILLUSTRATIONS: Landwehr, Wilde Orchideeën van Europa **2**: 51 (1977); Williams et al., Orchids of Britain and Europe, 145

(1978); Mossberg & Nilsson, Orchids of northern Europe, 31 (1979); Orchid Review **92**: 136 (1984); American Orchid Society Bulletin **54**: 681 (1985).

STEMS: 15–70 cm.

LEAVES: 3–6, 5–8 × 2–3 cm, ovate-lanceolate to oblong-ovate, the upper narrowest.

SPIKE: with 3–5 or more flowers.

FLOWERS: *c.* 2 cm, loosely arranged, creamy white.

BRACTS: as long as the ovary, lowermost much longer, ovate-lanceolate to lanceolate.

SEPALS AND PETALS: oblong to lanceolate, incurved, petals slightly shorter than sepals.

LIP: shorter than sepals and petals, yellowish orange at base; apical half heart-shaped, recurved, with 3–5 yellowish orange ridges.

DISTRIBUTION: Europe, N Africa, SW Asia.

HARDINESS: H2.

FLOWERING: spring–summer.

2. C. **longifolia** (Linnaeus) Fritsch

ILLUSTRATIONS: Landwehr, Wilde Orchideeën van Europa **2**: 515 (1977); Williams et al., Orchids of Britain and Europe, 145 (1978); Mossberg & Nilsson, Orchids of northern Europe, 29 (1979); American Orchid Society Bulletin **54**: 683 (1985), **55**: 593 (1986).

STEMS: 10–50 cm.

LEAVES: 4–12, 4–18 cm × 8–40 mm, linear-lanceolate to lanceolate.

SPIKE: with 10–20 or more flowers.

FLOWERS: *c.* 1.8 cm, white.

SEPALS AND PETALS: incurved but more open than in *C. damasonium*.

BRACTS: much shorter than ovary (except for the lowermost), mostly linear.

SEPALS: lanceolate.

PETALS: broader and shorter than sepals.

LIP: shorter than sepals and petals, yellowish orange at base; apical half heart-shaped, recurved, with 4–6 yellowish orange ridges.

DISTRIBUTION: Europe, N Africa, to the Middle East & temperate Asia.

HARDINESS: H2.

FLOWERING: spring.

3. C. **rubra** (Linnaeus) Richard

ILLUSTRATIONS: Danesch & Danesch, Orchideen Europas, Mitteleuropa, edn 3, 195 (1972); Landwehr, Wilde Orchideeën van Europa **2**: 517 (1978); Mossberg & Nilsson,

Orchids of Northern Europe, 27 (1979); American Orchid Society Bulletin **54**: 684 (1985).

STEMS: 10–60 cm, upper part hairy.

LEAVES: 2–8,3–14 × 1–3 cm, oblong to lanceolate.

SPIKE: with 2–15 flowers, hairy.

FLOWERS: 2–2.5 cm, opening fairly widely, pinkish red.

BRACTS: mostly longer than ovary, but the upper much shorter, linear.

SEPALS AND PETALS: to 2.5 cm, ovate to lanceolate.

LIP: as long as sepals, white, with several brownish ridges at base; apical half lanceolate, recurved, red-violet with *c.* 10 yellowish brown ridges.

DISTRIBUTION: Europe, N Africa, SW Asia east to Iran.

HARDINESS: H2.

FLOWERING: spring–early summer.

6. LISTERA R. Brown

A genus of about 10 species distributed throughout the temperate regions. Cultivation as for *Dactylorhiza* (p. 80).

HABIT: herbaceous perennials with short rhizomes; aerial growth dying down in winter.

LEAVES: 2 (rarely 3–4), ovate, almost opposite, arising at about the middle of the stem or below it.

FLOWERS: in a spike-like raceme.

SEPALS AND PETALS: more or less equal in size.

LIP: longer than the sepals and petals, strap-shaped, its apex 2-lobed; spur absent.

POLLINIA: 2, club-shaped, each longitudinally divided and made up of easily separable masses.

Key to species

1. Plants to 20 cm, leaves to 2.5 cm; raceme with 4–12 flowers **1. cordata**
2. Plants to 60 cm, leaves 5–20 cm; raceme with many flowers **2. ovata**

1. L. cordata (Linnaeus) R. Brown

ILLUSTRATIONS: Danesch & Danesch, Orchideen Europas, Mitteleuropa, edn 3, 135, 137 (1972); Landwehr, Wilde Orchideeën van Europa **2**: 533 (1977); Williams et al., Orchids of Britain and Europe, 141 (1978); Mossberg & Nilsson, Orchids of northern Europe, 49 (1979).

RHIZOME: slender, creeping.

STEMS: slender, sometimes copper-coloured, 4.5–20 cm, upper part minutely downy.

LEAVES: 2 (occasionally 4), ovate, shortly pointed, 1–2.5 cm.

RACEME: 1–6 cm with 4–12 flowers; bracts minute.

SEPALS: greenish, oblong-elliptic, 2–2.5 mm, obtuse.

PETALS: reddish.

LIP: purplish, linear, 3.5–4.5 mm, lateral lobes small and arising at the base, central lobe divided to about the middle, its segments diverging.

DISTRIBUTION: Europe, N America.

HARDINESS: H2.

FLOWERING: summer.

2. L. ovata (Linnaeus) R. Brown

ILLUSTRATIONS: Danesch & Danesch, Orchideen Europas, Mitteleuropa, edn 3, 134, 137 (1972); Landwehr, Wilde Orchideeën van Europa **2**: 531 (1977); Williams et al., Orchids of Britain and Europe, 141 (1978); Mossberg & Nilsson, Orchids of northern Europe, 47 (1979); Orchid Review **92**: 137 (1984).

RHIZOME: moderately thick.

STEMS: 20–60 cm, hairless below, downy above.

LEAVES: 2 (occasionally 3–4), ovate to broadly elliptic, distinctly 3–5-veined, obtuse, 5–20 cm.

RACEME: 7–25 cm, with numerous greenish or brownish green flowers; bracts minute.

SEPALS: incurved, ovate, obtuse, 4–5 mm.

PETALS: about the same length as sepals but narrower.

LIP: yellowish green, strap-shaped, 7–15 mm, the apex divided for one-third to half its length, lateral lobes vestigial or absent.

DISTRIBUTION: Eurasia.

HARDINESS: H1.

FLOWERING: summer.

7. CORYBAS Salisbury

A genus of about 60 species from the Himalaya and E Asia, extending to Australasia, of which only one is grown. Cultivation as for *Pleione* (p. 136), or they can be grown in peaty compost, with high humidity when growing. When dormant, they should be kept cool and dry.

HABIT: small, 1-leaved, terrestrial perennials; aerial growth dying off in winter; root tuberous, small and round.

LEAVES: ovate to heart-shaped, held flat against the ground.

FLOWER: 1 (rarely 2), large compared to the leaf size, shortly stalked (rarely not stalked), the stalk elongating in fruit.

UPPER SEPAL: conspicuous, broad, concave or hooded.

LATERAL SEPALS AND PETALS: thread-like, sometimes very small.

LIP: conspicuous, the basal part erect under the upper sepal, inrolled and tube-like around the column, the apical part expanded into a broad, often abruptly bent blade, the margin entire or fringed.

SPURS: 2, sometimes inconspicuous, placed on either side of the column.

POLLINIA: 2, granular.

1. C. **dilatatus** (Rupp & Nicholls) Rupp

ILLUSTRATION: American Orchid Society Bulletin **44**: 993 (1975), **46**: 989 (1977), **47**: 1120, 1121 (1978); Botanical Magazine n.s., 836 (1981).

HABIT: plant to 3 cm.

LEAF: to 3 cm wide, ovate to heart-shaped.

FLOWER: solitary, reddish purple.

UPPER SEPAL: to 2.5 cm.

LIP: with the apical part white or white tinged with crimson, papillose towards the apex; margins crimson, translucent, coarsely toothed, blotched and purple-veined.

DISTRIBUTION: SE Australia, including Tasmania.

HARDINESS: G1.

FLOWERING: summer.

8. SPIRANTHES Richard

A genus of about 30 species mainly in temperate zones, the majority occurring in N America. They should be grown in pots containing a well-drained compost and protected in a cold-frame or greenhouse.

HABIT: perennial herbs with slender or sturdy, fusiform tubers; aerial growth dying off in winter.

LEAVES: all basal in a rosette or evenly distributed on the stem, ovate-elliptic or lanceolate, rolled when young.

INFLORESCENCE: an elongate, twisted raceme or spike with many flowers which are close, spirally arranged and often scented.

SEPALS AND PETALS: almost equal, incurved, the tips free.

LIP: entire, its apical margins variously frilled or not; spur absent.

POLLINIA: 2, each made up of easily separable masses.

Key to species

1. Basal leaves ovate-elliptic, forming a flat rosette; raceme lateral

1.A. Inflorescence to 20 cm, with 7–20 flowers **1. spiralis**

1.B. Inflorescence to 40 cm, with up to 40 flowers **2. lacera**

2. Basal leaves linear-lanceolate, not forming a rosette; raceme terminal

2.A. Flowers in a dense raceme, not obviously arranged in rows, usually pink **5. sinensis**

2.B. Flowers in 1 spiral row in the raceme **4. aestivalis**
2.C. Flowers in 3–4 spiral rows in the raceme **3. cernua**

1. S. spiralis (Linnaeus) Chevallier
SYNONYM: *S. autumnalis* Richard.
ILLUSTRATIONS: Ross-Craig, Drawings of British plants, part 28: 47 (1971); Landwehr, Wilde Orchideeën van Europa **2**: 534, 535 (1977); Grey-Wilson & Mathew, Bulbs, t. 47 (1981).
TUBERS: stout, 5–13 mm thick.
LEAVES: 3–7, basal, 2–4 cm × 5–15 mm, ovate-elliptic, shortly and broadly stalked; upper leaves scale-like, sheathing the stem.
INFLORESCENCE: to 20 cm, glandular-hairy above, produced laterally from below the leaves.
FLOWERS: 7–20, white, scented, in 1 spiral row.
BRACTS: about as long as ovary.
SEPALS AND PETALS: 4–7 mm, the upper 3 and the lip forming a tube around the column.
LIP: apex truncate and scalloped.
DISTRIBUTION: Europe & N Africa to N Iran.
HARDINESS: H2.
FLOWERING: late summer–autumn.

2. S. lacera (Rafinesque) Rafinesque
SYNONYM: *S. gracilis* (Bigelow) Beck.
ILLUSTRATION: Luer, Native orchids of the United States and Canada, 111 (1975).
Like *S. spiralis* except as below.
TUBERS: slender, 3–6 mm thick.
INFLORESCENCE: to 50 cm, bearing up to 40 flowers.
DISTRIBUTION: eastern N America.
HARDINESS: H1.
FLOWERING: late summer.

3. S. cernua (Linnaeus) Richard
SYNONYM: *S. odorata* (Nuttall) Lindley.
ILLUSTRATION: Luer, Native orchids of the United States and Canada, 119, 121 (1975).
TUBERS: slender, elongate.
BASAL LEAVES: 2–6, 5–24 cm × 2–20 mm, linear-oblanceolate, the upper leaves small, sheath-like.
INFLORESCENCE: to 50 cm, with up to 60 flowers; upper parts downy, hairs sometimes glandular.
FLOWERS: white with yellowish centres, arranged in 3–4 rows.
BRACTS: shorter or longer than ovary.
SEPALS AND PETALS: 1.1–1.2 cm, sepals forming a tube with the lip.
LIP: apex rounded, scalloped and recurved.

DISTRIBUTION: eastern N America.
HARDINESS: H4.
FLOWERING: autumn.

4. **S. aestivalis** (Poiret) Richard
ILLUSTRATIONS: Ross-Craig, Drawings of British plants, part 28: 48 (1971); Landwehr, Wilde Orchideeën van Europa **2**: 537 (1977); Williams et al., Orchids of Britain and Europe, 139 (1978).
Like *S. cernua* except as below.
FLOWERS: 6–7 mm, arranged in 1 spiral row.
DISTRIBUTION: C & S Europe.
HARDINESS: H2.
FLOWERING: summer.

5. **S. sinensis** (Persoon) Ames
SYNONYM: *S. australis* Lindley.
ILLUSTRATIONS: Annals of the Royal Botanic Garden Calcutta 8: t. 369 (1898); Dansk Botanisk Arkiv **32**: 105 (1978); Williams et al., Orchids of Britain and Europe, 139 (1978).
Like *S. cernua* except as below.
FLOWERS: pink or rarely white, in a dense raceme.
LIP: with 2 basal calluses.
DISTRIBUTION: Asia to Australasia.
HARDINESS: G2.
FLOWERING: summer.

9. GOODYERA R. Brown

A genus of about 80 species from the north temperate regions, and SE Asia and Australasia; a few are grown, mainly for the sake of their coloured leaves. As most of them are of creeping habit, they are best grown in wide pans, using terrestrial compost which should be kept moist throughout the year. Most of the species require shade. A periodic dressing of pine-needles is beneficial (see Christian, P., Quarterly Bulletin of the Alpine Garden Society **43**: 322–4, 1975).

HABIT: terrestrial herbs, rhizomes creeping, sometimes above ground, aerial growth persistent through the winter.
LEAVES: membranous or fleshy, often asymmetric, mostly basal in rosettes, more rarely scattered on erect stems, often with differently coloured veins, rolled when young.
INFLORESCENCE: variable, erect, flowers few to many, arranged spirally or on 1 side, or uniformly in a cylindric spike or raceme.
SEPALS: directed forwards or spreading.
PETALS: directed forwards over the column.
LIP: saccate, the sac usually with hairs inside, entire, the tip usually pointed and reflexed.

COLUMN: short, blunt or with 2 long or short teeth on the rostellum.

POLLINIA: 2, club-shaped, each longitudinally divided and made up of easily separable masses.

Key to species

1. Flowers in a cylindric spike, not spirally arranged or in a 1-sided spike

1.A. Sepals with few hairs, reddish brown, tips white; leaves very dark, main veins red **1. colorata**

1.B. Sepals densely hairy, white with green midveins; leaves green with white veins **2. pubescens**

2. Flowers 5 mm or more; leaves green usually with broad, white mid-veins **3. oblongifolia**

3. Leaves green throughout or with a white network of veins **4. repens**

4. Leaves reddish with a pink or white network of veins **5. hispida**

1. **G. colorata** (Blume) Blume

ILLUSTRATIONS: Blume, Flore Javae **4**: t. 9b, f. 2 (1858); La Belgique Horticole **12**: 1, f. 4 (1862); J.J. Smith, Orchids of Java **2**: f. 93 (1909).

HABIT: stems *c.* 15 cm.

LEAVES: *c.* 6, scattered on the lower part of the stem, *c.* 6 × 2.5 cm, blackish green with a red main vein, narrowly ovate, pointed, sheathing at base.

INFLORESCENCE: *c.* 8 cm, hairy, with many flowers.

FLOWERS: *c.* 6 mm.

SEPALS: not spreading.

LIP: Base sac-like, hairy inside, the tip pointed and deflexed.

DISTRIBUTION: Indonesia (Java & Sumatra).

HARDINESS: G2.

2. **G. pubescens** (Willdenow) R. Brown

ILLUSTRATIONS: Botanical Magazine, 2540 (1825); Flore des Serres **15**: t. 1555 (1862–65); Luer, Native orchids of the United States and Canada, 141 (1975); American Orchid Society Bulletin **47**: 400 (1978), **48**: 1115 (1979).

RHIZOMES: creeping, sometimes above ground.

STEMS: to 50 cm, densely hairy.

LEAVES: in rosettes, to 9 × 4 cm, oblong-elliptic, blunt, net-veined, veins white.

INFLORESCENCE: cylindric, with up to 80 flowers, dense.

SEPALS: *c.* 5 × 3 mm, ovate, bluntly pointed.

PETALS: narrower than sepals, spathulate.

LIP: sac-like, the outer surface minutely warty, the tip pointed and recurved.

COLUMN: with a blunt rostellum.

DISTRIBUTION: eastern USA.
HARDINESS: H1.
FLOWERING: summer.

3. G. oblongifolia Rafinesque.
SYNONYMS: *G. decipiens* (Hooker) Hubbard; *G. menziesii* Lindley.
ILLUSTRATION: Luer, Native orchids of the United States and Canada, 143 (1975).
RHIZOMES: creeping, stoloniferous.
STEMS: to 48 cm, densely hairy.
LEAVES: to 11 × 4 cm, in rosettes, oblong-elliptic, blunt, mid-vein usually marked by a broad white line, the other veins forming a white, net-like pattern.
INFLORESCENCE: 1-sided, flowers loose or dense, scented.
SEPALS: 6–10 × 3–4 mm, oblong to narrowly ovate.
PETALS: spathulate.
LIP: sac-like, its tip oblong, blunt, recurved.
COLUMN: with a pointed rostellum.
DISTRIBUTION: western N America.
HARDINESS: H2.
FLOWERING: summer.

4. G. repens (Linnaeus) R. Brown
ILLUSTRATIONS: Danesch & Danesch, Orchideen Europas, Mitteleuropa, edn 3, 201, 202 (1972); Landwehr, Wilde Orchideeën van Europa 2: 541 (1977); Williams et al., Orchids of Britain and Europe, 137 (1978); Mossberg & Nilsson, Orchids of northern Europe, 52, 53 (1979).
RHIZOME: creeping, stoloniferous.
STEMS: to 25 cm, glandular-hairy.
LEAVES: 1.5–4 × 1–2 cm, ovate, blunt, forming a rosette, green or with a white, net-like pattern (var. **ophioides** Fernald).
INFLORESCENCE: spirally arranged or 1-sided, dense.
FLOWERS: glandular-hairy, cream-white, fragrant.
SEPALS: 3–4.5 × 3 mm, ovate.
PETALS: oblong-spathulate.
LIP: sac-like, its tip pointed, recurved.
COLUMN: with a blunt rostellum.
DISTRIBUTION: Eurasia, N America.
HARDINESS: H1.
FLOWERING: summer.

5. G. hispida Lindley
ILLUSTRATIONS: Annals of the Royal Botanic Garden Calcutta 8: t. 375 (1898); Bechtel et al., Manual of cultivated orchid species, 215 (1981).

Like *G. repens* except as below.

LEAVES: reddish, elliptic-lanceolate, with a pink or white net-like pattern.

COLUMN: with a pointed rostellum.

DISTRIBUTION: NE India, Malaysia, Thailand.

HARDINESS: G2.

FLOWERING: summer.

10. MACODES Lindley

A genus of about 10 species, (of which only a single one is grown), superficially similar to *Anoectochilus*, and distributed from Malaysia to New Guinea. Cultivation is as for tropical terrestrial species with the substitution of beech or oak leaf-mould for peat and bark and the addition of perlite to improve drainage. High humidity and protection from slugs are essential.

HABIT: terrestrial plants with short, fleshy, stems, rooting at basal nodes.

LEAVES: few, mostly basal, somewhat fleshy, attractively coloured and veined, rolled when young.

INFLORESCENCE: erect, flowers few to about 40.

FLOWERS: asymmetric due to the twisting in opposite directions of the lip and column.

LIP: uppermost, 3-lobed, its base concave and containing 2 glands.

COLUMN: with a single stigmatic area and with 2, thin, parallel wings on the front.

POLLINIA: 4, club-like, made up of small, easily separable masses of pollen.

1. M. petola (Blume) Lindley

ILLUSTRATIONS: Blume, Flore Javae 4: t. 31 (1858); Orchid Digest 38: 29 (1974); Bechtel et al., Manual of cultivated orchid species, 224 (1981).

LEAVES: *c.* 6 × 4 cm, ovate, velvety, dark green, the main veins and cross-veins beautifully marked with bright golden yellow.

INFLORESCENCE: *c.* 20 cm, downy with 10–40 flowers.

FLOWERS: reddish brown, lip white.

SEPALS, PETALS AND LIP: *c.* 5 mm.

DISTRIBUTION: Malaysia, Philippines to Indonesia (Sumatra).

FLOWERING: autumn–spring.

11. ANOECTOCHILUS Blume

A genus of perhaps 40 species distributed from India to NE Australia. Cultivation as for *Macodes* (p. 60).

HABIT: terrestrial; stems short, basal part rooting at nodes, upper part erect, fleshy.

LEAVES: few, mostly basal, thin, fleshy, with distinctively coloured veins, rolled when young.

INFLORESCENCE: erect, symmetric, flowers few.

LIP: lowermost, the apex widening abruptly to a transverse blade, the claw fringed, the base with a spur containing 2 glands.

COLUMN: with 2 wings on the front and 2 distinct stigmatic areas.

POLLINIA: 2, club-like, made up of small, easily separable masses of pollen on a common stipe.

Key to species

1. Lip inclined upwards; spur bent abruptly away from the ovary **1. roxburghii**
2. Lip pointing downwards; spur parallel to ovary **2. setaceus**

1. A. roxburghii (Wallich) Lindley

ILLUSTRATIONS: Annals of the Royal Botanic Garden Calcutta **8**: t. 390 (1898); Botanical Magazine, 9529 (1938); Orchid Digest **39**: 132 (1975); Dansk Botanisk Arkiv **32**: 55 (1978).

STEMS: to 30 cm.

LEAVES: 7 × 5.5 cm, ovate, velvety, brownish or pinkish green; veins marked in an attractive pink or yellowish net-like pattern, lower surface of leaf paler than upper.

INFLORESCENCE: with up to 15 flowers, glandular-hairy.

SEPALS: *c.* 1 cm, reddish or greenish red, reflexed.

PETALS: white or pinkish, forming a hood with the middle sepal, tips sickle-shaped.

LIP: *c.* 2 cm, inclined upwards.

SPUR: bent abruptly away from ovary.

DISTRIBUTION: NE India to Vietnam.

HARDINESS: G2.

FLOWERING: autumn–spring.

Anoectochilus roxburghii – lip.

2. A. setaceus (Blume) Lindley

SYNONYM: *A. regalis* Blume.

ILLUSTRATIONS: Edwards's Botanical Register **23**: t. 2010 (1837); Botanical Magazine, 4123 (1844); Flore de Serres, t. 6 (1846).

Like *A. roxburghii* except as below.

LIP: pointing downwards.

SPUR: parallel to ovary.

DISTRIBUTION: Sri Lanka, Java.

FLOWERING: spring.

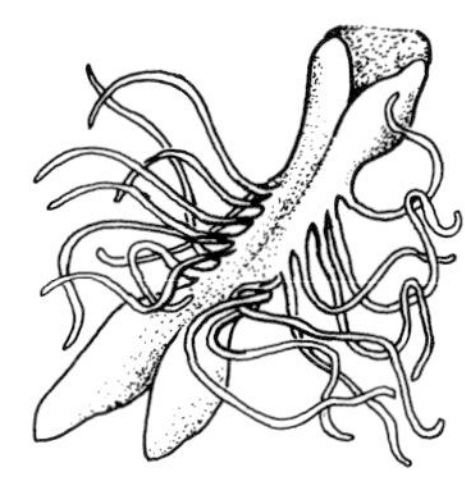

Anoectochilus setaceus – lip.

12. OPHRYS Linnaeus

A genus of between 40 and 50 species, many of which are superficially similar, thus making identification difficult. They are distributed from NW Europe throughout the Mediterranean region to the Middle East. There is some degree of specificity between pollinators (mostly bees and wasps) and the species of *Ophrys*. The male insects are attracted and sexually stimulated by the scent, shape, colour and the tactile stimuli of the hairs on the lip of the orchid. Fuller descriptions of these complex flowers are given in Danesch & Danesch (cited below) and Wood, *Orchid Review* **89**: 298 (1981). Literature: Nelson, E., *Gestaltwandel und Abbildung erörtert am Beispiel der Orchidaceen Europas und der Mittelmeerländer mit einer Monographie und Ikonographie der Gattung Ophrys* (1962)—all the species included in the following account are illustrated in this work, to which no further references are made; Danesch, E. & Danesch, O., *Orchideen Europas, Ophrys-Hybriden* (1972).

HABIT: perennial herbs.
TUBERS: 2 (occasionally 3), spherical, fleshy.
LEAVES: lanceolate, oblong or ovate, rolled when young, mostly in a basal rosette, sometimes spaced along the stem, when usually sheathing, usually appearing in autumn after the current season's growth has died back.
FLOWERS: few to *c*. 15 in a short or long, loose or rarely dense spike.
BRACTS: leaf-like, often inrolled.
SEPALS: spreading, usually greenish or pinkish, oblong or ovate, obtuse, all equal in length, the upper concave, erect or curved over the column.
PETALS: usually smaller and narrower than the sepals, often hairy or velvety.
LIP: spurless, hairy, complex in structure, often like a bee, wasp or other insect, oblong, square, rounded or diamond-shaped, usually marked with a coloured, hairless, often complex, patterned area (the speculum), lobed or not, with or without humps or horn-like basal protuberances; apex often with a short, tooth-like, deflexed or upcurved appendage.
COLUMN: erect or curved forwards; anther-connective beak-like, its apex obtuse or pointed.
POLLINIA: 2, club-like, made up of small, easily separable masses.

Synopsis of characters

Sepals. Pinkish or whitish: **6–9**.
Petals. Blackish or brownish purple: **1,2**; greenish-yellowish: **3–9**; brownish: 4,5; pink or purplish: 6–9. Hairy or velvety: **1,2,6–9**; hairless or papillose: 3–5. Rounded: **2**; triangular: 6–9; oblong to lanceolate: **2**–9; linear: **1,8,9**.
Lip. 3-lobed: **1–4,6–9**; entire: **5–7**. Mostly yellow: **3**. Densely fringed: **2**. Speculum bright, shiny blue: **2,4**; speculum H-shaped or of parallel lines: **5,8,9**. With an apical appendage: **5–9**; without an apical appendage: **1–5**.
Anther-connective. Obtuse: **1–4,6,7**; acute: 5–9.

Key to species

1. Lip with a tooth-like appendage at the apex

1.A. Speculum of 2 more or less parallel lines, or, if H-shaped, then petals hairless **5. sphegodes**

1.B. Lip distinctly 3-lobed

1.B.a. Sepals 1–1.5 cm **8. apifera**

1.B.b. Sepals 8–10 mm **9. scolopax**

1.C. Lip not or very indistinctly 3-lobed

1.C.a. Petals triangular to oblong-lanceolate; anther-connective acute **6. holoserica**

1.C.b. Petals broadly triangular; anther-connective obtuse **7. tenthredinifera**

2. Lip not lobed **5. sphegodes**

3. At least the margin of the lip bright yellow, and the whole usually hairless **3. lutea**

4. Speculum bright shining blue, covering almost the whole lip **2. vernixia**

5. Petals slender, antenna-like, blackish purple; lip sharply cleft at the apex **1. insectifera**

6. Petals oblong, greenish; lip merely notched at apex **4. fusca**

1. **O. insectifera** Linnaeus

SYNONYM: *O. muscifera* Hudson.

ILLUSTRATIONS: Danesch & Danesch, Orchideen Europas, Südeuropa, 128 (1969): Danesch & Danesch, Orchideen Europas, Mitteleuropa, edn 3, 222, 223 (1972); Ross-Craig, Drawings of British plants, part 28: t. 9 (1971); Mossberg & Nilsson, Orchids of northern Europe, 130, 131 (1979).

HABIT: plant to 60 cm.

LEAVES: linear to oblong.

SPIKE: with up to 10 flowers.

SEPALS: 6–8 mm, oblong-ovate, green.

PETALS: 4–6 mm, linear, antenna-like, blackish purple, velvety.

LIP: 9–10 × 6–7 mm, oblong, 3-lobed, central lobe cleft at apex, blackish purple or dark violet, papillose, lateral lobes oblong; speculum square or kidney-shaped, pale bluish violet.

COLUMN: with obtuse anther-connective.

DISTRIBUTION: most of Europe, uncommon in the extreme south and north and absent in the southeast.

HARDINESS: H2.

FLOWERING: spring–summer.

2. **O. vernixia** Brotero

SYNONYM: *O. speculum* Link.

ILLUSTRATIONS: Danesch & Danesch, Orchideen Europas, Südeuropa, 126 (1969); Sundermann, Europäische und mediterrane Orchideen, 82: (1975); Landwehr, Wilde

Orchideeën van Europa **2**: 377, 379 (1977); Grey-Wilson & Mathew, Bulbs, t. **37**: (1981).

HABIT: plant 7–30 cm.

LEAVES: oblong to lanceolate.

SPIKE: with up to 15 flowers.

SEPALS: 6–8 mm, oblong-ovate, greenish or yellowish green, sometimes lined with brown.

PETALS: one-third to half the length of the sepals, lanceolate, ovate-lanceolate or rounded, dark purple, hairy.

LIP: 1.3–1.5 cm, 3-lobed, margins densely fringed, hairs brownish purple to blackish purple, yellowish or brownish red, central lobe ovate, usually notched; lateral lobes oblong, yellowish; speculum large, shining blue, with a yellow margin.

COLUMN: with obtuse anther-connective.

DISTRIBUTION: Portugal, Mediterranean region.

HARDINESS: H3.

FLOWERING: spring.

3. **O. lutea** (Gouan) Cavanilles

ILLUSTRATIONS: Danesch & Danesch,Orchideen Europas, Südeuropa, 123, 124 (1969); Landwehr, Wilde Orchideeën van Europa **2**: 391 (1977); Williams et al., Orchids of Britain and Europe, **57**: (1978); Grey-Wilson & Mathew, Bulbs, t.37 (1981); American Orchid Society Bulletin **54**: 135 (1985); Kew Magazine **3**: 50 (1986).

HABIT: plant 7–30 cm.

BASAL LEAVES: ovate.

SPIKE: with up to 7 flowers.

SEPALS: *c.* 1 cm, green.

PETALS: one-third to half the length of the sepals, oblong, yellowish green, hairless or margins papillose.

LIP: 9–18 mm, oblong or rounded, 3-lobed, lateral lobes rounded, central lobe kidney-shaped,marginal zone broad, bright yellow, centre dark brown or blackish purple; speculum oblong, 2-lobed, greyish or bluish grey.

COLUMN: with obtuse anther-connective.

DISTRIBUTION: Mediterranean region.

HARDINESS: H3.

FLOWERING: spring.

4. **O. fusca** Link

ILLUSTRATIONS: Danesch & Danesch, Orchideen Europas, Südeuropa, 118 (1969); Landwehr, Wilde Orchideeën van Europa **2**: 383, 385, 387 (1977); Williams et al., Orchids of Britain and Europe, 55 (1978); Grey-Wilson & Mathew, Bulbs, t. 37 (1981); American Orchid Society Bulletin **54**: 284 (1985).

HABIT: plant 10–40 cm.

LEAVES: oblong or lanceolate.

SPIKE: with up to 8 flowers.

SEPALS: 9–11 mm, greenish or yellowish green.

PETALS: 6–8 mm, oblong, yellowish green or brownish green, indistinctly papillose.

LIP: 1.3–1.5 cm × 9–12 mm, oblong or obovate, purplish, purplish red or reddish brown, velvety, 3-lobed, lateral lobes not spreading, oblong, rounded, central lobe rounded, notched; speculum oblong, usually divided down the middle, bluish, often with a paler or yellowish margin.

COLUMN: with obtuse anther-connective.

DISTRIBUTION: Portugal, Mediterranean region, SW Romania.

HARDINESS: H3.

FLOWERING: spring.

Subsp. **iricolor** (Desfontaines) Swartz

SYNONYM: *O. iricolor* Desfontaines.

ILLUSTRATIONS: Danesch & Danesch, Orchideen Europas, Südeuropa, 119 (1969); Sundermann, Europäische und mediterrane Orchideen, 84 (1975); Landwehr, Wilde Orchideeën van Europa **2**: 383 (1977); Williams et al., Orchids of Britain and Europe, 55 (1978).

SPIKE: with up to 4 flowers.

PETALS: often papillose.

LIP: 2.5–3.1 cm, speculum bright blue.

DISTRIBUTION: C & E Mediterranean region.

5. **O. sphegodes** Miller

SYNONYM: *O. aranifera* Hudson.

ILLUSTRATIONS: Danesch & Danesch, Orchideen Europas, Südeuropa, 100–2 (1969); Ross-Craig, Drawings of British plants, part 28: t. 8 (1971); Landwehr, Wilde Orchideeën van Europa **2**: 397 (1977); Mossberg & Nilsson, Orchids of northern Europe, 132, 133 (1979); Orchid Review **89**: 295 (1981); American Orchid Society Bulletin **53**: 132 (1984), **54**: 138 (1985).

HABIT: plant 10–15 cm.

BASAL LEAVES: ovate-lanceolate.

SPIKE: with up to 10 flowers.

SEPALS: 6–12 mm, greenish.

PETALS: 4–8 mm, hairless, oblong-lanceolate, greenish or occasionally brownish, margins sometimes wavy.

LIP: 1–1.2 cm × 8–12 mm, entire, rounded, pale or dark brown or blackish brown, velvety, with or without basal protuberances; speculum usually H-shaped, the cross-line sometimes absent, greyish or purplish blue.

COLUMN: with acute anther-connective.

DISTRIBUTION: Europe.

HARDINESS: H3.

FLOWERING: spring–summer.

A variable species (see Wood, J.J., The subspecies of O. sphegodes in Cyprus and the eastern Mediterranean, Orchid Review **89**: 292–9, 1981).

Subsp. **litigiosa** (Camus) Becherer

SYNONYM: *O. litigiosa* Camus.

LIP: 5–10 × 5–10 mm, pale or yellowish brown, usually without basal protuberances.

Subsp. **mammosa** (Desfontaines) Soó

SYNONYM: *O. mammosa* Desfontaines.

LIP: 8–17 × 8–17 mm, blackish brown to blackish purple, usually with large basal protuberances.

Both of these subspecies are from southern Europe. The specific name is sometimes misspelled 'sphecodes'.

6. O. holoserica (Burmann) Greuter

SYNONYM: *O. fuciflora* (Schmidt) Moench.

ILLUSTRATIONS: Danesch & Danesch, Orchideen Europas, Südeuropa, 79a, b, c (1969); Danesch & Danesch, Orchideen Europas. Mitteleuropa, edn 3, 224–7 (1972); Ross-Craig, Drawings of British plants, part 28: t. 7(1971): Landwehr, Wilde Orchideeën van Europa **2**: 449, 461, 469 (1977); Mossberg & Nilsson, Orchids of northern Europe, 134, 135 (1979); Grey-Wilson & Mathew, Bulbs, t. 39 (1981); American Orchid Society Bulletin **54**: 133 (1985).

HABIT: plant 15–55 cm.

BASAL LEAVES: ovate-oblong.

SPIKE: with 2–6 flowers (occasionally more).

SEPALS: 9–13 mm, pinkish, whitish or greenish.

PETALS: about one-quarter to one-third of the length of the sepals, triangular to oblong-lanceolate, pink or purplish, hairy.

LIP: 9–13 mm, ovate or almost square, entire or rarely indistinctly 3-lobed, brownish or brownish purple,velvety; central part papillose, basal protuberances distinct, apical appendage upcurved, 3-toothed; speculum purplish or bluish grey bordered by a bold pattern of vertical and horizontal bands of pale green or yellow.

COLUMN: with acute anther-connective.

DISTRIBUTION: W, SW & C Europe.

HARDINESS: H3.

FLOWERING: spring–summer.

7. O. tenthredinifera Willdenow

ILLUSTRATIONS: Danesch & Danesch,Orchideen Europas, Südeuropa, 92, 93 (1969); Sundermann, Europäische und-mediterrane Orchideen, 98 (1975); Landwehr, Wilde Orchideeën van Europa **2**: 469, 471 (1977); Williams et al.,

Orchids of Britain and Europe, 43 (1978); Grey-Wilson & Mathew, Bulbs, t. 39 (1981); American Orchid Society Bulletin **53**: 133 (1984), **54**: 138 (1985).

Like *O. holoserica* except as below.

PETALS: broadly triangular.

LIP: almost square with a notched apex, dark brown, marginal area usually yellowish; speculum violet or bluish with a pale greenish border, pattern usually simple.

COLUMN: with obtuse anther-connective.

DISTRIBUTION: Portugal, Mediterranean region.

HARDINESS: H3.

FLOWERING: spring.

8. **O. apifera** Hudson

ILLUSTRATIONS: Danesch & Danesch, Orchideen Europas, Südeuropa, 94 (1969); Danesch & Danesch, Orchideen Europas, Mitteleuropa, edn 3, 228–33 (1972); Ross-Craig, Drawings of British plants, part 28: t. 6 (1971); Quarterly Bulletin of the Alpine Garden Society **44**: 59 (1976); Mossberg & Nilsson, Orchids of northern Europe, 136, 137 (1979); American Orchid Society Bulletin **54**: 16, 139 (1985).

HABIT: plants 15–50 cm.

BASAL LEAVES: lanceolate to ovate.

SPIKE: with up to 10 flowers.

SEPALS 1–1.5 cm, pink or purplish, rarely whitish.

PETALS: less than half as long as sepals, triangular to linear-lanceolate, greenish or rarely purplish, hairy.

LIP: 1–1.3 cm, ovate, 3-lobed, brownish or brownish purple; lateral lobes triangular, deflexed; basal protuberances *c.* 3 mm, apical appendage longer than wide, yellow, deflexed; speculum H-shaped or shield- or W-shaped, violet-purple with a pale green margin, sometimes indistinct.

COLUMN: with pointed anther-connective.

DISTRIBUTION: W, S & C Europe.

HARDINESS: H2.

FLOWERING: spring–summer.

9. **O. scolopax** Cavanilles

ILLUSTRATIONS: Danesch & Danesch, Orchideen Europas, Südeuropa, 72A (1969); Landwehr, Wilde Orchideeën van Europa **2**: 437, 438 (1977); Williams et al., Orchids of Britain and Europe, 35 (1978); Grey-Wilson & Mathew, Bulbs, t. 39 (1981).

Like *O. apifera* except as below.

SEPALS: 8–10 mm, the upper erect.

LIP: 8–12 mm, longer than wide, oblong-elliptic, with a narrow, hairless or papillose marginal zone, basal protuberances short, obtuse, apical appendage short, wider than long; speculum often X-shaped.

DISTRIBUTION: Portugal, W & C Mediterranean region.
HARDINESS: H3.
FLOWERING: spring.
A very variable species; several of its subspecies are occasionally grown in specialist collections

Subsp. **cornuta** (Steven) Camus
SYNONYM: *O.oestrifera* Bieberstein.
ILLUSTRATION: Kew Magazine **3**: pl. 49 (1986).
LIP: with a wide, hairless marginal zone and horn-like basal protuberances to 10 mm.
DISTRIBUTION: eastern Mediterranean area, the Caucasus and Iran.

Subsp. **heldreichii** (Schlechter) Nelson
SYNONYM: *O. oestrifera* subsp. *heldreichii* (Schlechter) Soó.
LIP: 1.3–1.5 cm.
DISTRIBUTION: eastern Mediterranean area.

Subsp. **orientalis** (Renz) Nelson
SYNONYM: *O.umbilicata* Desfontaines.
SEPALS: pale greenish, whitish or rarely reddish pink, the central hooded over the column.
DISTRIBUTION: Cyprus, Turkey, Syria and the Lebanon.

13. **SERAPIAS** Linnaeus

A genus of about 6 species between which the differences are not always clear, distributed from the Azores through the Mediterranean region to the Caucasus. Alpine house or cold-frame protection is necessary for plants grown in northern Europe. Pot culture and a compost incorporating coarse sand and limestone chips is recommended. New growth is produced during the autumn when water must be given sparingly and increased only when the plants are growing vigorously. As the growth dies down and the plants become dormant they require a warm, dry rest.

Literature: Nelson, E., *Monographie und Ikonographie der Orchidaceen Gattungen Serapias, Aceras, Loroglossum, Barlia* 3–45 (1968); Christian, P., The genus Serapias in the wild and in cultivation, *Quarterly Bulletin of the Alpine Garden Society* **43**: 188–98 (1975).

HABIT: terrestrial herbs; aerial growth dying off during the current year.
ROOTS: tuberous, tubers 2–4, occasionally 5, stalkless and in some species also produced at the end of root-like stolons, fleshy, ovoid.
STEMS: erect.
LEAVES: lanceolate, mostly basal or spaced along the stem, rolled when young, later often folded.
SPIKE: loose- or dense-flowered.

BRACTS: very conspicuous, boat-shaped, usually purplish or glaucous.

SEPALS AND PETALS: forming a pointed hood, usually coloured like the bracts, veins distinct; sepals lanceolate; petals ovate at base, tapering to a long point.

LIP: 3-lobed, composed of a basal and an apical part, usually hairy on the upper surface, lateral lobes upturned, central lobe usually abruptly bent downwards, tongue-like, the basal part with a swollen area or a double ridge at the base.

POLLINIA: 2, club-shaped,consisting of easily separable masses.

COLUMN APEX: beak-like.

Key to species

1. Lip with a single, black, hump-like swelling at the base **1. lingua**

2. Lip with 2 dark or pale ridges at the base

2.A. Apical part of the lip about as broad as the basal part when the whole lip is flattened out

2.A.a. Leaf-sheaths usually red-spotted; apical part of lip purplish to maroon **2. cordigera**

2.A.b. Leaf-sheaths not spotted; apical part of the lip yellow to orange **3. neglecta**

2.B. Spike with up to 10 flowers; tubers 2, stalkless **4. vomeracea**

2.C. Spikes with fewer flowers; tubers 3 or 4, one of them stalkless **5. olbia**

1. **S. lingua** Linnaeus

ILLUSTRATIONS: Danesch & Danesch, Orchideen Europas, Südeuropa, 196, 197 (1969); Sundermann, Europäische und mediterrane Orchideen, 116 (1975); Landwehr, Wilde Orchideeën van Europa **2**: 355 (1977); Williams et al., Orchids of Britain and Europe, 67 (1978); Grey-Wilson & Mathew, Bulbs, t. 40 (1981).

HABIT: plant to *c.* 30 cm.

TUBERS: 2–5, 1 stalkless.

LEAVES: 3–6, to *c.* 15 cm × 5–12 mm, linear to lanceolate, sheathing bases usually unspotted.

SPIKE: with 2–5 (sometimes as many as 9) flowers, loose.

BRACTS: *c.* 3–4 cm, about as long as the flowers.

HOOD: *c.* 2 cm.

LIP: lateral lobes purple; apical part to *c.* 1.9 cm × 7 mm, broadly lanceolate to narrowly ovate, hairless or shortly hairy, reddish, pinkish, violet or yellowish.

DISTRIBUTION: Portugal, N Africa, Mediterranean regions of S Europe east to Greece.

HARDINESS: H5.

FLOWERING: spring.

2. S. cordigera Linnaeus

ILLUSTRATIONS: Danesch & Danesch, Orchideen Europas, Südeuropa, 192, 193 (1969); Landwehr, Wilde Orchideeën van Europa **2**: 369 (1977); Williams et al., Orchids of Britain and Europe, 65 (1978); Grey-Wilson & Mathew, Bulbs, t. 40 (1981); American Orchid Society Bulletin **54**: 15 (1985).

HABIT: plant to 40 cm.

TUBERS: 2–3, 1 stalkless.

LEAVES: 3–8, to *c*. 15 × 1–1.5 cm, lanceolate, sheathing bases usually red-spotted.

SPIKE: 2–10-flowered, dense.

BRACTS: *c*. 3.4–4.5 cm, shorter or scarcely longer than flowers.

HOOD: *c*. 2.5 cm.

LIP: basal ridges diverging; lateral lobes dark purple or reddish; apical part 2–3 × 1.6–2 cm, heart-shaped, very hairy, purplish to maroon.

DISTRIBUTION: Spain, Portugal, Mediterranean region to W Turkey.

HARDINESS: H5.

FLOWERING: spring.

3. S. neglecta de Notaris

ILLUSTRATIONS: Danesch & Danesch, Orchideen Europas, Südeuropa, 189–91 (1969); Landwehr, Wilde Orchideeën van Europa, 363 (1977); Williams et al., Orchids of Britain and Europe, 63 (1978); Grey-Wilson & Mathew, Bulbs, t. 40 (1981); American Orchid Society Bulletin **53**: 139 (1984).

Like *S. cordigera* except as below.

HABIT: plant usually 15–25 cm, sturdy.

LEAVES: sheathing bases unspotted.

SPIKE: dense.

BRACTS: greenish.

LIP: basal ridges parallel; apical part heart-shaped, pale yellowish or orange.

DISTRIBUTION: C Mediterranean regions of Europe.

4. S. vomeracea (Burman) Briquet

SYNONYMS: *S. longipetala* (Tenore) Pollini; *S. pseudocordigera* Moricand.

ILLUSTRATIONS: Danesch & Danesch, Orchideen Europas, Südeuropa, 185 (1969); Sundermann, Europäische und mediterrane Orchideen, 113 (1975); Landwehr, Wilde Orchideeën van Europa **2**: 357, 359 (1977); Grey-Wilson & Mathew, Bulbs, t. 40 (1981); American Orchid Society Bulletin **53**: 139 (1984).

HABIT: plant 10–60 cm.

TUBERS: 2, stalkless.

LEAVES: 6–8, *c*. 6–11 cm × 5–12 mm, linear to lanceolate, sheathing bases green, unspotted.

SPIKE: with up to 10 flowers, loose.
BRACTS: to 5 cm, longer than the flowers.
LIP: reddish or brownish, distinctly hairy, basal ridges parallel, lateral lobes often blackish; apical part *c.* 1.5–2 cm × 6–10 mm, lanceolate.
DISTRIBUTION: S Europe.
HARDINESS: H5.
FLOWERING: spring.

Subsp. **orientalis** Greuter
SYNONYM: *S. orientalis* invalid.
ILLUSTRATIONS: Landwehr, Wilde Orchideeën van Europa **2**: 367 (1977); Williams et al., Orchids of Britain and Europe, 61 (1978); Kew Magazine 3: pl. 51 (1986).
HABIT: plant 15–35 cm.
SPIKE: 5–7-flowered, dense.
BRACTS: scarcely as long as or slightly longer than flowers, purplish.
LIP: apical part broadly lanceolate.

Serapias vomeracea subsp. *orientalis* – lip.

5. S. olbia Verguin
SYNONYM: *S. gregaria* Godfery.
ILLUSTRATIONS: Danesch & Danesch, Orchideen Europas, Südeuropa, 195 (1969); Sundermann, Europäische und mediterrane Orchideen, 116 (1975); Landwehr, Wilde Orchideeën van Europa **2**: 365, t. 3, 373 (1977); Williams et al., Orchids of Britain and Europe, 67 (1978).
Like *S. vomeracea* except as below.
HABIT: plant slender.
TUBERS: 3–4, 1 stalkless.
SPIKE: few flowered.
DISTRIBUTION: Southeast France.
Thought to be a natural hybrid between *S. lingua* and *S. parviflora*.

14. ACERAS R. Brown

A genus of a single species rarely seen in cultivation. It is similar to some species of *Orchis*, differing in the lack of a spur. In nature it occurs on chalk or limestone. Cultivation as for *Orchis*.

HABIT: herbaceous perennial; aerial growth usually dying down in winter or plant sometimes overwintering as a basal rosette.
STEMS: to *c.* 50 cm arising from 2 ovoid tubers.
LEAVES: unspotted, the lower crowded, spreading, 5–12 × *c.* 2.5 cm, lanceolate, obtuse, the upper smaller, erect, sheathing the stem.
SPIKE: to 20 cm, bearing 50 or more flowers.
SEPALS: obovate-lanceolate, 6–7 mm.

PETALS: linear, slightly shorter, than sepals, all yellowish or greenish yellow, hooded over the column.

LIP: *c.* 1.2 cm, shaped like a man, the lateral lobes being the arms and the divided central lobe the legs; spur absent.

POLLINIA: 2, club-shaped, composed of easily separable masses.

1. **A. anthropophorum** (Linnaeus) Aiton

ILLUSTRATIONS: Danesch & Danesch, Orchideen Europas, Mitteleuropa, edn 3, 131 (1972); Danesch & Danesch, Orchideen Europas, Südeuropa, 240 (1967); Williams et al., Orchids of Britain and Europe, 67 (1979); Mossberg & Nilsson, Wilde Orchideeën van Europa **2**: t. 153 (1977); American Orchid Society Bulletin **54**: 19 (1985).

DISTRIBUTION: W Europe, Mediterranean region.

HARDINESS: H5.

FLOWERING: summer.

15. HIMANTOGLOSSUM Koch

A genus of 3 to 5 species from Europe and SW Asia not often found in cultivation. Cultivation as for Orchis.

HABIT: herbaceous perennials with 2 ovoid tubers; aerial growth dying down in winter.

LEAVES: in a basal rosette and arranged along the stem.

SPIKE: elongate, bracts equalling the flowers.

SEPALS AND PETALS: free, forming a hood over the column.

LIP: long, strap-shaped, 3-lobed, the central lobe bifid and spirally twisted, 2 or more times longer than the lateral lobes; a single spur present, arising from the base of the lip.

POLLINIA: 2, club-shaped, composed of easily separable masses, attached to a single viscidium.

1. **H. hircinum** (Linnaeus) Sprengel

ILLUSTRATIONS: Danesch & Danesch, Orchideen Europas, Mitteleuropa, edn 3, 130 (1972); Danesch & Danesch, Orchideen Europas, Südeuropa, 241 (1967); Williams et al., Orchids of Britain and Europe, 69 (1978); Mossberg & Nilsson, Orchids of northern Europe, 125 (1979).

HABIT: stem to 90 cm, purplish-blotched.

LEAVES: the lower 4–6, 5–51 × 3–5 cm, elliptic-oblong, obtuse, the upper smaller, acute, their bases clasping the stem.

SPIKE: 10–50 cm, cylindric, with 15–18 loosely arranged flowers which emit a goat-like smell; bracts shorter than the flowers.

SEPALS AND PETALS: pale green, sometimes purplish-lined, hooded, ovate, obtuse, the petals linear, 7–10 mm.

LIP: 3–5 cm × 2 mm, lateral lobes strap-shaped, central lobe ribbon-like, coiled in bud, spirally twisted when open, apex bifid, pale green suffused with purple, purple-spotted at base.

SPUR: 2–2.5 mm, conical, downwardly pointing.

DISTRIBUTION: W, C & S Europe.

HARDINESS: H4.

FLOWERING: summer.

16. BARLIA Parlatore

A genus of a single species, requiring cultural conditions similar to those for *Orchis* (p. 74).

Like *Himantoglossum* except as below.

BRACTS: as long as or longer than the flowers

SEPALS: erect, not hooded.

LIP: much shorter, its central lobe not spirally twisted.

1. B. robertiana (Loiseleur) Greuter

SYNONYM: *Himantoglossum longibracteatum* (Bernardi) Schlechter.

ILLUSTRATIONS: Danesch & Danesch, Orchideen Europas, Südeuropa, 242, 243 (1967); Sundermann, Europäische und mediterrane Orchideen, 122 (1975); Landwehr, Wilde Orchideeën van Europa **2**: 345 (1977); Williams et al., Orchids of Britain and Europe, 97 (1978); American Orchid Society Bulletin **54**: 19 (1985).

DISTRIBUTION: Mediterranean area, Canary Islands.

HARDINESS: H5.

FLOWERING: summer.

17. ANACAMPTIS Richard

A genus of a single species, in nature occurring in open grassland and calcareous dunes. Although occurring naturally in northern Europe and therefore probably hardy enough for normal outdoor culture, the species is perhaps better suited for pot-culture as recommended for *Orchis*.

HABIT: perennial herb; aerial growth dying down in winter.

TUBERS: spherical or ovoid.

STEMS: to 75 cm with up to 8 leaves.

LEAVES: 15–25 × *c.* 2 cm, mostly basal, unspotted, lanceolate, upper leaves shorter, rolled when young.

SPIKE: to 8 cm, conical, becoming shortly cylindric, dense.

BRACTS: linear, slightly longer than ovary.

FLOWERS: pink, red or purple, occasionally white, emitting a faint fox-like smell.

SEPALS AND PETALS: 6–8 × *c.* 3 cm, upper sepal and petals incurved, lateral sepals spreading.

LIP: deeply 3-lobed, 6–9 × 8–10 mm, as broad as long or broader than long, lobes oblong, *c.* 4 mm, apex obtuse or truncate; base of lip with 2 erect ridges leading towards the mouth of the spur.

SPUR: 1–1.4 cm, slender, downwardly pointing.

POLLINIA: 2, club-like, made up of easily separated masses.

1. **A. pyramidalis** Richard

ILLUSTRATIONS: Danesch & Danesch, Orchideen Europas, Mitteleuropa, edn 3, 146, 147 (1972); Landwehr, Wilde Orchideeën van Europa **2**: 341, 343 (1977); Mossberg & Nilsson, Orchids of northern Europe, 123 (1979); Williams et al., Orchids of Britain and Europe, 81 (1978).

DISTRIBUTION: Europe, Mediterranean area, SW Asia.

HARDINESS: H1.

FLOWERING: summer.

18. **ORCHIS** Linnaeus

A genus of about 35 species from Europe to Asia. *Dactylorhiza* (p. 80) is often included under *Orchis* and most nurserymen still offer plants under the latter name. Most species, apart from those with northerly distributions, are best suited for growing in pots in a cold-frame or unheated greenhouse. The compost should be extremely well drained and kept moist only while the plants are in active growth. The tubers should be repotted annually and kept almost dry while dormant.

HABIT: perennial herbs; aerial growth dying down in winter.

TUBERS: 2–3, spherical or ovoid.

LEAVES: mostly in a basal rosette, spotted or unspotted, stem-leaves usually sheath-like.

FLOWERS: in short or long, dense- or loose-flowered spikes.

BRACTS: membranous.

SEPALS AND PETALS: almost equal, free, all curved forward over the column, or the lateral sepals spreading and the upper sepal and petals incurved and hooded over the column.

LIP: mostly 3-lobed, sometimes unlobed.

SPUR: 8–25 mm, arising from the base of the lip.

POLLINIA: 2, club-like, powdery.

Synopsis of characters

Leaves. Spotted: **1–4**; unspotted: **1–13**.
Bracts. Usually equal to ovary: **1–6**; shorter: **2,3,10–13**; longer: **1–5,8,9**.
Flower colour. Deep pink, purple, red or brownish: **1–13**; yellow: 1–3; greenish: **6,7,9**; white: **1–6,10–13**.
Spur. Length 2 mm: **10,11**; 5–12 mm: **1–9,12.13**; 15–25 mm: **4–6**. Erect: **1–6**; horizontal: **1–6**; downwardly pointing: **4,5,8–13**.
Lateral sepals. Spreading: **1–5**; hooded: **6–13**.
Lip. Unlobed: **1–8**; lobed: **1–6,9**.

Key to species

1. Lateral sepals spreading, only the petals and sometimes the upper sepal hooded over the column

1.A. Spur 1–2 cm, cylindric, not tapered to the apex, or flowers yellow

1.A.a. Flowers bright or pale yellow **2. provincialis**

1.A.b. Spur horizontal or erect **1. mascula**

1.A.c. Spur curved downwards **3. quadripunctata**

1.B. Bract as long as or shorter than the ovary; flowers in a cylindric spike **4. anatolica**

1.C. Bract longer than ovary; flowers in a 1-sided spike **5. joo-iokiana**

2. Lip unlobed, ovate to fan-shaped **8. papilionacea**

3. Spur pointing upwards or occasionally horizontal

3.A. Spur to 1 cm, shorter than to as long as lip **6. morio**

3.B. Spur to 1.6 cm, longer than lip **7. longicornu**

4. Central lobe of lip entire, at most indistinctly spotted **9. coriophora**

5. Sepals and petals much darker than lip; central lobe of lip divided into smaller lobes which are about as long as broad

5.A. Sepals to 3.5 mm; spur to 2 mm **11. ustulata**

5.B. Sepals to 1.4 cm; spur *c.* 5 mm **10. purpurea**

6. Sepals and petals not darker than lip; central lobe of lip divided into smaller lobes which are twice or more as long as broad

6.A. Leaves 1 or 2, margins not wavy **12. militaris**

6.B. Leaves 5–8, margins wavy **13. italica**

1. **O. mascula** Linnaeus

ILLUSTRATIONS: Sundermann, Europäische und mediterrane Orchideen, 146 (1975); Landwehr, Wilde Orchideeën van Europa 1: 261 (1977); Williams et al., Orchids of Britain and Europe, 101 (1978); American Orchid Society Bulletin **55**: 8 (1986).

HABIT: stem to 60 cm.

LEAVES: 3–11, 5–20 × 1.5–4.5 cm, lanceolate or oblanceolate to narrowly obovate, purplish-spotted or unspotted, mostly basal, upper leaves sheathing stem.

SPIKES: to 32 cm, few to many, loosely or densely flowered.

FLOWERS: purple, lilac, or sometimes white.

BRACT: about as long as the ovary.

SEPALS: upper sepal to 10 mm, oblong, curved forwards; lateral sepals to 1.3 cm, obliquely oblong or ovate, erect or reflexed.

PETALS: to 9 × 7 mm, obliquely ovate, hooded (with the upper sepal) over the column.

LIP: to 1.5 cm, 3-lobed or sometimes scarcely lobed, edges usually folded back, centre finely papillose, with purple lines or spots.

SPUR: to 1.2 cm, about equal to ovary, cylindric, horizontal or erect.
DISTRIBUTION: Europe to W Iran.
HARDINESS: H1.
FLOWERING: spring.

2. O. provincialis Balbis
ILLUSTRATIONS: Danesch & Danesch, Orchideen Europas, Südeuropa, 210, 211 (1967); Landwehr, Wilde Orchideeën van Europa 1: 279 (1977); Williams et al., Orchids of Britain and Europe, 103 (1978).
Like *O. mascula* except as below.
FLOWERS: bright or pale yellow.
DISTRIBUTION: S Europe, N Africa to E Mediterranean area.
HARDINESS: H5.
FLOWERING: spring.

3. O. quadripunctata Tenore
ILLUSTRATION: Danesch & Danesch, Orchideen Europas, Südeuropa, 206–8 (1967); Sundermann, Europäische und mediterrane Orchideen, 144 (1975); Landwehr, Wilde Orchideeën van Europa 2: 297 (1977); Williams et al., Orchids of Britain and Europe, 99 (1978); American Orchid Society Bulletin **55**: 12 (1986).
Like *O. mascula* except as below.
LIP: base with 2–4 small purple spots.
SPUR: to 1.2 cm, curved downwards, uniformly narrow.
DISTRIBUTION: E Mediterranean area.
HARDINESS: H5.
FLOWERING: spring.

4. O. anatolica Boissier
ILLUSTRATIONS: Sundermann, Europäische und mediterrane Orchideen, 146 (1975); Landwehr, Wilde Orchideeën van Europa 1: 283 (1977); Williams et al., Orchids of Britain and Europe, 99 (1978).
HABIT: stem to 40 cm, with 1–2 sheath-like leaves.
LEAVES: 3–8, to 13 × 2.5 cm, lanceolate to broadly lanceolate, unspotted, or blotched or spotted with purple, basal.
SPIKE: to 15 cm, loose, with few to many flowers.
FLOWERS: purplish to pale pink, sometimes white.
BRACT: as long as or shorter than ovary.
SEPALS: *c.* 10 × 4 mm, oblong to narrowly ovate; upper sepal slightly broader, hooded (with the slightly smaller petals) over the column.
LIP: to 1.4 cm, 3-lobed, edges often folded back, centre white with purple spots or lines.
SPUR: 1.5–2.5 cm, longer than ovary, slender, wider at mouth, narrow at apex, upwardly curved or horizontal.

DISTRIBUTION: E Mediterranean area to W Iran.
HARDINESS: H5.
FLOWERING: summer.

5. O. joo-iokiana (Makino) Maekawa
SYNONYM: *Ponerorchis joo-iokiana* (Makino) Maekawa.
ILLUSTRATIONS: Kitamura et al., Coloured illustrations of herbaceous plants of Japan **17**: t. 2 (1964); Maekawa, The wild orchids of Japan in colour, t. 19 (1971).
Like *O. anatolica* except as below.
LEAVES: 1–4, not in a basal rosette.
FLOWERS: mostly on one side of the inflorescence.
BRACTS: longer than ovary.
DISTRIBUTION: Japan.

6. O. morio Linnaeus
ILLUSTRATIONS: Danesch & Danesch, Orchideen Europas, Südeuropa, 222 (1967); Landwehr, Wilde Orchideeën van Europa **1**: 245, 247 (1977); Williams et al., Orchids of Britain and Europe, 93 (1978); American Orchid Society Bulletin **55**: 121 (1986).
HABIT: stem to 40 cm.
LEAVES: to 10, basal, to 16 × *c.* 1.3 cm, lanceolate to broadly oblong; 1–2 sheath-like leaves on stem.
SPIKE: 3–13 cm, loose or dense, with few to many flowers.
FLOWERS: reddish purple, mauve, or sometimes greenish or white.
BRACTS: about as long as ovaries.
SEPALS: 7–12 × 4–6 mm, oblong, apex rounded, often greenish- or purplish-veined, lateral sepals a little longer.
PETALS: to 7 × 3 mm, oblong.
LIP: *c.* 10 × 10 mm, often broader than long, edges often folded back.
SPUR: *c.* 8 mm, shorter than or as long as lip, cylindric, horizontal or upwardly curved.
DISTRIBUTION: Europe to W Iran.
HARDINESS: H1.
FLOWERING: spring.

7. O. longicornu Poiret
ILLUSTRATIONS: Danesch & Danesch, Orchideen Europas, Südeuropa, 225 (1967); Landwehr, Wilde Orchideeën van Europa **1**: 251 (1977); Williams et al., Orchids of Britain and Europe, 95 (1978).
Like *O. morio* except as below.
SPUR: to 1.6 cm, much longer than lip.
DISTRIBUTION: W Mediterranean area.
HARDINESS: H5.

8. O. papilionacea Linnaeus

ILLUSTRATIONS: Danesch & Danesch, Orchideen Europas, Südeuropa, 220 (1967); Sundermann, Europäische und mediterrane Orchideen, 138 (1975); Landwehr, Wilde Orchideeën van Europa 1: 233 (1977); American Orchid Society Bulletin **53**: 136 (1984), **55**: 117 (1986).

HABIT: stem to 40 cm.

LEAVES: up to 10, to *c.* 10 × 1 cm, lanceolate, mostly basal.

SPIKE: ovoid with up to 15 flowers.

FLOWERS: purplish, red or brownish.

BRACT: longer than ovary.

SEPALS AND PETALS: 1–1.8 cm, narrowly ovate, strongly veined, upper sepal and petals forming a hood.

LIP: 1.2–2.5 × 1.2–2.5 cm, entire, fan-shaped or ovate, distinctly lined, margin shallowly toothed.

SPUR: *c.* 1.2 cm, shorter than ovary, cylindric, downwardly curved.

DISTRIBUTION: S Europe to SW Asia.

HARDINESS: H5.

FLOWERING: spring.

9. O. coriophora Linnaeus

ILLUSTRATIONS: Danesch & Danesch, Orchideen Europas, Südeuropa, 226 (1967); Landwehr, Wilde Orchideeën van Europa **2**: 301, 303 (1977); Williams et al., Orchids of Britain and Europe, 83 (1978); American Orchid Society Bulletin **55**: 121 (1986).

HABIT: stem to *c.* 50 cm.

LEAVES: to 7, *c.* 15 × 2 cm, linear to lanceolate, erect, on basal part of stem, often turning brown before completion of flowering.

SPIKE: 3–12 cm, ovoid to cylindric, loose or dense.

FLOWERS: red, brownish or purplish, or sometimes greenish.

BRACTS: a little longer than ovaries.

SEPALS AND PETALS: hooded, tips free, sepals to 10 × 4 mm, ovate to lanceolate, the laterals wider than the upper; petals to 6 × 1–2 mm.

LIP: 5–11 mm, longer than broad, 3-lobed.

SPUR: 5–9 mm, shorter than or equalling ovary, conical or cylindric, downwardly curved.

DISTRIBUTION: S, C & E Europe.

HARDINESS: H2.

FLOWERING: spring.

10. O. purpurea Hudson

ILLUSTRATIONS: Danesch & Danesch, Orchideen Europas, Mitteleuropa, edn 3, 114–17 (1972); Landwehr, Wilde Orchideeën van Europa **2**: 325 (1977); Mossberg & Nilsson,

Orchids of northern Europe, 115 (1979); American Orchid Society Bulletin **54**: 17 (1985), **55**: 117 (1986).

HABIT: stem to 80 cm or more, robust.

LEAVES: to 6 or more, to 23 × 6 cm, broadly lanceolate to elliptic, shining, unspotted, forming a basal rosette, 1–2 leaves sheathing the lower part of the stem.

SPIKE: to 17 × 6 cm, broadly cylindric, many-flowered.

BRACTS: to 5 mm, much shorter than ovaries.

FLOWERS: brownish or reddish purple, rarely white.

SEPALS: to 1.4 cm, ovate, acute.

PETALS: shorter than sepals, linear-lanceolate.

LIP: to 1.5 cm, longer than broad, 4-lobed; lateral lobes oblong, central lobe divided into 2 smaller, oblong lobes which are about as long as broad and between which is a small tooth.

SPUR: *c.* 5 mm, less than half the length of ovary, cylindric, downwardly curved.

DISTRIBUTION: Europe to N Africa.

HARDINESS: H1.

FLOWERING: spring–summer.

11. **O. ustulata** Linnaeus

ILLUSTRATIONS: Danesch & Danesch, Orchideen Europas, Südeuropa, 230 (1967); Landwehr, Wilde Orchideeën van Europa **2**: 311 (1977); Mossberg & Nilsson, Orchids of northern Europe, 113 (1979).

Like *O. purpurea* except as below.

HABIT: stem to 25 cm.

SEPALS: to 3.5 mm.

LIP: with no tooth between the lobes of the middle lobe.

SPUR: 2 mm.

DISTRIBUTION: Europe to the Caucasus.

12. **O. militaris** Linnaeus

ILLUSTRATIONS: Danesch & Danesch, Orchideen Europas, Mitteleuropa, edn. 3, 119–24 (1972); Landwehr, Wilde Orchideeën van Europa **2**: 319 (1977); Williams et al., Orchids of Britain and Europe, 89 (1978); American Orchid Society Bulletin **54**: 17 (1985).

HABIT: stem to 45 cm, robust.

LEAVES: 3–5, to 13 × 4 cm, broadly elliptic to oblanceolate, obtuse, unspotted, basal; 1–2 leaves sheathing the lower part of the stem.

SPIKE: to 10 × 4.5 cm, broadly cylindric, many-flowered.

FLOWERS: whitish pink, rarely white, veins darker.

BRACTS: *c.* 3 mm, much shorter than ovaries.

SEPALS: to 1.5 cm, ovate-lanceolate, acute.

PETALS: shorter than sepals, linear.

LIP: to 1.5 cm, longer than broad, 4-lobed; lateral lobes linear, central lobe divided into 2, almost transverse, oblong lobes which are about twice as long as broad and between which is a small tooth.

SPUR: to 7 mm, about half the length of the ovary, cylindric, downwardly curved.

DISTRIBUTION: C Europe to the Caucasus.

HARDINESS: H1.

FLOWERING: spring–summer.

13. **O. italica** Poiret

ILLUSTRATIONS: Sundermann, Europäische und mediterrane Orchideen, 134 (1975); Landwehr, Wilde Orchideeën van Europe **2**: 321 (1977); Williams et al., Orchids of Britain and Europe, 91 (1978); American Orchid Society Bulletin **55**: 116 (1986).

Like *O. militaris* except as below.

LEAVES: 5–8, margins wavy.

LIP: lobes of central lobe linear, 4 or more times longer than broad.

DISTRIBUTION: Portugal & Mediterranean area.

HARDINESS: H5.

FLOWERING: spring.

19. DACTYLORHIZA Nevski

A temperate genus of 20–30 species from Europe, N Africa, Asia and N America. They are generally hardy and can be cultivated in full sun out-of-doors in a peat-bed. Cultivation conditions for the more southerly species are similar to those for *Serapias* and the less hardy species of *Ophrys* (see p. 61). The species comprising *Dactylorhiza* were long considered to be part of *Orchis*, and were separated as the genus *Dactylorchis* by Vermeulen in 1947 who was presumably unaware of the earlier name provided by Nevski.

Hybridisation is rife in *Dactylorhiza* and the resulting crosses are often as fertile and plentiful in the wild as their parents. In addition, many frequently used characters are very variable. As a result the delimitation of species and other categories is not well defined and the classification is confused. The absence or presence of spots on the leaves is not a reliable character when attempting to identify wild populations but should suffice for the species in cultivation.

Literature: Nelson, E., *Monographie und Ikonographie der Orchidaceen-Gattung Dactylorhiza* (1976).

HABIT: terrestrial; root-tubers palmately lobed or divided into finger-like sections; aerial parts dying down in winter.

LEAVES: 3–15, spirally arranged on the flowering stem, not forming a basal rosette at flowering time, sometimes spotted or blotched with brownish purple; upper leaves often small and bract-like.

FLOWERS: several to many, in compact spherical, conical or cylindric racemes.

BRACTS: leaf-like, green or purplish green.

SEPALS AND PETALS: free; lateral sepals spreading or erect; upper sepal and petals curved forwards forming a hood over the column.

LIP: simple or 3-lobed, spurred, usually marked with lines, dots or streaks.

COLUMN: short; rostellum 3-lobed.

Synopsis of characters

Height. Over 70 cm: **1–4**.
Stem. Hollow: **1–5**; solid: **6,7**.
Leaves. Yellowish green: **1–4**; mid to dark green: **1,2,5,7**. Spotted: **5–7**; unspotted: **1–4**. Hooded at tip: **3,4,6**; not hooded at tip: **1,2,5–7**. Not bract-like near top of stem: **3,4**. Overtopping inflorescence: **3,4**.
Bracts. Longer than flowers: 1–5; equal to or shorter than flowers: **1,2,6,7**.
Inflorescence. More than 15 cm long: **1,2**.
Flowers. White: **6,7**; pink: **1–4,6,7**; purplish pink: **1,2,5**;crimson, brownish pink or brick-red: **3,4**; yellow: **3,4**.
Lip. More than 1.2 cm wide: **1,2**.
Spur. 1.2–1.5 cm: **1–4**. Conical: **1–5**; cylindric: **3–7**.

Key to species

1. Leaves mid to dark green, spotted or blotched with brownish purple
1.A. Stem hollow **5. majalis**
1.B. Lip shallowly 3-lobed, central lobe shorter than lateral lobes **6. maculata**
1.C. Lip deeply 3-lobed, central lobe longer than lateral lobes **7. fuchsii**
2. Lip more than 1.1 cm wide; leaves 8–14
2.A. Inflorescence 15–25 cm; bracts longer than the flowers and protruding from the inflorescence **2. elata**
2.B. Inflorescence 5–13 cm; bracts at most as long as flowers, scarcely protruding from the inflorescence **1. foliosa**
3. Spur 5–8 mm **3. incarnata**
4. Spur 1.2–1.5 cm **4. sambucina**

1. D. foliosa (Solander) Soó

SYNONYM: *Orchis maderensis* Summerhayes.

ILLUSTRATIONS: Botanical Magazine, 5074 (1858); Landwehr, Wilde Orchideeën van Europa **2**: 219 (1977); Williams et al., Orchids of Britain and Europe, 119 (1978).

STEM: to 70 cm, hollow.

LEAVES: 8–14, to 20 cm, narrowly lanceolate to triangular, keeled, acute, stiff, ascending or arched, yellowish green, unspotted; upper leaves bract-like.

BRACTS: equal to or shorter than the flowers and scarcely protruding from the inflorescence.

INFLORESCENCE: 5–13 cm, cylindric, densely many-flowered.

FLOWERS: bright purplish pink.

SEPALS AND PETALS: 1–1.3 cm, lanceolate to ovate, rounded; lateral sepals a little broader than the rest.

LIP: 1–1.1 × 1.2–1.3 cm, always wider than long, transversely oblong or elliptic, shallowly 3-lobed, marked with lines and streaks; central lobe triangular.

SPUR: 6–8 mm, narrowly conical, almost straight.

DISTRIBUTION: Madeira.

HARDINESS: H5–G1.

FLOWERING: spring–summer.

2. **D. elata** (Poiret) Soó

ILLUSTRATIONS: Landwehr, Wilde Orchideeën van Europa **2**: 205–13 (1977); Williams et al., Orchids of Britain and Europe, 112 (1978); Grey-Wilson & Mathew, Bulbs, f. 44 (1981).

Like *D. foliosa* except as below.

STEM: to 110 cm.

BRACTS: longer than flowers and protruding from the inflorescence.

INFLORESCENCE: 5–25 cm.

FLOWERS: pink to purple.

SPUR: 1.2–1.5cm.

DISTRIBUTION: SW Europe.

3. **D. incarnata** (Linnaeus) Soó

ILLUSTRATIONS: Landwehr, Wilde Orchideeën van Europa **2**: 127–47 (1977); Williams et al., Orchids of Britain and Europe, 111 (1978); Grey-Wilson & Mathew, Bulbs, f. 44 (1981); American Orchid Society Bulletin **55**: 593 (1986).

STEM: to 80 cm, hollow.

LEAVES: 4–7, 10–20 cm, narrowly triangular to lanceolate, keeled, stiff, hooded at tip, ascending or arched, bright yellowish green, unspotted, often overtopping the inflorescence; upper 1 or 2 leaves bract-like.

LOWER BRACTS: longer than flowers and usually protruding from the inflorescence.

INFLORESCENCE: 4–10 cm, conical to cylindric, densely many-flowered.

FLOWERS: pale to bright pink or brick-red.

SEPALS AND PETALS: 5–8 mm, lanceolate to ovate, acute or obtuse, usually erect or ascending.

LIP: 5–7 × 5–9 mm, diamond-shaped, sometimes 3-lobed, marked with hieroglyphic-like lines of darker pink.

SPUR: 5–8 mm, broadly conical, slightly curved.

DISTRIBUTION: Europe & W Asia, uncommon in the south.

HARDINESS: H2.

FLOWERING: summer.

4. **D. sambucina** (Linnaeus) Soó

ILLUSTRATIONS: Landwehr, Wilde Orchideeën van Europa **2**: 59–61 (1977); Williams et al., Orchids of Britain and Europe, 109 (1978); Grey-Wilson & Mathew, Bulbs, f. 43 (1981).

Like *D. incarnata* except as below.

STEM: to 30 cm.

LEAVES: in lower half of the stem only, all much larger than bracts, never overtopping the inflorescence.

FLOWERS: yellow, crimson or reddish brown, in small, spherical to ovoid heads.

SPUR: 1.2–1.5 cm, equal to or longer than ovary, cylindric, obtuse.

DISTRIBUTION: Europe, N Africa & Turkey.

HARDINESS: H5–G1.

FLOWERING: summer.

5. **D. majalis** (Reichenbach) Hunt & Summerhayes

SYNONYM: *D. latifolia* (Linnaeus) Soó, confused name.

ILLUSTRATIONS: Landwehr, Wilde Orchideeën van Europa **2**: 173–81 (1977); Williams et al., Orchids of Britain and Europe, 113 (1978); Grey-Wilson & Mathew, Bulbs, f. 44 (1981).

STEM: to 60 cm, hollow.

LEAVES: 4–8, to 15 × 2.5 cm, broadly lanceolate to elliptic or oblong, acute, not hooded at tip, mid to dark green with irregular brownish purple spots and blotches; sometimes 1 or 2 upper leaves bract-like.

LOWER BRACTS: usually longer than flowers.

INFLORESCENCE: 6–12 cm, cylindric, dense, many-flowered.

FLOWERS: dark purplish pink.

SEPALS AND PETALS: 6–12 mm, lanceolate to oblong, obtuse, ascending or erect.

LIP: 6–9 × 9–12 mm, circular, diamond-shaped or transversely elliptic, simple or 3-lobed, with broad lateral lobes and a short, triangular central lobe, marked with lines and dots or dots alone.

SPUR: 7–10 mm, conical to cylindric, equal to or shorter than ovary.

DISTRIBUTION: Europe (except extreme south), N Russia, Turkey & the Baltic.

HARDINESS: H2.

FLOWERING: summer.

6. D. maculata (Linnaeus) Soó

SYNONYMS: *Orchis ericetorum* (Linden) Marshall; *O. elodes* Grisebach; *D. maculata* (Linnaeus) Soó subsp. *ericetorum* (Linden) Hunt & Summerhayes.

ILLUSTRATIONS: Landwehr, Wilde Orchideeën van Europa **2**: 73–5, 95–7 (1977); Williams et al., Orchids of Britain and Europe, 121 (1978); Grey-Wilson & Mathew, Bulbs, f. 44 (1981).

STEM: to 40 cm, solid.

LEAVES: 5–12, to 20 × 3 cm, linear-lanceolate to oblong, acute to rounded, sometimes slightly hooded at tip, mid to dark green with irregular spots and blotches of brownish purple; some small upper leaves bract-like.

BRACTS: about equal to flowers and not protruding from the inflorescence.

INFLORESCENCE: 5–10 cm, conical, ovoid or cylindric, dense, many-flowered.

FLOWERS: white to pink.

SEPALS AND PETALS: 7–11 mm, narrowly lanceolate, obtuse, usually horizontal.

LIP: 7–8 × 10–11 mm, diamond-shaped to transversely elliptic, usually marked with darker pink dots and short lines, shallowly 3-lobed; central lobe shorter than lateral lobes.

SPUR: 4–8 mm, cylindric, equal to or shorter than ovary.

DISTRIBUTION: Europe, N Africa & W Asia.

HARDINESS: H2.

FLOWERING: summer.

7. D. fuchsii (Druce) Soó

SYNONYM: *D. maculata* subsp. *fuchsii* (Druce) Hylander.

ILLUSTRATIONS: Landwehr, Wilde Orchideeën van Europa **2**: 103–5, 107–9 (1977); Williams et al., Orchids of Britain and Europe, 121 (1978); Grey-Wilson & Mathew, Bulbs, f. 44 (1981).

Like *D. maculata* except as below.

LIP: more deeply 3-lobed, marked with hieroglyphic-like lines; central lobe narrow and usually longer than lateral lobes.

DISTRIBUTION: Europe (except extreme south) & W Asia.

20. NIGRITELLA Richard

A genus of a single species, cultivated as *Dactylorhiza* (p. 80).

HABIT: herbaceous, terrestrial perennial; aerial growth dying down in winter.

TUBERS: 2, palmately lobed.

STEMS: to 28 cm.

LEAVES: 6–11, crowded, to 11 cm × 1–9 mm, linear to narrowly lanceolate, channelled, mostly basal, unspotted.

SPIKE: 1–3 cm × 7–25 mm, conical becoming ovoid or elongate, dense.
BRACTS: slender, as long as or longer than flowers.
FLOWERS: *c.* 5 mm, blackish crimson, red, yellowish or rarely white, vanilla-scented.
SEPALS AND PETALS: linear to lanceolate; tips acute.
LIP: uppermost, triangular to lanceolate-ovate, entire.
SPUR: *c.* 2 mm, conical, shorter than ovary.
POLLINIA: 2, granular.

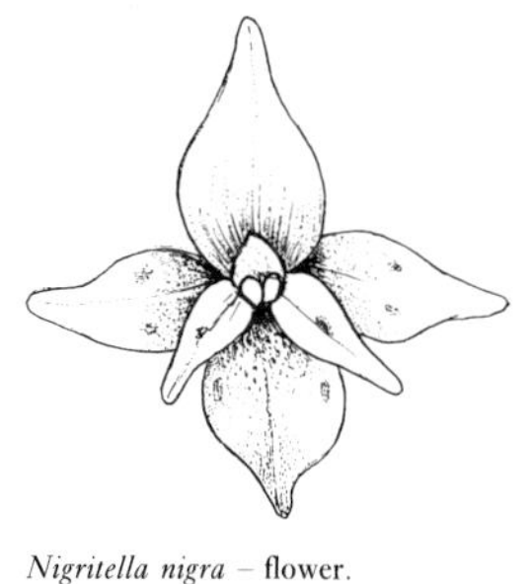

Nigritella nigra – flower.

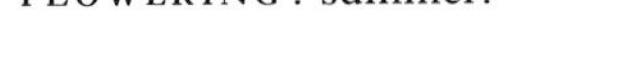

1. **N. nigra** (Linnaeus) Reichenbach
ILLUSTRATIONS: Sundermann, Europäische und mediterrane Orchideen, 178 (1975); Landwehr, Wilde Orchideeën van Europa 1: 229, 231 (1977); Williams et al., Orchids of Britain and Europe, 123 (1978); American Orchid Society Bulletin **54**: 837 (1985).
DISTRIBUTION: Scandinavia to N Spain & Greece.
HARDINESS: H1.
FLOWERING: summer.

21. **GYMNADENIA** R. Brown
A genus of perhaps 10 species from Europe and temperate Asia. Cultivation as for *Orchis* (p. 74).

HABIT: herbaceous, terrestrial perennials; aerial growth dying down in winter.
TUBERS: palmately lobed.
LEAVES: linear to lanceolate, arranged along lower part of stem, unspotted.
FLOWERS: in a cylindric spike, sweetly scented.
UPPER SEPAL AND PETALS: incurved to form a hood over the column.
LATERAL SEPALS: spreading almost horizontally.
LIP: shallowly 3-lobed.
SPUR: slender, shorter or longer than ovary.
POLLINIA: 2, club-like, powdery.

Key to species
1. Spur much longer than ovary; leaves oblong to lanceolate **1. conopsea**
2. Spur not longer than ovary; leaves linear **2. odoratissima**

1. **G. conopsea** (Linnaeus) R. Brown
ILLUSTRATIONS: Landwehr, Wilde Orchideeën van Europa 1: 221, 223 (1977); Williams et al., Orchids of Britain and Europe, 125 (1978); Mossberg & Nilsson, Orchids of northern Europe, 81 (1979); American Orchid Society Bulletin **54**: 942 (1985).

STEMS: to 50 cm.

LEAVES: to 7, arising along the lower third of the stem, to 25 cm × 6–35 mm, oblong to lanceolate.

SPIKE: to 25 cm.

BRACTS: lanceolate, as long as or longer than ovary.

FLOWERS: crowded, rose or purple, sometimes white.

SEPALS AND PETALS: 4–7 mm, upper sepal and petals elliptic to ovate, lateral sepals oblong, obtuse, spreading, margins slightly reflexed.

LIP: 5–7 mm, 3-lobed, lobes almost equal.

SPUR: to 2 cm, about twice as long as ovary, downwardly curved.

DISTRIBUTION: Eurasia.

HARDINESS: H1.

FLOWERING: spring–summer.

2. **G. odoratissima** (Linnaeus) Richard

ILLUSTRATIONS: Danesch & Danesch, Orchideen Europas, Mitteleuropa, edn 3, 150, 151 (1972); Landwehr, Wilde Orchideeën van Europa **1**: 223, 225 (1977); Mossberg & Nilsson, Orchids of northern Europe, 83 (1979); American Orchid Society Bulletin **54**: 943 (1985).

STEMS: to 40 cm.

LEAVES: to 7, mostly basal, to 18 × 1.6 cm, linear, slightly infolded.

SPIKE: to *c.* 10 cm.

BRACTS: lanceolate, longer than ovary.

FLOWERS: crowded, pale pink, purple, sometimes white.

SEPALS AND PETALS: to 3 mm, upper sepal and petals elliptic to ovate, lateral sepals oblong, obtuse, spreading.

LIP: *c.* 3 mm, 3-lobed, lateral lobes shorter than central lobe.

SPUR: to 5 mm, as long as or shorter than ovary, horizontal or downwardly curved.

DISTRIBUTION: Sweden to N Spain, N Italy & W Russia.

HARDINESS: H1.

FLOWERING: spring–summer.

22. COELOGLOSSUM Hartmann

A north temperate genus of 2 species. Cultivation as for *Dactylorhiza* (p. 80); in nature the 1 cultivated species occurs in calcareous grassland and at woodland edges.

HABIT: herbaceous, terrestrial perennials with 2 ovoid tubers; aerial growth dying down in winter.

LEAVES: unspotted, the lower ovate or oblong, obtuse, the upper smaller, lanceolate and pointed.

FLOWERS: slightly scented, 5–25 in a loose cylindric spike.

SEPALS AND PETALS: hooded over the column, greenish or

yellow-green often tinged reddish, sepals ovate, petals relatively narrower, almost linear.

LIP: strap-shaped, the apex shortly 2-lobed, with or without a short middle tooth, and with a short, rounded spur.

POLLINIA: 2, club-shaped, attached to separate viscidia.

1. **C. viride** (Linnaeus) Hartmann

ILLUSTRATIONS: Danesch & Danesch, Orchideen Europas, Mitteleuropa, edn 3, 138 (1972); Landwehr, Wilde Orchideeën van Europa **1**: 227 (1977); Williams et al., Orchids of Britain and Europe, 127 (1978); Mossberg & Nilsson, Orchids of northern Europe, 74 (1979); American Orchid Society Bulletin **54**: 941 (1985).

STEMS: 10–35 cm.

LEAVES: 2–6, 3–10 cm.

SPIKE: 2–10 cm.

SEPALS AND PETALS: 4.5–6 mm.

LIP: 3.5–6 mm.

DISTRIBUTION: Eurasia, N America.

HARDINESS: H1.

FLOWERING: summer.

23. PLATANTHERA Richard

A genus of between 80 and 100 species mostly from temperate areas of Europe, Asia and N & S America. Cultivation as for *Dactylorhiza* (p. 80).

HABIT: herbaceous, terrestrial perennials; tubers 2 or more, spindle-shaped, elongate; aerial growth dying down in winter.

STEMS: erect, unbranched.

LEAVES: basal or on the stem.

RACEME: usually with many small or medium-sized, white or greenish flowers.

SEPALS AND PETALS: free, upper sepal and petals usually incurved over the column, lateral sepals spreading or recurved.

LIP: entire with the base extended into a spur.

ANTHERS: distinctly separated.

POLLINIA: club-shaped, granular.

Key to species

1. Anther cells distinctly parallel, *c.* 2 mm apart at the base; spur 1.5–2.5 cm — 1. bifolia

2. Anther cells distinctly divergent towards the base, where they are *c.* 4 mm apart; spur 1.5–3.5 cm — **2. chlorantha**

1. **P. bifolia** (Linnaeus) Richard

ILLUSTRATIONS: Sundermann, Europäische und mediterrane

Orchideen, 182 (1975); Williams et al., Orchids of Britain and Europe, 129 (1978); Mossberg & Nilsson, Orchids of northern Europe, 68–70 (1979).

STEMS: to 50 cm.

TUBERS: 2, ovoid, the tips narrow and elongate.

LEAVES: 2 (rarely 3), basal, almost opposite, 7–25 × 2.5–8 cm, circular or elliptic or oblong-lanceolate, narrowing into a stalk; stem-leaves small and bract-like.

RACEME: 2.5–3.8 cm wide, cylindric, loose or dense, with many flowers.

BRACTS: lanceolate, as long as or slightly longer than the ovaries.

FLOWERS: 1.2–2 cm wide, white, scented.

LIP: 6–15 mm, strap-shaped, directed downwards, white, greenish towards the tip.

SPUR: 1.5–2.5 cm, narrowly cylindric, slightly inflated, greenish towards the apex, often straight.

ANTHER CELLS: parallel, *c.* 2 mm apart.

DISTRIBUTION: Europe, N Africa, SW Asia.

HARDINESS: H1.

FLOWERING: summer.

2. **P. chlorantha** (Custer) Reichenbach

ILLUSTRATIONS: Sundermann, Europäische und mediterrane Orchideen, 182 (1975); Landwehr, Wilde Orchideeën van Europa 1: 31–3 (1977); Mossberg & Nilsson, Orchids of northern Europe, 70, 71 (1979); Williams et al., Orchids of Britain and Europe, 129 (1978).

STEMS: to 60 cm.

TUBERS: 2, ovoid, the tips narrow, elongate.

LEAVES: 2 (rarely 3), basal, almost opposite, 6–20 × 2.5–8 cm, oblong to oblong-lanceolate or obovate, narrowing into a stalk; stem-leaves small and bract-like.

RACEME: 3–4.5 cm wide, cylindric, loose or dense, with many flowers.

BRACTS: lanceolate, the lower somewhat longer than the ovaries.

FLOWERS: to 2.5 cm wide, greenish white, scented.

LIP: 8–18 mm, strap-shaped, directed downwards, greenish, the tip darker.

SPUR: 1.5–3.5 cm, thickened towards the tip, greenish, darker at the tip.

ANTHER CELLS: divergent towards the base and *c.* 4 mm apart there.

DISTRIBUTION: Europe, SW Asia.

HARDINESS: H1.

FLOWERING: summer.

24. HABENARIA Willdenow

The genus contains between 600 and 800 species distributed throughout temperate N America and east Asia and the subtropics and tropics of the Old and New Worlds. Generic limits within this group are somewhat confused; Luer (reference below), for instance, place all the North American species under *Platanthera*. For the purpose of this book, *Bonatea* Willdenow and *Pecteilis* Rafinesque have been included in *Habenaria*. Most species require greenhouse or alpine-house protection and are best grown in terrestrial compost or in one containing peat, leaf-mould, loam and sand in equal parts. Adequate water is necessary while the plants are actively growing, with cool, almost dry conditions when dormant.

Literature: Luer, C., *Native orchids of the United States and Canada*, 175–242 (1975); Seidenfaden, G., Orchid genera in Thailand V, *Dansk Botanisk Arkiv* **31**: 22–6, 65–137 (1977).

HABIT: terrestrial herbs; aerial growth dying down in winter.
ROOTS: tuberous, fleshy, elongate or ovoid.
STEMS: erect.
LEAVES: spaced along stem or sometimes basal, linear to lanceolate, sheathing at the base, rolled when young.
RACEME: few- to many-flowered.
UPPER SEPAL: often forming a hood with the petals.
LATERAL SEPALS: spreading (in *H. bonatea* the basal parts of the sepals, petal-lobes, column and lip are united).
LIP: variously lobed or fringed.
SPUR: slender, long.
COLUMN: short or long, slender or thick; stigmas 2,borne on club-like arms, or stalkless.
POLLINIA: 2, club-like, breaking up into easily separable masses.

Key to species

1. Lobes of lip entire
1.A. Sepals and petals reddish; margins of lip papillose-ciliate **2. carnea**
1.B. Sepals and petals green; margins of lip not papillose-ciliate **3. rhodocheila**
2. Petals divided into 2 lobes; lateral lobes of lip linear to thread-like **1. bonatea**
3. Lip not lobed; flowers orange **5. ciliaris**
4. Leaves linear **4. radiata**
5. Leaves oblong to obovate **6. psycodes**

1. H. bonatea Reichenbach

SYNONYM: *Bonatea speciosa* (Linnaeus) Willdenow.
ILLUSTRATIONS: Botanical Magazine, 2926 (1829); Bolus, Orchids of South Africa **2**: t. 44, 45 (1911); Schelpe, An

introduction to the South African orchids, 75 (1966); South African Orchid Journal: cover (June 1971); American Orchid Society Bulletin **47**: 994 (1978).

HABIT: plant 30–80 cm.

LEAVES: many, often crowded along stem, to *c.* 15 × 5 cm, ovate-oblong or oblong, sometimes with purplish marks, obtuse.

RACEME: dense, with 15 or more flowers.

BRACTS: *c.* 3 cm, ovate-acuminate.

FLOWERS: scented, greenish.

SEPALS: ovate, 1.7–2.2 cm, lateral sepals longer than the upper.

PETALS: 2-partite, the upper segment as long as the upper sepal, sickle-shaped; the lower segment as long as or longer than the lateral sepal, linear or thread-like.

LIP: 3–3.5 cm, 3-lobed, the lobes thread-like, curved or bent, white tipped with green.

SPUR: 3–3.5 cm, toothed at the mouth, thickened towards tip, greenish.

DISTRIBUTION: South Africa.

HARDINESS: H5–G1.

FLOWERING: spring–summer.

2. H. carnea R. Brown

ILLUSTRATIONS: Gardeners' Chronicle **2**: 729 (1891); The Garden **47**: facing 182 (1895); Holttum, Flora of Malaya, edn 3, **1**: 86 (1964); Dansk Botanisk Arkiv **31**: 136 (1977).

HABIT: plant 15–30 cm.

LEAVES: 4–6, to 10 × 4 cm, lanceolate, mostly basal, olive green with pale spots.

RACEME: 4–15-flowered.

FLOWERS: pink or reddish.

SEPALS AND PETALS: *c.* 1 cm, lateral sepals broader and longer than the upper; petals broad, ovate, veins several.

LIP: 3-lobed, *c.* 3 × 2.5 cm, sometimes as wide as long, lobes oblong or ovate, central lobe notched, surface and margins finely papillose-ciliate.

SPUR: to 6 cm.

DISTRIBUTION: Malay Peninsula (Langkawi Islands), Thailand.

HARDINESS: G2.

FLOWERING: summer–autumn.

3. H. rhodocheila Hance

SYNONYM: *H. militaris* Reichenbach.

ILLUSTRATIONS: Botanical Magazine, 7571 (1897); American Orchid Society Bulletin **48**: 366, 367 (1979); Dansk Botanisk Arkiv **31**: 135 (1977); Bechtel et al., Manual of cultivated orchid species, 216 (1981).

Like *H. carnea* except as below.

SEPALS AND PETALS: green; petals slender, oblong-spathulate, 1-veined.

LIP: yellowish orange or bright red, margins not papillose-ciliate.

DISTRIBUTION: China, Thailand, Malay Peninsula, Laos, Vietnam, Philippines.

4. **H. radiata** Thunberg

SYNONYM: *Pecteilis radiata* (Thunberg) Rafinesque.

ILLUSTRATIONS: Australian Orchid Review **33**: 5 (1968); American Orchid Society Bulletin **44**: 809 (1975), **55**: 589 (1986); Maekawa, The wild orchids of Japan in colour, 90 (1971); Bechtel et al,, Manual of cultivated orchid species, 257 (1981).

HABIT: plant *c.* 30 cm.

LEAVES: 3–7, to 10 cm × *c.* 6 mm, linear, acute, the lowermost and uppermost shortest.

INFLORESCENCE: 1–2-flowered.

SEPALS AND PETALS: *c.* 1 cm, ovate, margins of petals finely and irregularly toothed.

LIP: *c.* 1.5 × 2.5 cm or larger, 3-lobed; lateral lobes obovate, their outer margins laciniate; central lobe linear or oblong, shorter than lateral lobes, margins entire.

DISTRIBUTION: Korea, Japan.

HARDINESS: H5–G1.

FLOWERING: summer.

5. **H. ciliaris** (Linnaeus) R. Brown

SYNONYMS: *Blephariglottis ciliaris* (Linnaeus) Linnaeus; *Platanthera ciliaris* (Linnaeus) Lindley.

ILLUSTRATIONS: Botanical Magazine, 1668 (1814); Shuttleworth et al., Orchids, 25 (1970); Luer, Native orchids of the United States and Canada, 181 (1975); Die Orchidee **27**: 155 (1976), **32**: 103 (1981).

Like *H. radiata* except as below.

LEAVES: lanceolate to oblong.

FLOWERS: orange.

LIP: oblong, unlobed.

DISTRIBUTION: Eastern USA.

HARDINESS H1.

6. **H. psycodes** (Linnaeus) Sprengel

SYNONYM: *Platanthera psycodes* (Linnaeus) Lindley.

ILLUSTRATIONS: American Orchid Society Bulletin **39**: 789 (1970); Luer, Native orchids of the United States and Canada, 196, 199 (1975); Die Orchidee **27**: 154, 155 (1976).

Like *H. radiata* except as below.

LEAVES: oblong to obovate.

FLOWERS: purplish, occasionally white.

LIP: 3-lobed, lobes obovate or fan-shaped.
DISTRIBUTION: SE Canada, Northeastern USA.
HARDINESS: H1.

25. STENOGLOTTIS Lindley

A genus of 3 species confined to eastern and southern Africa. Plants should be grown in terrestrial compost and given moist, shady conditions while growing vigorously.

HABIT: herbaceous perennials, terrestrial, growing on rocks or epiphytic.
ROOTS: several, tuberous, oblong, fleshy.
LEAVES: numerous, mostly basal and forming a rosette, membranous, rolled when young, usually dying back after flowering.
INFLORESCENCE: erect, many-flowered, bearing several sheaths, upper ones smaller than the rest and not clasping the stem.
SEPALS: basal part slightly joined to the lip and column.
PETALS: often slightly broader than sepals, hooded over the column.
LIP: joined to the column, longer than sepals and petals, oblong, widening towards the apex, which is 3- to 5-, or occasionally 7-lobed or -toothed.
SPUR: absent (in ours).
COLUMN: short.
POLLINIA: 2, club-like, attached to a short stalk, easily separable into distinct masses.

Key to species

1. Flowers to *c.* 20, arranged mostly to 1 side of the inflorescence; lip 3- or occasionally 4-toothed **1. fimbriata**
2. Flowers to *c.* 100, evenly arranged in a cylindric inflorescence; lip 5- or occasionally 7-toothed **2. longifolia**

1. S. fimbriata Lindley

ILLUSTRATIONS: Botanical Magazine, 5872 (1870); Flowering Plants of South Africa **15**: t. 585 (1935); American Orchid Society Bulletin **48**: 159 (1979); Stewart & Hennessy, Orchids of Africa, 51 (1981); Kew Magazine **6**: pl. 117 (1989).
LEAVES: 5–12, to 12 × *c.* 2 cm, lanceolate or oblanceolate, arranged in a basal rosette, purplish-spotted or not, margins often wavy.
RACEME: loose, 1-sided, to 38 cm.
FLOWERS: few to *c.* 20, pale pink to mauve, lip paler with darker spots; the ovary, sepals and petals are covered with small, crystal-like papillae.
SEPALS: 3–5 mm.

PETALS: slightly broader than sepals.

LIP: 5–9 mm, its apical third 3- or occasionally 4-lobed or toothed; lobes almost equal, 2–4 mm.

DISTRIBUTION: tropical E Africa to S Africa.

HARDINESS: G1.

FLOWERING: autumn.

2. **S. longifolia** Hooker

ILLUSTRATIONS: Botanical Magazine, 7186 (1891); Flowering Plants of South Africa **24**: t. 933 (1944); Orchid Review **87**: 62 (1979); Bechtel et al., Manual of cultivated orchid species, 275 (1981); American Orchid Society Bulletin **54**: 311 (1985); Kew Magazine **6**: pl. 20 (1989).

LEAVES: many, to 18 × 2.5 cm, lanceolate, gradually narrowed at either end, tips slightly recurved, occasionally blackish-spotted, margins often wavy.

FLOWERS: to *c.* 100, crowded in a cylindric spike to 70 cm, pale pink, occasionally white, with scattered, dark marks or dots.

SEPALS: *c.* 8 mm.

PETALS: smaller than sepals.

LIP: to *c.* 1.2cm, its apical part 5-, occasionally 7-toothed, teeth unequal, to *c.* 6 mm.

DISTRIBUTION: S Africa.

HARDINESS: G1.

FLOWERING: autumn.

26. DISA Bergius

A genus of over 100 species from tropical and southern Africa and Madagascar. Only a single species is widely grown, though some specialists have larger collections. Cultivation can be difficult, and full details are given in the papers by Vogelpoel and Stoutamire cited below.

Literature: Linder, H.P., A revision of Disa excluding Sect. Micranthae, *Contributions from the Bolus Herbarium* **9**: 1–370 (1981); Vogelpoel, L., Disa uniflora – its propagation and cultivation, *American Orchid Society Bulletin* **49**: 961–72 (1980); Vogelpoel, L., Disa species and their hybrids, *American Orchid Society Bulletin* **49**: 1084–92 (1980); Stoutamire, W., Cultivated Disas in Ohio, *American Orchid Society Bulletin* **50**: 1195–200 (1981).

HABIT: herbaceous, terrestrial perennials, occasionally evergreen, with tuberous roots, sometimes stoloniferous; sterile shoots usually rosette-like.

LEAVES: basal or on the stem, ovate to linear, though mostly lanceolate.

INFLORESCENCE: a dense or loose raceme or corymb.

FLOWERS: usually with lip lowermost (though this is not

obvious in single-flowered or corymbose inflorescences), less commonly with the lip uppermost.

MEDIAN SEPAL: concave and usually hood-like, and with a basal spur.

LATERAL SEPALS: spreading.

PETALS: smaller and narrower than sepals, often partially attached to the column.

LIP: small and narrow.

COLUMN: short.

POLLINIA: 2, granular.

1. D. uniflora Bergius

SYNONYM: *D. grandiflora* Linnaeus.

ILLUSTRATIONS: Botanical Magazine 4073 (1844); Orchid Review **78**: 95 (1970), **85**: 229 (1977), **86**: 127 (1978); Bechtel et al., Manual of cultivated orchid species, 204 (1981).

STEMS: to 60 cm.

LEAVES: those of sterile shoots in a loose basal rosette; leaves on flowering stems to 8, up to 15 × 1.25 cm, lanceolate, acuminate, the upper 3–4 smaller, sheathing.

INFLORESCENCE: with 1–3 (rarely to 10) flowers.

BRACTS: longer than ovaries.

FLOWERS: 8–12 cm wide, faintly scented, scarlet or carmine, orange to yellowish inside (pure white or yellow variants are occasionally recorded); lip lowermost

UPPER SEPAL: *c.* 5 × 3 cm, ovate, the hooded basal part forming a narrow conical, laterally flattened spur, 1–1.5 cm long.

LATERAL SEPALS: broadly elliptic, acuminate.

PETALS: *c.* 2 cm × 8 mm, narrowly obovate, erect.

DISTRIBUTION: South Africa (Cape Province).

HARDINESS: G1.

FLOWERING: winter.

27. PTEROSTYLIS R. Brown

A genus of about 95 species distributed from Australia to New Zealand and extending to New Guinea and New Caledonia. A compost of loam, sphagnum moss, peat and sand in equal parts is adequate for all species; it should be kept dry and frost-free while the plants are dormant.

The lip is sensitive to touch so that a small insect alighting at the tip is projected upwards against the column, where it is trapped between column and lip; in squeezing its way out it detaches the pollinia. The lip gradually returns to its original position in half an hour to two hours.

Literature: Cady, L., The genus Pterostylis in Australia and New Zealand, *The Orchadian* 8: 183–88 (1986).

HABIT: terrestrial herbs.

ROOTS: thread-like and fleshy with ovoid or spherical tubers.
STEMS: erect.
LEAVES: deciduous, mostly basal and forming a flat rosette, with 1–2 bract-like stem-leaves, or spaced along stem and linear to lanceolate; rolled when young.
FLOWERS: 1 or few, erect or pendent.
UPPER SEPAL AND PETALS: forming a hood, erect or curved forwards.
LATERAL SEPALS: joined along their basal inner margins, the tips tail-like or filament-like, diverging widely, erect or deflexed.
LIP: linear or oblong, apart from the tip which is mostly hidden inside hood; base with a small, erect appendage variously divided at its tip.
COLUMN: slender, variously winged towards apex and extended into a short foot at base.
POLLINIA: 4, powdery.

Key to species

1. Leaves linear to lanceolate, more than 5 times as long as broad, evenly spaced on the stem **1. banksii**
2. Leaves oblong to elliptic, less than 5 times as long as broad, mostly in a basal rosette
2.A. Flower to 3 cm, nodding, green **4. nutans**
2.B. Lip *c.* 5 mm, not twisted; tails of lateral sepals extending beyond the hood **3. pedunculata**
2.C. Lip *c.* 2 cm, twisted above the middle; tails of lateral sepals not extending beyond the hood **2. curta**

1. **P. banksii** Hooker

ILLUSTRATIONS: Botanical Magazine, 3172 (1832); Orchadian 1: 129 (1965); Salmon, New Zealand flowers and plants in colour, edn 2, 44 (1967); Cooper, A field guide to New Zealand native orchids, 60 (1981).
HABIT: plant robust, to 35 cm.
LEAVES: up to 6, to 25 × 1.5 cm, pointed, closely spaced along stem and extending above the flower.
FLOWER: solitary, to 7 cm long, erect, pale green with darker stripes.
SEPALS AND PETALS: tips pale orange to pink; tips of lateral sepals filament-like, bent around hood and diverging with age; tips of upper sepal and petals curving forwards and downwards.
LIP: with greenish margins and reddish central part.
DISTRIBUTION: New Zealand.
HARDINESS: H5–G1.
FLOWERING: summer–winter.

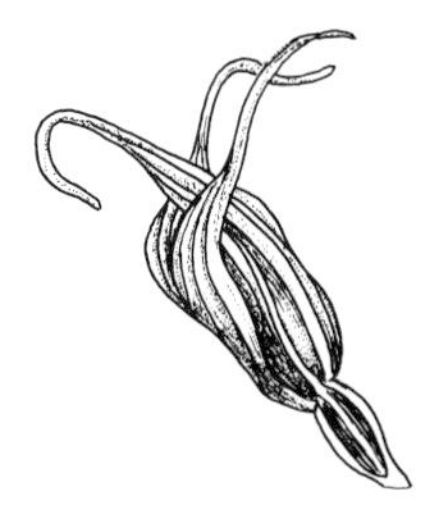

Pterostylis banksii – flower.

Pterostylis curta – flower.

2. P. curta R. Brown

ILLUSTRATIONS: Rupp, Orchids of New South Wales, edn 2, 81 (1969); Nicholls, Orchids of Australia, t. 301 (1969); American Orchid Society Bulletin **41**: 804, 805 (1972); Pocock, Ground orchids of Australia, t. 113 (1972).

HABIT: plant robust, to 30 cm.

LEAVES: usually 4, blade to 4 × 2 cm, ovate to elliptic, margins often wavy; stem-leaves 1 or 2, bract-like.

FLOWERS: usually solitary, *c.* 3 cm, erect, green flushed with brown.

LATERAL SEPALS: tips tailed, not extending beyond hood.

LIP: *c.* 2 cm, oblong-linear, characteristically twisted above the middle, reddish brown.

DISTRIBUTION: Australia, New Caledonia.

HARDINESS: H5–G1.

FLOWERING: summer–autumn.

3. P. pedunculata R. Brown

ILLUSTRATIONS: Australian Orchid Review **34**: 142 (1969); Nicholls, Orchids of Australia, t. 309 (1969): Pocock, Ground orchids of Australia, t. 125 (1972).

Like *P. curta* except as below.

FLOWER: to 2 cm, erect, green flushed with reddish brown or dark brown.

LATERAL SEPALS: filament-like, extending beyond the hood.

LIP: *c.* 5 mm, ovate, not twisted, dark reddish brown.

DISTRIBUTION: Australia.

FLOWERING: winter–spring.

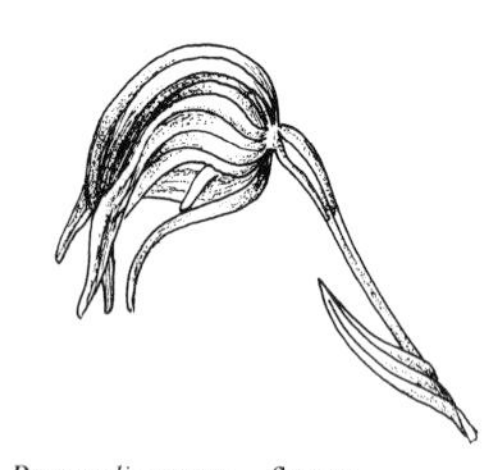

Pterostylis nutans – flower.

4. P. nutans R. Brown

ILLUSTRATIONS: Botanical Magazine, 3085 (1831); Australian Orchid Review **34**: 142 (1969); Nicholls, Orchids of Australia, t. 292 (1969); Pocock, Ground orchids of Australia, t. 123 (1972); American Orchid Society Bulletin **44**: 992 (1975).

Like *P. curta* except as below.

FLOWER: to 3 cm, conspicuously nodding, green.

SEPALS AND PETALS: tips pale brownish.

DISTRIBUTION: Australia & North Island of New Zealand.

28. SATYRIUM Swartz

A genus of perhaps 100 species, mainly from Africa, though 2 occur in Asia. Only one is generally grown. Crossland (*Quarterly Bulletin of the Alpine Garden Society* **46**: 301, 1978) recommends 'a compost providing humus and drainage, that normally used being one part each of loam, leaf mould, peat and sand. A position in semi-shade where the plant can remain cool and moist throughout growth, and dry but not arid when dormant is most suitable'. Annual potting should be done before growth commences. Apparently plants may remain dormant over several years.

HABIT: herbaceous, terrestrial perennials; aerial growth dying down in winter.
ROOTS: tuberous, ovoid.
STEMS: unbranched, leafy.
FLOWERS: several to many, in racemes or spikes, not resupinate, variously coloured.
LIP: uppermost, forming a hood.
SPURS: 2, attached to base of lip.
POLLINIA: 2, granular.

1. **S. nepalensis** Don
ILLUSTRATIONS: Annals of the Royal Botanic Garden Calcutta **8**: t. 443, 444 (1898); Orchid Review **85**: 149 (1977); Norwegian Journal of Botany **26**: 286 (1979).
TUBERS: ovoid, produced singly during the flowering season at the ends of vigorous roots.
STEM: 10–75 cm, bearing 2–3 leaves.
LEAVES: 7–25 × 3–9 cm, sheathing, oblong-ovate to lanceolate, obtuse, the upper shorter.
SPIKE: cylindric, 5–15 cm.
BRACTS: longer than the flowers, deflexed.
FLOWERS: numerous, pink or white, scented.
SEPALS AND PETALS: to 8 mm, linear-oblong, pointed, slightly recurved, petals narrower than sepals (which have hairy margins in var. **ciliata** (Lindley) King & Pantling).
LIP: to 8 mm, as broad as or broader than long, concave, keeled outside, the base with 2 spurs which are 8–12 mm.
DISTRIBUTION: N India to Sri Lanka & Burma.
HARDINESS: H5–G1.
FLOWERING: summer–autumn.
Plants offered under this name in the trade have occasionally proved to be *Herminium angustifolium*, *Habenaria arietina* or *Platanthera clavigera*.

29. VANILLA Miller

A genus of between 70 and 100 species distributed throughout the tropics. *V.planifolia* is the source of natural vanilla essence and is widely cultivated in the range of the genus. The long climbing stems of *Vanilla* species must be firmly supported, and can be attached to bark at various points along their length. Much of their moisture is obtained from the atmosphere so the pots, containing epiphytic compost, can be quite small. The plants should be frequently syringed, even in winter. Propagation is effected by cutting up the stem into small sections which can be planted.

HABIT: perennial branched climbers with long green stems, initially terrestrial, often becoming epiphytic, bearing a leaf and 1 or 2 roots at each node.

LEAVES: alternate, fleshy or leathery, not sheathing, sometimes reduced to scales, folded when young.

FLOWERS: in axillary spikes or racemes.

SEPALS AND PETALS: similar, fleshy, spreading; petals with distinct midrib.

LIP: with a narrow claw fused to the lower part of the column and an entire or obscurely lobed blade enclosing the upper part of the column.

POLLINIA: 2, powdery-granular.

FRUIT: a fleshy pod.

Key to species

1. Leaves reduced to small triangular scales **3. aphylla**
2. Sepals and petals *c.* 8 cm **2. pompona**
3. Sepals and petals *c.* 6 cm **1. planifolia**

1. **V. planifolia** Andrews

SYNONYMS: *Myrobroma fragrans* Salisbury; *V. fragrans* (Salisbury) Ames.

ILLUSTRATIONS: Botanical Magazine, 7167 (1891); Shuttleworth et al., Orchids, 30 (1970); Orchid Digest **41**: 181 (1977); Luer, Native orchids of Florida, 75 (1971).

STEM: fleshy, terete.

LEAVES: 15–20 × 3–5 cm, fleshy, ovate-elliptic to oblong, acute.

FLOWERS: 10–15 per spike, pale green or greenish yellow.

BRACTS: 7 mm, triangular to ovate.

SEPALS AND PETALS: to 6 × 1 cm, linear-oblong to narrowly obovate.

LIP: to 4.5 × 2 cm; blade triangular with longitudinal papillose veins in the mid-line and a blunt, reflexed apical lobe.

COLUMN: 3.5 cm, slender, downy.

OVARY: 4 cm, curved, maturing into a black fleshy cylindrical pod 15–25 cm.

DISTRIBUTION: C America and the West Indies south to Bolivia, Paraguay & N Argentina; naturalised in Madagascar, the Seychelles and elsewhere.

HARDINESS: G2.

FLOWERING: intermittent.

2. **V. pompona** Schiede

ILLUSTRATION: Bechtel et al., Manual of cultivated orchid species, 280 (1981).

Like *V. planifolia* except as below.

HABIT: generally more robust.

SEPALS AND PETALS: 8 cm, greenish yellow.

LIP: yellow or orange.

POD: 3-sided.

DISTRIBUTION: C America and the Lesser Antilles south through Peru & W Brazil to Bolivia.

3. V. aphylla Blume

Like *V. planifolia* except as below.

STEM: somewhat flattened.

LEAVES: reduced to small green triangular scales.

FLOWERS: 3 or 4 together.

SEPALS AND PETALS: pale green

LIP: white with pink hairs on the central lobe.

DISTRIBUTION: Burma (Tenasserim), N Malay peninsula & Java.

30. BLETILLA Reichenbach

A genus of about 10 species from E Asia, named for its apparent similarity to the genus *Bletia* (p. 108) from C & S America with which it is often confused. Species of *Bletilla* have sometimes been classified under *Limodorum*, *Arethusa* and *Cymbidium*. Pseudobulbs should be planted in terrestrial compost and kept in a cold-frame. In mild areas, plants will flourish in the open. Very little water should be given in winter.

HABIT: terrestrial, with corm-like pseudobulbs.

STEMS: erect, produced laterally from the bases of old pseudobulbs.

LEAVES: alternate, sheathing, thin and pleated, rolled in bud.

FLOWERS: several, in loose terminal racemes.

PETALS AND SEPALS: similar, free, pointing forward around lip.

LIP: 3-lobed; lateral lobes curled around column; central lobe with warts or ridges.

COLUMN: free, long and narrow.

POLLINIA: 8.

Key to species

1. Flowers magenta, sepals, petals and lip without dark tips **1. striata**
2. Flowers rose pink, sepals, petals and lip with dark tips **2. sinensis**

1. B. striata (Thunberg) Reichenbach

SYNONYM: *B. hyacinthina* (Smith) Pfitzer.

ILLUSTRATIONS: Botanical Magazine, 1492 (1812); Hay & Synge, Dictionary of garden plants, 126 (1969); Bechtel et al., Manual of cultivated orchid species, 171 (1981).

PLANT: to 60 cm.

PSEUDOBULBS: horizontally flattened.

LEAVES: 3–6, deciduous, 30–50 × 4–7 cm, elliptic, acute.
FLOWERS: 3–10 in a loose raceme, magenta.
SEPALS: 2–3 cm × 5–10 mm, elliptic, acute.
PETALS: similar to sepals but broader.
LIP: 2.5–3 cm; central lobe longitudinally ridged, shallowly notched at apex.
COLUMN: 2.5–3 cm, narrowly winged, purple at tip.
DISTRIBUTION: China & Japan.
HARDINESS: H5–G1.
FLOWERING: spring–summer.

2. **B. sinensis** (Rolfe) Schlechter
Like *B. striata* except as below.
FLOWERS: smaller, rose red.
SEPALS, PETALS AND LIP: with darker tips.
DISTRIBUTION: W China (Yunnan).
FLOWERING: early summer.

31. BOTHRIOCHILUS Lemaire

A genus of 4 species from C America frequently considered part of *Coelia* from which it is distinguished by the length of the column foot, the presence of a distinct mentum and the complicated lip. Cultivation requirements are similar to those for *Lycaste* (p. 299), though *B. machrostachyus*, having fairly hard pseudobulbs, benefits from a longer dry period.

Literature: Williams, L.O., The orchid genera Coelia & Bothriochilus, *Botanical Museum Leaflets, Harvard University* **8**: 145–8 (1940); Pridgeon, A., Revision of Coelia and Bothriochilus, *Orquidea* **7**: 57–94 (1978), reprinted in American Orchid Society Bulletin **56**: 601–609 (1987).

HABIT: epiphytic or terrestrial, often on rocks; rhizomes creeping.
PSEUDOBULBS: spherical to ovoid, smooth, simple, bearing several leaves.
LEAVES: linear to lanceolate, acute, pleated, rolled when young.
FLOWERS: few to many, in robust lateral racemes or spikes, on sheathed stems.
SEPALS: upper sepal free; laterals fused to column foot, forming a mentum.
PETALS: free, lip sharply bent or pouched at junction of blade and claw; blade shallowly 3-lobed; central lobe triangular, acute.
COLUMN: long and slender; column foot equalling it in length.
POLLINIA: 8, waxy.

Key to species

1. Pseudobulbs to 5 cm; flowers few, each *c.* 5 cm long **1. bellus**

2. Pseudobulbs 6 cm or more; flowers many, 1–1.5 cm long
2. macrostachyus

1. B. bellus Lemaire

SYNONYM: *Coelia bella* (Lemaire) Reichenbach.

ILLUSTRATIONS: Botanical Magazine, 6628 (1882); Orchid Review **82**: 314 (1974); American Orchid Society Bulletin **56**: 605 (1987).

HABIT: terrestrial, to 1 m.

PSEUDOBULBS: 3–5 × 1.5–2.5 cm, spherical to ovoid, each surmounted by a short stalk, bearing 3–5 leaves.

LEAVES: 15–60 × 1–2 cm, linear-lanceolate, acuminate, with 3–5 veins, pale green.

FLOWERING STEMS: to 20 cm, heavily sheathed, bearing 2–6 fragrant flowers

FLOWERS: 5 × 3.5–5 cm, funnel-shaped.

SEPALS: elliptic to oblong, obtuse, off-white with purple tips; the upper 3.5 × 1 cm, the laterals 5 × 1 cm, mentum 2 cm.

PETALS: 3.5 × 1 cm, oblong to spathulate, obtuse or rounded, off-white with a central patch of purple.

LIP: pouched; blade 4.5 × 1.5 cm with small lateral lobes; central lobe narrowly triangular, inner face with an orange, granular callus.

COLUMN 1.5 cm.

DISTRIBUTION: Mexico, Guatemala & Honduras.

HARDINESS: G2.

FLOWERING: summer.

2. B. macrostachyus (Lindley) Williams

SYNONYM: *Coelia macrostachya* Lindley.

ILLUSTRATIONS: Botanical Magazine, 4712 (1853); Shuttleworth et al., Orchids, 71 (1970); Ospina & Dressler, Orquideas de las Americas, t. 40 (1974); American Orchid Society Bulletin **56**: 603, 607 (1987).

HABIT: epiphytic, to 1 m.

PSEUDOBULBS: 6–10 cm, spherical to ovoid, each surmounted by a short stalk bearing 3–6 leaves.

LEAVES: to 100 × 2.5–4 cm, linear-lanceolate, acuminate, fleshy.

FLOWERING STEMS: to 60 cm, sheathed.

FLOWERS: many in a dense cylindric spike, each 10–15 × 5–7 mm, pale to dark pink or purple.

SEPALS: fleshy, warty on outer face; upper 8–10 mm, ovate-lanceolate; outer part of laterals 1.5 cm, semi-ovate to triangular, inner part forming a mentum.

PETALS: 9 mm, oblong-lanceolate, asymmetric, membranous.

LIP: pouched; blade to 12 mm, shallowly 3-lobed; central lobe narrowly triangular, acute.

COLUMN: 1.5 cm.

DISTRIBUTION: Mexico to Panama.
HARDINESS: G2.
FLOWERING: summer.

32. CALANTHE R. Brown

A genus of about 50 species with a distribution extending from Madagascar and the Mascarene Islands to the Himalaya, SE Asia generally, Japan and Australia. Many hybrids are cultivated, and Calanthes were among the first orchids to be deliberately hybridised.

Two types of plant occur, deciduous and evergreen. Both should be grown in terrestrial compost. Deciduous species should be given some shade and abundant moisture when growing; after the fall of the leaves they should be kept cool and dry. The evergreen species require more shade, and should not be dried out.

Literature: Seidenfaden, G., Orchid genera in Thailand I: Calanthe, *Dansk Botanisk Arkiv* **29**(2) (1975).

HABIT: usually terrestrial.
PSEUDOBULBS: usually inconspicuous, more rarely large and grooved.
LEAVES: large, thin, pleated, rolled when young, mostly evergreen, deciduous in some species with large pseudobulbs.
RACEMES: arising from the leaf axils, each with a long axis which is usually hairy.
BRACTS: conspicuous, persistent or soon falling, usually hairy.
SEPALS AND PETALS: similar, spreading, usually directed upwards, often hairy outside.
LIP: simple or 3-lobed, usually spurred, with warts or ridges at the base.
COLUMN: united to the lip for most of its length (except in No. 14).
POLLINIA: 8 in 2 groups of 4, attached to a common base.

Synopsis of characters

Pseudobulbs. Conspicuous: **1–3**; inconspicuous: **4–14**.
Leaves. Fallen at flowering: **1–3**; present at flowering: **4–14**.
Raceme. Arising from pseudobulb: **1–3**; arising from a rosette of leaves: **4–13**; arising from the lower part of an elongate leafy stem: **14**. Corymbose; **9**.
Lip. Simple: **1**; 3-lobed: **2–14**. Without spur: **12–14**; spur 2–12 mm: **10,11**; spur 1.5–2.9 cm: **1–3,6–9**; spur 3–5 cm:**4**. With central lobe itself 2-lobed or notched: **2–7,9–11,13,14**; with central lobe entire: **8–12**. Free from column: **14**.

Key to species

1. Bracts falling as soon as the flower opens; lip free from column; plant with conspicuous, leafy stems **14. gracilis**

2. Plant with large pseudobulbs, leafless when flowering
2.A. Lip not lobed **1. rosea**
2.B. Sepals and petals 2 cm or more, white; lip usually white **2. vestita**
2.C. Sepals and petals to 1.5 cm, pink; lip pink **3. rubens**
3. Spur absent
3.A. Sepals and petals spreading; upper surface of lip with 3 conspicuous ridges **12. tricarinata**
3.B. Sepals and petals reflexed; upper surface of lip without ridges **13. reflexa**
4. Spur at most 1.2 cm
4.A. Spur to 3 mm, oblong, straight **11. brevicornu**
4.B. Spur 5–12 mm, pointed, tapering, curved **10. discolor**
5. Lip with its central lobe entire, not notched or further lobed **8. cardioglossa**
6. Flower entirely white except for yellow warts on the base of the lip; raceme corymbose **9. triplicata**
7. Lip white, its central lobe deeply lobed with lobes oblong, diverging at a wide angle
7.A. Sepals and petals greenish or yellowish; lip white with yellow warts at the base; spur 2–3 cm **6. herbacea**
7.B. Sepals and petals pink; lip white flushed violet; spur 1.6–2 cm **7. izu-insularis**
8. Spur 2–2.7 cm, broadening towards its apex **5. sylvatica**
9. Spur 3–5 cm, not broadening towards its apex **4. masuca**

1. C. **rosea** (Lindley) Bentham & Hooker
ILLUSTRATION: Dansk Botanisk Arkiv **29**(2): 41 (1975).
PSEUDOBULBS: 10–15 cm, conspicuous, ovoid to oblong, grooved, narrowed just below the apex.
LEAVES: lanceolate, fallen by flowering time.
RACEME: to 45 cm with 7–12 flowers.
BRACTS: persistent, 1.5–2 cm.
FLOWERS: 4–5 cm in diameter, red-pink.
SEPALS AND PETALS: lanceolate, acute.
LIP: simple, larger than the petals, its base rolled around the column.
SPUR: *c.* 1.5cm, straight, oblong.
DISTRIBUTION: Thailand.
HARDINESS: G1–2.
FLOWERING: winter.

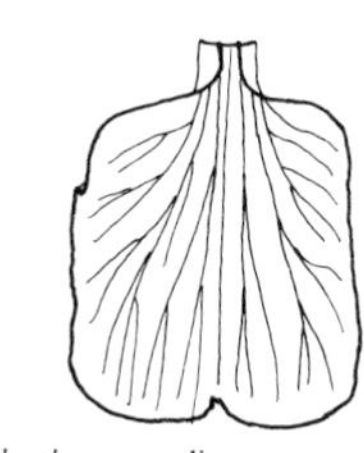
Calanthe rosea – lip.

2. C. **vestita** Lindley.
ILLUSTRATIONS: Botanical Magazine, 4671 (1852); Dansk Botanisk Arkiv **29**(2): 29 & t. 4 (1975); Bechtel et al., Manual of cultivated orchid species, 175 (1981).
PSEUDOBULBS: to 8 cm, ovoid, grooved.
LEAVES: lanceolate, fallen by flowering time.
RACEME: to 70 cm with 6–12 flowers.

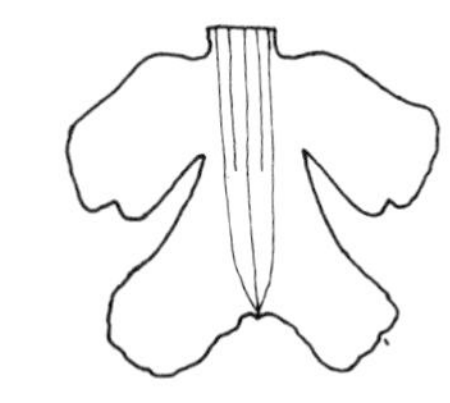
Calanthe vestita – lip.

BRACTS: 2–3 cm, persistent.
FLOWERS: 6–6.5 cm in diameter, white or cream.
LIP: white blotched with yellow or rarely entirely pink, 3-lobed, the lateral lobes about the same size as the divergently notched central lobe.
SPUR: 1.8–2.5 cm, tapering, curved.
DISTRIBUTION: SE Asia, from Thailand to Sulawesi.
HARDINESS: G2.
FLOWERING: winter–spring.
Variable in the precise coloration of the lip, a character formerly used in the recognition of varieties (cultivars). 'Regnieri' (*C. regnieri* Reichenbach), which has a pink lip, is frequently grown.

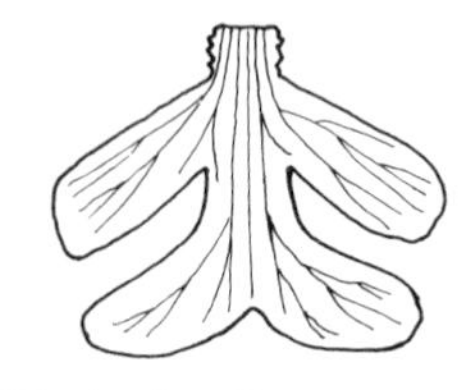
Calanthe rubens – lip.

3. C. rubens Ridley
ILLUSTRATIONS: Dansk Botanisk Arkiv **29**:(2): 33 & t. 5 (1975); Bechtel et al., Manual of cultivated orchid species, 175 (1981).
Like *C. vestita* except as below.
FLOWERS: *c.* 3 cm in diameter, pink.
DISTRIBUTION: Thailand, NE Malaysia.

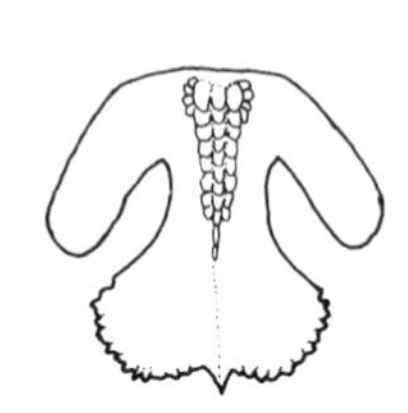
Calanthe masuca – lip.

4. C. masuca (Don) Lindley
ILLUSTRATIONS: Botanical Magazine, 4541 (1850); Dansk Botanisk Arkiv **29**(2): t. 2 (1975).
PSEUDOBULBS: inconspicuous.
LEAVES: 20–30 × 7–11 cm, narrowly elliptic.
RACEME: arising from the leaf rosette, to 70 cm, with many flowers.
BRACTS: persistent, 1.3–2 cm.
FLOWERS: to 6 cm in diameter, pale reddish purple.
LIP: as long as the petals, 3-lobed, with oblong lateral lobes and a somewhat notched central lobe, deep reddish violet with yellow warts towards the base.
SPUR: 3–5 cm, curved.
DISTRIBUTION: Himalaya, SW China, Thailand & Malaysia.
HARDINESS: G1–2.
FLOWERING: summer.
This species is now usually considered part of *C. sylvatica*. However, the distinctions between them, though imprecise, have some horticultural importance, so the 2 are maintained here separately.

5. C. sylvatica (Thouars) Lindley
ILLUSTRATION: Bechtel et al., Manual of cultivated orchid species, 175 (1981).
PSEUDOBULBS: inconspicuous.
LEAVES: 15–25 × 5–7 cm, narrowly elliptic.

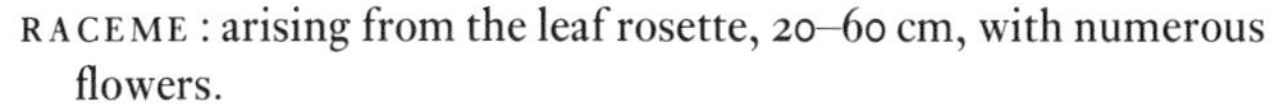

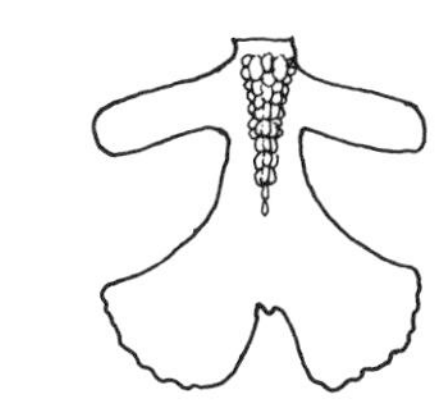

Calanthe sylvatica – lip.

RACEME: arising from the leaf rosette, 20–60 cm, with numerous flowers.

BRACTS: persistent 1.5–3 cm.

FLOWERS: 3–4 cm in diameter, usually reddish purple, more rarely greenish with the lip tinged with purple.

LIP: as long as the petals, 3-lobed, the lateral lobes small, oblong, the central lobe reversed- heart-shaped, notched at the apex.

SPUR: 2–2.7 cm, curved, broadening towards the apex.

DISTRIBUTION: N Madagascar, Mascarene Islands.

HARDINESS: G2.

FLOWERING: spring–summer.

6. C. **herbacea** Lindley

ILLUSTRATION: Annals of the Royal Botanic Garden Calcutta **5**: t. 44 (1895).

PSEUDOBULBS: inconspicuous.

LEAVES: 20–25 × 5–8 cm, elliptic.

RACEME: arising from the leaf rosette, 30–40 cm, with numerous flowers.

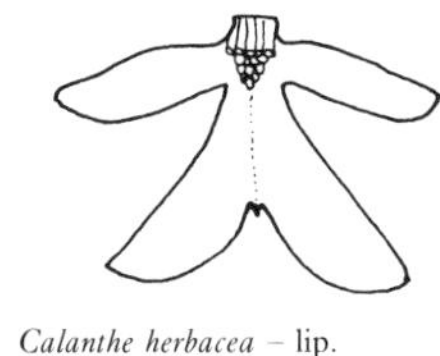

Calanthe herbacea – lip.

BRACTS: persistent, 1.8–2.5 cm.

SEPALS AND PETALS: greenish or yellowish.

LIP: longer than the petals, white with yellow warts at the base, 3-lobed, the lateral lobes oblong, the central deeply and divergently 2-lobed.

SPUR: straight or S-shaped, 2–3 cm.

DISTRIBUTION: N India.

HARDINESS: G1–2.

7. C. **izu-insularis** (Satomi) Ohwi

ILLUSTRATION: Maekawa, Wild orchids of Japan in colour, t. 142 (1979).

Like *C. herbacea* except as below.

SEPALS AND PETALS: pink.

LIP: white flushed with violet

SPUR: 1.6–2 cm.

DISTRIBUTION: Japan.

8. C. **cardioglossa** Schlechter

ILLUSTRATION: Dansk Botanisk Arkiv **29**(2): 37 & t. 6 (1975).

PSEUDOBULBS: inconspicuous.

LEAVES: to 30 × 11 cm, lanceolate to elliptic.

RACEME: arising from the leaf rosette, to 70 cm, with 10–20 flowers.

BRACTS: persistent, 1.5–2 cm.

FLOWERS: to 2 cm in diameter, pale reddish purple.

LIP: about as long as the petals, 3-lobed, with reversed-triangular, upcurving lateral lobes which are pale violet

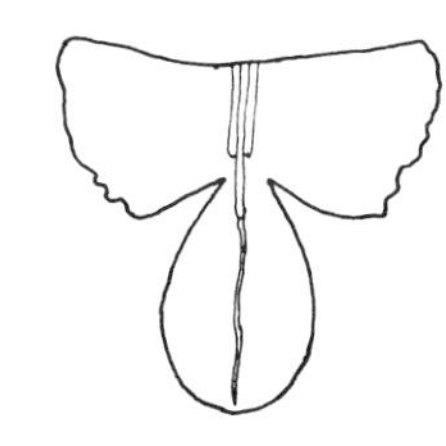

Calanthe cardioglossa – lip.

spotted darker, and an entire, diamond-shaped or elliptic central lobe.

DISTRIBUTION: Vietnam, Laos, Kampuchea & Thailand.

HARDINESS: G2.

FLOWERING: winter.

9. C. triplicata (Willemet) Ames

SYNONYMS: *C. veratrifolia* (Willdenow) R. Brown; *C. furcata* Lindley.

ILLUSTRATIONS: Dansk Botanisk Arkiv **29**(2): 16 & t. 1 (1975); Maekawa, Wild orchids of Japan in colour, t. 150 (1979); Bechtel et al., Manual of cultivated orchid species, 175 (1981); Orchid Review **92**: 243 (1984).

PSEUDOBULBS: inconspicuous.

LEAVES: elliptic, to 40 × 13 cm.

RACEME: arising from the leaf rosette, to 1 m, the flowers corymbosely arranged at the top.

BRACTS: persistent, 8–15 mm.

Calanthe triplicata – lip.

FLOWERS: 3–5 cm in diameter, white except for the yellow warts on the lip.

LIP: longer than the petals, 3-lobed, the lateral lobes oblong, the central deeply and divergently lobed.

SPUR: 1.5–2.5 cm, curved, tapering.

DISTRIBUTION: SE Asia, Japan & E Australia.

HARDINESS: G1–2.

FLOWERING: spring.

10. C. discolor Lindley

SYNONYMS: *C. lurida* Decaisne; *C. amamiana* Fukuyama.

ILLUSTRATIONS: Botanical Magazine, 7026 (1888); Lin, Native orchids of Taiwan, t. 46 (1975); Maekawa, Wild orchids of Japan in colour, t. 141, 143 (1975); Bechtel et al., Manual of cultivated orchid species, 174 (1981); American Orchid Society Bulletin **55**: 589 (1986).

PSEUDOBULBS: inconspicuous.

LEAVES: narrowly elliptic, 17–23 × 4–7 cm.

RACEME: arising from the leaf rosette, to 40 cm, with many flowers.

BRACTS: persistent, 4–8 mm.

FLOWERS: 3–5 cm in diameter, reddish purple or yellow.

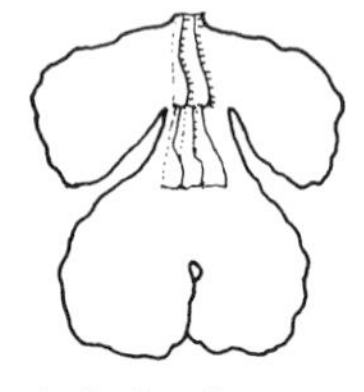

Calanthe discolor – lip.

LIP: about as long as the petals, 3-lobed, the lateral lobes oblong or broadening towards their apices, the central reversed-heart-shaped, deeply notched, bearing 3 prominent ridges towards the base.

SPUR: 5–12 mm, curved, tapering.

DISTRIBUTION: Japan, Korea & Taiwan.

HARDINESS: G1–2.

FLOWERING: spring.

The yellow-flowered variant is called var. **flava** Yatabe in recent

Japanese Floras. It has also been called *C. striata* (Banks) R. Brown, *C. bicolor* Lindley, *C. sieboldii* Decaisne and various combinations of these names at the level of variety or forma.

11. C. brevicornu Lindley

ILLUSTRATION: Annals of the Royal Botanic Garden Calcutta 8: t. 227 (1898).

PSEUDOBULBS: inconspicuous.

LEAVES: 20–30 × 5–6 cm, narrowly elliptic.

RACEME: arising from the leaf rosette, to 50 cm, with many flowers.

BRACTS: persistent, *c.* 1.5 cm.

FLOWERS: 2.5–4 cm in diameter, sepals and petals brown.

LIP: a little shorter than the petals, 3-lobed, the lateral lobes rounded, the central reversed-heart-shaped, deeply notched, purplish red with white margins, and with 3 yellow ridges near the base.

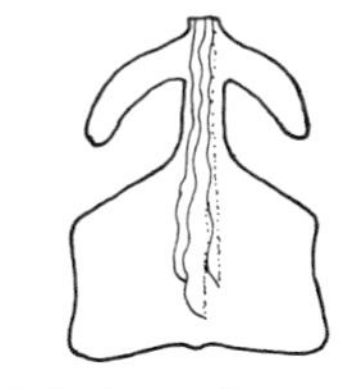

Calanthe brevicornu – lip.

SPUR: to 3 mm, oblong, blunt, straight.

DISTRIBUTION: Nepal, NE India (Sikkim).

HARDINESS: G1.

FLOWERING: summer.

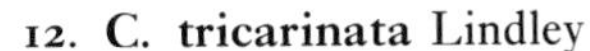

12. C. tricarinata Lindley

ILLUSTRATIONS: Annals of the Royal Botanic Garden Calcutta 8: t. 223 (1898); Botanical Magazine, 8803 (1919); Lin, Native orchids of Taiwan, t. 49, 50 (1975); Maekawa, Wild orchids of Japan in colour, t. 138 (1979).

PSEUDOBULBS: inconspicuous.

LEAVES: 18–30 × 5–9 cm, elliptic.

RACEME: arising from the leaf rosette, 30–40 cm, with several flowers.

BRACTS: 8–20 mm, persistent.

FLOWERS: 2.5–4 cm in diameter, sepals and petals greenish yellow.

LIP: about as long as the petals, red-brown, 3-lobed, the lateral lobes small, rounded, the central reversed-heart-shaped, notched, margins wavy, bearing 3 very pronounced ridges.

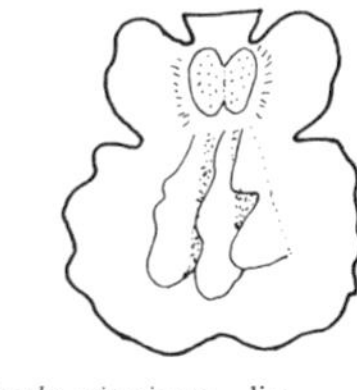

Calanthe tricarinata – lip.

SPUR: absent.

DISTRIBUTION: Himalaya, China, Japan & Taiwan.

HARDINESS: G1–2.

13. C. reflexa Maximowicz

ILLUSTRATIONS: Botanical Magazine, 9648 (1943); Lin, Native orchids of Taiwan, t. 44, 45 (1975); Maekawa, Wild orchids of Japan in colour, t. 137 (1979).

Like *C. tricarinata* except as below.

HABIT: generally smaller.

SEPALS AND PETALS: reflexed, each narrowed into a bristle-like point.

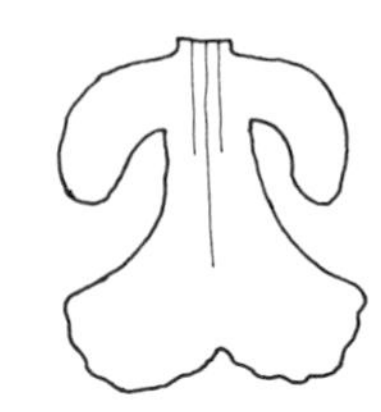
Calanthe reflexa – lip.

LIP: pale purple or white with oblong lateral lobes and an obovate, scarcely wavy, irregularly toothed but not notched central lobe which bears no ridges.
DISTRIBUTION: Japan, Korea (Quelpaert Islands) & Taiwan.
FLOWERING: summer.

14. C. gracilis (Lindley) Lindley
ILLUSTRATIONS: Botanical Magazine, 4714 (1853); Annals of the Royal Botanic Garden Calcutta 8: t. 222 (1898).
PSEUDOBULBS: inconspicuous.
LEAFY STEMS: to 40 cm.
LEAVES: 15–20 × 3–4 cm, narrowly elliptic.
RACEMES: arising laterally near the bases of the leafy stems and shorter than them, with 10–15 flowers.
FLOWERS: 2–3 cm in diameter, greenish yellow.
SEPALS AND PETALS: spreading.
LIP: free from the column, white with a yellow band, 3-lobed, the lateral lobes obliquely triangular, the central finely irregularly toothed and deeply notched.
SPUR: absent.
DISTRIBUTION: Himalaya.
HARDINESS: G1–2.
FLOWERING: autumn.

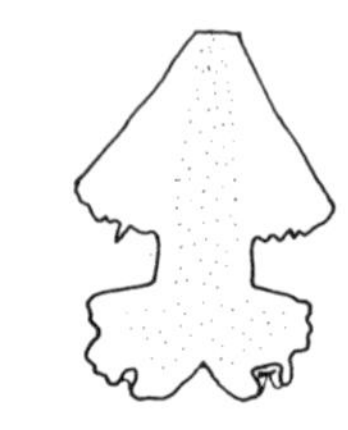
Calanthe gracilis – lip.

Var. **venusta** (Schlechter) Maekawa
SYNONYM: *C. venusta* Schlechter.
ILLUSTRATION: Maekawa, Wild orchids of Japan in colour, t. 136 (1979).
HABIT: larger, with stems to 1 m.
LEAVES: 20–40 × 4–7 cm.
SEPALS: longer than the petals.
LIP: with lateral lobes ovate.
DISTRIBUTION: Japan.

33. BLETIA Ruiz & Pavon

Between 25 and 50 species ranging from Mexico, Florida and the West Indies south to Argentina. Several species now placed in other genera were originally described in *Bletia* and the name is still sometimes used for plants quite unlike the genus as currently recognised. *Bletia* species should be grown in terrestrial compost and repotted when new growth first appears; the pseudobulbs should be partially covered. Little water should be given once growth has ceased. Plants can be divided after flowering.

Literature: Dressler, R.L., Notes on Bletia, *Brittonia* **20**: 182–90 (1968).

HABIT: terrestrial, rarely epiphytic.
PSEUDOBULBS: almost spherical, slightly flattened, corm-like, bearing 1–several leaves.

LEAVES: linear to lanceolate, pleated, rolled when young, deciduous.

FLOWERING STEMS: produced from lower nodes of pseudobulbs.

FLOWERS: many, in loose racemes or panicles

SEPALS AND PETALS: similar, free, spreading; lateral sepals closely pressed together at the base but not fused.

LIP: free, simple or 3-lobed; lateral lobes erect or curved around the column, central lobe broad, spreading, notched or more deeply 2-lobed, with 3–7 longitudinal toothed or scalloped keels.

COLUMN: arched, apically winged.

POLLINIA: 8, hard and compressed.

Key to species

1. Lip 3.5 × 2–2.5 cm, central lobe deeply notched or 2-lobed **1. catenulata**
2. Lip 1.5–2 × 1.2–1.5 cm, central lobe almost square **2. purpurea**

1. B. catenulata Ruiz & Pavon

SYNONYMS: *B. sanguinea* Poeppig & Endlicher; *B. sherrattiana* Lindley; *B. watsonii* Hooker.

ILLUSTRATIONS: Botanical Magazine, 5646 (1867); Pabst & Dungs, Orchidaceae Brasilienses 1: 229 (1975); Bechtel et al., Manual of cultivated orchid species, 170 (1981).

HABIT: to 2 m.

PSEUDOBULBS: 2–6 cm, each bearing 1–6 leaves in 2 opposite ranks.

LEAVES: 20–100 × 1–10 cm, variable in shape, linear to lanceolate or elliptic, acuminate.

FLOWERING STEM: to 2 m, bearing 3–15 flowers.

FLOWERS: 5–6.5 cm wide.

SEPALS: 3–3.5 cm × 7–10 mm, oblong, acute, red to pinkish purple.

PETALS: 3–3.5 × 1.7–2 cm, ovate, rounded, darker than sepals.

LIP: 3.5 × 2–2.5 cm, circular to oblong, deeply 3-lobed; lateral lobes semicircular to oblong, erect to spreading, purple with darker veins; central lobe deeply notched to 2-lobed, rounded, pink or pinkish purple with 3–5 pale to bright yellow ridges.

COLUMN: club-shaped.

DISTRIBUTION: Colombia, Ecuador, W Brazil & Bolivia.

HARDINESS: G2.

FLOWERING: summer.

2. B. purpurea (Lamarck) de Candolle

SYNONYMS: *B. verecunda* (Salisbury) R. Brown; *B. shepherdii* Hooker.

ILLUSTRATIONS: Botanical Magazine, 3319 (1834); Dunsterville & Garay, Venezuelan orchids illustrated **2**: 41 (1961); Luer, Native orchids of Florida, 227 (1972); American Orchid Society Bulletin **53**: 18, 19 (1984), **54**: 269 (1985), **55**: 234 (1986).

HABIT: terrestrial, rarely epiphytic, to 1.5 m.

PSEUDOBULBS: 2–4 cm, each bearing 3–5 leaves.

LEAVES: to 100 × 5 cm, linear-lanceolate, tapering, pale green.

FLOWERING STEM: to 1.5 m, sometimes branched, bearing 3–80 flowers.

FLOWERS: 2–2.5 cm wide, opening in succession.

SEPALS AND PETALS: 1.5–2.5 cm × 6–8 mm, pink to pinkish purple; upper sepal oblong to lanceolate; lateral sepals and petals ovate-lanceolate, acuminate, asymmetric; lateral sepals often with a central yellow stripe.

LIP: 1.5–2 × 1.2–1.5 cm, 3-lobed; lateral lobes triangular to diamond-shaped, scalloped, curved over column, purple; central lobe almost square, scalloped, purple, with 5 yellow scalloped ridges.

COLUMN: white, 1–1.3 cm.

DISTRIBUTION: USA (Florida), Bahamas & Mexico, south through C America and the West Indies to Colombia, Venezuela & Guyana.

HARDINESS: G2.

FLOWERING: throughout the year.

34. **PHAIUS** Loureiro

A widespread genus of 30–50 species ranging from tropical Africa eastwards to the Himalaya, N India, China & Japan, and south through SE Asia to N Australia and the Pacific islands. The spurless species, formerly treated as the separate genus *Gastrorchis* Schlechter, are now usually considered to be part of *Phaius*. The terrestrial species should be grown in large, well-drained pots and kept moist and partially shaded during the growing season, though overhead spraying should be avoided. During the resting period small amounts of moisture can still be given and plants can be exposed to stronger light. The epiphytic species require more shading and are easily harmed by draughts.

HABIT: terrestrial, rarely epiphytic, robust.

PSEUDOBULBS: spherical to ovoid, (sometimes stem-like), bearing 3–10 leaves.

LEAVES: large, lanceolate to elliptic, alternate, stalked, pleated, persistent, rolled when young.

FLOWERING STEMS: lateral, arising from pseudobulb bases or lower axils.

FLOWERS: few to many in loose or dense racemes.

SEPALS AND PETALS: similar, fleshy, free.

LIP: simple or 3-lobed with lateral lobes erect or curved around column, spurred or pouched below (except Nos. 3 & 4), shortly fused to column at base.

COLUMN: short.

POLLINIA: 8, in 2 groups of 4.

Key to species

1. Pseudobulbs stem-like, narrowly fusiform to cylindric **1. mishmensis**
2. Flowers at least 10 cm wide; sepals and petals lanceolate, more than 3 times as long as broad **2. tankervilleae**
3. Leaves with white spots; sepals and petals yellow to greenish yellow; lip spurred **5. flavus**
4. Sepals and petals white **4. tuberculosus**
5. Sepals and petals pale pink with darker purple streaks **3. humblotii**

1. P. mishmensis (Lindley) Reichenbach

ILLUSTRATIONS: Botanical Magazine, 7497 (1896); Bose & Bhattacharjee, Orchids of India, 431 (1980); Bechtel et al., Manual of cultivated orchid species, 258 (1981).

HABIT: terrestrial.

STEMS: to 140 cm, narrowly fusiform to cylindric, fleshy, bearing 6–8 leaves.

LEAVES: 15–30 × 8–14 cm, lanceolate to ovate or oblong, acuminate.

FLOWERING STEMS: 1 or more from lower axils of pseudobulbs, to 60 cm, bearing 7–10 flowers.

FLOWERS: 5–6 cm wide.

SEPALS AND PETALS: similar, pale pink to darker reddish brown; sepals to 3.5 × 1 cm, oblong, acuminate; petals somewhat smaller.

LIP: 3 × 2.5 cm, 3-lobed; lateral lobes oblong, rounded, pink speckled with red or purple, curved around column; central lobe square to oblong, shallowly lobed, with a central hairy ridge, pink speckled with red or purple, margins yellow.

SPUR: 1.5–1.7 cm, tapering, curved.

COLUMN: 1.2 cm, straight.

DISTRIBUTION: NE India, Burma, Thailand, Taiwan & Philippines.

HARDINESS: G2.

FLOWERING: summer–autumn.

2. P. tankervilleae (Banks) Blume

SYNONYMS: *P. bicolor* Lindley; *P.grandifolius* Loureiro; *P. blumei* Lindley; *P. wallichii* Hooker.

ILLUSTRATIONS: Botanical Magazine, 4078 (1844), 7023 (1888); Sheehan & Sheehan, Orchid genera illustrated, 133 (1979); Bechtel et al., Manual of cultivated orchid species, 258 (1981).

HABIT: terrestrial.

PEUDOBULBS: 2.5–6 cm, conical to ovoid, few-leaved.

LEAVES: 30–120 × 9–20 cm, elliptic to lanceolate, acuminate, membranous.

FLOWERING STEM: lateral, from base of pseudobulb, to 2 m, bearing 10–20 flowers.

FLOWERS: 10–12.5 cm wide.

SEPALS AND PETALS: similar, 4.5–6.5 × 1–1.3 cm, lanceolate, acuminate, reddish or yellowish brown, margin paler, greenish red on back.

LIP: 5 × 4.5 cm, 3-lobed; lateral lobes oblong, rounded, curved around column, pink to purplish red, yellow below; central lobe broadly ovate, acute or truncate, pink and yellow; throat yellow and purplish red; callus oblong.

SPUR: 1 cm, cylindric, curved.

COLUMN: 2 cm, club-shaped.

DISTRIBUTION: C China, N India, Sri Lanka, through SE Asia to N Australia.

HARDINESS: G2.

FLOWERING: spring–summer.

3. **P. humblotii** Reichenbach

SYNONYM: *Gastrorchis humblotii* (Reichenbach) Schlechter.

ILLUSTRATIONS: Cogniaux & Goossens, Dictionnaire Iconographique des Orchidées, Phaius, t. 1 (1898); Veitch, Manual of orchidaceous plants, part 6: 12 (1890).

HABIT: terrestrial.

PSEUDOBULBS: 4–6 × 3–4 cm, spherical to conical, ridged, few-leaved.

LEAVES: 30–60 × 6–10 cm, broadly lanceolate, acuminate, membranous.

FLOWERING STEM: robust, 30–80 cm, from base of pseudobulb, bearing 7–12 flowers.

FLOWERS: 5–6 cm wide.

SEPALS AND PETALS: similar, 2.5–3 × 1.5–2 cm, obovate to elliptic, acuminate, pale pink with darker streaks of purple.

LIP: 3-lobed, lacking a spur, strongly crisped; lateral lobes oblong, erect to spreading, pinkish purple to reddish brown; central lobe semicircular to oblong, pink with white streaks; callus yellow, divided into 2 tooth-like ridges.

COLUMN: club-shaped.

DISTRIBUTION: Madagascar.

HARDINESS: G2.

FLOWERING: summer.

4. **P. tuberculosus** (Thouars) Blume

SYNONYM: *Gastrorchis tuberculosa* (Thouars) Schlechter.

ILLUSTRATIONS: Botanical Magazine, 7307 (1893); Proceedings of 8th world orchid conference, t. 7 (1976); Bechtel et al., Manual of cultivated orchid species, 258 (1981).

Like *P. humblotii* except as below.

PSEUDOBULBS: *c.* 2.5 cm × 5 mm.

SEPALS AND PETALS: white.

LIP: with 3 yellow keels; lateral lobes yellow with red spots.

DISTRIBUTION: Madagascar.

HARDINESS: G1.

5. **P. flavus** (Blume) Lindley

SYNONYM: *P. maculatus* Lindley.

ILLUSTRATIONS: Botanical Magazine, 3960 (1842); Cogniaux & Goossens, Dictionnaire Iconographique des Orchidées, Phaius, t. 3 (1904).

Like *P. humblotii* except as below.

LEAVES: with white spots.

SEPALS AND PETALS: yellow to greenish yellow, oblong, obtuse.

LIP: spurred, greenish yellow with reddish brown streaks in the heavily crisped margin.

DISTRIBUTION: SE Asia.

HARDINESS: G2.

FLOWERING: spring.

35. SPATHOGLOTTIS Blume

A genus of about 40 species distributed from N India through SE Asia and C China to Australia and the Pacific Islands. *Spathoglottis* species perform best if repotted every spring in well-drained terrestrial compost. Plenty of water can be supplied through the summer but this should be decreased when the plants are dormant. Growing temperatures depend on the country of origin. Propagation is effected by dividing sections of rhizome with pseudobulbs.

HABIT: terrestrial.

PSEUDOBULBS: conical to ovoid, sheathed, bearing several leaves.

LEAVES: narrowly lanceolate, tapering gradually to the stalk, pleated, prominently veined, rolled when young.

FLOWERING STEMS: robust, borne laterally from basal leaf axils, unbranched, basally sheathed, many-flowered.

SEPALS: spreading, free.

PETALS: similar to sepals but often wider.

LIP: deeply 3-lobed; lateral lobes erect or spreading; central lobe lanceolate to spathulate, with a long narrow claw and a prominent basal callus and small basal teeth.

COLUMN: slender, arched.

POLLINIA: club-shaped, 8, in 2 groups of 4.

Key to species

1. Flowers yellow

1.A. Flowers no more than 4 cm wide **3. pubescens**

1.B. Flowers deep yellow; lateral lobes of lip broadest at their bases **1. aurea**

1.C. Flowers pale yellow; lateral lobes of lip spathulate **2. kimballiana**

2. Flowers 3.5–4.5 cm wide **4. plicata**

3. Flowers more than 4.5 cm wide **5. vieillardii**

1. S. **aurea** Lindley

ILLUSTRATIONS: Holttum, Flora of Malaya I: Orchids, 162 (1964); Shuttleworth et al., Orchids, 86 (1970).

PSEUDOBULBS: to 20 cm, ovoid, with 2–4 leaves.

LEAVES: to 100 × 5 cm.

FLOWERING STEM: to 60 cm or more, bearing 4–10 flowers.

FLOWERS: 6–7 cm wide, deep yellow.

BRACTS: 1–2 cm, spathulate, concave, spreading.

SEPALS AND PETALS: 2.5–3 × 1.2–1.5 cm, elliptic to ovate, obtuse, slightly concave.

LIP: 2.5–3 × 2.5–3 cm, deeply divided into 3 main lobes; lateral lobes oblong, rounded, broadest at base, erect; central lobe 3–4 mm wide, lanceolate, rounded, with 2 small triangular lobes or teeth at base; callus of 2 erect triangular teeth; lateral lobes, callus and base of central lobe spotted with crimson.

COLUMN: 1–5 cm, narrowly club-shaped, arched.

DISTRIBUTION: Malay peninsula.

HARDINESS: G2.

FLOWERING: summer.

2. S. **kimballiana** Hooker

ILLUSTRATIONS: Botanical Magazine, 7443 (1895); Gardeners' Chronicle **4**: 93 (1888).

Like *S. aurea* except as below.

FLOWERS: to 7.5 cm wide, pale yellow.

SEPALS: flecked with purple.

LIP: lateral lobes broadly spathulate.

DISTRIBUTION: Borneo.

FLOWERING: spring.

3. S. **pubescens** Lindley

SYNONYM: *S. fortunei* Lindley.

ILLUSTRATIONS: Edwards's Botanical Register **31**: t. 19 (1845); Schlechter, Die Orchideen, edn 2, 308 (1927).

Like *S. aurea* except as below.

LEAVES: 15–20 × 1.5–2 cm.

BRACTS ON FLOWERING STEM: narrowly ovate, attenuate, not spreading.

FLOWERS: 3–4 cm wide.
LIP: lateral lobes oblong, rounded,strongly sickle-shaped.
DISTRIBUTION: Hong Kong.
FLOWERING: winter.

4. S. plicata Blume
ILLUSTRATIONS: Holttum, Flora of Malaya I: Orchids, 6 (1964); Shuttleworth et al., Orchids, 86 (1970); Bechtel et al., Manual of cultivated orchid species, 274 (1981).
PSEUDOBULBS: to 7 × 5 cm, with 4–7 leaves.
LEAVES: 30–120 × 2–7 cm.
FLOWERING STEM: to 150 cm, bearing 5–25 flowers.
BRACTS: lanceolate, acuminate.
FLOWERS: 3.5–4.5 cm wide, purple to pinkish purple.
SEPALS: 2.3 × 1.2 cm, elliptic, acute; lateral sepals asymmetric.
PETALS: 2 × 1.4 cm, elliptic to ovate, obtuse.
LIP: 1.7 × 1.8 cm, 3-lobed; lateral lobes ovate to narrowly spathulate, rounded to truncate; central lobe spathulate, long-clawed, with 2 small teeth at base; callus triangular to heart-shaped, yellow.
COLUMN: 1.2 cm, club-shaped, arched.
DISTRIBUTION: SE Asia from Sumatra to the Philippines (?India); naturalised in Florida, Hawaii and parts of tropical Africa.
HARDINESS: G2.
FLOWERING: more or less continuously.

5. S. vieillardii Reichenbach
ILLUSTRATIONS: Botanical Magazine, 7013 (1888); Graf, Tropica, 754 (1978).
Like *S. plicata* except as below.
HABIT: larger in all parts.
FLOWERS: pale purple, to 6 cm wide.
LIP: lateral lobes pinkish or brownish purple.
DISTRIBUTION: Malaysian & Indonesian islands.
FLOWERING: autumn.

36. CHYSIS Lindley

A genus of about 6 species from Mexico, south through Central America to Venezuela and Peru. They should be grown in epiphytic compost in pots or baskets. A warm shady humid environment gives best results when the plants are actively growing though lower temperatures and much less water should be applied once growth ceases and the leaves fall.

HABIT: epiphytic or on rocks.
PSEUDOBULBS: cylindric to fusiform or narrowly club-shaped, compound, fleshy, bearing several leaves in upper half, sheathed below.

LEAVES: in 2 opposite rows, pleated, rolled when young, prominently veined, deciduous.

FLOWERING STEMS: lateral from lower leaf axils.

FLOWERS: several in loose racemes.

SEPALS AND PETALS: free, spreading; lateral sepals arising obliquely from column foot forming a mentum.

LIP: 3-lobed, basally fused to column foot; lateral lobes broad, erect; central lobe spreading or reflexed, entire or notched; base of lip with keels or thickened veins.

COLUMN: short and thick, 2-winged, with a prominent foot.

POLLINIA: 8, often fused to column, causing self-fertilisation.

Key to species

1. Sepals and petals white with purplish streaks **2. limminghii**
2. Bracts almost as long as ovaries; lip callus of 5 hairy ridges **1. aurea**
3. Bracts at most half as long as ovaries; lip callus ridges not hairy **3. laevis**

1. C. **aurea** Lindley

ILLUSTRATIONS: Botanical Magazine, 3617 (1837); Skelsey, Orchids, 101 (1979); Shuttleworth et al., Orchids, 87 (1979); Bechtel et al., Manual of cultivated orchid species, 183 (1981).

HABIT: epiphytic.

PSEUDOBULBS: to 50 × 4 cm, cylindric to fusiform, pendent, bearing 4–15 leaves.

LEAVES: to 45 × 7 cm, lanceolate, acuminate (when young) to attenuate.

FLOWERING STEM: to 45 cm, bearing 3–12 flowers.

BRACTS: *c.* 2 cm, nearly equal to ovary.

FLOWERS: to 6 cm wide, yellow with reddish lines and spots on petals and lip.

UPPER SEPAL: 3–3.5 × 1.5 cm, oblong, bluntly tapering.

LATERAL SEPALS: 2.5 × 1.7 cm, triangular, asymmetric, obtuse.

PETALS: 2.5–3 × 1–1.5 cm, spathulate to narrowly diamond-shaped, obtuse or rounded.

LIP: 2–2.5 × 2.5–3 cm, 3-lobed, very fleshy and crisped, lateral lobes oblong to ovate, sickle-shaped, erect; central lobe ovate to circular; callus of 5 fleshy, hairy ridges.

COLUMN: white, with a prominent foot.

DISTRIBUTION: Mexico to Venezuela, Colombia & Peru.

HARDINESS: G2.

FLOWERING: spring.

Var. **bractescens** (Lindley) Allen

SYNONYM: *C. bractescens* Lindley.

ILLUSTRATIONS: Botanical Magazine, 5186 (1860); Williams et al., Orchids for everyone, 184 (1980); Bechtel et al., Manual of cultivated orchid species, 183 (1981); American Orchid Society Bulletin **55**: 813 (1986).

BRACTS ON FLOWERING STEM: 2.5 cm or more, leaf-like.

FLOWERS: 6–8 cm wide.

SEPALS AND PETALS: white with yellow streaks.

DISTRIBUTION: Mexico, Guatemala & Belize.

FLOWERING: winter–spring.

2. **C. limminghii** Linden & Reichenbach

ILLUSTRATIONS: Botanical Magazine, 5265 (1861); Cogniaux & Goossens, Dictionnaire Iconographique des Orchidées, Chysis, t. 3 (1901).

Like *C. aurea* except as below.

SEPALS AND PETALS: white with purple or purplish brown streaks.

LIP: margin not crisped.

DISTRIBUTION: C Mexico.

FLOWERING: summer.

3. **C. laevis** Lindley

ILLUSTRATIONS: Cogniaux & Goossens, Dictionnaire Iconographique des Orchidées, Chysis, t. 2 (1901); Ospina & Dressler, Orquideas de las Americas, t. 42 (1974); American Orchid Society Bulletin **55**: 813 (1986).

Like *C. aurea* except as below.

BRACTS: about half the ovary length.

FLOWERS: yellow to orange with streaks of reddish purple.

LIP: pale yellow with pinkish purple spots and blotches; callus ridges hairless.

DISTRIBUTION: Mexico & Costa Rica.

FLOWERING: summer.

37. THUNIA Reichenbach

A genus of about 6 species from India, Burma and SE Asia, which could all be regarded as colour variants of a single species. They perform best if repotted annually on the appearance of new growth, and should then be kept shaded and slightly moist. Once the new growth is well established more water can be given. After leaf-fall plants should be kept cool and dry. Propagation is by division.

HABIT: terrestrial.

STEMS: tall, tufted, cane-like, sheathed below, densely leafy above.

LEAVES: long, acuminate, slightly bluish, spreading, in 2 opposite ranks, rolled when young.

FLOWERS: densely clustered in pendent, terminal racemes.

BRACTS: large, boat-shaped, white to pale green.

SEPALS AND PETALS: free, similar.

LIP: unlobed, rolled around column, with 5–7 narrow fringed keels and a short, obtuse spur.

COLUMN: 2-winged at apex.

POLLINIA: 8.

Key to species

1. Sepals and petals pale purple with darker tips — **3. bensoniae**
2. Flowers to 12.5 cm wide — **1. alba**
3. Flowers to 7.5 cm wide — **2. bracteata**

1. T. alba (Lindley) Reichenbach

SYNONYM: *T. marshalliana* Reichenbach.

ILLUSTRATIONS: Shuttleworth et al., Orchids, 43 (1970); Skelsey, Orchids, 140 (1979); Bechtel et al., Manual of cultivated orchid species, 276 (1981).

HABIT: to 150 cm.

LEAVES: 15–20 cm, lanceolate, acuminate, pale green.

FLOWERS: 5–12, to 12.5 cm wide, fragrant.

BRACTS: to 5 cm, white, boat-shaped.

SEPALS AND PETALS: 6.5–7.5 × 3.3–1.5 cm, lanceolate to ovate, acuminate, white.

LIP: 5–6 × 4.5 cm, broadly elliptic to oblong, yellow with a white border, margin irregular and fringed; blade with orange, branching, hairy ridges.

SPUR: *c.* 1 cm, shallowly 2-lobed.

COLUMN: 2–2.5 cm.

DISTRIBUTION: Burma, Thailand & S China.

HARDINESS: G2.

FLOWERING: summer.

2. T. bracteata (Roxburgh) Schlechter

ILLUSTRATIONS: Botanical Magazine, 3991 (1843); Graf, Tropica, 754 (1978).

Like *T. alba* except as below.

HABIT: robust, to 60 cm.

FLOWERS: 4–10, to 7.5 cm wide.

LIP: less than 4 cm wide, oblong to ovate; ridges orange or purple.

DISTRIBUTION: N.India, Burma & Thailand.

Thunia bensoniae – lip.

3. T. bensoniae Hooker

ILLUSTRATION: Botanical Magazine, 5694 (1868).

Like *T. alba* except as below.

SEPALS AND PETALS: pale purple with darker tips.

LIP: white at base; blade dark purple with bright yellow, hairy ridges.

DISTRIBUTION: Burma.

38. ARUNDINA Blume

Formerly considered to contain several species, *Arundina* is now usually regarded as a single widespread and highly variable species. Distinctive variants from the extremes of the range appear to be interconnected by intermediates. Cultivation conditions are similar to those for *Sobralia* though a slightly higher minimum temperature is required.

HABIT: terrestrial, to 3 m.

STEMS: erect, slender and rigid, tufted, unbranched, with many leaves.

LEAVES: 12–30 × 1.5–4 cm, linear to lanceolate, grass-like, in 2 opposite ranks, folded when young; sheaths overlapping at base.

FLOWERS: many, 6–7 cm wide, in simple or branched terminal spikes, fragrant.

BRACTS: prominent.

SEPALS AND PETALS: *c.* 4 cm, white to pale pinkish purple, acute; sepals *c.* 1 cm wide, lanceolate to elliptic; lateral sepals lying close together under lip; petals *c.* 2 cm wide, ovate to circular, wavy.

LIP: 4 × 3.5 cm, shallowly 3-lobed, oblong, rounded, base rolled around the column, crisped, deeply notched at apex, 3-keeled; throat pale, usually with a yellow patch; blade white to bright pink.

COLUMN: slender, without foot.

POLLINIA: 8, flattened.

1. A. graminifolia (Don) Hochreutiner

SYNONYMS: *Bletia graminifolia* Don; *A. speciosa* Blume; *A. chinensis* Blume; *A. bambusifolia* Lindley; *A. revoluta* Hooker.

ILLUSTRATIONS: Botanical Magazine, 7284 (1893); Shuttleworth et al., Orchids, 43 (1970); Graf, Tropica, 709 (1978); Opera Botanica **89**: f. 4 & pl. 1d (1986).

DISTRIBUTION: Himalaya & S China, Sri Lanka, SE Asia & Pacific islands.

HARDINESS: G1.

FLOWERING: more or less continuously.

39. SOBRALIA Ruiz & Pavon

A genus of 35–50 species from C America and tropical S America. They perform best if repotted very infrequently and should be allowed to become well established in large pots or beds. When actively growing, plants should be kept well shaded and watered. Once growth ceases watering can be considerably reduced.

HABIT: terrestrial, rarely epiphytic.

STEMS: erect, slender and unbranched, with many leaves.

LEAVES: leathery, pleated, rolled when young, long-sheathed, in 2 opposite ranks.

FLOWERS: few or solitary, terminal or in the upper leaf axils.

BRACTS: usually well developed, sometimes several to each flower.

SEPALS: fused at base.

PETALS: free, broader than sepals.

LIP: slightly fused to base of column, simple or 2-lobed, rolled around column; blade expanded, wavy; small calluses and ridges often present.

COLUMN: slender, without a foot, 3-lobed at apex.

POLLINIA: 8, in 2 groups of 4, granular.

Key to species

1. Flowers less than 12 cm wide
1.A. Leaves with warts and short, stiff hairs beneath **2. decora**
1.B. Flowers in axillary panicles **3. cattleya**
1.C. Flowers solitary **1. sessilis**
2. Bracts 2.5–5 cm; lip to 8 cm **4. leucoxantha**
3. Flower stalked, purple or pinkish purple **5. macrantha**
4. Flowers almost stalkless, yellow **6. xantholeuca**

1. **S. sessilis** Lindley

ILLUSTRATIONS: Dunsterville & Garay, Venezuelan orchids illustrated 1: 397 (1959); American Orchid Society Bulletin 42: 395 (1973).

STEMS: to 90 cm.

LEAVES: to 22 × 8 cm, stiff, narrowly elliptic to ovate, acute or obtuse.

FLOWERS: solitary, terminal, tapering smoothly into ovary and stalk.

BRACTS: several, dark brown.

SEPALS: 4.5–5 × 1.5–2 cm, lanceolate, acute.

PETALS: 4–5 × 1–2 cm, lanceolate to ovate, acute or obtuse, strongly recurved; petals and sepals white.

LIP: 4.2–5 × 2.8–3.5 cm, ovate to circular, shallowly notched, slightly scalloped, rolled around column, white with yellow stripe in throat and patch on blade; callus of 2 small basal projections.

COLUMN: 2.5 cm.

DISTRIBUTION: Brazil, Guyana & Venezuela.

HARDINESS: G2.

FLOWERING: summer.

2. **S. decora** Bateman

ILLUSTRATIONS: Hamer, Las Orquideas de El Salvador, t. 34 (1974); Graf, Tropica, 759 (1978).

Like *S. sessilis* except as below.

HABIT: occasionally epiphytic.
LEAVES: with warts and short stiff hairs beneath and on sheaths.
FLOWERS: to 10 cm wide, pale to dark purple.
DISTRIBUTION: Mexico to Honduras.

3. **S. cattleya** Reichenbach.
ILLUSTRATION: American Orchid Society Bulletin **44**: 195 (1975).
Like *S. sessilis* except as below.
FLOWERS: in several axillary panicles, each with 5–12 flowers, pinkish brown inside, white or cream outside.
LIP: maroon with yellow ridges.
DISTRIBUTION: Venezuela & Colombia.
HARDINESS: G2.

4. **S. leucoxantha** Reichenbach
ILLUSTRATIONS: Botanical Magazine, 7058 (1889); Graf, Tropica, 759 (1978); Bechtel et al., Manual of cultivated orchid species, 273 (1981).
HABIT: terrestrial.
STEMS: to 150 cm, clustered.
LEAVES: 20–30 × 3–4 cm, elliptic to lanceolate, acuminate, scurfy beneath; sheaths warty.
FLOWERS: solitary, terminal, 15–18 cm wide.
BRACTS: 2.5–5 cm.
SEPALS: to 7 × 2.5 cm, linear to lanceolate, white.
PETALS: to 7 × 3–3.5 cm, lanceolate to oblong, white.
LIP: 6–8 × 5–6 cm, ovate, notched, wavy, rolled around column, white with a yellow to reddish orange throat.
COLUMN: 2.5 cm.
DISTRIBUTION: Costa Rica & Panama.
HARDINESS: G2.
FLOWERING: summer–autumn.

5. **S. macrantha** Lindley
ILLUSTRATIONS: Botanical Magazine, 4446 (1849); Skelsey, Orchids, 137 (1979); Bechtel et al., Manual of cultivated orchid species, 273 (1981); Orchid Review **92**: 315 (1984).
HABIT: terrestrial or on rocks.
STEMS: to 2 m, erect, clustered.
LEAVES: 13–30 × 2–7.5 cm, lanceolate, acuminate, dark green above, paler beneath.
FLOWERS: solitary, 15–18 cm wide, fragrant.
BRACTS: to 11 cm, green.
SEPALS: to 9 × 2.5 cm, lanceolate to elliptic, obtuse or acute, with curled tips, pinkish purple.
PETALS: to 9 × 3–3.5 cm, elliptic, asymmetric, obtuse or rounded, pinkish purple.

LIP: to 10 × 7cm, ovate, rolled around column, deeply scalloped and notched, slightly darker than petals, paler in the throat, with 7 darker ridges; callus of 2 white, fleshy teeth near lip base.

COLUMN: 3–4 cm, club-shaped.

DISTRIBUTION: Mexico to Costa Rica.

HARDINESS: G2.

FLOWERING: summer.

6. S. xantholeuca Reichenbach

ILLUSTRATIONS: Botanical Magazine, 7332 (1894); Hamer, Las Orquideas de El Salvador, t. 34 (1974); Graf, Tropica, 759 (1978).

HABIT: epiphytic or on rocks.

STEMS: to 180 cm, thick.

LEAVES: 15–28 × 3–7 cm, lanceolate to oblong, acuminate; sheaths speckled with reddish brown.

FLOWERS: solitary, terminal, *c.* 15 cm wide, almost stalkless, bright yellow.

SEPALS: 8–11 × 1.5–2 cm, lanceolate, acuminate.

PETALS: 8–10 × 2–2.5 cm, lanceolate to elliptic, acute or acuminate.

LIP: 8–11 × 6–7 cm, ovate to oblong, rolled around column, deeply notched, scalloped; throat streaked with deeper yellow.

COLUMN: 2.5 cm.

DISTRIBUTION: Guatemala & El Salvador.

HARDINESS: G2.

FLOWERING: spring–summer.

40. ISOCHILUS R. Brown

This genus is now usually regarded as consisting of 2 species from tropical America with rather variable floral characters interconnecting them. This variation has resulted in the use of many different names at species level. The degree of sepal fusion may vary considerably, even on a single plant. Cultivation conditions are generally similar to those for *Cattleya* (p. 167), though water should be given throughout the year.

HABIT: epiphytic, terrestrial or on rocks.

STEMS: slender, erect and clustered or creeping.

LEAVES: many, in 2 opposite ranks, linear to oblong, folded when young, with sheathing bases, often notched.

FLOWERS: terminal in 1- or 2-ranked racemes, sometimes solitary, white to pinkish purple.

SEPALS: upper free; laterals almost always at least half-fused, inflated below, keeled or occasionally winged.

PETALS: free, shorter and broader than sepals, short-clawed.

LIP: fused to base of column or column-foot, short-clawed, narrow, often S-shaped.
COLUMN: erect, sometimes with a small foot.
POLLINIA: 4, waxy.

1. **I. linearis** (Jacquin) R. Brown
ILLUSTRATIONS: Dunsterville & Garay, Venezuelan orchids illustrated 1: 178 (1959); Shuttleworth et al., Orchids, 71 (1970); Skelsey, Orchids, 115 (1979).
HABIT: epiphytic.
STEMS: to 50 cm, erect, covered in grey leaf-sheaths.
LEAVES: to 6 cm × 3 mm, linear, slightly tapering, asymmetrically notched, with a prominent midrib.
FLOWERS: 1–several, in dense 1- or 2-sided racemes, *c.* 1 cm long, pale pink to dark pinkish purple (rarely orange).
SEPALS: 8–9 × 3 mm, triangular to narrowly ovate; lateral sepals fused to about half their length, inflated and forming a small mentum with column foot.
PETALS: 6–8 × 3–4 mm, lanceolate, irregular, short-clawed.
LIP: 8 × 2 mm, linear to lanceolate, constricted in the middle, short-clawed.
COLUMN: 3–4 mm, with a short foot.
DISTRIBUTION: Mexico to Peru & Brazil, West Indies.
HARDINESS: G2.
FLOWERING: summer.

41. ARPOPHYLLUM Llave & Lexarza

A small genus of 2 to 5 species from tropical America. Floral differences between the species are slight and specific distinctions are made largely on vegetative characters. Cultivation conditions are generally similar to those for Cattleya (p. 167), though a simpler epiphytic compost of equal parts of sphagnum and tree-fern fibre is adequate. Good ventilation is important.

Literature: Garay, L., Synopsis of the genus Arpophyllum, *Orquidea (Mexico)* 4: 16–19 (1974).

HABIT: epiphytic or on rocks, with creeping, simple or branched rhizomes.
STEMS: cane-like, sheathed, 1-leaved.
LEAVES: fleshy or leathery, folded when young.
FLOWERS: many, small, in dense terminal racemes; flowering stem with large basal sheath.
SEPALS: spreading; lateral sepals inflated below and fused to column.
PETALS: smaller than sepals.
LIP: uppermost (flower non-resupinate), usually longer than sepals and petals, inflated and pouched below.
COLUMN: slightly arched, with a small foot.
POLLINIA: 8, waxy, pear-shaped.

Key to species

1. Leaf fleshy, strongly folded along the midrib **1. spicatum**
2. Leaf leathery, folded only at the base **2. giganteum**

1. A. **spicatum** Llave & Lexarza

ILLUSTRATIONS: Botanical Magazine, 6022 (1873); Shuttleworth et al., Orchids, 70 (1970); Graf, Tropica, 710 (1978); American Orchid Society Bulletin **54**: 869 (1985).

STEMS: to 75 cm, flattened, almost concealed in sheaths.

LEAF: to 50 × 4 cm, oblong, slightly sickle-shaped, obtuse, strongly folded along midrib, fleshy.

INFLORESCENCE: to 30 × 3 cm, cylindric; stalks and ovaries with black warts; basal sheath to 15 × 2 cm.

FLOWERS: 8–10 mm wide, purplish pink.

UPPER SEPAL: 4.5–5.5 × 1.5–3 mm, oblong to ovate, obtuse or with a small point.

LATERAL SEPALS: 5–6 × 2–3 mm, elliptic to oblong, acute to rounded, 3-veined, asymmetric, inflated at base.

PETALS: 5–5.5 × 1.5–2 mm, linear to narrowly elliptic, obtuse or rounded, 1-veined.

LIP: 5.5–6 × 3.5 mm, constricted above pouch; blade ovate, curved over column, 7-veined.

COLUMN: 3.5–4 mm.

DISTRIBUTION: Mexico.

HARDINESS: G2.

FLOWERING: spring.

2. A. **giganteum** Lindley

ILLUSTRATIONS: Dunsterville & Garay, Venezuelan orchids illustrated **6**: 50 (1976); Bechtel et al., Manual of cultivated orchid species, 168 (1981).

Like *A. spicatum* except as below.

STEMS: to 1 m.

LEAF: leathery, flattened, folded only towards the base, straight-sided.

INFLORESCENCE: to 18 cm long.

DISTRIBUTION: Mexico to Costa Rica; Jamaica & Colombia.

42. COELOGYNE Lindley

A genus of 100–150 species from India, China, SE Asia and the Pacific islands, many of which have showy and long-lasting flowers. Himalayan species such as *C. cristata* and *C. barbata* perform best in cool conditions never exceeding 25°C. They have a very definite resting period when no water should be given; this may last for several months. If the pseudobulbs shrivel during this time, light syringing will supply all the moisture required until growth begins again. Tropical species, e.g. *C. speciosa* and *C. dayana*, grow throughout the year and should be kept constantly moist with a minimum temperature of

15°C. Good drainage is essential. The spreading habit of many of these species is best accommodated on bark or in hanging baskets. Well-established plants flower heavily and repotting should be avoided unless absolutely necessary.

Literature: Seidenfaden, G., Orchid genera in Thailand 3, Coelogyne, *Dansk Botanisk Arkiv* **29**(4) (1975).

HABIT: epiphytic.

PSEUDOBULBS: cylindric, ovoid or fusiform, clustered or distant, with 1 or 2 leaves.

LEAVES: linear to elliptic, leathery and pleated, rolled when young.

FLOWERING STEMS: erect to pendent, variously sheathed, terminal on pseudobulbs or basal and sometimes later developing into pseudobulbs.

FLOWERS: 1–many in loose racemes, often fragrant.

SEPALS AND PETALS: free; sepals often keeled; petals usually narrower, sometimes similar.

LIP: 3-lobed; lateral lobes erect; disc with entire, warty, toothed or fringed keels which may extend on to the central lobe.

COLUMN: arched, exposed.

POLLINIA: 4.

Synopsis of characters

Pseudobulbs. Clustered: **1–4,10–12,14,15,18,23**; distant: **5–9,13,16,21,22,24–26**. One-leaved: **1**.

Leaves. 3 cm wide or less: **2–7,23**.

Flowering stem. Basal: **1,10–26**; on mature pseudobulbs: **2–8,16**. Erect or arched (majority of flowers held at or above level of pseudobulbs): **1–8,16–26**; pendent (majority of flowers held below level of pseudobulbs): **10–15,24–26**. 5 cm or less: **21**; more than 50 cm: **8–10,12,14,16,17**; more than 100 cm: **12**. Densely hairy:**11**; with a few scattered hairs: **12,14**. Sheathed at base only: **1.5–7,10,11,13–21,23–26**; with a cluster of scales just below flowers: **8,9**; sheathed from base up to lowest flowers: **2–4,12,22**.

Flowers. Solitary: **1–7**; 10 or fewer per stem: **1,5–11,13,14, 16–21,23–26**; more than 25 per stem: **10,12,14,22**. About 2 cm wide: **2–4,10**; 3–4.5 cm wide: **2–4,5–7,11,13,15,18,23–26**; more than 5 cm wide: **1–9,12,14–26**.

Bracts. Persistent: **10–12,14,16,21,22**; deciduous: **1–9,13, 18–20**.

Petals. 2 cm or less: **5–7,10,11,18–20,21**; 5 cm or more: **16–21**. Narrower than sepals: **1–9,13,15,18–20,22–26**; about as wide as sepals: **10–12,14,16–21**.

Lip. 4.5 cm or more: **1–4,16,17**. Margin entire:**2–4,10–15, 18,22,23**; margin toothed: **1,8,9**; margin fringed:**5–8,21**; margin scalloped: **16,17**. With a distinct waist: **16,17**. With ringed patches: **18**. With an even number of keels: **2–8, 7,12,22**; with an odd number of keels: **1–11,13–15,18–21,23**. Keels entire: **5–7,10,18,22**; keels wavy: **2–4,13,23–26**; keels papillose: **1,11,14,16,17**; keels toothed: **12,21**; keels fringed: **1–4,8,9,15,21**.

Key to species

1. Flower solitary, or if more then not all the flowers open at the same time — *Go to Subkey 1*
2. Flowering stems pendent, most of the flowers held below the level of the pseudobulbs — *Go to Subkey 2*
3. Flowering stem erect or arched, most flowers held above the level of the pseudobulbs — *Go to Subkey 3*

Subkey 1. Flowers solitary, or if more than one, then not all the flowers open at the same time.

1. Lip margin entire, not fringed
1.A. Pseudobulbs 1-leaved — **1. speciosa**
1.B. Flower to 2 cm wide, bright red — **4. miniata**
1.C. Sepals and petals yellow — **2. lawrenceana**
1.D. Sepals and petals white — **3. holochila**
2. Flowering stem short, with scales at the base only
2.A. Flowers at most 4 cm wide; lip with 5 keels, 2 longer, 3 shorter — **6. fimbriata**
2.B. Sepals and petals pinkish brown; lip yellow to pinkish brown with purplish brown patches — **5. ovalis**
2.C. Sepals and petals syellowish orange; lip dark brown — **7. fuliginosa**
3. Lip 3.5 cm, with 3 keels — **8. barbata**
4. Lip less than 3 cm, with 2 keels — **9. stricta**

Subkey 2. Flowers numerous, several open at the same time; flowering stem pendent, most of the flowers held below the level of the pseudobulbs.

1. Flowers to 2 cm wide; keels on lip not reaching the base of the central lobe — **10. veitchii**
2. Inflorescence-stalk densely hairy — **11. tomentosa**
3. Main keels on lip 2; raceme to 1 m, with up to 100 flowers — **12. dayana**
4. Flowers 10 or fewer, each less than 4 cm wide — **13. flaccida**
5. Sepals and petals pale yellowish — **14. massangeana**
6. Sepals and petals whitish — **15. swaniana**

Subkey 3. Flowers numerous, several open at the same time; flowering stems erect or arching, most flowers held above the level of the pseudobulbs.

1. Flowers predominantly green
1.A. Flowering stem with more than 5 flowers which are *c.* 10 cm wide — **16. pandurata**

1.B. Flowering stem with up to 10 flowers which are 6–8 cm wide **17. parishii**
2. Lip with ringed spots
2.A. Lip with 3 keels, 2 long and 1 short
2.A.a. Flowers 3–4 cm wide, lip to 2 cm **18. nitida**
2.A.b. Flowers *c.* 7 cm wide, lip to 3 cm **19. corymbosa**
2.B. Lip with 3 equal keels **20. corrugata**
3. Petals and lateral sepals more or less equal in width **21. cristata**
4. Flowers 6–7 cm wide; flowering stem with 10–18 flowers **22. asperata**
5. Leaves less than 2 cm wide **23. viscosa**
6. Flowers white **26. lactea**
7. Flowers *c.* 3 cm wide; lip to 1.8 cm **24. trinervis**
8. Flowers *c.* 5 cm wide; lip to 4 cm **25. fuscescens**

1. **C. speciosa** (Blume) Lindley

ILLUSTRATIONS: Botanical Magazine, 4889 (1855); Die Orchidee **29**: 18 (1978); Bechtel et al., Manual of cultivated orchid species, 187 (1981).

PSEUDOBULBS: 2–8 cm, ovoid, with 4 slightly concave sides, clustered, 1-leaved.

LEAF-BLADE: to 35 × 9 cm, narrowly elliptic, acuminate.

LEAF-STALK: *c.* 5 cm.

FLOWERING STEM: to 20 cm, basal, erect or arched, sheathed at base, bearing 1–3 flowers.

BRACTS: deciduous.

FLOWERS: *c.* 6 cm wide, drooping.

SEPALS: 5–7 × 1–1.8 cm, narrowly ovate, acuminate, yellowish green.

PETALS: 5–6 cm × 2–3 mm, linear, yellowish green.

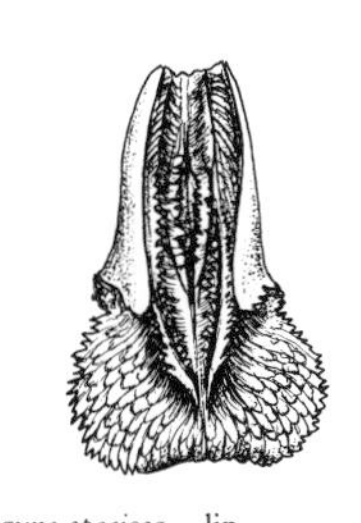

Coleogyne speciosa – lip.

LIP: 5 × 3–3.5 cm, 3-lobed, dark reddish brown with a white or yellow tip; keels 3, 2 long and 1 short, papillose, fringed; central lobe with a short broad claw.

COLUMN: 3.5 cm, arched.

DISTRIBUTION: Java & Sumatra (Malaysian records refer to *C. xyrekes* and those from Vietnam to *C. lawrenceana*).

HARDINESS: G2.

FLOWERING: continuously.

2. **C. lawrenceana** Rolfe

ILLUSTRATIONS: Botanical Magazine, 8164 (1907); Gardeners' Chronicle **47**: 335 (1910); Bechtel et al., Manual of cultivated orchid species, 186 (1981).

PSEUDOBULBS: 5–8 × 2.5–3 cm, ovoid to cylindric, clustered, 2-leaved.

LEAVES: 20–30 × 3 cm, elliptic, acuminate, tapering into stalk.

FLOWERING STEM: 15–20 cm, erect, on mature pseudobulb, sheathed throughout, 1-flowered.

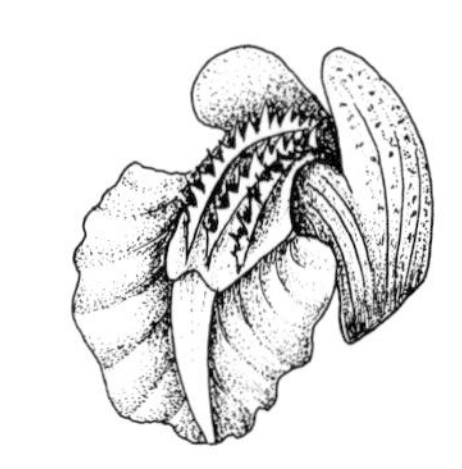

Coleogyne lawrenceana – lip.

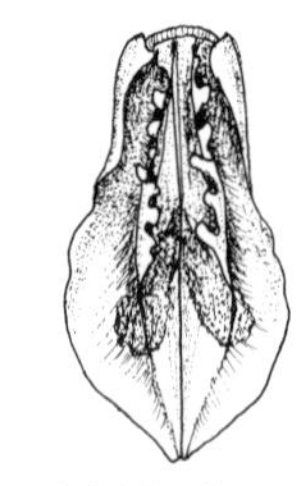

Coleogyne holochila – lip.

FLOWER: 5–7 cm wide.

SEPALS: 5 × 1.5 cm, triangular to ovate, obtuse, yellow.

PETALS: 4.5 cm × 3–4 mm, linear, drooping, yellow.

LIP: 5 × 2.5–3 cm, 3-lobed, with 3 long, fringed keels and 2 shorter basal ones; disc and lateral lobes purplish red; central lobe white with a yellow stripe.

COLUMN: 3–4 cm, club-shaped.

DISTRIBUTION: Vietnam.

HARDINESS: G2.

FLOWERING: spring.

3. C. **holochila** Hunt & Summerhayes

SYNONYMS: *C. elata* misapplied; *C.stricta* misapplied.

ILLUSTRATION: Botanical Magazine, 5001 (1857).

Like *C. lawrenceana* except as below.

FLOWERS: 6–10 per raceme, *c.* 4.5 cm wide.

SEPALS AND PETALS: white.

LIP: obscurely 3-lobed, yellow or orange with 2 sinuous, fringed keels.

DISTRIBUTION: NE India & Burma.

HARDINESS: G2.

FLOWERING: spring–summer.

4. C. **miniata** (Blume) Lindley

SYNONYM: *Hologyne miniata* (Blume) Pfitzer.

ILLUSTRATIONS: Cogniaux & Goossens, Dictionnaire Iconographique des Orchidées, Coelogyne, t. 7 (1907); Van Steenis, Mountain flora of Java, t. 35 (1972).

Like *C. lawrenceana* except as below.

PSEUDOBULBS: oblong to fusiform, distant, enclosed in scaly sheaths.

FLOWERS: several, *c.* 2 cm wide, bright red.

COLUMN: with a small foot.

DISTRIBUTION: Indonesia (Java & Bali; ?Sumatra).

HARDINESS: G2.

FLOWERING: summer.

5. C. **ovalis** Lindley

ILLUSTRATIONS: Botanical Magazine, 9255 (1931); Orchid Review **80**: 98 (1976); Bechtel et al., Manual of cultivated orchid species, 187 (1981).

PSEUDOBULBS: 3–9 × 2 cm, fusiform to ovoid, sheathed below, distant, 2-leaved.

LEAVES: 7–17 × 1.5–4.5 cm, lanceolate, acute; leaf-stalk short.

FLOWERING STEM: to 12 cm, erect, on mature pseudobulb, 1–3 flowered, sheathed below.

FLOWERS: to 6 cm wide.

BRACTS: deciduous.

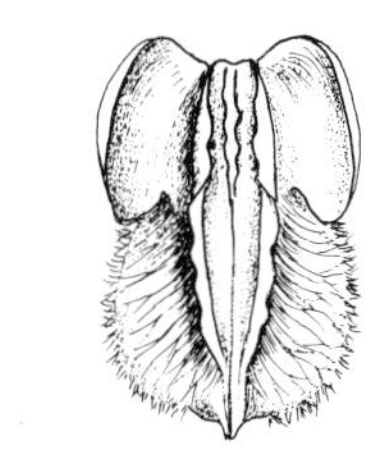

Coleogyne ovalis – lip.

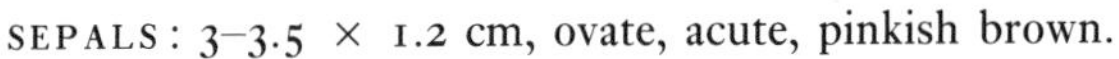

SEPALS: 3–3.5 × 1.2 cm, ovate, acute, pinkish brown.
PETALS: 3 cm × 1 mm, linear, acute, pinkish brown.
LIP: 3–3.5 × 2–2.5 cm, oblong, 3-lobed, finely fringed, yellow to pinkish brown with purplish brown patches; keels 3, wavy, 2 long and 1 short.
COLUMN: 1.5–1.8 cm, apically winged.
DISTRIBUTION: W Himalaya to Burma, China & Thailand.
HARDINESS: G2.
FLOWERING: summer.

6. C. fimbriata Lindley
ILLUSTRATIONS: Edwards's Botanical Register **11**: t. 868 (1825); Dansk Botanisk Arkiv **29**(4): 18 (1975); Bechtel et al., Manual of cultivated orchid species, 186 (1981).
Like *C. ovalis* except as below.
FLOWERING STEM: to 5 cm.
FLOWERS: 3–4 cm wide, pale yellow.
UPPER SEPAL: 1.5–2.5 mm.
LIP: marked with purplish brown, with 2 extra, short keels on the central lobe.
DISTRIBUTION: NE India to China & Malaysia.
FLOWERING: summer–autumn.

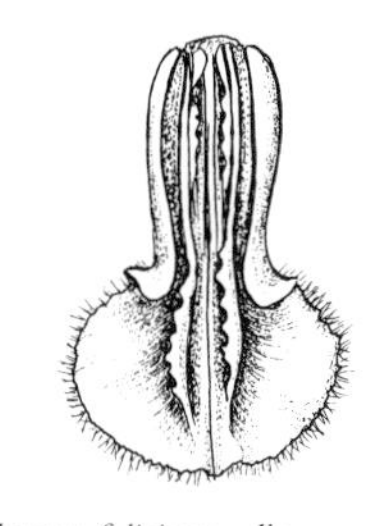

Coleogyne fuliginosa – lip.

7. C. fuliginosa Lindley
ILLUSTRATIONS: Botanical Magazine, 4440 (1849); Dansk Botanisk Arkiv **29**:(4): 21 (1975); Bechtel et al., Manual of cultivated orchid species, 186 (1981).
Like *C. ovalis* except as below.
FLOWERS: 5–6 cm wide.
SEPALS AND PETALS: yellowish orange.
LIP: dark brown; keels 4, slightly wavy, 2 long and 2 short.
DISTRIBUTION: Burma; ?Java.
FLOWERING: winter.

8. C. barbata Griffith
ILLUSTRATIONS: Veitch, Manual of orchidaceous plants: Eriae, 32 (1890); Lindenia **16**: 735 (1901); Orchid World **6**: 274 (1916).
PSEUDOBULBS: to 9 × 2.5 cm, conical to ovoid, sheathed below when young, distant, 2-leaved.
LEAVES: to 40 × 5 cm, oblong to lanceolate, acute or acuminate; leaf-stalk *c.* 5 cm.
FLOWERING STEM: to 60 cm, erect, on mature pseudobulb, up to 10-flowered, with a cluster of overlapping scales just below the flowers.
BRACTS: deciduous.
SEPALS: 3.5–4 × 1.5 cm, ovate to oblong, acute, white.
PETALS: 3.5–4 cm × 2–3 mm, linear, acute, white.

LIP: 3.5 × 3 cm, 3-lobed; central lobe laciniate; disc and central lobe dark brown to black, lateral lobes white; keels 3, fringed, middle one shorter.

COLUMN: 2.5–3 cm.

DISTRIBUTION: Bhutan & NE India.

HARDINESS: G2.

FLOWERING: winter.

9. C. stricta (D. Don) Schlechter

SYNONYM: *C. elata* Lindley.

ILLUSTRATION: Orchid Review **80**: 210 (1972).

Like *C. barbata* except as below.

LIP: less than 3 cm, keels 2, margin finely and shortly toothed.

DISTRIBUTION: Himalaya.

FLOWERING: spring–summer.

The illustration under this name in Botanical Magazine, 5001 (1857) is not of this species.

10. C. veitchii Rolfe

ILLUSTRATIONS: Botanical Magazine, 7764 (1901); Orchid Review **33**: 219 (1925).

PSEUDOBULBS: 7.5–10 × 2–4 cm, fusiform, becoming grooved, at first almost enclosed in sheathing scales, clustered, 2-leaved.

LEAF-BLADE: to 30 × 8 cm, broadly elliptic, acute, distinctly 5-veined beneath; stalk *c.*. 10 cm.

FLOWERING STEM: 15–60 cm, basal, pendent, sheathed at base, bearing 30 flowers or more.

BRACTS: persistent.

FLOWERS: *c.* 2 cm wide, white.

SEPALS: 12–14 × 5 mm, oblong-lanceolate, acute, incurved and concave.

PETALS: 10–13 × 3 mm, lanceolate, acute, incurved and concave.

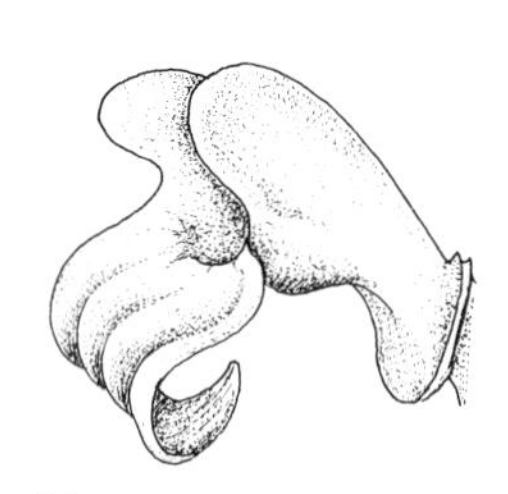

Coleogyne veitchii – lip.

LIP: 1.5 × 1.3 cm, shallowly 3-lobed, with 3 short basal keels.

COLUMN: 7 mm, widening apically with a winged slightly wavy margin.

DISTRIBUTION: New Guinea.

HARDINESS: G2.

FLOWERING: summer–autumn.

11. C. tomentosa Lindley

ILLUSTRATION: Dansk Botanisk Arkiv **29**:(4): 61 (1975).

PSEUDOBULBS: to 6 × 3 cm, ovoid, clustered, 2-leaved.

LEAF-BLADE: to 38 × 10 cm, elliptic; stalk 8 cm.

FLOWERING STEM: to 40 cm, basal, pendent, densely covered in short black hairs, base loosely sheathed, bearing 20 flowers or more.

BRACTS: persistent, hairy.

FLOWERS: 4–4.5 cm wide, pale orange or salmon-coloured.

SEPALS: 2.5–3 cm × 7–9 mm, lanceolate.

PETALS: slightly shorter and narrower.

LIP: 2.5 × 1.8 cm, 3-lobed, yellowish; lateral lobes brown with paler veins, keels 5, 3 long and 2 short, papillose at base, warty at apex.

COLUMN: *c.* 6 mm.

DISTRIBUTION: Malaysia, Thailand; ?Borneo.

HARDINESS: G2.

FLOWERING: spring.

12. C. **dayana** Reichenbach

ILLUSTRATIONS: Schlechter, Die Orchideen, edn 2, 135 (1927); Botanical Magazine n.s., 309 (1958); Graf, Tropica, 715 (1978); Bechtel et al., Manual of cultivated orchid species, 186 (1981).

PSEUDOBULBS: 10–22 × 2–5 cm, fusiform to cylindric, clustered, 2-leaved.

LEAF-BLADE: 25–75 × 5–10 cm, elliptic, acute, with 7–9 prominent veins beneath; stalk *c.* 9 cm.

FLOWERING STEM: to 100 cm, basal, pendent, with short scattered hairs, sheathed at base and above, bearing up to 50 flowers.

BRACTS: moderately persistent.

FLOWERS: *c.* 7 cm wide, pale yellowish, brownish yellow or whitish.

SEPALS: 3–3.5 cm × 5–7 mm, lanceolate, acute.

PETALS: a little smaller.

LIP: *c.* 3.5 × 2.5–3 cm, 3-lobed; lateral lobes brown with white veins; central lobe brown with a narrow white margin, keels 6, 2 long and 4 short.

DISTRIBUTION: Malay peninsula, Sumatra, Borneo; Java.

HARDINESS: G2.

FLOWERING: winter–spring.

13. C. **flaccida** Lindley

ILLUSTRATIONS: Botanical Magazine, 3318 (1834); Schlechter, Die Orchideen, edn 2, 136 (1927); Orchid Review **80**: 190 (1972); Bechtel et al., Manual of cultivated orchid species, 186 (1981).

PSEUDOBULBS: to 12 × 2.5 cm, cylindric to fusiform, grooved, sheathed in shining purple scales below, 2-leaved.

LEAF-BLADE: to 16 × 3.6 cm or more, lanceolate, narrowing gradually to base, apex acuminate.

FLOWERING STEM: 20–25 cm, basal, pendent, sheathed at base, bearing *c.* 8 flowers.

BRACTS: deciduous.

FLOWERS: to 4 cm wide, white.

SEPALS: 3–4 cm × 9 mm, lanceolate.

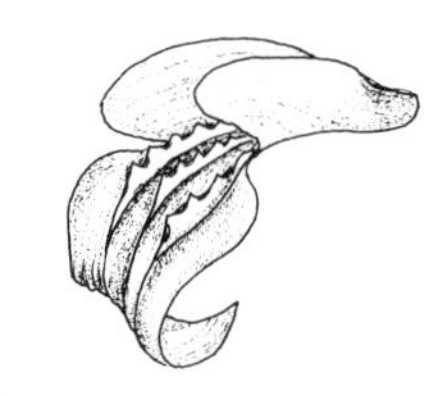
Coleogyne flaccida – lip.

PETALS: 2.7–3.5 cm × 3 mm, linear.

LIP: 1.7 × 1.2 cm, 3-lobed, white; base of central lobe yellow; lateral lobes veined with reddish brown; disc with 3 wavy keels.

DISTRIBUTION: Himalaya.

HARDINESS: G2.

FLOWERING: winter–spring.

14. C. massangeana Reichenbach

ILLUSTRATIONS: Botanical Magazine, 6979 (1888); Dansk Botanisk Arkiv **29**:(4): 62 (1975); Williams et al., Orchids for everyone, 75 (1980); Bechtel et al., Manual of cultivated orchid species, 187 (1981).

PSEUDOBULBS: 5–10 × 2.5 cm, fusiform to ovoid, clustered, 2-leaved.

LEAF-BLADE: to 50 × 12 cm, elliptic; stalk 5–10 cm.

FLOWERING STEM: to 60 cm, basal, pendent, with scattered short black hairs, sheathed at base, bearing up to 30 flowers.

BRACTS: persistent.

FLOWERS: to 7 cm wide, pale yellowish.

SEPALS: to 3.5 × 1 cm, lanceolate.

PETALS: slightly smaller.

Coleogyne massangeana – lip.

LIP: 2–2.8 × 1.9 cm, 3-lobed; lateral lobes brown with white or yellowish veins inside, whitish outside; central lobe brown and yellow; keels 5, 3 long and 2 short, warted except for the crested basal area.

DISTRIBUTION: Thailand, Malay peninsula, Java; ?Sumatra.

HARDINESS: G2.

FLOWERING: spring–summer.

15. C. swaniana Rolfe

ILLUSTRATIONS: Botanical Magazine, 7602 (1898); Orchid Review **83**: 150 (1975).

PSEUDOBULBS: to 10 × 3.5 cm, fusiform, 4- to 6-angled, clustered, 2-leaved.

LEAF-BLADE: to 25 × 6 cm, elliptic; stalk 5–7.5 cm.

FLOWERING STEM: to 40 cm, basal, pendent, with minute purple spots, loosely sheathed at base, bearing *c.* 20 flowers.

BRACTS: apparently deciduous.

FLOWERS: *c.* 5 cm wide.

SEPALS: 2.5 cm × 7–10 mm, lanceolate, white, or white tinged with pale brown.

PETALS: narrower, white.

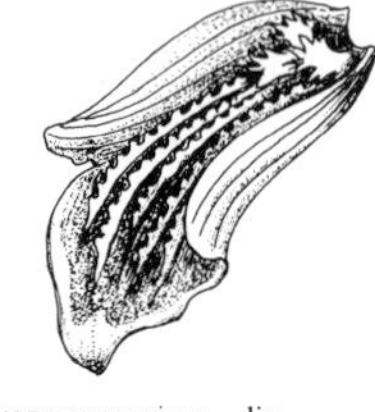
Coleogyne swaniana – lip.

LIP: 2.5 × 1.6 cm, 3-lobed, pale brown, darker at the margins and lobe-tips; lateral lobes white-veined; keels 5, 3 long and 2 short.

DISTRIBUTION: Malay peninsula, Sumatra, Borneo; ?Philippines.

HARDINESS: G2.
FLOWERING: spring–summer.

16. C. pandurata Lindley.

ILLUSTRATIONS: Botanical Magazine, 5084 (1858); Shuttleworth et al., Orchids, 41 (1970); Skelsey, Orchids, 103 (1979).
PSEUDOBULBS: 7–12 × 2–3 cm, cylindric to fusiform or ovoid, compressed, 2-leaved, distant.
LEAF-BLADE: 20–50 × 7 cm, lanceolate to elliptic, acute to acuminate; stalk *c.* 6 cm.
FLOWERING STEM: to 50 cm, basal, erect or arched, sheathed below, with up to 15 flowers.
BRACTS: persistent.
FLOWERS: *c.* 10 cm wide, fragrant, pale green.
SEPALS AND PETALS: 3.5–5.5 × 1.2–1.5 cm; sepals lanceolate to ovate, acute; petals lanceolate to slightly spathulate, acute.
LIP: 4.5 × 1.5–2.5 cm, 3-lobed with a long waist, mottled with purplish brown or black, 2-keeled at base; disc with warty ridges.
DISTRIBUTION: Malaysia, Sumatra & Borneo.
HARDINESS: G2.
FLOWERING: summer.

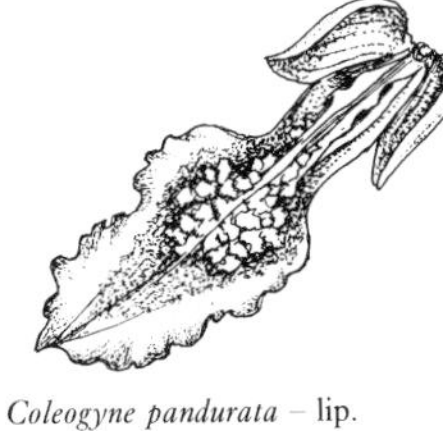

Coleogyne pandurata – lip.

17. C. parishii Hooker

ILLUSTRATIONS: Botanical Magazine, 5323 (1862); Graf, Tropica, 715 (1978); Bechtel et al., Manual of cultivated orchid species, 187 (1981).

Like *C. pandurata* except as below.

FLOWERING STEM: with 3–5 flowers.
FLOWERS: 6–8 cm wide.
LIP: to 2.5 cm.
DSITRIBUTION: Burma.
FLOWERING: spring–summer.

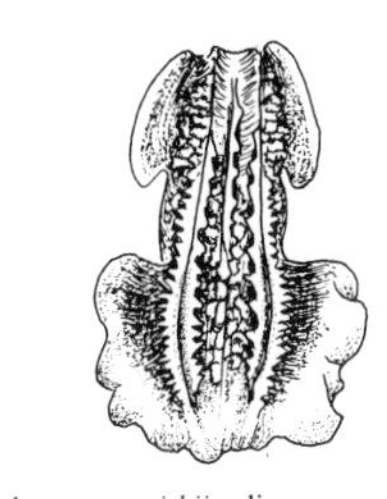

Coleogyne parishii – lip.

18. C. nitida (Wallich) Lindley

SYNONYMS: *C. ochracea* Lindley; not *C. nitida* misapplied, which is probably *C. punctulata* Lindley.
ILLUSTRATIONS: Botanical Magazine, 4661 (1852); Hay & Synge, Dictionary of garden plants in colour, 60 (1969); Bechtel et al., Manual of cultivated orchid species, 187 (1981).
PSEUDOBULBS: 2–8 × 1–1.5 cm, cylindric to fusiform or narrowly ovoid, clustered, 2-leaved.
LEAVES: 8–30 × 2–6 cm, lanceolate to elliptic, acute to acuminate; stalk short or absent.
FLOWERING STEM: to 20 cm, erect or arched, basal, with 2–8 flowers, sheathed below.

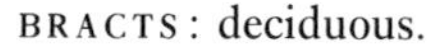

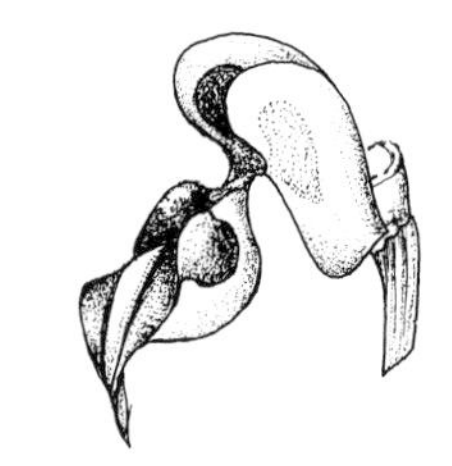

Coleogyne nitida – lip.

BRACTS: deciduous.

FLOWERS: 3–4 cm wide, fragrant.

SEPALS: 2–3 cm × 7–9 mm, white, lanceolate, acuminate, upper sepal almost spathulate.

PETALS: 2–2.5 cm × 5 mm, white, lanceolate, acute.

LIP: 2 × 1–1.5 cm, 3-lobed, ovate, white with large yellow or orange patches ringed with reddish orange; keels 3, 2 long and 1 short.

COLUMN: *c.* 1 cm.

DISTRIBUTION: W Himalaya to Burma, China, Thailand & Laos.

HARDINESS: G2.

FLOWERING: spring–summer.

19. C. corymbosa Lindley

ILLUSTRATIONS: Botanical Magazine, 6955 (1887); Northen, Miniature orchids, 57 (1980); Bechtel et al., Manual of cultivated orchid species, 185 (1981).

Like *C. nitida* except as below.

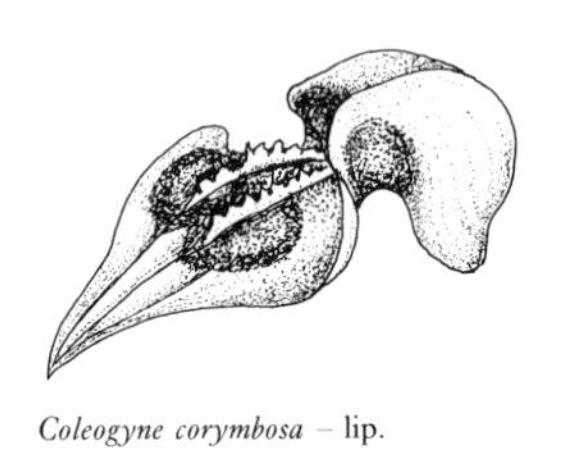

Coleogyne corymbosa – lip.

PSEUDOBULBS: 2.5–4 × 1.5–2.5 cm, ovoid to fusiform.

FLOWERS: to 7 cm wide.

LIP: 3 cm.

DISTRIBUTION: E Himalaya & NE India.

FLOWERING: summer.

20. C. corrugata Wight

SYNONYM: *C. nervosa* Richard.

ILLUSTRATIONS: Wight, Icones Plantarum Indiae Orientalis, t. 1639 (1851); Botanical Magazine, 5601 (1866).

Like *C. nitida* except as below.

PSEUDOBULBS: wrinkled.

SEPALS AND PETALS: more than 1 cm wide, elliptic.

LIP: with 3 equal wavy keels; central lobe longer than broad, mostly yellow with orange streaks.

DISTRIBUTION: S India.

FLOWERING: autumn.

21. C. cristata Lindley

ILLUSTRATIONS: Botanical Magazine, 8477 (1913); Williams et al., Orchids for everyone, 185 (1980); Bechtel et al., Manual of cultivated orchid species, 185 (1981).

PSEUDOBULBS: 3.5–7.5 × 2–4 cm, cylindric to fusiform or ovoid, irregularly ridged, distant, 2-leaved.

LEAVES: 12–30 × 2–3 cm, lanceolate, acute; stalk short or absent.

FLOWERING STEM: 15–30 cm, erect or arched, basal, with 3–10 flowers, sheathed at base.

BRACTS: persistent.

FLOWERS: 7–9 cm wide, white, very fragrant.

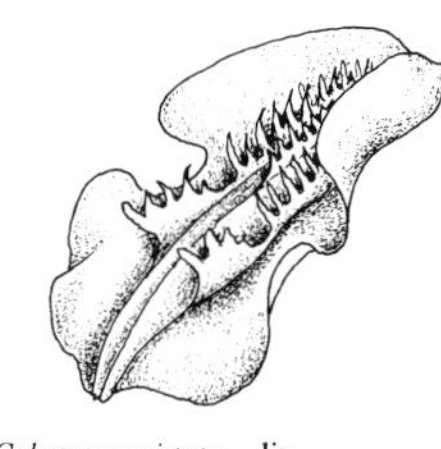

Coleogyne cristata – lip.

SEPALS AND PETALS: similar, 3.5–5.5 × 1.5–2 cm, oblong to lanceolate, acute or obtuse.

LIP: 3–4 × 3–3.5 cm, 3-lobed; keels 3, yellow and fringed, 2 long and 1 short.

COLUMN: 2.5 cm.

DISTRIBUTION: E Himalaya.

HARDINESS: G2.

FLOWERING: winter–spring.

22. C. asperata Lindley

ILLUSTRATIONS: Orchid Review **84**: 338 (1976); Millar, Orchids of Papua New Guinea, 74 (1978).

Like *C. pandurata* except as below.

FLOWERS: 6–7 cm wide, pale cream-coloured.

LIP: *c.* 3 cm, lined and speckled with brownish orange.

DISTRIBUTION: Malaysia, Sumatra to New Guinea.

FLOWERING: spring–summer.

The illustration under this name in Sheehan & Sheehan, Orchid genera illustrated, 67 (1979) is probably of *C. pandurata*.

23. C. viscosa Reichenbach

SYNONYM: *C. graminifolia* Parish & Reichenbach.

ILLUSTRATIONS: Botanical Magazine, 7006 (1885); Dansk Botanisk Arkiv **29**(4): 39 (1975).

PSEUDOBULBS: 2.5–8 cm, ovoid, ribbed when mature, clustered, 2-leaved.

LEAVES: 20–45 × 1–3 cm, linear.

FLOWERING STEM: 10–15 cm, basal, erect, sheathed at base, bearing 2–7 flowers.

BRACTS: deciduous.

FLOWERS: *c.* 5 cm wide, white, apparently fragrant.

SEPALS: 2–3.5 cm × 6–9 mm, lanceolate.

PETALS: narrower.

LIP: 2 × 1–1.5 cm, 3-lobed, white; lateral lobes brownish-veined; central lobe dark yellow; keels 3, finely wavy, the middle one less distinct, purplish apically.

DISTRIBUTION: SW China, N India, Burma, Vietnam & Malaysia.

HARDINESS: G2.

FLOWERING: winter.

24. C. trinervis Lindley

SYNONYM: *C. cinnamomea* Teijsmann & Binnendijk.

ILLUSTRATIONS: Smith, Orchideen von Java **2**: t. 109 (1905); Dansk Botanisk Arkiv **29**(4): 47 (1975).

PSEUDOBULBS: to 6 × 3 cm, ovoid to narrowly ovoid, *c.* 2 cm apart, 2-leaved.

LEAF-BLADE: to 40 × 4.5 cm, lanceolate, acute, distinctly 3-veined beneath, stalk *c.* 7 cm.

Coleogyne trinervis – lip.

FLOWERING STEM: to 18 cm, basal, erect, sheathed at base, bearing 6–8 flowers.

BRACTS: deciduous.

FLOWERS: *c.* 3 cm wide, cream or pale brown.

SEPALS: 2.2 cm × 6 mm, lanceolate.

PETALS: 22 cm × 3 mm, linear.

LIP: to 1.8 × 1.4 cm, 3-lobed; keels 7, 3 long, 4 short.

COLUMN: 1.4 cm.

DISTRIBUTION: Burma, Cambodia, Vietnam, Thailand, Malaysia & Java.

HARDINESS: G2.

FLOWERING: winter.

25. C. fuscescens Lindley

ILLUSTRATIONS: Annals of the Royal Botanical Garden Calcutta **8**: t. 181 (1898); Orchid Review **80**: 188 (1972); Dansk Botanisk Arkiv **29**(4): 27 (1975).

Like *C. trinervis* except as below.

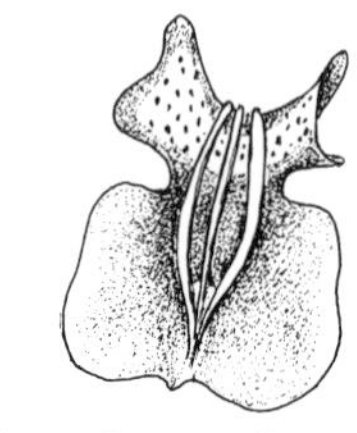

Coleogyne fuscescens – lip.

FLOWERS: to 5 cm wide, pale brown.

LIP: *c.* 4 × 1.5 cm; lateral lobes forward pointing, less than 5 mm long.

DISTRIBUTION: NE India.

Var. **brunnea** (Lindley) Lindley.

ILLUSTRATIONS: Botanical Magazine, 5494 (1865); Dansk Botanisk Arkiv **29**(4): 29 (1975).

Like *C. trinervis* except as below.

LIP: forwardly pointing lateral lobes more than 5 mm.

DISTRIBUTION: Burma, Thailand, Laos & Vietnam.

26. C. lactea Reichenbach

SYNONYM: *C. huettneriana* misapplied.

ILLUSTRATIONS: Cogniaux & Goossens, Dictionnaire Iconographique des Orchidées, Coelogyne, t. 5 (1902); Dansk Botanisk Arkiv **29**:(4): 42 (1975).

Like *C. trinervis* except as below.

LEAVES: to 30 × 3–5 cm.

FLOWERS: to 4 cm wide, white, sweetly scented.

LIP: 2 × 1.4–1.7 cm.

DISTRIBUTION: Burma, Thailand, Laos & Vietnam.

Further study may show this to be merely a variety of *C. flaccida* (No. 13).

43. PLEIONE D. Don

A genus of 16–20 Asian alpines, growing mostly between 1000 and 1500 m altitude in the wild; most of them were originally described in *Coelogyne*. They should be grown in epiphytic compost in well-drained, shallow pots or wide pans in cool conditions with temperatures never exceeding 20–25°C. In winter they shoud be protected from frost. They perform best if repotted annually; this should be done just after flowering has finished when new growth has just appeared. They should be kept in shady, humid conditions, though no water should be given until the new roots are actively growing. Once growth ceases and the leaves drop, temperatures should be lowered and water withheld.

The classification of the genus is in a state of flux, especially in connection with the species which form the *P. bulbocodioides* complex. Because the genus is popular, the species being marginally hardy in Europe, and as new introductions from China become available, it is likely that the details of specific recognition will change.

Literature. Hunt, P.F. & Vosa, C.G., The cytology and taxonomy of the genus Pleione, *Kew Bulletin* **25**: 432–32 (1971); Hunt, P.F. & Butterfield, I., The genus Pleione, *The Plantsman* 1(2): 112–23 (1979); Cribb, P. & Butterfield, I., *The genus Pleione* (1988); Bailes, C., Pleione: a neglected genus, *American Orchid Society Bulletin* **57**: 493–499 (1988); Stergianou, K.K., Habit differentiation and chromosome evolution in Pleione, *Plant Systematics and Evolution* **166**: 253–264 (1989); Stergianou, K.K. & Harberd, D.J., A cytotaxonomic study of some species of Pleione, *Botanical Journal of the Linnean Society* **101**: 213–228 (1989).

HABIT: epiphytic or terrestrial, often on rocks.

PSEUDOBULBS: clustered, spherical, ovoid or barrel-shaped, often sheathed, formed from bases of previous season's flowering stems.

LEAVES: membranous, pleated, rolled when young, deciduous, 1 or 2 per flowering stem and produced with or soon after the flowers, or, in autumn-flowering species, several months later on mature pseudobulbs (Nos. 1 and 2).

FLOWERING STEMS: usually 1-flowered (occasionally 2- or 3-flowered) sheathed below.

SEPALS AND PETALS: free, similar.

LIP: entire or very shallowly 3-lobed; lateral lobes or basal portion rolled or upturned around column, often concealing it; central lobe irregularly toothed, lacerate or fringed, with 2–9 longitudinal keels.

COLUMN: slender, hooded, often winged at the apex.

POLLINIA: 4.

Key to species

1. Pseudobulbs barrel-shaped, 2-leaved

1.A. Sheaths on flowering stems somewhat inflated, not warty; sepals and petals white **1. maculata**

1.B. Lip with purple blotches on the margin towards the apex; callus not extending to the apex of the lip; pseudobulbs green **2. × lagenaria**

1.C. Lip without purple blotches; callus extending to the lip apex or nearly so; pseudobulbs mottled with red-brown or purple **3. praecox**

2. Callus of lip consisting of keels barbed with fine, backwardly curving teeth

2.A. Leaves produced with the flowers; flowering stems more than 3 times as long as the pseudobulbs **4. hookeriana**

2.B. Leaves produced after the flowers; flowering stems to twice as long as the pseudobulbs **5. humilis**

3. Keels of the lip-callus with entire, straight margins

3.A. Flowers pale yellow to orange-yellow, rarely white, with red or brown spots on the lip; column 2.6–3.2 cm; flower-stalk 2–5 cm **6. forrestii**

3.B. Flowers pale lavender to rose-pink, rarely white, with red or purple spots on the lip; column 1.8–2.2 cm; flower-stalk 7–15 cm **8. yunnanensis**

4. Flowers pale yellow; lip spotted with red **7. × confusa**

5. Flowers white, lip with no markings or with very pale markings **10. formosana**

6. Petals at most 3.5 cm **11. limprichtii**

7. Keels of lip unbroken, lacerate towards the apex; lip of the same rose-pink to magenta as the sepals and petals **9. bulbocodioides**

8. Keels of lip broken, irregularly toothed or lacerate; lip whitish with pink margins, paler than sepals and petals **10. formosana**

1. **P. maculata** (Lindley) Lindley

ILLUSTRATIONS: Botanical Magazine, 4691 (1853) & n.s., 860 (1983); The Plantsman 1(2): 113 (1979); Bechtel et al., Manual of cultivated orchid species, 263 (1981); Cribb & Butterfield, The genus Pleione, pl. 1 & 15 (1988).

HABIT: on rocks.

PSEUDOBULBS: 3–4 cm, barrel-shaped, narrowing abruptly above, warty, 2-leaved.

LEAVES: 15–30 × 2.5–3 cm, lanceolate to elliptic, acute.

FLOWERING STEMS: 5–10 cm, sheaths not warty, with a leafy bract above.

FLOWERS: 5–6 cm wide.

SEPALS AND PETALS: 3–3.5 cm × 5–7 mm, lanceolate, acute, white or pale cream; petals sometimes streaked with pink.

LIP: 2–3.5 × 2.5–3.5 cm, shallowly 3-lobed; lateral lobes triangular, curled around column; central lobe semicircular to oblong, wavy, margin irregular, slightly notched at tip, white with purple blotches and a central yellow patch; keels 5–7, papillose, running almost to tip.

DISTRIBUTION: NE India, Bhutan, Burma & Thailand.
HARDINESS: H5–G1.
FLOWERING: winter.

2. P. × **lagenaria** Lindley

ILLUSTRATIONS: Botanical Magazine, 5370 (1863) & n.s., 860 (1983); Cribb & Butterfield, The genus Pleione, pl. 1 (1988).

Like *P. maculata* except as below.

LEAF SHEATHS: mottled with purple.
FLOWERING STEM: with warty sheaths.
SEPALS AND PETALS: pink to rose-purple.
LIP: heavily blotched.
DISTRIBUTION: NE India (Assam), W China (Yunnan).

Thought to be a naturally-occurring hybrid between *P. maculata* and *P. praecox*.

3. **P. praecox** (J.E. Smith) D. Don

SYNONYMS: *Coelogyne wallichiana* Lindley; *P. reichenbachiana* (Moore & Veitch) Williams.
ILLUSTRATIONS: The Plantsman 1(2): 121 (1979); Bechtel et al., Manual of cultivated orchid species, 263 (1981); Botanical Magazine, n.s., 861 (1983); Cribb & Butterfield, The genus Pleione, pl. 2 & 15 (1988); American Orchid Society Bulletin **57**: 494- 495 (1989).
HABIT: epiphytic or on rocks.
PSEUDOBULBS: 2–2.5 cm, barrel-shaped, narrowing abruptly above, purple with small greenish warts, 2-leaved.
LEAVES: to 20 × 3.5 cm, lanceolate to narrowly elliptic, acute.
FLOWERING STEMS: to 15 cm, basal sheaths warty.
FLOWERS: *c.* 8 cm wide.
SEPALS AND PETALS: 4–6 cm × 5–15 mm, lanceolate, acute, pink to bluish purple.
LIP: 5–6 × 3 cm, shallowly 3-lobed; lateral lobes oblong to triangular, rolled around column; central lobe oblong, irregularly toothed, strongly notched, pinkish white with large marginal purple blotches and a central patch of yellow; keels 5, papillose; lateral keels running about two-thirds of lip length.
DISTRIBUTION: Nepal, NE India, China & Burma.
HARDINESS: H5–G1.
FLOWERING: autumn.

4. **P. hookeriana** (Lindley) Williams

ILLUSTRATIONS: Botanical Magazine, 6388 (1878) & n.s., 862 (1983); The Plantsman 1(2): 114 (1979); Bechtel et al., Manual of cultivated orchid species, 263 (1981); Cribb & Butterfield, The genus Pleione, pl. 3 & 15 (1988).
HABIT: epiphytic or on rocks.
PSEUDOBULBS: 2–3.5 cm, conical to ovoid, not sheathed.

LEAVES: 1 per flowering stem, 3–15 × 3–4 cm, lanceolate to elliptic, acuminate.

FLOWERING STEMS: to 15 cm, bearing closely adpressed, smooth bracts below the leaf.

FLOWERS: 5–6.5 cm wide.

SEPALS AND PETALS: 2.5–3.5 cm, pink to pale purple, sepals broader than petals, lateral sepals sickle-shaped.

LIP: 2–2.5 × 3–4 cm, shallowly 3-lobed; lateral lobes bluntly triangular, not concealing column, white to pale pink; central lobe semicircular to oblong, notched, finely fringed, pink with brown blotches and a central yellow patch; keels 5–7, finely barbed.

COLUMN: winged.

DISTRIBUTION: NE India, Nepal, Bhutan, Thailand, Laos.

HARDINESS: H5–G1.

FLOWERING: spring.

5. **P. humilis** (J.E. Smith) D. Don

ILLUSTRATIONS: Botanical Magazine, 5674 (1867) & n.s., 863 (1983); The Plantsman 1(2): 117 (1979); American Orchid Society Bulletin **57**: 493 (1988); Cribb & Butterfield, The genus Pleione, pl. 4 & 15 (1988).

HABIT: epiphytic or on rocks.

PSEUDOBULBS: 2.5–5 cm, pear-shaped, sheathed.

LEAVES: 1 per flowering stem, 20–30 × 4–5 cm, produced after flowering, lanceolate.

FLOWERING STEM: *c.* 4 cm, sheathed at base.

FLOWERS: 7–9 cm wide.

SEPALS AND PETALS: 3.5–4.5 cm, lanceolate to narrowly oblong, pale pinkish white, sepals broader than petals.

LIP: 3–4 × 2.5–3 cm, oblong, more or less unlobed, base rolled around column, apical portion finely fringed, pale pink strongly marked with crimson or orange; keels 5–7, barbed.

DISTRIBUTION: NE India, Nepal & Burma.

HARDINESS: H5–G1.

FLOWERING: winter–spring.

6. **P. forrestii** Schlechter

ILLUSTRATIONS: Botanical Magazine, n.s., 864 (1983); Cribb & Butterfield, The genus Pleione, pl. 6 & 16 (1988).

HABIT: usually on rocks, more rarely epiphytic.

PSEUDOBULBS: conical, 1.5–3 cm × 6–15 mm, 1-leaved.

LEAVES: narrowly elliptic-lanceolate, to 15 x 1.5–4 cm.

FLOWERING STEM: 5–11 cm, appearing before the leaf.

FLOWERS: pale yellow to orange, rarely white, marked with brown or crimson spots on the lip.

SEPALS AND PETALS: 3–4.2 cm, petals somewhat sickle-shaped.

LIP: 3.2–3.7 cm, apical part lacerate; keels 5–7, entire.
DISTRIBUTION: NE Burma and SW China (Yunnan).
HARDINESS: G1–H5.
FLOWERING: spring.
Since the beginning of this century, genuine *P. forrestii* has been lost to cultivation, though it has more recently been re-introduced. In the interim, the name has been applied to hybrid plants (see *P.* × *confusa*, below).

7. P. × **confusa** Cribb & Tang
SYNONYM: *P. forrestii* misapplied.
ILLUSTRATIONS: Botanical Magazine n.s., 501 (1967); Orchid Review 83: 10 (1975); Cribb & Butterfield, The genus Pleione, pl. 7 & 16 (1988).
HABIT: on rocks.
PSEUDOBULBS: 2.5 cm, pear-shaped, sheathed.
LEAVES: 1 per flowering stem, to 15 × 4 cm, lanceolate to elliptic, produced after flowering.
FLOWERING STEM: 2–5 cm, sheathed at base.
FLOWERS: 5–6 cm wide.
SEPALS AND PETALS: *c.* 4 cm, lanceolate to elliptic, yellow or orange.
LIP: almost circular, obscurely 3-lobed, its base rolled around column; blade finely fringed, yellow with purple blotches; keels 5 or 6, irregularly and bluntly toothed.
DISTRIBUTION: Burma & China (Yunnan).
HARDINESS: H5–G1.
FLOWERING: spring.
Recent cytological and morphological studies have shown that this plant, formerly grown as *P. forrestii*, is a natural hybrid between *P. forrestii* Schlechter and *P. albiflora* Cribb & Tang; this last species is not in general cultivation.

8. **P. yunnanensis** (Rolfe) Rolfe
ILLUSTRATIONS: Botanical Magazine, 8106 (1906) & n.s., 865 (1983); Orchid Review 83: 9 (1975); Cribb & Butterfield, The genus Pleione, pl. 8 (1988).
HABIT: on rocks.
PSEUDOBULBS: 2–3 cm, spherical, flattened, shiny, sheathed.
LEAVES: 1 per flowering stem, 20–30 × 2–3 cm, lanceolate to elliptic, produced after flowering.
FLOWERING STEM: 10–20 cm, sheathed at base.
FLOWERS: 5–7 cm wide.
SEPALS AND PETALS: 3–4 cm, oblong to lanceolate, obtuse, pinkish purple; sepals broader than petals.
LIP: oblong to circular; base rolled around column; apical portion finely fringed, purplish pink with purple blotches; keels 5, finely toothed or papillose.

DISTRIBUTION: China (Yunnan).
HARDINESS: H5–G1.
FLOWERING: spring.

9. P. bulbocodioides (Franchet) Rolfe
SYNONYMS: *Coelogyne pogonioides* Rolfe; *P. delavayi* (Rolfe) Rolfe; *P. henryi* (Rolfe) Schlechter.
ILLUSTRATIONS: Botanical Magazine, n.s., 866 (1983); Cribb & Butterfield, The genus Pleione, pl. 9, 16 (1988).
HABIT: epiphytic or on rocks.
PSEUDOBULBS: 2–2.6 cm, conical, ovoid or pear-shaped, pale or dark green to purple, sheathed.
LEAVES: 1 per flowering stem, 14 × 2.5 cm or more, lanceolate to elliptic, acute, reaching full size after flowering.
FLOWERING STEMS: 10–20 cm, sheathed, with 1 flower.
FLOWERS: 5–12 cm wide.
SEPALS AND PETALS: 3.3–4.6 cm × 4–10 mm, lanceolate to asymmetrically elliptic, acute, pink to pale purple.
LIP: 3.2–4.5 × 2.5–3.5 cm, shallowly 3-lobed to bluntly diamond-shaped or almost circular, base rolled around column, apical portion finely toothed to lacerate, shallowly notched, white to pink, heavily spotted with pale brown or purplish pink or both; keels 2–7, wavy or toothed.
DISTRIBUTION: W China, Burma & Taiwan.
HARDINESS: H5–G1.
FLOWERING: spring.
Variable. The two following species are often included within *P. bulbocodioides*, and are not always easy to distinguish; for further details, see the monograph by Cribb and Butterfield cited above.

10. P. formosana Hayata
SYNONYM: *P. pricei* Rolfe.
ILLUSTRATIONS: Botanical Magazine, 8729 (1917) & n.s., 421 (1962); Cribb & Butterfield, The genus Pleione, 69 (1988).
Like *P. bulbocodioides* except as below.
LIP: whitish with pink margins, paler than sepals and petals; keels broken, irregularly toothed or lacerate.
DISTRIBUTION: E China.
HARDINESS: G1–H5.
FLOWERING: spring.

11. P. limprichtii Schlechter
ILLUSTRATIONS: Botanical Magazine, n.s., 397 (1962), 868 (1983); Cribb & Butterfield, The genus Pleione, pl. 11 (1988).
Like *P. bulbocodioides* except as below.
PETALS: at most 3.5 cm, pink to pink-magenta.
LIP: spotted with rose red.
DISTRIBUTION: SW China (Sichuan).

HARDINESS: G1–H5.
FLOWERING: spring.

44. DENDROCHILUM Blume

A genus of 120–150 species from SE Asia. Most of the commonly cultivated species are in section *Platyclinis* which has sometimes been recognised as a distinct genus. Cultivation conditions are similar to those for the tropical species of *Coelogyne* (p. 124).

HABIT: epiphytic, rarely on rocks.
PSEUDOBULBS: fusiform, conical or ovoid, clustered or distant, 1- or 2-leaved.
LEAVES: flat, narrow, leathery, stalked, tapered at both ends, rolled in bud.
FLOWERING STEMS: lateral, long and slender, arched, bearing many flowers in dense, 2-ranked racemes or spikes.
FLOWERS: small, fragrant.
SEPALS AND PETALS: spreading, similar; upper sepal often keeled; lateral sepals slightly fused to column foot.
LIP: simple or 3-lobed, fleshy below, loosely hinged to column foot, often with 2 or 3 keels.
COLUMN: short, with narrow side lobes and apical wings.
POLLINIA: 4, pear-shaped.

Key to species

1. Flowers less than 1 cm wide **1. filiforme**
2. Lip about equal to sepals and petals, lateral lobes very small or absent, central lobe wedge- or fan-shaped **4. cobbianum**
3. Flowers off-white to pale yellow; lip pale green, yellow or white, lateral lobes broadly triangular, rounded or obtuse **2. glumaceum**
4. Flowers greenish yellow; lip striped with brown, lateral lobes linear to narrowly triangular, acute **3. latifolium**

1. D. filiforme Lindley

SYNONYM: *Platyclinis filiformis* (Lindley) Bentham.
ILLUSTRATIONS: Orchid Review 77: 154 (1969); Skelsey, Orchids, 109 (1979); Sander, Orchids and their cultivation, t. 39 (1979).
HABIT: epiphytic.
PSEUDOBULBS: 2.5 × 1 cm, ovoid, densely clustered, 1- or 2-leaved.
LEAVES: 12–20 × 1–2 cm, linear-lanceolate, acute.
FLOWERING STEM: 20–45 cm, thread-like, arched, bearing up to 100 flowers or more.
FLOWERS: *c.* 6 mm wide in 2 ranks, fragrant, pale yellow or greenish yellow.

SEPALS AND PETALS: 2–3 × 1–1.5 mm, lanceolate to ovate or elliptic.

LIP: 2 × 2 mm, heart-shaped, shallowly notched, 2-ridged.

COLUMN: *c.* 1 mm.

DISTRIBUTION: Philippines.

HARDINESS: G2.

FLOWERING: summer.

2. D. glumaceum Lindley

SYNONYM: *Platyclinis glumacea* (Lindley) Bentham.

ILLUSTRATIONS: Botanical Magazine, 4853 (1855); Graf, Tropica, 724 (1978); Bechtel et al., Manual of cultivated orchid species, 203 (1981).

HABIT: epiphytic.

PSEUDOBULBS: 2–5 × 1–2 cm, ovoid, densely clustered, 1-leaved, fusiform, covered in red sheaths when young.

LEAVES: 25–30 × 2–4 cm, narrowly lanceolate, acute, with a short stalk.

FLOWERING STEM: to 40 cm, thread-like, arched, bearing many flowers.

BRACTS: spreading, papery, acute.

FLOWERS: 1.2–1.8 cm wide, 2-ranked, fragrant, off-white to pale yellow.

SEPALS: 7–10 × 1–2 mm, linear-lanceolate, acute.

PETALS: 4–7 × 1 mm, linear-lanceolate.

LIP: 3–4 × 2 mm, 3-lobed, pale green, yellow or white, with 2 thick basal ridges; lateral lobes broadly triangular, rounded or obtuse, central lobe circular to broadly ovate, curved downwards.

COLUMN: erect.

DISTRIBUTION: Philippines.

HARDINESS: G2.

FLOWERING: spring.

3. D. latifolium Lindley

SYNONYM: *Platyclinis latifolia* (Lindley) Hemsley.

ILLUSTRATION: Orchid Review 77: 155 (1969).

Like *D. glumaceum* except as below.

FLOWERS: greenish yellow.

LIP: striped with brown; lateral lobes linear to narrowly triangular, acute, deeply and finely toothed; central lobe obovate, obtuse.

DISTRIBUTION: Philippines.

4. D. cobbianum Reichenbach

SYNONYM: *Platyclinis cobbiana* (Reichenbach) Hemsley.

ILLUSTRATIONS: Sheehan & Sheehan, Orchid genera illustrated, 83 (1979); Williams et al., Orchids for everyone, 187

(1980); Bechtel et al., Manual of cultivated orchid species, 203 (1981).

HABIT: epiphytic.

PSEUDOBULBS: 2.5–8 × 1–2 cm, narrowly conical, tapering above, 1-leaved.

LEAVES: 6–35 × 2.5–6 cm, lanceolate to oblong, acuminate, softly leathery.

FLOWERING STEM: 35–50 cm, slender, arched.

BRACTS: 4–7 mm, papery.

FLOWERS: 1.2–1.8 cm wide, in 2 twisted ranks, slightly spaced out, fragrant.

SEPALS AND PETALS: 7–10 × 2–3 mm, oblong to elliptic, obtuse, white.

LIP: 6–8 × 3 mm, wedge- or fan-shaped, yellow to orange; callus oblong.

DISTRIBUTION: Philippines.

HARDINESS: G2.

FLOWERING: autumn.

45. **PHOLIDOTA** Hooker

About 40 species from India, China and SE Asia. Cultivation conditions are generally similar to those for the tropical species of *Coelogyne* (p. 124) though less water should be given from when the pseudobulbs are fully formed until new growth appears.

HABIT: epiphytic with creeping rhizomes.

PSEUDOBULBS: clustered, distant, or in branching chains, 1- or 2-leaved.

LEAVES: rolled when young.

FLOWERING STEMS: terminal, slender, erect, arched or pendent, several- to many-flowered; axis often zig-zag.

BRACTS: conspicuous, sometimes falling early.

FLOWERS: small, alternate in 2 ranks.

SEPALS AND PETALS: spreading or pointing forwards; sepals concave, lateral sepals often keeled; petals similar to sepals or narrower.

LIP: fused to column base, pouched below, with a small deflexed blade, often 2- or 3-lobed.

COLUMN: short, winged around anther.

POLLINIA: 4.

Key to species

1. Pseudobulbs borne one on top of another, forming long, pendent chains — **1. articulata**
2. Flowering stem erect — **3. ventricosa**
3. Pseudobulbs 1-leaved — **2. pallida**
4. Pseudobulbs 2-leaves — **4. chinensis**

1. P. articulata Lindley.

ILLUSTRATIONS: Smith, Orchideen von Java **2**: t. 115 (1909); Orchid Review **85**: 300 (1977); Banerjee & Thapa, Orchids of Nepal, 122 (1978).

PSEUDOBULBS: to 10 × 1.5 cm, cylindric, 2-leaved, developed on top of older pseudobulbs, forming long pendent, sometimes branched chains.

LEAVES: 8–16 × 3–5 cm, elliptic, acute; stalk *c.* 1 cm.

FLOWERING STEM: arched, *c.* 15 cm.

BRACTS: *c.* 1 cm, falling as flowers open.

FLOWERS: 10–20, *c.* 1.2 cm wide, fragrant, pink to dull brown.

SEPALS AND PETALS: similar, 7–8 mm, spreading, ovate to elliptic, acute or obtuse.

LIP: *c.* 5 mm; pouch with 5 yellow ridges, tapering to junction with blade; blade 3 × 5 mm, kidney-shaped, 2-lobed, orange at base.

COLUMN: 3–4 mm.

DISTRIBUTION: Himalaya through SE Asia to Java.

HARDINESS: G2.

FLOWERING: spring–summer.

2. P. pallida Lindley

SYNONYM: *P. imbricata* Lindley.

ILLUSTRATIONS: Shuttleworth et al., Orchids, 42 (1970); Sheehan & Sheehan, Orchid genera illustrated, 141 (1979); Bechtel et al., Manual of cultivated orchid species, 261 (1981).

PSEUDOBULBS: 3–6 cm conical, 1-leaved.

LEAVES: 30 × 6 cm, lanceolate to elliptic, acute; stalk to 5 cm.

FLOWERING STEM: to 30 cm, pendent, many-flowered.

BRACTS: 8–10 mm, persistent.

FLOWERS: 6–7 mm wide, off-white to pale pink.

SEPALS: upper 5 × 3 mm, curved over column; laterals to 6 mm, fused below, keeled.

PETALS: 4 × 1.5 mm, linear to oblong, curved over column.

COLUMN: almost circular, 3–4 mm.

DISTRIBUTION: Burma & S China to Australia.

HARDINESS: G2.

FLOWERING: spring–summer.

3. P. ventricosa (Blume) Reichenbach

ILLUSTRATION: Smith, Orchideen von Java **2**: t. 114 (1909).

Like *P. pallida* except as below.

PSEUDOBULBS: 2-leaved.

LEAF-STALKS: to 15 cm.

FLOWERING STEM: erect.

FLOWERS: 1–1.2 cm wide, greenish white.

DISTRIBUTION: Sumatra, Java & Borneo.

4. P. chinensis Lindley

ILLUSTRATIONS: Sander, Orchids and their cultivation, t. 35 (1979); Williams et al., Orchids for everyone, 192 (1980).

Like *P. pallida* except as below.

PSEUDOBULBS: spherical to oblong, 2-leaved.

FLOWERING STEM: pendent.

FLOWERS: 1.2–2 cm.

SEPALS AND PETALS: pinkish brown, petals linear.

LIP: white.

DISTRIBUTION: China & Hong Kong.

FLOWERING: spring–summer.

46. HEXISEA Lindley

A genus of 5 species from tropical America, similar to *Scaphyglottis* but lacking a movable lip. Cultural requirements are similar to those for *Cattleya* (p. 167), though the plants can also be grown on bark slabs, and generally require shadier conditions throughout the year. Propagation is best effected by separating the stems when repotting.

Literature: Dressler, R.L., The genus Hexisea, *Orquidea (Mexico)* **4**: 197–200 (1974).

HABIT: epiphytic.

STEMS: sheathed, sometimes branched, composed of chains of successive pseudobulbs.

LEAVES: linear to strap-shaped, folded when young, in opposite pairs at the apex and upper nodes of the stems.

FLOWERS: few, in short terminal racemes on short scale-bearing stalks.

SEPALS AND PETALS: similar, spreading; lateral sepals fused to column foot, forming a small mentum.

LIP: entire or shallowly lobed, divided into a narrow basal claw and a sharply deflexed blade; claw pressed against or fused to column forming a nectary.

COLUMN: club-shaped, with a broad hollow foot, trifid.

POLLINIA: 4 (in ours).

1. H. bidentata Lindley

ILLUSTRATIONS: Botanical Magazine, 7031 (1888); Sheehan & Sheehan, Orchid genera illustrated, 105 (1979); Bechtel et al., Manual of cultivated orchid species, 216 (1981).

STEMS: to 50 cm.

PSEUDOBULBS: to 10 × 1 cm, brownish purple, grooved.

LEAVES: to 15 × 1.5 cm, strap-shaped, dark green, bifid or trifid at apex.

FLOWERS: 2.5–3 cm wide, bright reddish orange.

SEPALS: 1.5–2 cm × 5 mm, lanceolate, acute or obtuse.

PETALS: similar but smaller.

LIP: to 1.5 cm × 5 mm, oblong, obtuse with a narrow claw fused to the column.
COLUMN: 3–4 mm.
DISTRIBUTION: C America to Venezuela, Colombia & Peru.
HARDINESS: G2.
FLOWERING: intermittent.

47. EPIDENDRUM Linnaeus

A genus of about 750 species from the New World. Its relationships with other genera are controversial: here, *Epidendrum* is restricted to plants with the column united for all or most of its length to the lip. Species with the lip and column mostly free from each other are included in *Encyclia* (p. 154) and *Barkeria* (p. 165).

Plants should be potted in early spring in an epiphytic compost, using the smallest size of pot possible. Propagation is by division or by means of the young plantlets produced by some species after the flowering spikes have been cut back.

Literature: Ames, O., Hubbard, F.T. & Schweinfurth, C., *The genus Epidendrum in the United States and Middle America* (1936).

HABIT: terrestrial or epiphytic.
STEMS: usually elongate and leafy, more rarely pseudobulb-like with 1–2 leaves, erect or hanging.
LEAVES: acute or notched at the apex, folded when young.
FLOWERS: in racemes or panicles which are sometimes umbel-like.
BRACTS: small or conspicuous.
FLOWERS: resupinate or not, variable in size.
SEPALS AND PETALS: similar, spreading.
LIP: joined to the column for most of the length of the latter, variable in shape, entire or 3-lobed.
POLLINIA: 4, laterally flattened, parallel.

Synopsis of characters

Main axis. Stem-like: **4–8,12–16**; pseudobulb-like: **1–3,9–11**.
Inflorescence. Panicle with many flowers: **1–3**; spherical raceme with many flowers:**4–7**; elongate raceme with up to 5 flowers: **8–16**.
Sepals and petals. To 3 cm: **1–7**; more than 3 cm: **8–16**.
Lip. Not lobed: **12–16**; lobes more or less equal, the lip forming a cross-shaped structure at the top of the column: **4–7**; lobes unequal, not as above: **1–3,8–11**.

Key to species

1. Lip not lobed, though sometimes fringed or notched *Go to Subkey 1*

2. Central lobe of lip narrowly oblanceolate or linear, projecting between and beyond the lateral lobes *Go to Subkey* 2
3. Inflorescence an almost spherical raceme borne on a long stalk covered with bladeless sheaths, which terminates the leafy shoot; lobes of the lip approximately equal in size, forming a cross at the apex of the column, often all the lobes fringed *Go to Subkey* 3
4. Inflorescence a panicle, rarely an elongate raceme, borne directly on the leafy shoot or on an entirely leafless shoot; lobes of the lip various, usually unequal, rarely fringed, not forming a cross at the top of the column *Go to Subkey* 4

Subkey 1. *Lip not lobed, though sometimes fringed or notched.*
1. Sepals and petals at least 3 cm **12. medusae**
2. Stems hanging, not erect and cane-like **16. loefgrenii**
3. Lip deeply notched **15. latilabrum**
4. Sepals and petals *c.* 8 mm **13. diffusum**
5. Sepals and petals *c.* 2.5 cm **14. eximium**

Subkey 2. *Lip lobed, central lobe narrowly oblanceolate or linear, projecting between and beyond the lateral lobes.*
1. Stem elongate, bearing several leaves along its length **8. nocturnum**
2. Lateral lobes of lip deeply fringed along their outer margins **9. ciliare**
3. Leaves linear-lanceolate to lanceolate **10. parkinsonianum**
4. Leaves elliptic-oblong **11. oerstedtii**

Subkey 3. *Lip lobed, lobes almost equal, forming a cross at the apex of the column, often all fringed; inflorescence an almost spherical raceme borne on a long stalk covered with bladeless sheaths which terminates the leafy shoot.*
1. Stem arching or scrambling, rooting from many of the nodes **4. ibaguense**
2. Flower yellow except for the bright red apex of the column **6. xanthinum**
3. Raceme compact; at least the central lobe of the lip fringed **5. arachnoglossum**
4. Raceme elongate; lobes of lip entire **7. cnemidophorum**

Subkey 4. *Lip lobed, lobes various, usually unequal, rarely fringed, not forming a cross at the apex of the column; raceme borne directly on a leafy shoot or on an entirely leafless shoot.*
1. Stem to 25 cm, increasing in thickness upwards, then tapering, not tall and cane-like **1. stamfordianum**
2. Leaf-sheaths warty; sepals, petals and lip lilac-purple **3. myrianthum**
3. Leaf sheaths not warty; sepals and petals greenish brown, lip white **2. paniculatum**

1. E. stamfordianum Bateman

ILLUSTRATIONS: Fieldiana **26**: 381 (1953); Bechtel et al., Manual of cultivated orchid species, 210 (1981).

STEMS: to 25 cm, erect, increasing in thickness upwards, then tapering, bearing 4–6 leaves near the apex.

LEAVES: 13–24 × 3–7 cm, linear-oblong to linear-elliptic, slightly notched at the apex.

INFLORESCENCE: a panicle (rarely a raceme) arising on a leafless stalk from the base of the leafy stem or, more rarely, terminating a leafy shoot.

FLOWERS: greenish yellow spotted with red.

SEPALS AND PETALS: similar, 1.4–2.5 cm.

LIP: deeply 3-lobed, the lateral lobes oblong, recurved, the central deeply notched with divergent lobes, margins fringed.

DISTRIBUTION: Guatemala.

HARDINESS: G2.

FLOWERING: spring.

2. E. paniculatum Ruiz & Pavon

SYNONYM: *E. floribundum* Humboldt, Bonpland & Kunth.

ILLUSTRATION: Botanical Magazine, 3637 (1838).

Like *E. stamfordianum* except as below.

STEMS: tall, cane-like.

PANICLE: terminating the stem.

SEPALS: greenish brown, reflexed.

PETALS: very narrow, greenish brown, reflexed.

LIP: white with very divergently lobed central lobe.

DISTRIBUTION: C America, northern S America, West Indies.

3. E. myrianthum Lindley

SYNONYM: *E. verrucosum* Swartz var. *myrianthum* (Lindley) Ames & Correll.

ILLUSTRATION: Bateman, Second century of orchidaceous plants, t. 163 (1867).

Like *E. paniculatum* except as below.

LEAF-SHEATHS: warty.

SEPALS, PETALS AND LIP: lilac-purple.

DISTRIBUTION: Guatemala.

4. E. ibaguense Humboldt, Bonpland & Kunth

SYNONYM: *E. radicans* Ruiz & Pavon.

ILLUSTRATIONS: Hamer, Orquideas de El Salvador **1**: 211 (1974); Bechtel et al., Manual of cultivated orchid species, 209 (1981).

STEMS: to 1 m, scrambling or arching, producing roots from many of the nodes.

LEAVES: ovate-oblong to oblong, notched at the apex.

RACEME: more or less spherical, borne on a long stalk covered with bladeless sheaths, which terminates the leafy shoot.

FLOWERS: variable in colour, orange, scarlet red, purplish red or yellow.

SEPALS AND PETALS: erect, 1–2.2 cm.

LIP: forming a cross at the top of the column, 3-lobed, the lobes all similar or the central larger, deeply fringed, the central 2-lobed.

DISTRIBUTION: C America, northern S America, West Indies.

HARDINESS: G2.

FLOWERING: most of the year.

E. × o'brienianum Rolfe (*E. ibaguense* × *E. evectum* Hooker) is similar but has a longer inflorescence, the sepals and petals orange and the column tipped with yellow.

5. E. arachnoglossum Reichenbach

ILLUSTRATION: Revue Horticole 1882: 554.

Like *E. ibaguense* except as below.

STEMS: erect, not rooting.

FLOWERS: purplish pink throughout.

DISTRIBUTION: Colombia.

HARDINESS: G2.

FLOWERING: summer.

6. E. xanthinum Lindley

ILLUSTRATION: Botanical Magazine, 7586 (1898). Like *E. arachnoglossum* except as below.

FLOWERS: yellow, the apex of the column bright red.

DISTRIBUTION: Brazil.

7. E. cnemidophorum Lindley

ILLUSTRATION: Botanical Magazine, 5656 (1867).

Like *E. arachnoglossum* except as below.

INFLORESCENCE: more elongate.

LIP: more obscurely 3-lobed, with entire lobes.

DISTRIBUTION: C America.

8. E. nocturnum Jacquin

ILLUSTRATIONS: Fieldiana **26**: 347 (1953); Bechtel et al., Manual of cultivated orchid species, 210 (1981); American Orchid Society Bulletin **55**: 237 (1986).

STEMS: to 1 m, erect, leafy.

LEAVES: oblong-lanceolate to elliptic, unequally notched at the apex.

RACEME: terminal with up to 5 flowers, borne directly on the leafy stems, with conspicuous, boat-shaped bracts.

SEPALS: greenish white, 3.5–9 cm.

PETALS: white, 3.5–9 cm, spreading.

LIP: white, 3-lobed, the lateral lobes reversed-triangular with finely toothed outer margins, the central lobe 2–2.5 cm, very narrowly triangular, projecting between and beyond the lateral lobes.

DISTRIBUTION: USA (Florida), C America, northern S America, West Indies.

HARDINESS: G2.

FLOWERING: summer–autumn.

9. E. ciliare Linnaeus

ILLUSTRATIONS: Botanical Magazine, 463 (1799); Cogniaux & Goossens, Dictionnaire Iconographique des Orchidées, Epidendrum, t. 6 (1898); Fieldiana **26**: 319 (1953); Bechtel et al., Manual of cultivated orchid species, 209 (1981).

STEMS: short, pseudobulb-like, bearing 1–2 leaves at the apex.

LEAVES: oblong-elliptic, somewhat unequally notched at the apex.

RACEME: terminal, with few flowers and conspicuous, boat-shaped bracts.

SEPALS AND PETALS: 4–9 cm, erect, pale green or yellowish.

LIP: white, 3-lobed, the lateral lobes narrowly reversed-triangular with deeply fringed outer margins, the central lobe 2.5–6 cm, linear, projecting between and beyond the lateral lobes.

DISTRIBUTION: C America, northern S America, West Indies.

HARDINESS: G2.

FLOWERING: winter.

10. E. parkinsonianum Hooker

SYNONYM: *E. falcatum* Lindley.

ILLUSTRATION: Botanical Magazine, 3778 (1840).

STEMS: short, pseudobulb-like, usually hanging, bearing 1–2 or rarely more leaves.

LEAVES: linear-lanceolate to lanceolate, acute.

RACEME: terminal with flowers and conspicuous, boat-shaped bracts.

SEPALS AND PETALS: 5.5–8.5 cm, spreading, white, yellowish or yellowish green, becoming pale orange if not fertilised.

LIP: white or yellowish, deeply 3-lobed, the lateral lobes reversed-triangular with finely toothed outer margins, the central lobe 3.5–5.5 cm, projecting between and beyond the lateral lobes.

DISTRIBUTION: C America.

HARDINESS: G2.

FLOWERING: autumn.

11. E. oerstedtii Reichenbach

SYNONYM: *E. costaricense* Reichenbach.

ILLUSTRATIONS: Cogniaux & Goossens, Dictionnaire Iconographique des Orchidées, Epidendrum, t. 3 (1897); Botanical Magazine, n.s,, 375, (1962).

Like *E. parkinsonianum* except as below.

LEAVES: elliptic-oblong.

LIP: central lobe narrowly oblanceolate.
DISTRIBUTION: Costa Rica.

12. E. medusae (Reichenbach) Veitch
SYNONYM: *Nanodes medusae* Reichenbach.
ILLUSTRATIONS: Botanical Magazine, 5723 (1868); Bechtel et al., Manual of cultivated orchid species, 210 (1981).
STEMS: to 20 cm, usually leafy, hanging.
LEAVES: greyish green, lanceolate, tapering to an acute apex.
RACEME: with 1–3 large flowers.
SEPALS AND PETALS: 3–4 cm, spreading, yellowish green flushed with red.
LIP: broadly kidney-shaped, broader than long (to 6 cm broad), purplish brown, deeply fringed all round.
DISTRIBUTION: Ecuador.
HARDINESS: G2.
FLOWERING: summer.
There are several other rarely cultivated species, which, though generally unlike each other and *E. medusae*, are described here because of their simple lips.

13. E. diffusum Swartz
ILLUSTRATION: Botanical Magazine, 3565 (1837).
STEMS: erect.
LEAVES: ovate.
PANICLE: wide, of small flowers.
SEPALS AND PETALS: *c.* 8 mm, entirely yellowish green.
DISTRIBUTION: C America.

14. E. eximium Williams
ILLUSTRATION: Orquidea **2**: 245 (1972).
STEMS: erect, cane-like.
RACEME: of several flowers.
SEPALS AND PETALS: yellow, *c.* 2.5 cm.
LIP: oblong, entire, yellow, red-lined.
DISTRIBUTION: Mexico, El Salvador.

15. E. latilabrum Lindley
ILLUSTRATION: Martius, Flora Brasiliensis 3(5): t. 31 (1898).
STEMS: erect, zig-zag, cane-like.
RACEME: few-flowered, umbel-like.
SEPALS AND PETALS: 1.8–3.3 cm.
LIP: entire, broader than long, deeply notched.
DISTRIBUTION: Brazil.

16. E. loefgrenii Cogniaux
ILLUSTRATION: Martius, Flora Brasiliensis 3(5): t. 49 (1898).
STEMS: hanging, with broad, overlapping leaves.

RACEME: with few, very small (*c.* 1 cm in diameter) greenish flowers.
LIP: broadly obovate, unfringed, notched.
DISTRIBUTION: Brazil.

48. ENCYCLIA Hooker

A genus of about 150 species from C & S America, most of which were originally described in *Epidendrum*. Cultivation as for *Epidendrum* (p. 148).

Literature: Dressler, R.L. & Pollard, G.E., *The genus Encyclia in Mexico* (1974).

Key to species

1. Lip uppermost, flowers not resupinate *Go to Subkey 1*
2. Lip-blade or central lobe more or less isodiametric, not tapering to a point *Go to Subkey 2*
3. Lip-blade or central lobe usually longer than broad, tapering to a point *Go to Subkey 3*

Subkey 1. Lip uppermost, flowers not resupinate.

1. Pseudobulbs each with a single leaf **3. fragrans**
2. Sepals firm and erect, at most 3 times as long as broad
2.A. Lip 1–1.3 cm, cream or greenish white with radiating purple lines **5. radiata**
2.B. Lip pale yellowish green, smaller **6. vespa**
3. Petals at most 6 times as long as broad; flowers 2–3 per raceme; lip cream with radiating purple lines **4. baculus**
4. Sepals and petals pale green **1. cochleata**
5. Sepals and petals white streaked with pink **2. glumacea**

Subkey 2. Lip lowermost, flowers resupinate; lip blade or central lobe more or less idsodiametric, not tapering to a point.

1. Lip (including claw) 3 cm or more
1.A. Flowers erect or spreading, 3–15 per raceme; lip 4 cm or less, deeply 3-lobed, purple **19. cordigera**
1.B. Flowers green and white; lip notched **20. mariae**
1.C. Flowers yellow; lip truncate **21. citrina**
2. Lip without lateral lobes on blade or claw **25. polybulbon**
3. Petals hooded at their tips
3.A. Sepals and petals reddish purple, lip pink **23. hanburii**
3.B. Lip 1.7–2.2 cm, cream, pink or lavender **22. selligera**
3.C. Lip more than 3 cm, white with pink streaks **24. advena**
4. Lateral lobes of lip 12 × 6 mm or more, strongly constricted at base **13. alata**
5. Lip papillose or warty, notched **18. varicosa**
6. Lateral lobes of lip arising from the base of the blade rather than from the claw **17. michuacana**
7. Leaves at most to 25 cm **16. bractescens**
8. Lip 1.5–2 cm, white **14. tampensis**

9. Lip 1–1.5 cm, cream or yellow, with red or brown veins **15. aromatica**

Subkey 3. Lip lowermost, flowers resupinate; lip blade or central lobe longer than broad, tapering to a point

1. Lip with lateral lobes on the claw

1.A. Lip (including claw) 3 cm or more; sepals and petals linear, 3 cm or more, yellowish green **9. brassavolae**

1.B. Sepals and petals 8–14 mm, yellowish green heavily spotted with brown **8. boothiana**

1.C. Sepals and petals more than 1.4 cm, orange to bright scarlet **7. vitellina**

2. Flower-stalk and ovary covered in warts **12. adenocaula**

3. Petals unspotted, greenish brown; lip 1–1.5 cm **11. patens**

4. Petals with purple spots, the background pale greenish yellow; lip 2–2.5 cm **10. prismatocarpa**

1. E. **cochleata** (Linnaeus) Lemée

SYNONYM: *Hormidium cochleatum* (Linnaeus) Brieger.

ILLUSTRATIONS: Botanical Magazine, 572 (1801); Luer, Native orchids of Florida, t. 60 (1972); Dressler & Pollard, The genus Encyclia in Mexico, t. 7 (1974); American Orchid Society Bulletin **54** 441 (1985) & **55**: 237 (1986); Orchid Review **94**: 84 (1986).

PSEUDOBULBS: 5–26 cm, ellipsoid to pear-shaped, sometimes stalked, loosely clustered.

LEAVES: 2–4 per pseudobulb, 20–33 × 3–4 cm, elliptic to lanceolate, acute.

RACEME: to 50 cm, many-flowered.

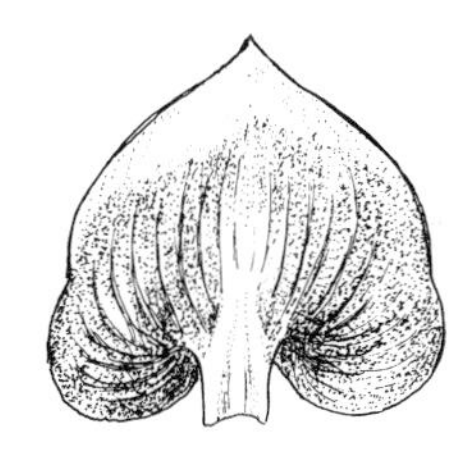

Encyclia cochleata – lip.

SEPALS AND PETALS: 3–7.5 cm × 3–7 mm, narrowly lanceolate, acute, twisted and drooping, pale green.

LIP: uppermost, 1–2.1 × 1.3–2.6 cm, triangular to heart-shaped, concave, deep purple flushed with yellowish green; base white with deep purple veins.

COLUMN: 7–9 mm, green with purple spots; base fused to claw of lip.

DISTRIBUTION: USA (Florida); S & W Mexico to Colombia & Venezuela.

HARDINESS: G2.

FLOWERING: continuously.

2. E. **glumacea** (Lindley) Pabst

SYNONYM: *Hormidium glumaceum* (Lindley) Brieger.

ILLUSTRATION: Edwards's Botanical Register **26**: t. 6 (1840).
Like *E. cochleata* except as below.

RACEME: base enclosed in long, chaffy scales.

FLOWERS: white streaked with pink.

Encyclia glumacea – lip.

LIP: narrower, acuminate, 1.8–2.5 cm × 8–12 mm.

DISTRIBUTION: E Brazil & Ecuador.

3. E. fragrans (Swartz) Lemée

SYNONYM: *Hormidium fragrans* (Swartz) Brieger.

ILLUSTRATIONS: Botanical Magazine, 1669 (1814); Dunsterville & Garay, Venezuelan orchids illustrated 4: 87 (1966); Dressler & Pollard, The genus Encyclia in Mexico, t. 4 (1974).

PSEUDOBULBS: 4.5–11 cm, narrowly ovoid to ellipsoid, 1–4 cm apart.

LEAVES: 9–35 cm, solitary, elliptic to strap-shaped, obtuse or acute.

RACEME: 5–20 cm, with 2–6 white, cream or greenish, very fragrant flowers.

SEPALS: 2–3.5 cm × 4–9 mm, lanceolate to elliptic, acuminate.

PETALS: 2–3 cm × 7–12 mm, elliptic to diamond-shaped, acuminate.

LIP: uppermost, 1.7–2.4 cm × 9–17 mm, ovate-acuminate with purple radial lines.

COLUMN: 6–8 mm, basal half fused to lip.

DISTRIBUTION: Greater Antilles, S. Mexico to Brazil.

HARDINESS: G2.

FLOWERING: spring–summer.

Encyclia fragrans – lip.

4. E. baculus (Reichenbach) Dressler & Pollard

SYNONYMS: *Epidendrum pentotis* Reichenbach; *Hormidium baculus* (Reichenbach) Brieger.

ILLUSTRATIONS: Botanical Magazine, 702 (1975); Dressler & Pollard, The genus Encyclia in Mexico, t. 1 (1974); Graf, Tropica, 729 (1978).

Like *E. fragrans* except as below.

PSEUDOBULBS: 15–35 cm, bearing 2 or 3 leaves, gradually tapering.

RACEMES: short 2- or 3-flowered.

SEPALS AND PETALS: more than 3.5 cm.

LIP: blade with prominent basal auricles.

DISTRIBUTION: S Mexico to Brazil.

HARDINESS: G2.

FLOWERING: spring–summer.

Encyclia baculus – lip.

5. E. radiata (Lindley) Dressler

SYNONYM: *Hormidium radiatum* (Lindley) Brieger.

ILLUSTRATIONS: Botanical Magazine n.s., 662 (1974); Dressler & Pollard, The genus Encyclia in Mexico, t. 8 (1974); Graf, Tropica, 729 (1978).

PSEUDOBULBS: 7–11 cm, ellipsoid to ovoid, grooved, sometimes stalked, about 2.5 cm apart.

LEAVES: 2–4 per pseudobulb, 14–35 × 1.5–3 cm, lanceolate to elliptic, obtuse or acute.

RACEME: 7–20 cm, with 4–12 cream or greenish white flowers.

SEPALS: 1.5–2 cm × 5–7 mm, elliptic.

Encyclia radiata – lip.

PETALS: 1.5–2 cm × 8–11 mm, broadly elliptic to diamond-shaped.

LIP: uppermost, 1–1.3 × 1.3–2 cm, semicircular to kidney-shaped, notched, with purple radial lines.

COLUMN: 8–10 mm, basal half fused to lip.

DISTRIBUTION: C & S Mexico, Guatemala, Honduras & Costa Rica.

HARDINESS: G2.

FLOWERING: autumn–winter.

6. E. vespa (Vellozo) Dressler

SYNONYMS: *Epidendrum variegatum* Hooker; *Hormidium variegatum* (Hooker) Brieger.

ILLUSTRATIONS: Botanical Magazine, 3151 (1832); Pabst & Dungs, Orchidaceae Brasilienses **1**: 187 (1975); Graf, Tropica, 729 (1978).

Like *E. radiata* except as below.

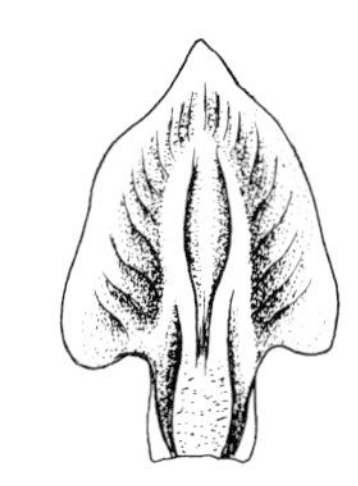

Encyclia vespa – lip.

LEAVES: faintly mottled.

SEPALS AND PETALS: yellowish green densely spotted with blackish purple.

LIP: small, pale yellowish green, scarcely longer than the column.

DISTRIBUTION: West Indies, Venezuela, Guyana, Surinam, Ecuador, Peru Brazil.

HARDINESS: G2.

FLOWERING: autumn–summer.

7. E. vitellina (Lindley) Dressler

ILLUSTRATIONS: Botanical Magazine, 4107 (1844); Die Orchidee **22**: 45 (1971), **35**: centre pull-out (1984); Dressler & Pollard, The genus Encyclia in Mexico, t. 31 (1974); American Orchid Society Bulletin **54**: 831 (1985).

PSEUDOBULBS: 2.5–5 cm, conical to ovoid, clustered.

LEAVES: 1–3, 7–22 × 1–4 cm, lanceolate to elliptic, obtuse or acute.

RACEME OR PANICLE: 12–30 cm, with 4–12 flowers.

SEPALS AND PETALS: lanceolate to elliptic, acute, orange to bright scarlet; sepals 1.5–2.5 × 3–8 mm; petals 1.5–2.5 cm × 6–11 mm.

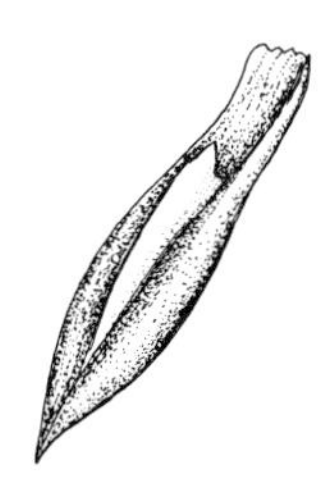

Encyclia vitellina – lip.

LIP: 1.1–1.6 cm, yellow or orange with a red tip; blade elliptic to oblong, acute with deflexed sides.

COLUMN: 6–7 mm; base fused to lip.

DISTRIBUTION: S Mexico & Guatemala.

HARDINESS: G2.

FLOWERING: spring–summer.

Var. **majus** Veitch has shorter, denser racemes of larger, more brightly coloured flowers.

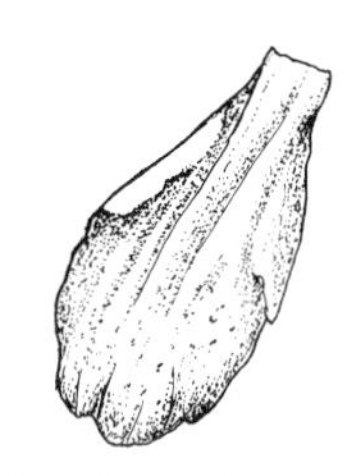
Encyclia boothiana – lip.

8. E. boothiana (Lindley) Dressler

SYNONYMS: *Diacrium bidentatum* (Lindley) Hemsley; *Hormidium boothianum* (Lindley) Brieger.

ILLUSTRATIONS: Luer, Native orchids of Florida, t. 61 (1972); Dressler & Pollard, The genus Encyclia in Mexico, t. 29 (1974).

Like *E. vitellina* except as below.

FLOWERS: smaller and fleshier.

SEPALS AND PETALS: 8–14 × 2–5 mm, rounded, yellowish green heavily spotted with purplish brown.

LIP: 7–10 mm, diamond-shaped or weakly 3-lobed, with down-curled margins, white to greenish yellow.

DISTRIBUTION: USA (Florida), S Mexico, Belize, Guatemala & West Indies.

FLOWERING: winter.

9. E. brassavolae (Reichenbach) Dressler

SYNONYM: *Hormidium brassavolae* (Reichenbach) Brieger.

ILLUSTRATIONS: Botanical Magazine, 5664 (1867); Dressler & Pollard, The genus Encyclia in Mexico, t. 32 (1974); Rittershausen & Rittershausen, Orchids in colour, 177 (1979); Orchid Review **94**: 85 (1986).

PSEUDOBULBS: 9–18 cm, conical, ovoid or fusiform, clustered or up to 4.5 cm apart.

LEAVES: 2 or 3 per pseudobulb, 15–30 × 3.5–5 cm, oblong to lanceolate, obtuse.

RACEME: 15–100 cm with 3–15 flowers.

SEPALS AND PETALS: linear-lanceolate, acuminate, pale yellowish green to brown; sepals 3.5–5.5 cm × 3–6 mm; petals 3.2–4.7 cm × 2–4 mm.

LIP: 3–4.5 cm, cream-coloured with a purple tip; blade ovate to triangular, acuminate.

COLUMN: 1.2–1.5 cm; basal half fused to lip.

DISTRIBUTION: S Mexico to W Panama.

HARDINESS: G2.

FLOWERING: summer–autumn.

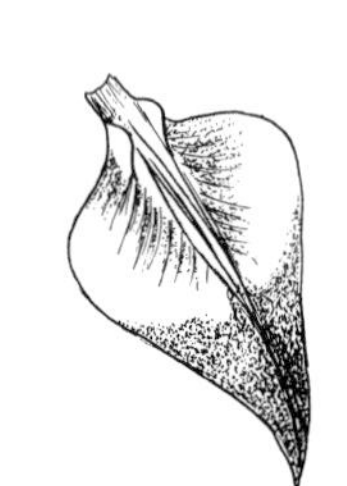
Encyclia brassavolae – lip.

10. E prismatocarpa (Reichenbach) Dressler

SYNONYM: *Hormidium prismatocarpum* (Reichenbach) Brieger.

ILLUSTRATIONS: Botanical Magazine, 5336 (1862); Shuttleworth et al., Orchids, 48 (1970); Graf, Tropica, 729 (1978).

PSEUDOBULBS: to 30 cm, ovoid with a long 'neck'.

LEAVES: 2–4 per pseudobulb, 25–40 × 3–6 cm, strap-shaped, obtuse.

RACEME: to 40 cm, many-flowered.

SEPALS AND PETALS: similar, 2–2.5 cm × 3–6 mm, linear-lanceolate to oblong, acute, pale greenish yellow with large purple blotches.

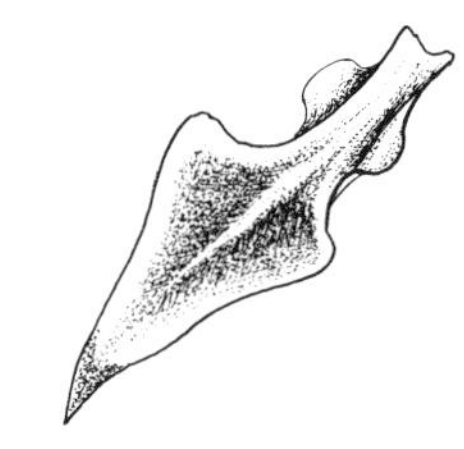
Encyclia prismatocarpa – lip.

LIP: 2–2.5 cm, 3-lobed; lateral lobes *c.* 5 mm, obtuse and rounded, yellow or greenish yellow; central lobe narrowly triangular, acute, pale yellow or white with a large central patch of purple and a yellow tip.
COLUMN: with basal third fused to lip, lobed and fringed at tip.
DISTRIBUTION: Costa Rica, Panama.
HARDINESS: G2.
FLOWERING: intermittently.

11. **E. patens** Hooker
SYNONYM: *Epidendrum odoratissimum* Lindley not *Epidendrum patens* Swartz.
ILLUSTRATIONS: Botanical Magazine, 3013 (1830); Pabst & Dungs, Orchidaceae Brasilienses 1: 197 & 298 (1975).
Like *E. prismatocarpa* except as below.

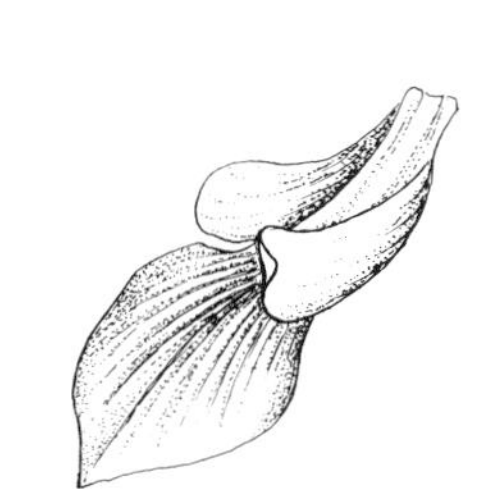
Encyclia patens – lip.

PETALS: unspotted, greenish brown, spathulate.
LIP: shorter (1–1.5 cm), cream-coloured, with oblong to obovate lateral lobes and an obovate central lobe.
DISTRIBUTION: SW Brazil.
HARDINESS: G2.
FLOWERING: summer.
The Botanical Magazine plate shows the raceme growing from beside the pseudobulb, which would exclude this species from *Encyclia*. Subsequent authors make no reference to this; presumably the illustration is inaccurate.

12. **E. adenocaula** (Llave & Lexarza) Schlechter
SYNONYMS: *Epidendrum nemorale* Lindley; *Epidendrum verrucosum* Lindley not Swartz.
ILLUSTRATIONS: Botanical Magazine, 4606 (1851); Orchid Digest 37: 45 (1973); Rittershausen & Rittershausen, Orchids in colour, 177 (1979).
PSEUDOBULBS: 5–8 cm, conical to ovoid, clustered.
LEAVES: 2 or 3 per pseudobulb, 11–35 cm × 5–30 mm, strap-shaped, acute or obtuse.
PANICLES: 30–100 cm, many-flowered; stem, branches and ovaries covered in warts.
SEPALS AND PETALS: linear to lanceolate, acute, pale purplish pink; sepals 2.5–5.5 cm × 5–7 mm; petals 3–5 cm × 4–8 mm.

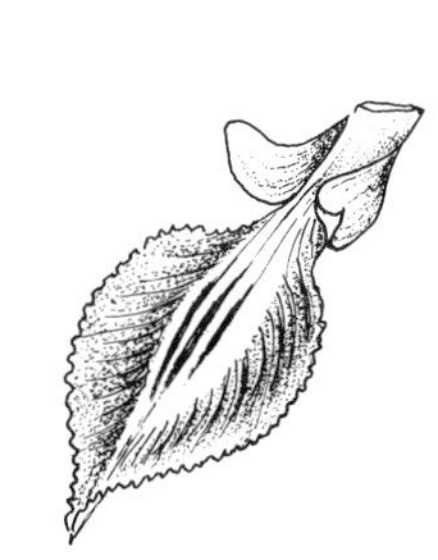
Encyclia adenocaula – lip.

LIP: 3.5–4.5 cm, 3-lobed; lateral lobes lanceolate to oblong, *c.* 10 × 5 mm; central lobe circular to ovate, regularly scalloped, mucronate, with 1 or more streaks of dark red.
COLUMN: 1.3–1.5 cm, club-shaped, apically winged; basal quarter fused to lip.
DISTRIBUTION: W Mexico.
HARDINESS: G2.
FLOWERING: spring–summer.

13. E. alata (Bateman) Schlechter

ILLUSTRATIONS: Botanical Magazine, 3898 (1842); Dressler & Pollard, The genus Encyclia in Mexico, t. 53 (1974); Hamer, Las Orquideas de El Salvador **1**: t. 14 (1974).

PSEUDOBULBS: 6–12 cm, conical to ovoid, clustered.

LEAVES: 2 or 3 per pseudobulb, 20–60 × 1.5–3.5 cm, strap-shaped to narrowly elliptic, acute.

PANICLES: 30–200 cm, many-flowered.

SEPALS AND PETALS: 2.2–3 cm × 5–8 mm, greenish yellow with a central patch of reddish brown near the tip; margins curled back; sepals lanceolate to elliptic, obtuse; petals elliptic to spathulate, obtuse or mucronate.

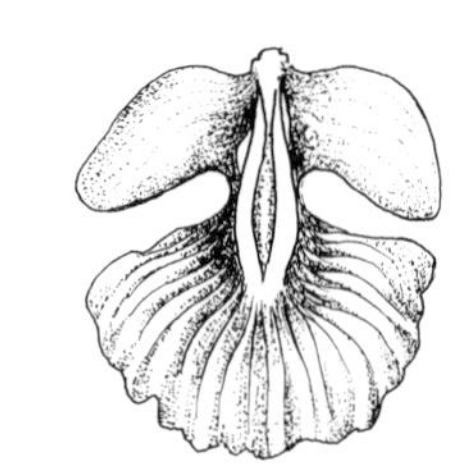

Encyclia alata – lip.

LIP: 1.8–2.3 cm, 3-lobed, pale greenish yellow, lateral lobes 1.2–1.6 cm, oblong to obovate, rounded, flared; central lobe circular to transversely diamond-shaped, fleshy, veined with dark red.

COLUMN: 1–1.2 cm, apically winged; basal quarter fused to lip.

DISTRIBUTION: S Mexico to Costa Rica.

HARDINESS: G2.

FLOWERING: summer.

14. E. tampensis (Lindley) Small

ILLUSTRATIONS: Shuttleworth et al., Orchids, 46 (1970); Luer, Native orchids of Florida, t. 57 & 58 (1972); American Orchid Society Bulletin **43**: 808 (1974), **53**: 25 (1984), **54**: 267, 981 (1985).

PSEUDOBULBS: 1–7 cm, ovoid, slightly separated.

LEAVES: 1–3 per pseudobulb, 30–40 × 1.5–2 cm, linear to narrowly lanceolate.

PANICLES: to 1 m, few- or many-flowered.

SEPALS AND PETALS: 1.2–2.2 cm × 4–6.5 mm, lanceolate to obovate, greenish yellow to brown suffused with purple.

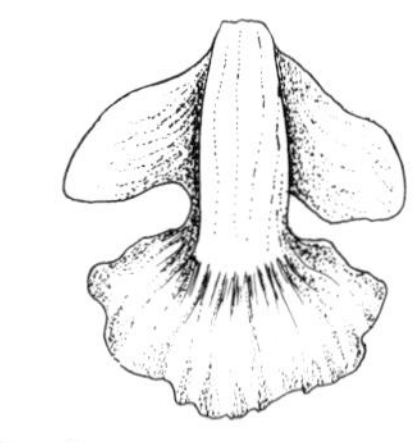

Encyclia tampensis – lip.

LIP: 1.5–2 cm, white, 3-lobed; lateral lobes triangular, upright or slightly flared; central lobe semicircular with a central patch of pink or pink streaks.

COLUMN: 10 mm, apically winged; base fused to claw of lip.

DISTRIBUTION: USA (Florida).

HARDINESS: G2.

FLOWERING: intermittently.

15. E. aromatica (Bateman) Schlechter

ILLUSTRATIONS: Dressler & Pollard, The genus Encyclia in Mexico, t. 63 (1974); Hamer, Las Orquideas de El Salvador **1**: t. 15 (1974).

Like *E. tampensis* except as below.

Encyclia aromatica – lip.

LIP: smaller (1–1.5 cm), cream or yellow with red or brown veins and an oblong to diamond-shaped, wavy-edged central lobe.

DISTRIBUTION: SE Mexico.

FLOWERING: spring–summer.

16. E. bractescens (Lindley) Hoehne

ILLUSTRATIONS: Botanical Magazine, 4572 (1851); Dressler & Pollard, The genus Encyclia in Mexico, t. 46 (1974); Northen, Miniature orchids, C-31 (1980).

Like *E. tampensis* except as below.

PSEUDOBULBS: 3 cm or less.

LEAVES: to 25 × 1 cm.

SEPALS AND PETALS: narrowly lanceolate, purplish brown with greenish yellow tips.

LIP: with long narrow lateral lobes and circular central lobe.

DISTRIBUTION: S Mexico.

FLOWERING: spring–summer.

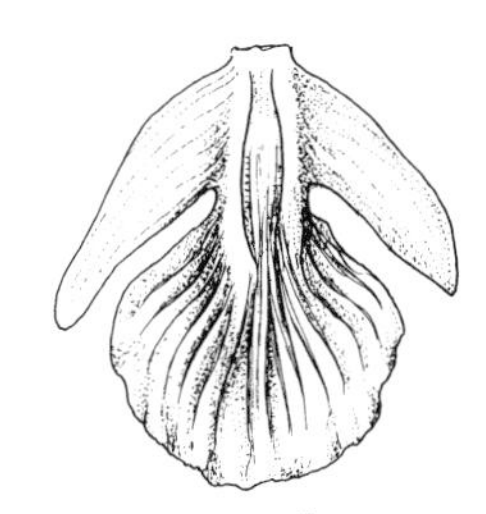

Encyclia bractescens – lip.

17. E. michuacana (Llave & Lexarza) Schlechter

SYNONYM: *Epidendrum virgatum* Lindley.

ILLUSTRATIONS: Dressler & Pollard, The genus Encyclia in Mexico, t. 20 (1974); Hamer, Las Orquideas de El Salvador 1: t. 23 (1974).

Like *E. tampensis* except as below.

SEPALS AND PETALS: dark brown.

LIP: 10–11 mm, with narrow lateral lobes arising from base of blade rather than from claw, cream or pale yellow often spotted with purple.

DISTRIBUTION: S Mexico & Guatemala.

HARDINESS: G2.

FLOWERING: intermittently.

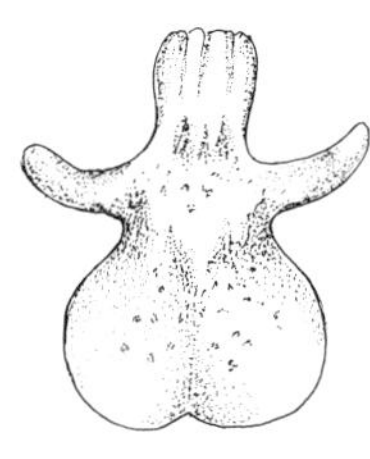

Encyclia michuacana – lip.

18. E. varicosa (Lindley) Schlechter

ILLUSTRATIONS: Dressler & Pollard, The genus Encyclia in Mexico, t. 22a (1974); Hamer, Las Orquideas de El Salvador 1: t. 23 (1974).

Like *E. tampensis* except as below.

PSEUDOBULBS: with long slender necks.

SEPALS AND PETALS: brown.

LIP: papillose or warty, notched, with triangular lateral lobes arising from claw, and blade of 2 oblong truncate lobes.

DISTRIBUTION: S & W Mexico to Panama

FLOWERING: mainly winter.

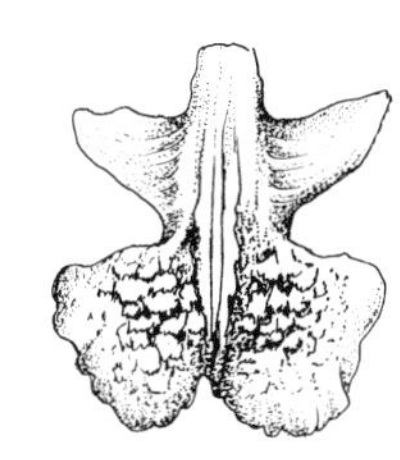

Encyclia varicosa – lip.

19. E. cordigera (Humboldt, Bonpland & Kunth) Dressler

SYNONYMS: *Epidendrum atropurpureum* misapplied: *Epidendrum macrochilum* Hooker.

ILLUSTRATIONS: Botanical Magazine, 3534 (1836); Dressler & Pollard, The genus Encyclia in Mexico, t. 57 (1974); Sheehan & Sheehan, Orchid genera illustrated, 89 (1979).

PSEUDOBULBS: 3–11 cm, conical to ovoid, clustered.

LEAVES: 2 or 3 per pseudobulb, 12.5–45 × 1.5–4.5 cm, narrowly elliptic.

RACEMES OR PANICLES: 15–75 cm with 3–15 flowers.

SEPALS AND PETALS: fleshy, 2.5–3.5 cm × 5–11 mm, green

Encyclia cordigera – lip.

streaked with brown; sepals obovate to elliptic, obtuse or mucronate; petals elliptic to spathulate.

LIP: 3.3–4 cm, cream or pink, centrally streaked with magenta, 3-lobed, lateral lobes oblong to lanceolate, central lobe broadly obovate.

COLUMN: 1.5–1.7 cm, base fused to lip, white.

DISTRIBUTION: S Mexico to Colombia & Venezuela.

HARDINESS: G2.

FLOWERING: spring–summer.

20. E. mariae (Ames) Hoehne

SYNONYM: *Hormidium mariae* (Ames) Brieger.

ILLUSTRATIONS: American Orchid Society Bulletin **41**: 7 (1972); Dressler & Pollard, The genus Encyclia in Mexico, t 36 (1974); Graf, Tropica, 729 (1978); Kew Magazine **2**: pl. 25 (1985).

FLOWERS: pendent, 2–4 per inflorescence.

SEPALS AND PETALS: green, spreading.

LIP: white with green or yellowish green throat and unlobed or shallowly 3-lobed, apex notched.

DISTRIBUTION: E Mexico.

HARDINESS: G2.

FLOWERING: summer.

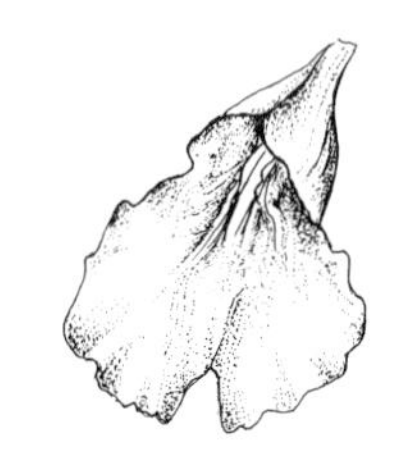

Encyclia mariae – lip.

21. E. citrina (Llave & Lexarza) Dressler

SYNONYMS: *Cattleya citrina* (Llave & Lexarza) Lindley; *Epidendrum citrinum* (Llave & Lexarza) Reichenbach.

ILLUSTRATIONS: Botanical Magazine, 3742 (1839); Dressler & Pollard, The genus Encyclia in Mexico, t. 37 (1974); Bechtel et al., Manual of cultivated orchid species, 207 (1981); Orchid Review **97**: 42 (1989).

FLOWERS: bright yellow, pendent, 1 or 2 per inflorescence.

LIP: shallowly 3-lobed, central lobe oblong, truncate.

DISTRIBUTION: Mexico.

HARDINESS: G1.

FLOWERING: spring.

Encyclia citrina – lip.

22. E. selligera (Lindley) Schlechter

SYNONYM: *E. oncidioides* misapplied.

ILLUSTRATIONS: Reichenbach, Xenia Orchidacearum **3**: t. 233 (1892); Hamer, Las Orquideas de El Salvador **1**: t. 17 (1974); Dressler & Pollard, The genus Encyclia in Mexico, t. 59 (1974).

PSEUDOBULBS: 4.5–8 cm, conical to ovoid, clustered.

LEAVES: 2 per pseudobulb, 25 × 1.5–4 cm, elliptic to strap-shaped, obtuse.

PANICLES: 25–80 cm, many-flowered.

SEPALS AND PETALS: pale green, strongly flushed with purplish brown; sepals 1.8–2.4 cm × 5–9 mm, lanceolate to

Encyclia selligera – lip.

spathulate, obtuse; petals 1.8–2.1 cm × 7–9 mm, spathulate, hooded, obtuse or mucronate.

LIP: 1.7–2.2 cm, 3-lobed, cream, pink or lavender; lateral lobes to 10 × 5 mm, broad-tipped, constricted at base; central lobe ovate to circular; callus elliptic, running into 2 ridges on central lobe of lip.

COLUMN: 1.1–1.3 cm, base fused to claw of lip.

DISTRIBUTION: S Mexico & Guatemala.

HARDINESS: G2.

FLOWERING: winter–spring.

Frequently confused with *E. oncidioides* (Lindley) Schlechter, which has smaller narrowly ellipsoid pseudobulbs and smaller yellow flowers flushed with brown. Dressler considers the latter to be very rare in cultivation. The following 2 species are also very similar.

23. E. **hanburii** (Lindley) Schlechter

ILLUSTRATIONS: Orchid Digest **34**: 307 (1970); Dressler & Pollard, The genus Encyclia in Mexico, t. 60 (1974).

Like *E. selligera* except as below.

SEPALS AND PETALS: reddish purple.

LIP: pink, heavily veined; callus ovate to rhombic, running into a single ridge on the central lobe of the lip.

DISTRIBUTION: S Mexico.

HARDINESS: G2.

FLOWERING: spring–summer.

24. E. **advena** (Reichenbach) Dressler

SYNONYMS: *Epidendrum capartianum* Lindley; *Epidendrum osmanthum* Barbosa-Rodrigues.

ILLUSTRATION: Botanical Magazine, 7792 (1901).

Like *E. selligera* except larger in all its parts, and as below.

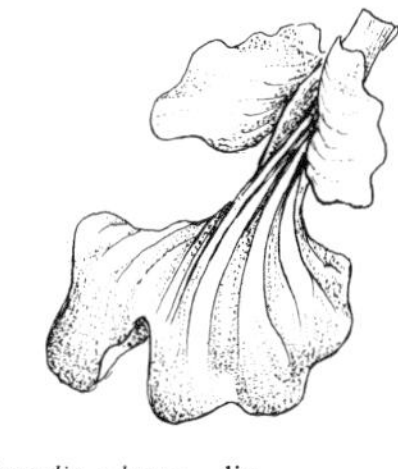

Encyclia advena – lip.

PSEUDOBULBS: 7.5–13 cm.

PETALS: spathulate to circular, about 1.5 cm wide, yellow streaked with pink.

LIP: more than 3 cm, wavy and scalloped, white with pink streaks.

DISTRIBUTION: Brazil.

FLOWERING: autumn–winter.

25. E. **polybulbon** (Swartz) Dressler

SYNONYM: *Dinema polybulbon* (Swartz) Lindley.

ILLUSTRATIONS: Botanical Magazine, 4067 (1844); Die Orchidee 29: lxxxvii (1978); Dressler & Pollard, The genus Encyclia in Mexico, t. 38 (1974).

PSEUDOBULBS: 1–1.5 cm, ovoid, 5–40 mm apart.

LEAVES: 2 or 3 per pseudobulb, 1–5 cm × 5–10 mm, elliptic to ovate, notched.

FLOWERS: solitary.

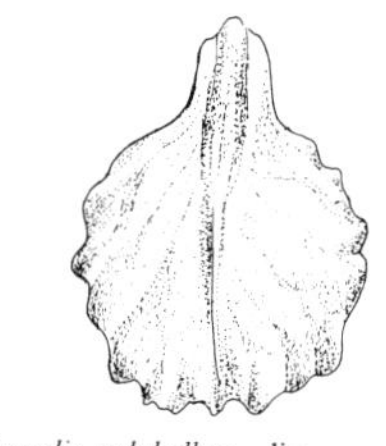
Encyclia polybulbon – lip.

SEPALS: 9–17 × 1.5–3 mm, linear-lanceolate, acute, brownish yellow.

PETALS: similar, narrower with slightly hooked tips.

LIP: 1.5 cm, white, with a wavy, more or less circular, sometimes mucronate blade.

COLUMN: *c.* 4 mm, fused to lip, winged, with two 3 mm apical processes.

DISTRIBUTION: S Mexico, Guatemala, Honduras, Cuba & Jamaica.

HARDINESS: G2.

FLOWERING: intermittently.

49. BROUGHTONIA R. Brown

A single species from the West Indies which grows best on bark or rafts suspended near the roof of a glasshouse. General conditions are similar to those for *Epidendrum* (p. 148); it should not be heavily shaded, nor should too much water be given in winter.

HABIT: epiphytic, to 50 cm.

PSEUDOBULBS: 3–6 cm, spherical to ovoid, slightly flattened, with 1 or 2 apical leaves.

LEAVES: 7.5–15 × 2–4 cm, oblong, stiff and leathery, folded when young.

FLOWERING STEMS: to 50 cm, borne on top of pseudobulbs, with a small scale at each node, bearing 5–12 flowers.

FLOWERS: 2–3 cm wide, purplish red.

SEPALS: 1.5–2.5 cm × 5 mm, lanceolate, acute, extending into a long narrow spur which is fused to the ovary.

PETALS: 1.5–2.5 × 1 cm, elliptic to ovate, rounded.

LIP: 2.5 × 2.5 cm, more or less circular, notched and irregularly toothed.

COLUMN: free, exposed.

POLLINIA: 4.

1. B. sanguinea (Swartz) R. Brown

SYNONYM: *B. coccinea* Lindley.

ILLUSTRATIONS: Botanical Magazine, 3536 (1836); Skelsey, Orchids, 96 (1979); Bechtel et al., Manual of cultivated orchid species, 173 (1981).

DISTRIBUTION: Jamaica & Cuba.

HARDINESSS: G2.

FLOWERING: intermittent.

50. LAELIOPSIS Lindley

Two species from the West Indies, of which only one is commonly grown. Cultivation as for *Broughtonia*.

PSEUDOBULBS: ovoid to fusiform.

LEAVES: 1–3, apical, rigid, sharply toothed.
FLOWERS: in terminal spikes.
SPUR: absent.
LIP: rolled around column.
COLUMN: without foot, concealed.
POLLINIA: 4.

1. L. domingensis Lindley
SYNONYMS: *Broughtonia domingensis* (Lindley) Rolfe; *B. lilacina* Henfrey.
ILLUSTRATIONS: Shuttleworth et al., Orchids, 52 (1970); Orchid Digest **40**: 212 (1976).
PSEUDOBULBS: 5–8 cm.
FLOWERS: 3–8, 3.5 cm wide, pale pinkish purple.
LIP: yellow-throated, finely toothed.
DISTRIBUTION: Dominican Republic.
HARDINESS: G2.
FLOWERING: spring–summer.

51. BARKERIA Knowles & Westcott

A genus of about 10 species from C America, closely related to and often included in *Epidendrum*, but differing in that at least half of the column is free from the lip; differing from *Encyclia* in the lack of true pseudobulbs. Cultivation as for *Epidendrum* (p. 148), but the plants are better grown on rafts or bark, as their elongate rhizomes are restricted in pots.

Literature: Halbinger, F., The genus Barkeria, *American Orchid Society Bulletin* **42**: 620–6 (1973); Halbinger, F. & Kennedy, J., Barkeria, *Orchid Digest* **44**: 56–62 (1980); Marsh, R., Barkerias — the shape of things to come, *American Orchid Society Bulletin* **56**: 580–587 (1987).

HABIT: epiphytes.
STEMS: erect, elongate, cane-like, sometimes somewhat swollen at the base, leafy.
LEAVES: leathery, folded when young.
RACEME (RARELY PANICLE): with few to many flowers.
FLOWERS: resupinate.
SEPALS AND PETALS: similar or the petals somewhat broader.
LIP: united to the column for up to half the length of the latter, entire.
COLUMN: winged.
POLLINIA: 4, laterally flattened, parallel.

Key to species

1. Lip joined to the column for about half the length of the latter, its blade 1.3–1.8 cm, bearing 3 yellow ridges **3. skinneri**

2. Lip without ridges, though the main vein sometimes thickened **1. elegans**
3. Lip with 3–5 ridges **2. spectabilis**

1. B. elegans Knowles & Westcott

SYNONYM: *Epidendrum elegans* (Knowles & Westcott) Reichenbach.

ILLUSTRATIONS: Botanical Magazine, 4784 (1854); American Orchid Society Bulletin **42**: 621 (1973).

STEMS: to 25 cm, erect, thick, leafy.

LEAVES: linear-lanceolate to elliptic-lanceolate, acute.

RACEME: loose, with few flowers.

SEPALS AND PETALS: similar or the petals broader, all narrowly elliptic, acute, 2–3.5 cm.

LIP: 2–3.5 cm, obovate-elliptic to oblong, broadly rounded but with a small point, white with a large purple blotch and with a callus but without ridges.

COLUMN: joined to the lip for up to one-third of its length.

DISTRIBUTION: Mexico, Guatemala.

HARDINESS: G2.

FLOWERING: spring.

2. B. spectabilis Lindley

ILLUSTRATIONS: Botanical Magazine, 6098 (1874); Fieldiana **26**: 342 (1953); American Orchid Society Bulletin **42**: 619 (1973), **56**: 585 (1987); Bechtel et al., Manual of cultivated orchid species, 170 (1981).

Like *B. elegans* except as below.

PLANT: often larger.

RACEME: often with many flowers.

FLOWERS: white, pink or deep purple.

LIP: often spotted, bearing 3–5 ridges which are the same colour as the rest.

DISTRIBUTION: C America.

HARDINESS: G2.

FLOWERING: winter.

Variable in flower colour and the precise shape of the lip. Halbinger recognises *B. lindleyana*, which is usually sunk in *B. spectabilis*, as a distinct species on the basis of lip shape and column length (15–18 mm in *B. spectabilis*, 6–14 mm in *B. lindleyana*).

3. B. skinneri (Lindley) Richard & Galeotti

SYNONYM: *Epidendrum skinneri* Lindley.

ILLUSTRATIONS: Botanical Magazine, 3951 (1843); American Orchid Society Bulletin **42**: 624 (1973); Bechtel et al., Manual of cultivated orchid species, 170 (1981).

Like *B. elegans* except as below.

PLANT: smaller.

INFLORESCENCE: sometimes paniculate.

FLOWERS: red to purple.
LIP: 1.3–1.8 cm with 3 yellow ridges.
COLUMN: joined to the lip for almost half its length.
DISTRIBUTION: Guatemala.
HARDINESS: G2.
FLOWERING: winter.

52. CATTLEYA Lindley

A genus of about 40 species from C & S America; often used in bi- and trigeneric hybrids. They require maximum light throughout the year though new growth should be protected from strong sunlight. In the growing season water should be applied freely and the atmosphere kept moist with some air movement. Watering should be reduced during the winter but stems should not be allowed to shrink. Clusters of stems can be divided when repotting.

Literature: Withner, C., The genus Cattleya, *American Orchid Society Bulletin* **17**: 296–306, 363–9 (1948); Fowlie, J.A., *Brazilian bifoliate Cattleyas and their color varieties* (1977); Braem, G., *The Brazilian bifoliate Cattleyas* (1984); Braem, G., *The unifoliate Cattleyas* (1986); Withner, C., *The Cattleyas and their relatives. I. The Cattleyas* (1988); Braem, G., *Cattleya*, 2 volumes (1985, 1986).

HABIT: epiphytic or terrestrial, often on rocks; rhizomes creeping, bearing erect, swollen or cane-like stems with several nodes.
LEAVES: 1 or 2 (rarely 3 or 4) borne on top of the stems, folded when young.
FLOWERS: solitary or in terminal racemes (except Nos. **19** & **20**).
SEPALS AND PETALS: free.
LIP: 3-lobed or entire; lateral lobes or basal part rolled around the column, often concealing it.
COLUMN: free from lip, slightly flattened, often winged.
POLLINIA: 4 cohering in pairs, 6 or 8 (with some smaller than others) in one species.

Key to species

1. Flowers lateral, on slender, leafless shoots
1.A. Pseudobulbs bearing 1 leaf **19. walkeriana**
1.B. Pseudobulbs bearing 2 or 3 leaves **20. nobilior**
2. Pollinia 6 or 8 (2 or 4 of them usually smaller than the others) **5. dormaniana**
3. Stems or pseudobulbs bearing 2 or more leaves
3.A. Lip simple, shallowly lobed or with very small basal lateral lobes, never with deep sinuses *Go to Subkey* 1
3.B. Lip with deeply incised lobes *Go to Subkey* 2
4. Stems or pseudobulbs bearing a single leaf *Go to Subkey* 3

Subkey 1. Stems or pseudobulbs bearing 2 or more leaves; lip simple, shallowly lobed or with very small basal lateral lobes, never with deep sinuses.

1. Flowers red to orange, to 4.5 cm across; lip less than 2.5 × 1.5 cm, obtuse or acute, not pale in the throat **1. aurantiaca**
2. Lip clearly divided into a claw and a fan-shaped blade **18. bicolor**
3. Flowers 6–8 cm across; stems 1 cm or less in diameter, swollen at the base **2. bowringiana**
4. Flowers 9–12 cm across; stems 2–3 cm in diameter, not swollen at the base **3. skinneri**

Subkey 2. Stems or pseudobulbs bearing 2 or more leaves; lip with conspicuous, deeply incised lateral lobes.

1. Claw of central lobe of lip at least one quarter of the central lobe length

1.A. Claw of central lobe of lip more than one-third of the total lip length **15. granulosa**

1.B. Sepals and petals copper-coloured, without spots **17. elongata**

1.C. Racemes 5–10-flowered; sepals and petals green with red-brown spots **14. guttata**

1.D. Racemes 10–20-flowered; sepals and petals bronze-green or deep brown with faint to conspicuous darker spots **16. leopoldii**

2. Lateral lobes of lip strongly angled, not rounded

2.A. Claw of central lobe of lip not less than half the width of the central lobed **13. violacea**

2.B. Flowers 6–12 per raceme; sepals and petals pale pinkish purple **11. amethystoglossa**

2.C. Flowers 1–2 per raceme; sepals and petals yellow to brown **12. schilleriana**

3. Lip widest across the central lobe (rather than across the lateral lobes); sinus broadly triangular or oblong

3.A. Lip white tinged with yellow and with darker veining **10. velutina**

3.B. Lip pink **9. aclandiae**

4. Petals narrower than sepals, usually less than 1 cm wide **4. forbesii**

5. Petals as broad as sepals, usually 1.5 cm wide or more

5.A. Sepals and petals 2 cm wide or more, usually with purple spots near the bases **7. loddigesi**

5.B. Lip yellow towards the tip **8. harrisoniana**

5.C. Lip without any yellow **6. intermedia**

Subkey 3. Stems or pseudobulbs bearing a single leaf.

1. Flowers up to 10 cm wide

1.A. Flowers yellow **22. luteola**

1.B. Flowers pale to dark purple, lip sometimes almost white **21. lawrenceana**
2. Lip with a central yellow stripe **34. maxima**
3. Sepals and petals yellow
3.A. Lip crimson to purple **24. dowiana**
3.B. Lip yellow in the basal part, pink with a white border towards the apex, all purple-veined **33. rex**
4. Sepals and petals white to pale pink; lip white to pale pink with a yellow throat and an indistinct central patch of purplish pink **25. gaskelliana**
5. Sepals and petals white to pale pink; lip white to pale pink with a distinct central patch of purplish pink **26. mendelii**
6. Sepals and petals pink; lip white with a yellow throat with purple veins which merge to a large central patch of purplish pink **27. mossiae**
7. Sepals and petals white to pale pink; lip pale pink with a purple tip **28. schroederae**
8. Sepals and petals pink; lip bright reddish purple with a pink margin, throat white and yellow **23. labiata**
9. Sepals and petals pink to white, lip purple **29. trianae**
10. Sepals and petals purplish pink, lip darker with a white and yellow throat **30. warneri**
11. Flower very large, lip mostly deep purplish crimson, throat greenish yellow veined with crimson **31. warscewiczii**
12. Sepals and petals pinkish purple; lip maroon with a yellow throat and pink margins **32. percivalliana**

1. C. **aurantiaca** (Bateman) P.N. Don

ILLUSTRATIONS: Shuttleworth et al., Orchids, 58 (1970); Williams et al., Orchids for everyone, 105 (1980); Bechtel et al., Manual of cultivated orchid species, 178 (1981); American Orchid Society Bulletin **55**: 25 (1986); Withner, The Cattleyas, f. 5 (1988).
HABIT: epiphytic.
STEMS: 20–35 cm, club-shaped, 2-leaved.
LEAVES: 5–20 × 2.5–5.5 cm, ovate to elliptic, notched.
FLOWERS: several per raceme, 3–4 cm wide, orange to scarlet.
SEPALS: 1.5–2.5 cm × 5 mm, lanceolate, acute.
PETALS: 1.5–2.5 cm × 5 mm, ovate to lanceolate, acute.
LIP: 1.5–2.5 × 1 cm, ovate to elliptic; base rolled around column.
COLUMN: 1 cm.
DISTRIBUTION: Mexico, Guatemala, Honduras & El Salvador.
HARDINESS: G2.
FLOWERING: summer.

Cattleya aurantiaca – lip.

2. C. **bowringiana** Veitch

ILLUSTRATIONS: Botanical Magazine n.s., 451 (1964); Shuttleworth et al., Orchids, 60 (1970); Williams et al., Orchids for everyone, 105 (1980); American Orchid Society Bulletin **53**: 691 (1984); Orchid Review **92**: 39 (1984); Withner, The Cattleyas, f. 9 (1988).

HABIT: terrestrial, on rock or sand.

STEMS: 20–35 cm × 5–10 mm, 4-or 5-noded, swollen at base, bearing 2 (rarely 3) leaves.

LEAVES: 13–20 × 5–6 cm, oblong to elliptic, leathery, dark green.

FLOWERS: 5–20 (rarely to 45) per raceme, 6–7.5 cm wide, purple to magenta; lip darker, pale-throated.

SEPALS: 3.8–5 × 1–1.4 cm, oblong to elliptic.

PETALS: 4–5 × 2.5–3 cm, oblong to ovate.

LIP: 3.5–4 × 2.8–3.5 cm, oblong to ovate, base rolled around column.

COLUMN: 1–1.2 cm, winged.

DISTRIBUTION: Belize & Guatemala.

HARDINESS: G2.

FLOWERING: autumn–winter.

Cattleya skinneri – lip.

3. C. skinneri Bateman

ILLUSTRATIONS: Botanical Magazine, 4270 (1846); Williams et al., Orchids for everyone, 106 (1980); Bechtel et al., Manual of cultivated orchid species, 181 (1981); Withner, The Cattleyas, f. 64 (1988).

Like *C. bowringiana* except as below.

HABIT: epiphytic or terrestrial.

STEMS: pseudobulbous (2–3 cm in diameter).

FLOWERS: larger (7.5–10 cm wide), 4–12 per raceme.

DISTRIBUTION: Mexico to Costa Rica; ?Panama.

FLOWERING: winter–spring.

4. C. forbesii Lindley

SYNONYM: *C. pauper* (Vellozo) Stellfeld.

ILLUSTRATIONS: Botanical Magazine, 3265 (1833); Williams et al., Orchids for everyone, 105 (1980); Bechtel et al., Manual of cultivated orchid species, 179 (1981); Withner, The Cattleyas, f. 19 (1988).

HABIT: epiphytic.

STEMS: 10–20 × 1–1.5 cm, 3- or 4-noded, slightly swollen above, 2-leaved.

LEAVES: 9–15 × 3–7 cm, oblong, leathery.

FLOWERS: 1–5, 9–11 cm wide.

SEPALS AND PETALS: 5–6 cm, pinkish yellow to pale brown; sepals to 1.5 cm wide; lateral sepals sickle-shaped; petals to 1 cm wide.

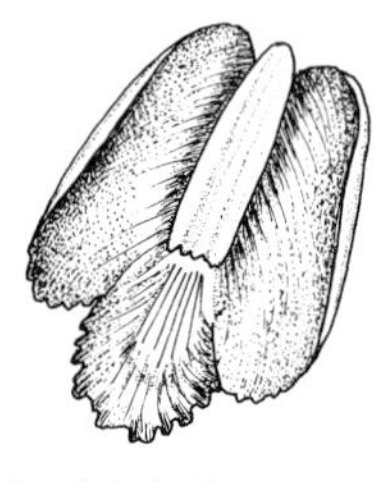

Cattleya forbesii – lip.

LIP: 4.5–5.5 × 3–3.5 cm, 3-lobed, off-white with pinkish brown veins and a central yellow stripe; lateral lobes semicircular to oblong, rolled around column, separated from the small central lobe by a short narrow sinus; central lobe 1.4–1.6 cm wide, white, almost circular.

COLUMN: 4 cm.

POLLINIA: 4.

DISTRIBUTION: S Brazil.

HARDINESS: G2.

FLOWERING: spring–summer.

5. **C. dormaniana** Reichenbach

ILLUSTRATIONS: American Orchid Society Bulletin **45**: 978 (1976); Fowlie, Brazilian bifoliate Cattleyas, 17 (1977); Bechtel et al., Manual of cultivated orchid species, 178 (1981); Withner, The Cattleyas, f. 13 (1988).

Generally like *C. forbesii* but differing as below.

LIP: central lobe of lip larger (2.3–2.8 cm wide), dark purple, overlapped by lateral lobes when flattened out.

POLLINIA: 6 or 8, 2 or 4 of them smaller than the others and probably sterile.

DISTRIBUTION: S Brazil.

FLOWERING: summer–autumn.

6. **C. intermedia** Hooker

ILLUSTRATIONS: Botanical Magazine, 2851 (1821); Fowlie Brazilian bifoliate Cattleyas, 110 (1977); Bechtel et al., Manual of cultivated orchid species, 179 (1981); American Orchid Society Bulletin **53**: 696 (1984); Withner, The Cattleyas, f. 32, 33 (1988).

HABIT: epiphytic or on rocks.

STEMS: 15–40 × 1.5 cm, club-shaped, 2-leaved.

LEAVES: 7–15 × 3.5–7 cm, elliptic to oblong, obtuse, fleshy.

FLOWERS: 3–10, 10 × 12.5 cm wide.

SEPALS: off-white to pale pink; upper 6–7 × 1.5 cm, laterals shorter and broader.

PETALS: 5.5–7 × 1.5–1.8 cm, off-white to pale pink.

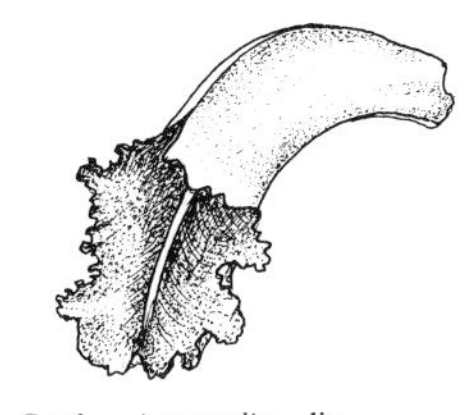

Cattleya intermedia – lip.

LIP: 5–6 × 4–4.6 cm, 3-lobed; lateral lobes asymmetrically ovate, rolled around column, off-white to pale pink with irregularly toothed tips, separated from the central lobe by a narrow rounded sinus; central lobe fishtail-shaped, purplish pink, minutely toothed.

COLUMN: 2.5–3.5 cm, winged.

DISTRIBUTION: S Brazil.

HARDINESS: G2.

FLOWERING: autumn.

7. **C. loddigesi** Lindley

ILLUSTRATIONS: Fowlie, Brazilian bifoliate Cattleyas, 123 (1977); Williams et al., Orchids for everyone, 106 (1980);

Bechtel et al., Manual of cultivated orchid species, 180 (1981); Withner, The Cattleyas, f. 43 (1988).

Like *C. intermedia* except as below.

SEPALS AND PETALS: 2 cm wide or more, sparsely and finely spotted.

LIP: lateral lobes broadly triangular, overlapping the semi-circular central lobe.

DISTRIBUTION: S Brazil.

FLOWERING: summer–winter.

8. C. harrisoniana Lindley

SYNONYMS: *C. loddigesii* var. *harrisoniana* Rolfe; *C. harrisoniana* Bateman).

ILLUSTRATIONS: Botanical Magazine, 4085 (1844); Fowlie, Brazilian bifoliate Cattleyas, 126 (1977); American Orchid Society Bulletin **54**: 1121 (1985); Withner, The Cattleyas, f. 31 (1988).

Like *C. intermedia* except as below.

STEMS: slightly thicker above.

FLOWERS: deep pink, unspotted.

PETALS: strongly sickle-shaped.

LIP: lateral lobes rounded, pink with yellow veins; central lobe pale, wavy, slightly notched.

DISTRIBUTION: Brazil.

FLOWERING: spring & autumn

Cattleya harrisoniana – lip.

9. C. aclandiae Lindley

ILLUSTRATIONS: Botanical Magazine, 5039 (1858); Fowlie, Brazilian bifoliate Cattleyas, 32 (1977); Bechtel et al., Manual of cultivated orchid species, 178 (1981); Withner, The Cattleyas, f. 2 (1988).

HABIT: epiphytic.

STEMS: to 12 cm, 1- or 2-noded, slightly thicker above, 2-leaved.

LEAVES: 4–10 × 2–3.5 cm, elliptic, fleshy, sparsely spotted with purple.

FLOWERS: 1 or 2 per raceme, 7–10 cm wide.

SEPALS AND PETALS: 4.5 × 1.5–2 cm, elliptic, yellowish green to brown, densely spotted with purplish brown; sepals acuminate; petals obtuse or rounded, wavy.

LIP: 4.5–5 × 3–3.5 cm, 3-lobed; lateral lobes semicircular, white to pale pink, not concealing column, separated from central lobe by a broad triangular sinus; central lobe kidney-shaped with a deep pointed notch, pink, minutely toothed, wavy.

Cattleya aclandiae – lip.

COLUMN: 2.5 cm, winged, deep pinkish purple.

DISTRIBUTION: E Brazil.

HARDINESS: G2.

FLOWERING: spring.

10. C. velutina Reichenbach

ILLUSTRATIONS: Shuttleworth et al., Orchids, 62 (1970); Fowlie, Brazilian bifoliate Cattleyas, 20 (1977); Bechtel et al., Manual of cultivated orchid species, 182 (1981); American Orchid Society Bulletin **54**: 1321 (1985); Withner, the Cattleyas, f. 70 (1988).

Like *C. aclandiae* except as below.

STEMS: 20–30 cm, cane-like.

SEPALS AND PETALS: yellow with fine purple spots, wavy.

LIP: central lobe fan-shaped, pale pink with darker veins and a broad yellow margin.

DISTRIBUTION: S Brazil.

FLOWERING: summer.

11. C. amethystoglossa Linden & Reichenbach

ILLUSTRATIONS: Botanical Magazine, 5683 (1868); Williams et al., Orchids for everyone, 105 (1980); Bechtel et al., Manual of cultivated orchid species, 178 (1981); Withner, The Cattleyas, f. 3 (1988).

HABIT: on rocks.

STEMS: to 1 m, cane-like, slightly swollen above, with 4–6 nodes, 2-leaved.

LEAVES: 16–23 × 5–7.5 cm, oblong to narrowly elliptic, obtuse, leathery, dark green.

FLOWERS: 6–12 per raceme, 7–10 cm wide, very pale pinkish purple spotted with dark purple; lip with darker lobes.

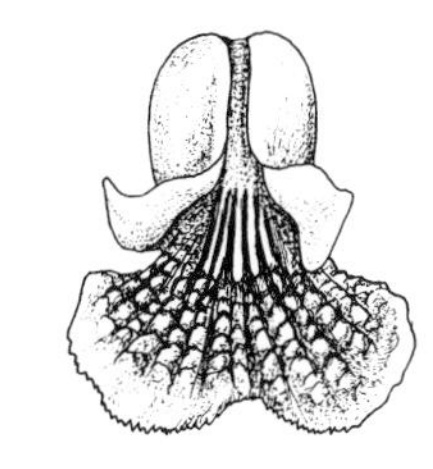

Cattleya amethystoglossa – lip.

SEPALS: upper 3.8–4.5 × 1.4–1.7 cm, ovate to oblong; laterals shorter and broader, curved downwards.

PETALS: 3.5–4.5 × 2–2.5 cm, lanceolate to obovate.

LIP: 3–4 × 2.7–3.5 cm, 3-lobed; lateral lobes ovate to triangular, with warty, slightly spathulate tips, rolled around column, separated from the central lobe by a narrow triangular sinus; central lobe fishtail-shaped with warty ridges.

COLUMN: 2 cm.

DISTRIBUTION: E Brazil.

HARDINESS: G2.

FLOWERING: spring–summer.

12. C. schilleriana Reichenbach

ILLUSTRATIONS: Shuttleworth et al., Orchids, 61 (1970); Fowlie, Brazilian bifoliate Cattleyas, 36 (1977); Bechtel et al., Manual of cultivated orchid species, 181 (1981); Withner, The Cattleyas, f. 60, 61 (1988).

Like *C. amethystoglossa* except as below.

STEMS: 15–20 cm, swollen.

LEAVES: 6–10 cm, elliptic, spotted.

FLOWERS: 1 or 2 per raceme, to 9 cm wide.

Cattleya schilleriana – lip.

SEPALS AND PETALS: yellow to brown, spotted.

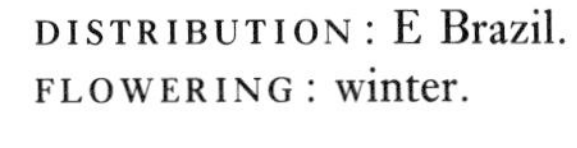

DISTRIBUTION: E Brazil.
FLOWERING: winter.

13. C. violacea (Kunth) Rolfe
SYNONYM: *C. superba* Lindley.
ILLUSTRATIONS: Botanical Magazine, 4083 (1944); Fowlie, Brazilian bifoliate Cattleyas, 48 (1977); Bechtel et al., Manual of cultivated orchid species, 182 (1981); Withner, The Cattleyas, f. 71 (1988).

Like *C. amethystoglossa* except as below.

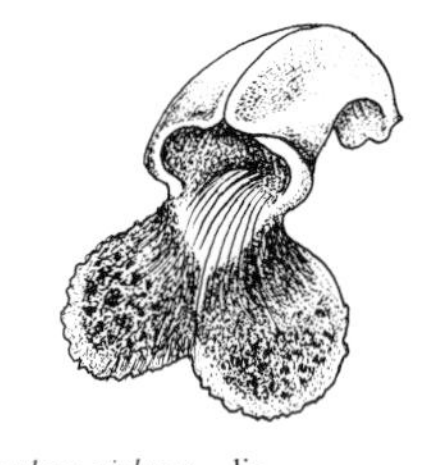

Cattleya violacea – lip.

FLOWERS: 1–5 per raceme.
SEPALS AND PETALS: pink, unspotted.
DISTRIBUTION: Venezuela, Guyana, Peru & Brazil.
FLOWERING: spring–summer.

14. C. guttata Lindley
SYNONYM: *C. elatior* Lindley.
ILLUSTRATIONS: Botanical Magazine, 3693 (1839); American Orchid Society Bulletin **45**: 979 (1976), **54**: 1316 (1985), **55**: 227 (1986); Fowlie, Brazilian bifoliate Cattleyas, 102 (1977); Withner, The Cattleyas, f. 27, 28 (1988).
HABIT: epiphytic or on rocks.
STEMS: 20–80 (rarely to 150) × 1–3 cm, slightly swollen above, with 3–8 nodes, 2- or 3-leaved.
LEAVES: 15–25 × 5–7 cm, elliptic, leathery.
FLOWERS: 3–20 per raceme, 7.5–10 cm wide.
SEPALS AND PETALS: 3.5–4.5 cm × 8–10 mm, yellowish green to brown, boldly spotted with purplish red; lateral sepals shorter, sickle-shaped; petals slightly wavy.
LIP: 2.6–3.3 × 2.5–3 cm, 3-lobed; lateral lobes diamond-shaped, rounded at base, rolled around column, pale pink; central lobe fishtail-shaped, tapering gradually into a long claw, ridged, reddish purple.
COLUMN: 2–2.5 cm.
DISTRIBUTION: S Brazil.
HARDINESS: G2.
FLOWERING: winter.

15. C. granulosa Lindley
ILLUSTRATIONS: Botanical Magazine, 5048 (1858); Fowlie, Brazilian bifoliate Cattleyas, 71 (1977); Bechtel et al., Manual of cultivated orchid species, 179 (1981); American Orchid Society Bulletin **55**: 227 (1986); Withner, The Cattleyas, f. 21 (1988).

Generally like *C. guttata* except as below.

LEAVES: 7–12 × 3.5–5 cm.
SEPALS AND PETALS: green, yellow or orange, finely spotted with red.

Cattleya granulosa – lip.

LIP: off-white to pale yellow; lateral lobes more or less triangular, not fully concealing column; central lobe kidney-shaped, tapering abruptly into a long narrow claw, pale pink, densely and finely spotted with purple.

DISTRIBUTION: E Brazil.

FLOWERING: spring.

16. C. **leopoldii** Lemaire

SYNONYM: *C. guttata* var. *leopoldii* (Lemaire) Rolfe.

ILLUSTRATIONS: American Orchid Society Bulletin **45**: 978 (1976); Fowlie, Brazilian bifoliate Cattleyas, 98 (1977); Bechtel et al., Manual of cultivated orchid species, 180 (1981); Withner, The Cattleyas, f. 42 (1988).

Generally like *C. guttata* except as below.

Cattleya leopoldii – lip.

FLOWERING STEM: 10–25 cm.

LIP: central lobe broad, fishtail-shaped, almost touching lateral lobes when flattened, abruptly contracted into claw, purple, ridged.

DISTRIBUTION: S Brazil.

FLOWERING: summer.

17. C. **elongata** Barbosa Rodrigues

ILLUSTRATIONS: Botanical Magazine, 7543 (1897); Fowlie, Brazilian bifoliate Cattleyas, 87 (1977); Withner, The Cattleyas, f. 18 (1988).

Very similar to *C. leopoldii*, differing as below.

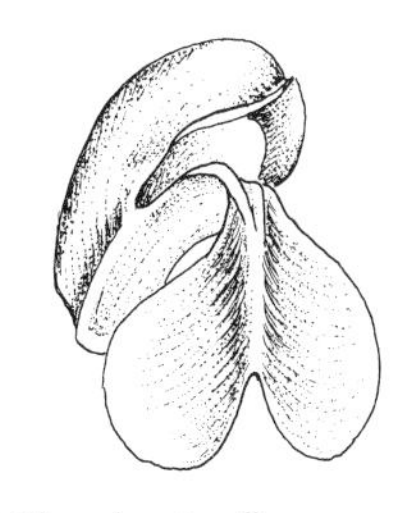
Cattleya elongata – lip.

FLOWERING STEM: 35–60 cm.

SEPALS AND PETALS: unspotted, copper-coloured.

DISTRIBUTION: E Brazil.

FLOWERING: autumn–winter.

18. C. **bicolor** Lindley

ILLUSTRATIONS: Botanical Magazine, 4909 (1856); Fowlie, Brazilian bifoliate Cattleyas, 23 (1977); Bechtel et al., Manual of cultivated orchid species, 178 (1981); Withner, The Cattleyas, f. 8 (1988).

HABIT: epiphytic.

STEMS: 30–80 cm, slender, slightly thicker above, with 4–6 nodes, 2-leaved.

LEAVES: 12–20 × 2–2.5 cm, oblong, leathery.

FLOWERS: 2–9 per raceme, 8–9 cm wide.

SEPALS: 4.5–6 × 1–1.5 cm; laterals broader, sickle-shaped, coppery to greenish brown.

PETALS: 4.5–5 × 1.5–2 cm, wavy, coppery to greenish brown.

LIP: 3.5 × 1.5 cm, without lobes, or with small basal flaps, long-clawed; blade fan-shaped, shallowly notched, wavy, minutely toothed, pink.

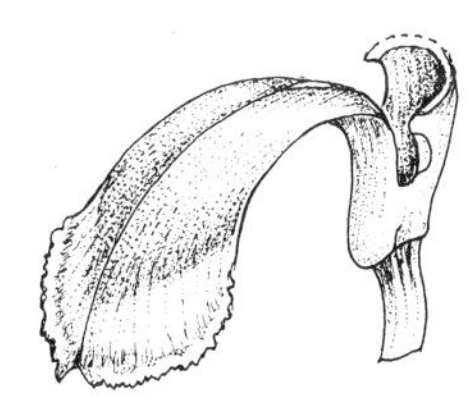
Cattleya bicolor – lip.

COLUMN: 3 × 1.2 cm, pale pink, winged.
DISTRIBUTION: S Brazil.
HARDINESS: G2.
FLOWERING: summer–autumn.

19. C. walkeriana Gardner
SYNONYM: *C. bulbosa* Lindley.
ILLUSTRATIONS: Shuttleworth et al., Orchids, 62 (1970); Fowlie, Brazilian bifoliate Cattleyas, 56 (1977); Bechtel et al., Manual of cultivated orchid species, 182 (1981); Withner, The Cattleyas, f. 72, 73 (1988).
HABIT: epiphytic or on rocks.
PSEUDOBULBS: 3–9 × 1–2 cm, 2- or 3-noded, bearing a single leaf (rarely 2-leaved).
LEAVES: 5–10 × 3–5 cm, elliptic to ovate, rounded or notched.
FLOWERS: 1–3 per raceme, 8–9 cm wide, borne on slender, leafless, lateral shoots.
SEPALS AND PETALS: 4.5–5.5 cm, pinkish purple; sepals 1.2–1.5 cm wide, lanceolate, acuminate; petals 2.5–3.5 cm wide, diamond-shaped to obovate, rounded.

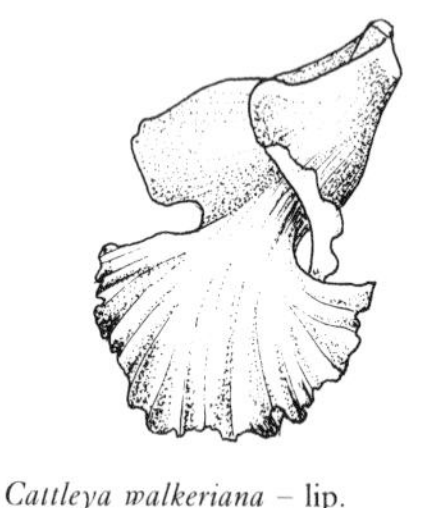
Cattleya walkeriana – lip.

LIP: 3.5–4.5 × 2.5–3.5 cm, 3-lobed, broadest across central lobe; lateral lobes triangular to obovate, pink, asymmetrically notched; sinus broad and rounded; central lobe almost circular, notched, dark purplish red with a narrow central patch of yellow.
COLUMN: 2.5–3 cm, pink, winged, exposed.
DISTRIBUTION: E Brazil.
HARDINESS: G2.
FLOWERING: spring.

20. C. nobilior Reichenbach
ILLUSTRATIONS: Orchid Digest **36**: 33 (1972); Fowlie, Brazilian bifoliate Cattleyas, 68 (1977); Pabst & Dungs, Orchidaceae Brasilienses 1: 192 (1975); Withner, The Cattleyas, f. 53 (1988).
Like *C. walkeriana* except as below.
LEAVES: 2 or 3.
FLOWERS: 10–12 cm wide.
SEPALS AND PETALS: pink.
LIP: central lobe pink with a large central yellow patch; lateral lobes of lip almost concealing column.
DISTRIBUTION: Brazil.
FLOWERING: summer.

21. C. lawrenceana Reichenbach
ILLUSTRATIONS: Botanical Magazine, 7133 (1890); Bechtel et al., Manual of cultivated orchid species, 180 (1981); Withner, The Cattleyas, f. 40 (1988).
HABIT: epiphytic.

Cattleya lawrenceana – lip.

STEMS: to 25 cm, swollen above, 2- or 3-noded, bearing a single leaf.

LEAVES: 8–20 × 4–5 cm, oblong, hard and leathery, sometimes purple-spotted.

FLOWERS: 3–8 per raceme, to 10 cm wide, pale to dark reddish purple, sometimes almost white; lip with a pale throat and a central reddish brown patch.

SEPALS: 6.5–7 × 1.5–2 cm, elliptic to oblong.

PETALS: 6.5 × 2.5–4 cm, oblong to broadly elliptic, obtuse, wavy.

LIP: 5.5–6.3 × 4 cm, oblong to ovate, rolled around column.

COLUMN: 2 cm.

DISTRIBUTION: Venezuela & Guyana.

HARDINESS: G2.

FLOWERING: spring–summer.

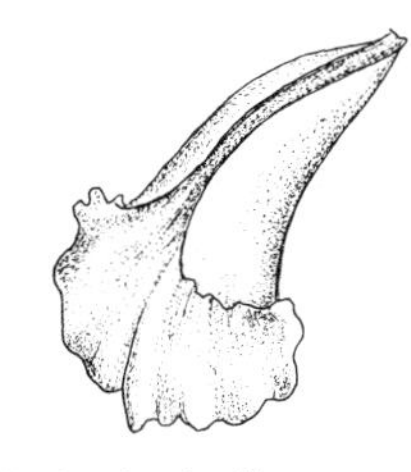

Cattleya luteola – lip.

22. C. luteola Lindley

ILLUSTRATIONS: Botanical Magazine, 5032 (1858); Bechtel et al., Manual of cultivated orchid species, 180 (1981); Withner, The Cattleyas, f. 46 (1988).

Generally like *C. lawrenceana*, differing as below.

FLOWERS: yellow.

SEPALS AND PETALS: 5 cm or less.

LIP: to 3 cm, obscurely 3-lobed.

DISTRIBUTION: Ecuador, Peru, Brazil & Bolivia.

FLOWERING: autumn.

23. C. labiata Lindley

ILLUSTRATIONS: Botanical Magazine, 3998 (1843); Bechtel et al., Manual of cultivated orchid species, 180 (1981); Withner, The Cattleyas, f. 39 (1988).

HABIT: epiphytic.

STEMS: to 30 cm, club-shaped, bearing a single leaf.

LEAVES: 15–30 × 4–7 cm, oblong, obtuse, leathery.

FLOWERS: 3–5 per raceme, *c.* 15 cm wide.

SEPALS AND PETALS: 7–8 cm, pale pink; sepals 2–2.5 cm wide, lanceolate, acute; petals 5–7 cm wide, elliptic, wavy.

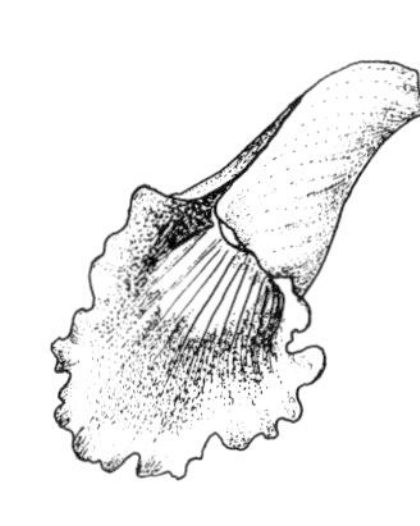

Cattleya labiata – lip.

LIP: 7–9 × 4.5–7 cm, ovate, shallowly 3-lobed, rolled around column, bright reddish purple with a pink margin; throat white and yellow.

COLUMN: 3–3.5 cm.

DISTRIBUTION: E Brazil.

HARDINESS: G2.

FLOWERING: autumn.

The following 9 species are recognised as such by Withner (1988). However, they seem to be merely colour-variants of *C. labiata*, and may be better considered as varieties of that species; see also Menezes, S.C., *Orchid Digest* **51**: 105–153 (1987).

24. C. dowiana Bateman

SYNONYM: *C. labiata* var. *dowiana* (Bateman & Reichenbach) Veitch.

ILLUSTRATIONS: Botanical Magazine, 5618 (1867); Bechtel et al., Manual of cultivated orchid species, 178 (1981); Withner, The Cattleyas, f. 14 (1988).

SEPALS AND PETALS: bright yellow.

LIP: crimson to purple.

DISTRIBUTION: Costa Rica & Colombia.

25. C. gaskelliana Reichenbach

SYNONYM: *C. labiata* var. *gaskelliana* (Reichenbach) Veitch.

ILLUSTRATIONS: Bechtel et al., Manual of cultivated orchid species, 179 (1981); Withner, The Cattleyas, f. 20 (1988).

STEMS: fusiform.

FLOWERS: 15–18 cm wide.

SEPALS AND PETALS: white to pale pink.

LIP: white to pale pink with a yellow throat and a small indistinct central patch of purplish pink separated from the yellow by an area of white or pale pink.

DISTRIBUTION: Venezuela.

FLOWERING: summer.

26. C. mendelii Backhouse

SYNONYM: *C. labiata* var. *mendelii* (Backhouse) Reichenbach.

ILLUSTRATIONS: Bechtel et al., Manual of cultivated orchid species, 181 (1981); Withner, The Cattleyas, f. 48 (1988).

Very like *C. mendelii*, differing as below.

LIP: with red-veined throat and larger central patch of purplish pink on the lip.

DISTRIBUTION: Colombia.

FLOWERING: spring–summer.

27. C. mossiae Hooker

SYNONYM: *C. labiata* var. *mossiae* (Hooker) Lindley.

ILLUSTRATIONS: Bechtel et al., Manual of cultivated orchid species, 181 (1981); Withner, The Cattleyas, f. 51 (1988).

Very similar to *C. gaskelliana* and *mendelii*, differing as below.

SEPALS AND PETALS: pink.

LIP: white; throat yellow with purple veins which merge with a large central patch of purplish pink.

DISTRIBUTION: Venezuela.

FLOWERING: spring–summer.

28. C. schroederae Reichenbach

SYNONYM: *C. labiata* var. *schroederae* (Reichenbach) Anon.

ILLUSTRATIONS: Bechtel et al., Manual of cultivated orchid species, 181 (1981); Withner, The Cattleyas, f. 63 (1988).

FLOWERS: very fragrant.
SEPALS AND PETALS: white or very pale pink.
LIP: pale pink with a purple tip.
DISTRIBUTION: Colombia.
FLOWERING: winter–spring.

29. C. trianae Linden & Reichenbach
SYNONYM: *C. labiata* var. *trianae* Linden & Reichenbach.
ILLUSTRATIONS: Bechtel et al., Manual of cultivated orchid species, 182 (1981); Withner, The Cattleyas, f. 68 (1988).
Like *C. schroederae* differing as below.
FLOWERS: less fragrant.
LIP: purple.
DISTRIBUTION: Colombia.
FLOWERING: winter–spring.

30. C. warneri Moore
SYNONYM: *C. labiata* var. *warneri* (Moore) Veitch.
ILLUSTRATIONS: Bechtel et al., Manual of cultivated orchid species, 182 (1981); Withner, The Cattleyas, f. 74 (1988).
FLOWERS: purplish pink.
LIP: darker with a white and yellow throat.
DISTRIBUTION: S Brazil.
FLOWERING: summer.

31. C. warscewiczii Reichenbach
SYNONYM: *C. labiata* var. *warscewiczii* (Reichenbach) Reichenbach.
ILLUSTRATIONS: Bechtel et al., Manual of cultivated orchid species, 182 (1981); Withner, The Cattleyas, f. 76 (1988).
FLOWERS: 17–20 cm wide.
LIP: mostly deep purplish crimson; throat greenish yellow, veined with crimson.
DISTRIBUTION: Colombia.

32. C. percivalliana O'Brien
SYNONYM: *C. labiata* var. *percivalliana* (O'Brien) Reichenbach.
ILLUSTRATIONS: Bechtel et al., Manual of cultivated orchid species, 181 (1981); Withner, The Cattleyas, f. 54 (1988).
SEPALS AND PETALS: pinkish purple.
LIP: maroon, with a yellow throat, margin pink.
DISTRIBUTION: Venezuela.
FLOWERING: winter–spring.

33. C. rex O'Brien
ILLUSTRATIONS: Cogniaux & Goossens, Dictionnaire Iconographique des Orchidées, Cattleya, t. 22 (1899); Botanical Magazine, 8377 (1911); Withner, The Cattleyas, f. 59 (1988).
Like *C. labiata* except as below.

Cattleya rex – lip.

Cattleya maxima – lip.

SEPALS AND PETALS: white to very pale yellow.
LIP: purple-veined, yellow in basal half, apically pink with a white border.
DISTRIBUTION: Peru & Colombia.
FLOWERING: summer.

34. C. **maxima** Lindley

ILLUSTRATIONS: Bechtel et al., Manual of cultivated orchid species, 180 (1981); Withner, The Cattleyas, f. 47 (1988).
Like *C. labiata* except as below.
FLOWERS: pale to deep pink.
PETALS: about twice the width of the sepals.
LIP: strongly veined with purple, with a central yellow stripe.
DISTRIBUTION: Ecuador, Colombia & Peru.
HARDINESS: G2.

53. LAELIA Lindley

A genus of about 50 species extending from C America and the West Indies south to Peru and Brazil, often used in bi- and trigeneric hybrids with *Cattleya* and *Brassavola*. Many of the species were originally described in *Bletia* or *Cattleya* and can only reliably be distinguished from the latter on the number of pollinia. Cultural treatment should be as for *Cattleya* (p. 167), though the Mexican species need a longer and cooler resting period.

Literature: Jones, H.G., Review of sectional division in the genus Laelia of the Orchidaceae, *Botanischer Jahrbücher* **97**: 309–16 (1976); Withner, C., *The Cattleyas and their relatives. II. The Laelias* (1988).

HABIT: epiphytic or terrestrial, often on rocks; rhizomes creeping, bearing erect pseudobulbous or cane-like stems of several nodes.
LEAVES: 1–3 per stem, folded when young.
FLOWERS: in terminal racemes.
SEPALS AND PETALS: free, rarely wavy.
LIP: 3-lobed or simple, usually internally ridged or keeled; lateral lobes or basal part curved around column, often concealing it.
COLUMN: free from lip, apical teeth lacking or adpressed to anther cap.
POLLINIA: 8, all equal, fertile.

Synopsis of characters:

Stem. Club-shaped: **1–7**; swollen at base: **8–10**; conical, ovoid or fusiform: **11–22**; very slender throughout: **8,9**. Ribbed: **14,15,17**. Of 1 internode: **20–22**. 7 cm or less: **10–22**; 7 cm or

more: **1–16.** 1-leaved: **1–16,10,19,20–22**; 2-leaved: **8,9,14,16,17,20–22**; 3-leaved: **14.**
Leaves. 20 cm or more: **1–9,14–16**; less than 10 cm: **11–13.** Less than 5 mm wide: **20–22**; 1.5–2.5 cm wide: **10,17,18,20–22**; 2.5–4.5 cm wide: **1–5,8–22**; more than 4.5 cm wide: **6,7.**
Flowering stem. 10 cm or less: **1–3,19–22**; 10–25 cm: **1–7,10,11–13,17–22**; 25–50 cm: **4–10,14–17**; more than 50 cm: **14–16.** Unsheathed: **19**; with large basal sheaths: **1–13,20–22**; with several small sheathing scales: **14–17.**
Flowers. 1 per stem: **14,15,19–22**; 2–7 per stem; **1–22**; 8–15 per stem: **8–15,17.** 6 cm wide or less: **1–3,8–13,16–18**; 7–12 cm wide: **1–5,14–16,20–22**; 13 cm wide or more: **1–7,19.**
Ovary. Glandular: **14,15.**
Sepals and petals. Pink to purple: **1–3,6,7,10–22**; white: **1–3,6,7,10–13,17,18,20–22**; yellow, orange or red: **1–5,8–13.** 4.5 cm or less: **1–3,8–13,17,20–22**; 5–7 cm: **1–5,14–16,20–22**; more than 7.5 cm: **4–6,19.**
Petals. Wider than sepals: **1–7,14–22**; as wide as sepals: **8–13,20–22.** Less than 1 cm wide: **8–13,17,20–22**; more than 4 cm wide: **6,7,19.** With recurved margins: **1–3.** Crisped: **1–3.**
Lip. Clearly 3-lobed: **1–5,8–22**; unlobed or obscurely 3-lobed: **6,7,20–22** Crisped: **11–13.** With reflexed central lobe: **14,15,17.** 3 × 2 cm or less: **8–13,17**; 4 × 2.5 cm or more: **1–7,14–16,19–22.** Lacking keels or ridges: **1–7**; 1-ridged: **16,19**; 2-ridged: **11–13,19**; 3-ridged: **8–10,14,15,17–22**; 4 or 5 ridged: **20–22**; 7-ridged: **11.**
Column. Concealed: **6,7,10–13,15–18,20–22**; exposed or partly exposed: **1–5,8–10,14,15,19.**

Key to species

1. Lip without internal keels or ridges — *Go to Subkey 1*
2. Flowering stem sheathed at base or with small sheathing scales at each node — *Go to Subkey 2*
3. Flowering stem without sheaths or scales — *Go to Subkey 3*

Subkey 1. Lip without internal keels or ridges.

1. Column distinctly less than half as long as lip
1.A. Petals more than twice as long as broad
1.A.a. Flowers 10–13 cm wide; sepals and petals greenish yellow to orange — **4. grandis**
1.A.b. Flowers 14–19 cm wide; sepals and petals greenish purple — **5. tenebrosa**
1.B. Flowers 15–20 cm wide; sepals and petals white to very pale pink — **6. purpurata**
1.C. Flowers 13–15 cm wide; sepals and petals pink — **7. lobata**
2. Sepals and petals yellow; flower less than 8 cm wide — **2. xanthina**
3. Margins of petals and lip strongly crisped — **3. crispa**

4. Margins of petals and lip not strongly crisped **1. perrinii**

Subkey 2. Lip with ridges or keels; flowering stem sheathed at base or with a small sheathing scale at each node.

1. Sepals distinctly narrower than petals
1.A. Flowers to 6 cm wide; lip to 2.5 × 1.6 cm
1.A.a. Leaves 2 per stem, 1–1.5 cm wide **18. albida**
1.A.b. Leaves 1 per stem, more than 2.5 cm wide **17. rubescens**
1.B. Lip not lobed, or very obscurely 3-lobed
1.B.a. Sepals and petals 3.5–4.5 cm; central lobe of lip dark purple **20. pumila**
1.B.b. Sepals and petals 6–7 cm; central lobe of lip white **21. jongheana**
1.C. Stems flattened; lateral lobes of lip pinkish purple outside, yellowish inside **16. anceps**
1.D. Stem with a single leaf **15. furfuracea**
1.E. Sepals and petals 5–6.5 cm, pinkish purple **14. autumnalis**
1.F. Sepals and petals 2–2.5 cm, white; plant of extensively creeping habit **22. lundii**
2. Flowers less than 5 cm wide; sepals and petals orange to red
2.A. Flowering stems exceeding leaves **8. cinnabarina**
2.B. Flowering stems shorter than leaves **9. harpophylla**
3. Lip more than 2 × 1.5 cm; sepals and petals yellow **10. flava**
4. Sepals and petals white or very pale yellow or very pale pink **13. longipes**
5. Sepals and petals pink with pale centres **11. crispata**
6. Sepals and petals pinkish purple with darker veins **12. crispilabia**

Subkey 3. Lip with ridges or keels; flowering stem without sheaths or scales.

1. Lateral lobes of lip spreading, exposing the column **19. speciosa**
2. Lateral lobes of lip covering and almost completely concealing the column **20. pumila**

1. L. **perrinii** Lindley

ILLUSTRATIONS: Botanical Magazine, 3711 (1840); Orchid Digest **41**: 149 (1977); Bechtel et al., Manual of cultivated orchid species, 219 (1981); Withner, The Laelias, f. 64 (1988).

HABIT: epiphytic.

STEM: 7–30 cm, ovoid to club-shaped, 1-leaved.

LEAVES: 20–30 × 3–4.5 cm, oblong to strap-shaped, obtuse or slightly notched, leathery, often spotted underneath.

FLOWERING STEM: 10–25 cm, robust, 2- or 3-flowered, basally sheathed.

FLOWERS: 10–13 cm wide.

SEPALS: upper 6.5 × 1.5 cm, lanceolate; laterals smaller, sickle-shaped, pink.

PETALS: 6.5 × 2 cm, ovate, pink.

LIP: 5 × 3.5 cm, 3-lobed; lateral lobes obtuse, white to pale pink, sometimes edged with purple; central lobe oblong to circular, crisped, with a broad purple margin and a pale yellow throat, lacking keels or ridges.

COLUMN: 3–4 cm, white to pale pink, partially concealed.

DISTRIBUTION: SE Brazil.

HARDINESS: G2.

FLOWERING: autumn.

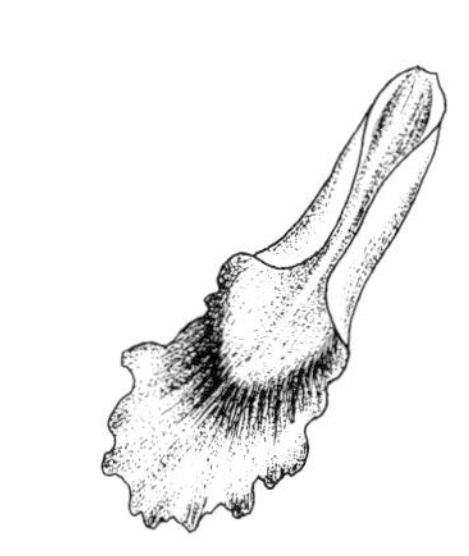

Laelia perrinii – lip.

2. L. xanthina Hooker

ILLUSTRATIONS: Botanical Magazine, 5144 (1859); Pabst & Dungs, Orchidaceae Brasilienses 1: 210 (1975); Bechtel et al., Manual of cultivated orchid species, 220 (1981); Withner, The Laelias, f. 91 (1988).

Similar to *L. perrinii*, differing as below.

FLOWERS: 4 or 5 per flowering stem, less than 8 cm wide.

SEPALS AND PETALS: yellow; petal margins recurved.

LIP: white and yellow, striped with pink.

DISTRIBUTION: E Brazil.

FLOWERING: spring–summer.

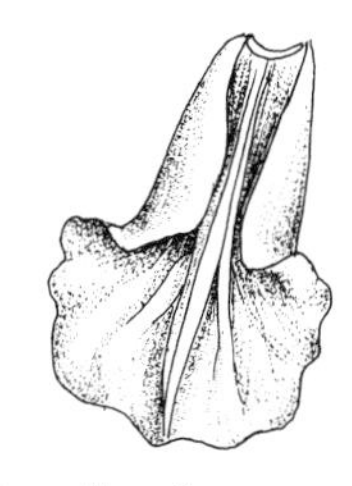

Laelia xanthina – lip.

3. L. crispa Reichenbach

ILLUSTRATIONS: Botanical Magazine, 3910 (1842); Orchid Digest 41: 144 (1977); Bechtel et al., Manual of cultivated orchid species, 218 (1981); Withner, The Laelias, f. 24, 25 (1988).

Like *L. perrinii* except as below.

SEPALS AND PETALS: white or very pale purple, tinged with pink at base; petals crisped and wavy.

LIP: strongly crisped, lateral lobes white, central lobe purple with darker veins and a white margin, throat yellow.

DISTRIBUTION: S Brazil.

FLOWERING: summer.

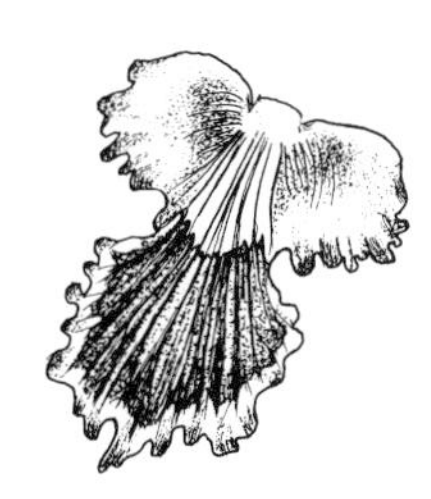

Laelia crispa – lip.

4. L. grandis Lindley

ILLUSTRATIONS: Botanical Magazine, 5553 (1866); Graf, Tropica, 733 (1978); Bechtel et al., Manual of cultivated orchid species, 219 (1981); Withner, The Laelias, f. 45 (1988).

HABIT: epiphytic.

STEM: 20–30 cm, club-shaped, 1-leaved.

LEAVES: 20–35 × 3–4.5 cm, strap-shaped, obtuse.

FLOWERING STEM: to 30 cm, basally sheathed, bearing 2–5 flowers.

FLOWERS: 10–13 cm wide.

SEPALS: 5–7 × 1–1.5 cm, lanceolate, greenish yellow to orange.

PETALS: 5–5.5 × 1.8–2 cm, diamond-shaped to elliptic, wavy, greenish yellow to orange.

Laelia grandis – lip.

LIP: 5.5–6 × 3.5 cm, 3-lobed, white with purple veins; lateral lobes obtuse; central lobe oblong, obtuse, crisped and wavy, lacking keels or ridges.

COLUMN: 2 cm, partially concealed.

DISTRIBUTION: E Brazil.

HARDINESS: G2.

FLOWERING: summer.

5. **L. tenebrosa** (Gower) Rolfe

SYNONYM: *L. grandis* var. *tenebrosa* Gower.

ILLUSTRATIONS: Ospina & Dressler, Orquideas de las Americas, t. 64 (1974); Orchid Digest **41**: 144 (1977); Bechtel et al., Manual of cultivated orchid species, 220 (1981); Withner, The Laelias, f. 87 (1988).

Like *L. grandis* except as below.

FLOWERS: 14–19 cm wide.

SEPALS: 8–10 × 1.6–2 cm, greenish purple.

PETALS: 7–10 × 2.5–3.5 cm, greenish purple.

LIP: 6.5–8.5 × 4–6 cm, pale pinkish purple, darker inside with a deep purple throat.

DISTRIBUTION: E Brazil.

6. **L. purpurata** Lindley

ILLUSTRATIONS: Graf, Tropica, 737 (1978); Williams et al., Orchids for everyone, 107 (1980); Bechtel et al., Manual of cultivated orchid species, 220 (1981); Withner, The Laelias, f. 70, 71 (1988).

HABIT: epiphytic.

STEM: to 30 × 3 cm, fusiform to club-shaped, 1-leaved.

LEAVES: 20–30 × 5 cm, oblong, rounded, leathery.

FLOWERING STEM: to 30 cm, basally sheathed, bearing 2–7 flowers.

FLOWERS: 15–20 cm wide.

SEPALS: 7.5–11 × 2 cm, lanceolate to strap-shaped, white to very pale pink.

PETALS: 8–10 × 4–6 cm, oblong to ovate, white to very pale pink.

Laelia purpurata – lip.

LIP: 7–10 × 6.5–8 cm, obscurely 3-lobed, wavy; lateral lobes curled around column; central lobe more or less circular; throat yellow, lateral lobes deep purple, margins paler, all veined with darker purple, lacking keels or ridges.

COLUMN: 2.5–3 cm, pale green, concealed.

DISTRIBUTION: S Brazil.

HARDINESS: G2.
FLOWERING: early summer.

7. L. lobata (Lindley) Veitch

SYNONYM: *L. boothiana* Reichenbach.
ILLUSTRATIONS: Pabst & Dungs, Orchidaceae Brasilienses **1**: 208 (1975); Orchid Digest **41**: 144 (1977); Withner, The Laelias, f. 55, 56 (1988).

Like *L. purpurata* except as below.

FLOWERS: 13–15 cm wide, pink.
LIP: 6 × 5 cm with dark veins, throat deep pink.
DISTRIBUTION: Brazil.
FLOWERING: spring.

8. L. cinnabarina Lindley

ILLUSTRATIONS: Botanical Magazine, 4302 (1847); Williams et al., Orchids for everyone, 191 (1980); Bechtel et al., Manual of cultivated orchid species, 218 (1981); Withner, The Laelias, f. 22, 23 (1988).
HABIT: on rocks.
STEMS: 10–25 cm, swollen below, 1- or 2-leaved.
LEAVES: 10–25 × 3–3.5 cm, linear to oblong, acute, often purplish green.
FLOWERING STEMS: 35–50 cm, basally sheathed, bearing 5–15 flowers.
FLOWERS: 5–6 cm wide, orange to red.
SEPALS AND PETALS: 3–3.5 cm × 5 mm, linear to lanceolate, acute.
LIP: 2–2.5 × 1 cm, 3-lobed, lateral lobes acute, curled around column; central lobe narrowly diamond-shaped to ovate, acute, with 3 ridges, crisped and reflexed.
COLUMN: club-shaped, mostly concealed.

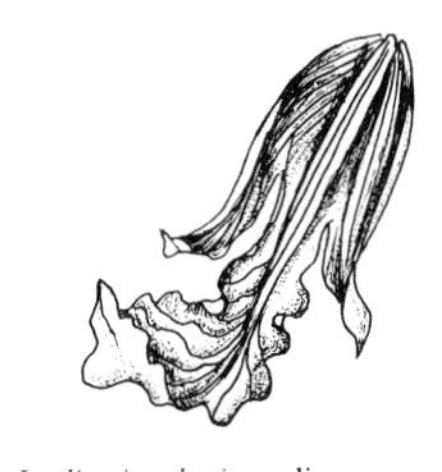

Laelia cinnabarina – lip.

DISTRIBUTION: SE Brazil.
HARDINSS: G2.
FLOWERING: spring.

9. L. harpophylla Reichenbach

ILLUSTRATIONS: Ospina & Dressler, Orquideas de las Americas, t. 63 (1974); Graf, Tropica, 733 (1978); Bechtel et al., Manual of cultivated orchid species, 219 (1981); Withner, The Laelias, f. 46, 47 (1988).

Like *L. cinnabarina* except as below.

HABIT: epiphytic.
STEMS: very slender; flowering stem shorter than leaves, bearing 4–7 flowers.
DISTRIBUTION: SE Brazil.
FLOWERING: winter–spring.

10. L. flava Lindley

ILLUSTRATIONS: Orchid Digest **38**: 69 (1974); Graf, Tropica, 733 (1978); Bechtel et al., Manual of cultivated orchid species, 218 (1981); American Orchid Society Bulletin **54**: 825 (1985); Withner, The Laelias, f. 37 (1988).

HABIT: on rocks.

STEMS: 5–20 cm, clustered, swollen below, purplish green, 1-leaved.

LEAVES: 8–15 × 2.5–3 cm, lanceolate to oblong, obtuse, leathery, purple beneath.

FLOWERING STEM: 20–45 cm, basally sheathed, bearing 5–10 flowers.

FLOWERS: 3–4.5 cm wide, yellow.

SEPALS AND PETALS: 2–2.5 cm × 6 mm, linear to narrowly lanceolate, acute.

LIP: 2.5–3 × 2 cm, 3-lobed; lateral lobes blunt, curled around column; central lobe oblong, obtuse, crisped, wavy and slightly reflexed, 3-ridged.

COLUMN: 9 mm, mostly concealed.

DISTRIBUTION: SE Brazil.

HARDINESS: G2.

FLOWERING: winter, spring & early summer.

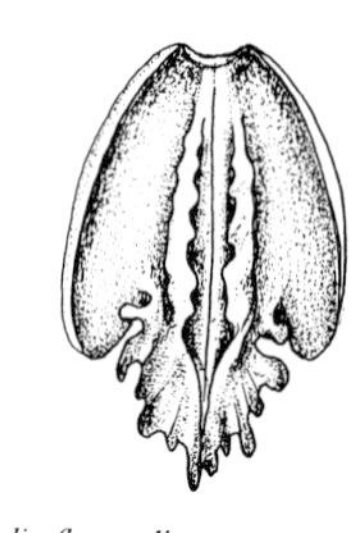

Laelia flava – lip.

11. L. crispata (Thunberg) Garay

SYNONYM: *L. rupestris* Lindley.

ILLUSTRATIONS: American Orchid Society Bulletin **43**: 992 (1974); Pabst & Dungs, Orchidaceae Brasilienses 1: 212 (1975); Northen, Miniature orchids, C-37 (1980); Withner, The Laelias, f. 26 (1988).

HABIT: on rocks.

STEMS: 4–10 cm, not swollen, 1-leaved.

LEAVES: 10–15 × 3–3.5 cm, strap-shaped to oblong, obtuse, leathery.

FLOWERING STEM: 15–25 cm, with a single basal sheath, bearing 2–10 flowers.

FLOWERS: 4–5 cm wide.

SEPALS AND PETALS: 2–2.5 cm × 8–10 mm, oblong to elliptic, obtuse to rounded, pink with pale centres.

LIP: 1.5–2 × 1.2 cm, 3-lobed; lateral lobes oblong, curled around column, pale pink central lobe ovate to oblong, crisped and wavy, yellow with a pink margin, 2-ridged.

COLUMN: concealed.

DISTRIBUTION: SE Brazil.

HARDINESS: G2.

FLOWERING: winter.

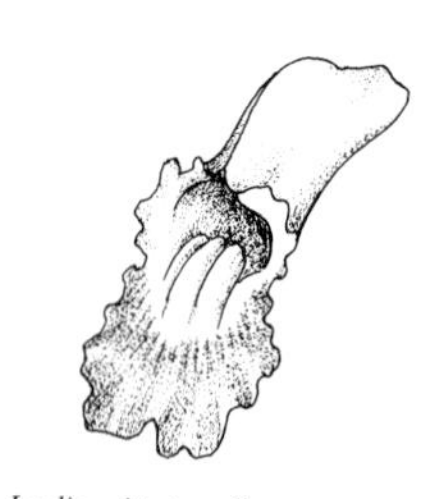

Laelia crispata – lip.

12. L. crispilabia Warner

ILLUSTRATIONS: American Orchid Society Bulletin **43**: 993

(1974); Orchid Digest 38: 69 (1974); Withner, The Laelias, f. 27 (1988).

Like *L. crispata* except as below.

STEMS: pear-shaped.

LEAVES: 6–8 cm.

SEPALS AND PETALS: pinkish purple with darker veins.

LIP: similar in colour to sepals and petals but with dark margins.

DISTRIBUTION: SE Brazil.

FLOWERING: summer.

13. L. longipes Reichenbach

ILLUSTRATIONS: Botanical Magazine, 7541 (1897); American Orchid Society Bulletin 43: 993 (1974); Bechtel et al., Manual of cultivated orchid species, 218 (1981); Withner, The Laelias, f. 57 (1988).

Like *L. crispata* except as below.

STEMS: conical to ovoid.

SEPALS AND PETALS: white, very pale pink or pale yellow.

LIP: pale yellow with a deep yellow throat.

DISTRIBUTION: SE Brazil.

FLOWERING: summer.

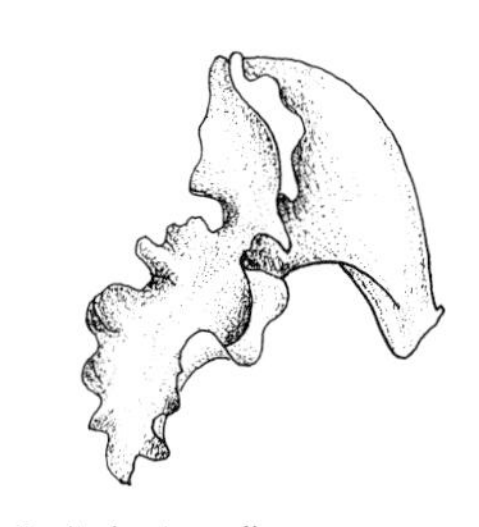

Laelia longipes – lip.

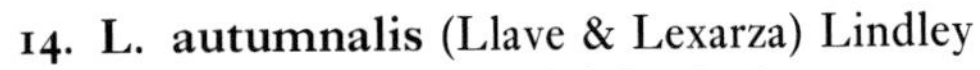

14. L. autumnalis (Llave & Lexarza) Lindley

SYNONYM: *L. gouldiana* Reichenbach.

ILLUSTRATIONS: Botanical Magazine, 3817 (1841); Orchid Digest 42: 20 (1978); Bechtel et al., Manual of cultivated orchid species, 218 (1981); Withner, The Laelias, f. 11 (1988).

HABIT: epiphytic or on rocks.

STEMS: 10–15 cm, conical to pear-shaped, 2- or 3-leaved, ribbed.

LEAVES: 12–20 × 3–4 cm, oblong to lanceolate, acute or obtuse, leathery.

FLOWERING STEMS: 30–100 cm, with several small sheathing scales, bearing 4–10 flowers.

FLOWERS: 7.5–12 cm wide.

SEPALS AND PETALS: 5–6.5 cm, acuminate, pinkish purple with darker streaks; sepals 1.3 cm wide, lanceolate; petals 1.8 cm wide, ovate.

LIP: 4–5 × 2.5–3 cm, 3-lobed; lateral lobes rounded, partially enclosing column, white, central lobe oblong, acuminate, reflexed, purple with a white and yellow base, 3-ridged.

COLUMN: 2.5–3 cm.

DISTRIBUTION: S Mexico.

HARDINESS: G2.

FLOWERING: winter.

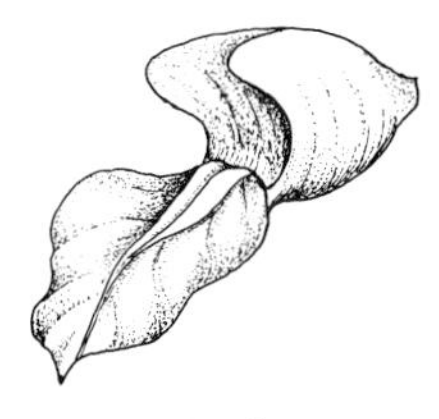

Laelia autumnalis – lip.

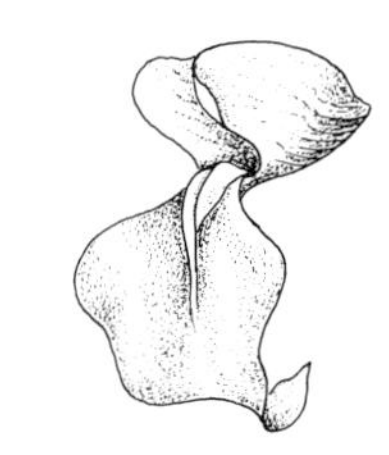

Laelia furfuracea – lip.

15. L. furfuracea (Lindley) Rolfe

SYNONYM: *L. autumnalis* var. *furfuracea* Lindley.

ILLUSTRATIONS: Botanical Magazine, 3810 (1841); Northen, Miniature orchids, C-40 (1980); Withner, The Laelias, f. 39 (1988).

Like *L. autumnalis* except as below.

STEMS: 4–6 cm, ovoid, deeply grooved, 1-leaved.

FLOWERS: 1–3 per flowering stem.

OVARY: mealy-glandular.

DISTRIBUTION: S Mexico.

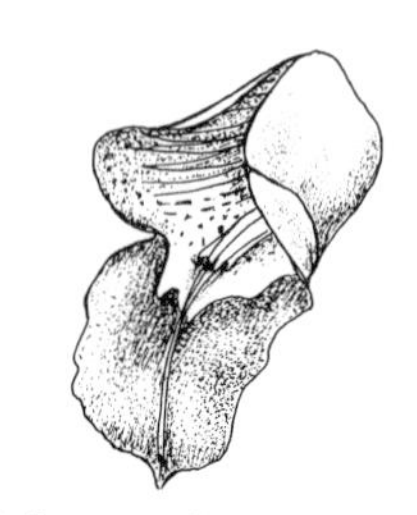

Laelia anceps – lip.

16. L. anceps Lindley

ILLUSTRATIONS: Botanical Magazine, 3804 (1841); Williams et al., Orchids for everyone, 107 (1980); Bechtel et al., Manual of cultivated orchid species, 218 (1981); Withner, The Laelias, f. 7 (1988).

Similar to *L. autumnalis*, differing as below.

STEMS: 7–12 cm, flattened, 1- or 2-leaved.

FLOWERING STEM: bearing 2–5 flowers.

LIP: lateral lobes pinkish purple outside, yellow with purple veins and border inside; central lobe blunt, dark reddish purple, 1-ridged.

DISTRIBUTION: C Mexico.

17. L. rubescens Lindley

SYNONYMS: *L. acuminata* Lindley; *L. peduncularis* Lindley.

ILLUSTRATIONS: Botanical Magazine, 4099 (1844); Orchid Digest **42**: 20 (1978); Graf, Tropica, 733 (1978); Withner, The Laelias, f. 76, 77 (1988).

HABIT: epiphytic or on rocks.

STEMS: 2.5–4.5 × 1.5–2.5 cm, clustered, ovoid, flattened, smooth, becoming furrowed, 1-leaved.

LEAVES: 10–14 × 3–3.5 cm, oblong to lanceolate, obtuse, channelled, leathery.

FLOWERING STEM: 25–50 cm, with several small sheathing scales, bearing 2–10 flowers.

FLOWERS: 5–6 cm wide.

SEPALS: 3.5 cm × 5 mm, lanceolate, acute, white to pinkish purple.

PETALS: 2.5–3 cm × 8–10 mm, ovate to oblong, acute, wavy, white to pinkish purple.

Laelia rubescens – lip.

LIP: 2.5–3 × 1.5 cm, 3-lobed; lateral lobes oblong, tips flared; central lobe oblong, acute, reflexed, wavy, white to pinkish purple; throat with dark purple patch surrounded by yellowish green.

COLUMN: 8 mm, concealed.

DISTRIBUTION: S Mexico, Guatemala, Nicaragua & Costa Rica.

HARDINESS: G2.
FLOWERING: winter.

Laelia albida – lip.

18. L. albida Lindley
ILLUSTRATIONS: Botanical Magazine, 3957 (1842); Graf, Tropica, 733 (1978); Bechtel et al., Manual of cultivated orchid species, 217 (1981); Withner, The Laelias, f. 5 (1988).
Like *L. rubescens* except as below.
LEAVES: 2 per stem, less than 1.5 cm wide, linear to lanceolate.
FLOWERS: white, faintly flushed with pink.
LIP: 3-ridged.
DISTRIBUTION: Mexico.

19. L. speciosa (Humboldt, Bonpland & Kunth) Schlechter
SYNONYMS: *L. grandiflora* Lindley; *L. majalis* Lindley.
ILLUSTRATIONS: Botanical Magazine, 5667 (1867); Die Orchidee **22**: 239 (1974); Orchid Digest **42**: 21 (1978); Withner, The Laelias, f. 83, 84 (1988).
HABIT: epiphytic.
STEMS: 3–5 cm, ovoid to fusiform, clustered, 1-leaved.
LEAVES: 10–15 cm, oblong to lanceolate, acute or obtuse, stiff.
FLOWERING STEMS: 10–20 cm, lacking sheathing scales, bearing 1 or 2 flowers.
FLOWERS: 15–20 cm wide.
SEPALS: 10–12 × 1.5–2.5 cm, lanceolate, acute, pale pinkish purple.
PETALS: 8–10 × 4.5–6 cm, ovate to diamond-shaped, obtuse, wavy, pale pinkish purple.
LIP: 7.5–9 × 4.5–5.5 cm, deeply 3-lobed; lateral lobes ovate to oblong, rounded, flared away from column, white to pale pinkish purple, edged with darker pink; central lobe oblong to circular, with 1–3 yellow keels, notched, white with pink streaks and a broad pink or pinkish purple border.
COLUMN: exposed.
DISTRIBUTION: C Mexico.
HARDINESS: G2.
FLOWERING: spring–summer.

Laelia speciosa – lip.

20. L. pumila (Hooker) Reichenbach
ILLUSTRATIONS: Botanical Magazine, 3656 (1839); Northen, Miniature orchids, C-40 (1980); Bechtel et al., Manual of cultivated orchid species, 220 (1981); Withner, The Laelias, f. 67–69 (1988).
HABIT: epiphytic.
STEMS: 2–3 cm, ovoid to fusiform, 1-leaved.
LEAVES: 10–13 × 2–2.5 cm, linear to oblong, obtuse.
FLOWERING STEM: 4–10 cm, 1 (rarely 2)-flowered.
FLOWERS: 8–12 cm wide.

SEPALS: 4.5 × 1–1.5 cm, oblong to lanceolate, acute, pinkish purple.

PETALS: 3.5–4.5 × 2–3 cm, elliptic, acute or obtuse, wavy, pinkish purple.

LIP: 4–4.5 × 3 cm, shallowly 3-lobed or unlobed, basal part pinkish purple, curled around column; central lobe oblong to circular, with 3–5 ridges, wavy, sometimes notched, dark purple with a white or pale yellowish green throat.

COLUMN: 2–2.5 cm, concealed.

DISTRIBUTION: SE Brazil.

HARDINESS: G2.

FLOWERING: autumn.

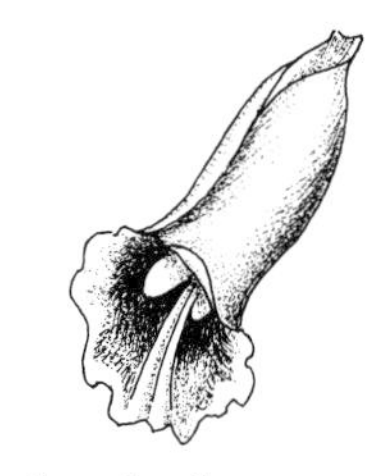

Laelia pumila – lip.

21. L. jongheana Reichenbach

ILLUSTRATIONS: Botanical Magazine, 6038 (1873); Northen, Miniature orchids, C-41 (1980); Bechtel et al., Manual of cultivated orchid species, 285 (1981); Withner, The Laelias, f. 50, 51 (1988).

Generally similar to *L. pumila* except as below.

LEAVES: 3.5–4 cm wide.

SEPALS AND PETALS: 6–7 cm.

LIP: 5.5 × 3 cm; central lobe white, throat white and yellow with 7 wavy keels.

DISTRIBUTION: SE Brazil.

FLOWERING: spring.

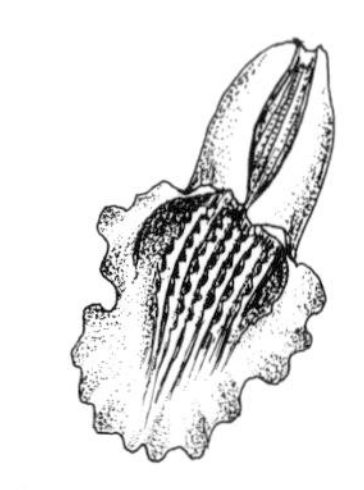

Laelia jongheana – lip.

22. L. lundii Reichenbach

SYNONYM: *L. regnellii* Rodrigues.

ILLUSTRATION: Northen, Miniature orchids, C-42 (1980); Skelsey, Orchids, 117 (1980); Bechtel et al., Manual of cultivated orchid species, 219 (1981); Withner, The Laelias, f. 60 (1988).

HABIT: rhizomes extensively creeping.

LEAVES: 2 per stem, 8–10 cm × 5 mm, linear, channelled and pointed.

FLOWERING STEM: usually 1-flowered.

SEPALS AND PETALS: 2–2.5 cm × 3–4 mm, white.

LIP: clearly 3-lobed, white with dense pink or red veins, wavy.

COLUMN: concealed.

DISTRIBUTION: SE Brazil.

HARDINESS: G2.

54. SCHOMBURGKIA Lindley

A genus of 12–15 species from the American tropics, many of which were originally described in *Cattleya* or *Laelia*. Cultural requirements are similar to those for *Cattleya* (p. 000) though the plants perform best if repotted less frequently.

HABIT: epiphytic or on rocks; rhizomes branched, creeping.

PSEUDOBULBS: erect, cylindric to fusiform, few-noded.

LEAVES: 2 or 3 (rarely 4) per pseudobulb, folded when young, narrowly ovate to oblong.

FLOWERS: several, in tall terminal racemes or panicles.

FLOWERING STEMS: with a sheathing scale at each node.

FLOWER-STALKS: with long or short bracts.

SEPALS AND PETALS: similar, free, spreading, wavy.

LIP: 3-lobed; lateral lobes loosely enclosing column, becoming erect or spreading; central lobe rounded or notched, often curled downwards, sometimes with keels or ridges.

COLUMN: stout, with a single, adpressed, apical tooth.

POLLINIA: 8.

Key to species

1. Bracts less than half the length of the flower-stalks **5. tibicinis**
2. Flowers more than 9 cm wide **1. superbiens**
3. Flowers 7.5–8.5 cm wide **2. gloriosa**
4. Sepals and petals cream with purplish spots **4. lyonsii**
5. Sepals and petals purplish brown **3. undulata**

1. S. superbiens (Lindley) Rolfe

ILLUSTRATIONS: Botanical Magazine, 4090 (1844); Orchid Digest **42**: 21 (1978); Bechtel et al., Manual of cultivated orchid species, 272 (1981).

PSEUDOBULBS: to 1 m or more, fusiform, flattened, 1- or 2-leaved.

LEAVES: to 35 × 6 cm, oblong, acute, leathery and stiff.

FLOWERING STEM: to 1 m or more, robust, bearing 12–20 flowers.

FLOWERS: 10–12 cm wide, fragrant; ovary plus flower-stalk 6–8 cm.

BRACTS: 5–7 cm.

SEPALS AND PETALS: 5–7 × 1.3–1.8 cm, oblong to narrowly lanceolate, acute, pink to pinkish purple, paler at base; petals slightly wavy.

LIP: 4–5.5 × 3–3.5 cm; lateral lobes rounded, crisped, greenish yellow with dark purple veins and a pink margin; central lobe ovate, notched, wavy, with 5 yellow ridges and a pink blade.

COLUMN: 2.5–3 cm, club-shaped, curved.

DISTRIBUTION: Mexico, Guatemala & Honduras.

HARDINESS: G2.

FLOWERING: winter.

2. S. gloriosa Reichenbach

SYNONYMS: *S. crispa* Lindley; *S. crispa* misapplied, not *Laelia crispa* (Lindley) Reichenbach.

ILLUSTRATIONS: Pabst & Dungs, Orchidaceae Brasilienses **1**: 218 (1975); Dunsterville & Garay, Venezuelan orchids illustrated **6**: 398 (1976); Bechtel et al., Manual of cultivated orchid species, 383 (1981).

PSEUDOBULBS: to 30 × 5 cm, club-shaped with long tapering bases, ridged, 2-or 3-leaved.

LEAVES: to 35 × 6 cm, oblong to lanceolate, acute or obtuse.

FLOWERING STEM: to 75 cm, slender, bearing 8–15 flowers.

FLOWERS: 7.5–8.5 cm wide; ovary plus flower-stalk to 9 cm.

BRACTS: to 6 cm.

SEPALS AND PETALS: 3.5–4 × 1 cm, oblong, obtuse (sepals each with a small point), wavy, yellowish brown with darker veins.

LIP: 2 × 2 cm, white to pale pink with yellowish brown margins and central lobe; lateral lobes kidney-shaped to ovate; central lobe broadly triangular, rounded, crisped, curled downwards, 3-ridged.

COLUMN: 2 cm.

DISTRIBUTION: Venezuela, Guyana & Surinam.

HARDINESS: G2.

FLOWERING: winter.

3. S. **undulata** Lindley

ILLUSTRATIONS: Dunsterville & Garay, Venezuelan orchids illustrated **2**: 256 (1961); Skelsey, Orchids, 38 (1979); Bechtel et al., Manual of cultivated orchid species, 272 (1981).

PSEUDOBULBS: to 25 cm, club-shaped with long tapering bases, ridged, 2-leaved.

LEAVES: to 30 × 5 cm, lanceolate to oblong, obtuse, leathery.

FLOWERING STEM: to 50 cm, slender, bearing 5–20 flowers.

FLOWERS: 5–6 cm wide; ovary plus flower-stalk 5–7 cm, purplish green.

BRACTS: *c.* 5 cm.

SEPALS AND PETALS: 3–3.5 × 1.5 cm, oblong, acute, crisped and very wavy, purplish brown.

LIP: 3 × 1.5–2 cm, purple; lateral lobes broad and shallow, erect; central lobe oblong, shallowly notched, slightly curled downwards, wavy, 5-ridged.

COLUMN: 1.5 cm.

DISTRIBUTION: Venezuela & Trinidad.

HARDINESS: G2.

FLOWERING: summer.

4. S. **lyonsii** Lindley

ILLUSTRATIONS: Botanical Magazine, 5172 (1860); Gardeners' Chronicle **26**: 203 (1899).

Like *S. undulata* except as below.

FLOWERS: *c.* 5 cm wide, cream-coloured with purple spots.

DISTRIBUTION: Jamaica.

FLOWERING: summer–autumn.

5. S. tibicinis Lindley

ILLUSTRATIONS: Botanical Magazine, 4476 (1849); Shuttleworth et al., Orchids, 66 (1970); Skelsey, Orchids 137 (1979).

HABIT: epiphytic, very large and robust.

PSEUDOBULBS: to 60 × 8 cm or more, fusiform, hollow, 2- or 3-leaved.

LEAVES: to 45 × 8 cm, elliptic to oblong, rounded or obtuse.

FLOWERING STEMS: to 5 m, branched, bearing 10–20 flowers.

FLOWERS: 6–8 cm wide, opening in succession; ovary plus flower-stalk 3–5 cm.

BRACTS: 1 cm.

SEPALS AND PETALS: 4–5 × 1.5–2 cm, oblong to spathulate, obtuse or rounded, purplish blue flushed with brown, paler beneath.

LIP: 4–5 × 3–4 cm, 3-lobed, yellow, white and purple; lateral lobes semicircular to oblong; central lobe diamond-shaped, curled downwards.

DISTRIBUTION: Mexico to Panama.

HARDINESS: G2.

FLOWERING: spring.

55. BRASSAVOLA R. Brown

A genus of 15–20 species extending from C America and the West Indies south to Brazil and Argentina, often used as a parent in bi- and trigeneric hybrids with *Cattleya* and *Laelia*. Species Nos. **1** and **2** are sometimes treated as the separate genus *Rhyncholaelia* Schlechter, as they are distinct in stem and leaf shape. Cultural treatment should be as for *Epidendrum* (p. 148) though the *Brassavola* growth habit is more easily accommodated on blocks of tree fern or cork bark. They should have a definite winter rest.

Literature: Jones, H.G., Nomenclatural revision of the genus Brassavola of the Orchidaceae, *Annalen des Naturhistorischen Museums in Wien* **79**: 9–22 (1975).

HABIT: epiphytic or on rocks; rhizomes creeping, bearing erect, narrow stems or pseudobulbs.

STEMS OR PSEUDOBULBS: sheathed, 1-leaved.

LEAVES: fleshy, mostly cylindric or linear, sometimes flattened and leathery, folded when young.

FLOWERS: terminal (except No. **9**), solitary or in racemes.

SEPALS AND PETALS: free and spreading, similar, linear to narrowly lanceolate.

LIP: unlobed, or obscurely 3-lobed, base or claw rolled around and often concealing column, blade or central lobe oblong, circular or heart-shaped.

COLUMN: free from lip, with acute projecting teeth at tip.

POLLINIA: 8 (often a few extra abortive).

Synopsis of characters

Plant. Erect: **1–8**; pendent: **9**.
Stems. Pseudobulbous: 1,2; cane-like: 3–9.
Leaves. Flat: **1,2**; cylindric or linear: 3–9. Glaucous: **1,2**. Less than 12 cm long: **1,4–6**; 12–40 cm long: **1–8**; more than 40 cm: **4–6.9**. Less than 1 cm wide: **3–8**; 1–2 cm wide: **9**; 2–4 cm wide: **1–6**; 5 cm wide or more: 2.
Flowers. Lateral: **9**; terminal: 1–9. 1–3 per raceme: **1–3,7–9**; **4** or more per raceme: **4–8**. Less than 7 cm wide: **4–8**; more than 7.5 cm wide: **1–3,7–9**.
Petals and sepals. Spotted or warty: **4–8**. Less than 5 cm long: 4–6; 5 cm long or more: **1–3,7–9**. Less than 1 cm wide: 3–9; more than 1 cm wide: **1–3**.
Lip. Unclawed or with very short claw: **1–6**; with claw about one-third of total lip length: **7–9**. Lacerate: **2,3**. Long and gradually tapering: **3**. 3–4 cm long: **4–6**; 3.5–5.5 cm long: **1,7,8**; 6 cm long or more: **2,3,9**. Less than 4.5 cm wide: **1,3–8**; more than 4.5 cm wide: **2,9**.

Key to species

1. Leaves flattened; stems pseudobulb-like
1.A. Lip margin entire; column less than 2 cm **1. glauca**
1.B. Lip margin deeply and irregularly toothed; column more than 2 cm **2. digbyana**
2. Lip 6 cm or more, long and gradually tapering **3. cucullata**
3. Lip without a claw or with a very short claw
3.A. Leaves 35–50 cm × 5 mm; petals 2–3 mm wide **6. flagellaris**
3.B. Flowers 1–2; sepals greenish yellow **4. perrinii**
3.C. Flowers 1–12; sepals cream with red spots **5. tuberculata**
4. Plant pendent; flowering stem lateral **9. acaulis**
5. Flower 7.5–8.5 cm wide, not spotted **7. nodosa**
6. Flower 5–7 cm wide, spotted with red **8. subulifolia**

1. **B. glauca** Lindley

SYNONYMS: *Rhyncholaelia glauca* (Lindley) Schlechter; *Laelia glauca* (Lindley) Bentham.
ILLUSTRATIONS: Botanical Magazine, 4033 (1843); Shuttleworth et al., Orchids, 68 (1970); Bechtel et al., Manual of cultivated orchid species, 172 (1981).
HABIT: epiphytic or terrestrial.
PSEUDOBULBS: 2–10 cm, fusiform to club-shaped, 1-leaved.
LEAVES: 5–12.5 × 2.5–3.5 cm, oblong to elliptic, obtuse, leathery and glaucous.
FLOWERING STEMS: to 10 cm, terminal, 1-flowered.
FLOWER: 8–10 cm wide, very fragrant.
SEPALS AND PETALS: similar, 5–6.5 cm, linear to lanceolate,

obtuse, white to pale green or purple; sepals 1–1.5 cm wide, margins reflexed; petals 1.5–2 cm wide, margins wavy.

LIP: 5–5.5 × 4 cm, obscurely 3-lobed, wavy; lateral lobes rounded, bases rolled around column; central lobe oblong, abruptly pointed, throat with a narrow purple patch.

COLUMN: 5 mm, partly exposed.

DISTRIBUTION: Mexico to Panama.

HARDINESS: G2.

FLOWERING: winter–spring.

2. **B. digbyana** Lindley

SYNONYMS: *Rhyncholaelia digbyana* (Lindley) Schlechter; *Laelia digbyana* (Lindley) Bentham.

ILLUSTRATIONS: Botanical Magazine, 4474 (1849); Williams et al., Orchids for everyone, 107 (1980); Bechtel et al., Manual of cultivated orchid species, 171 (1981); American Orchid Society Bulletin **57**: 269 (1988).

HABIT: epiphytic.

PSEUDOBULBS: 12–20 cm, club- shaped, 3- or 4-noded, 1-leaved.

LEAVES: 12–20 × 5–6 cm, elliptic, obtuse, leathery and slightly glaucous.

FLOWERING STEM: to 15 cm, terminal, 1-flowered.

FLOWER: 10–12 cm wide, fragrant.

SEPALS: 8–9.5 × 2–2.5 cm, lanceolate, obtuse, very pale green to greenish yellow, margins wavy.

PETALS: similar to sepals but somewhat wider.

LIP: 7–8 × 7.5 cm, obscurely 3-lobed, almost circular, margins deeply and finely lacerate; lateral lobes semicircular, bases rolled around but not concealing column; central lobe rectangular to oblong.

COLUMN: 3–3.5 cm.

DISTRIBUTION: Mexico, Belize & Honduras.

HARDINESS: G2.

FLOWERING: summer.

3. **B. cucullata** (Linnaeus) R. Brown

SYNONYM: *B cuspidata* Hooker.

ILLUSTRATIONS: Botanical Magazine, 3722 (1839); Dunsterville & Garay, Venezuelan orchids illustrated **2**: 43 (1961); Bechtel et al., Manual of cultivated orchid species, 171 (1981).

HABIT: epiphytic.

STEMS: to 25 cm, cylindric, thin, several-noded, 1-leaved.

LEAVES: 20–40 cm × 5–10 mm, cylindric to linear, fleshy.

FLOWERING STEMS: short, bearing 1–3 flowers.

FLOWERS: 8–10 cm wide, fragrant, white and greenish yellow tinged with pink or brown.

OVARY PLUS FLOWER-STALK: 15–25 cm.

SEPALS AND PETALS: similar, 4 cm × 5–10 mm, lanceolate, tapering.

LIP: 6.5–10 × 1.5–2.5 cm, base circular to heart-shaped with lacerate margin; tip long and tapering.

COLUMN: 1.5–2 cm, partly exposed.

DISTRIBUTION: Mexico & West Indies south to Ecuador.

HARDINESS: G2.

FLOWERING: autumn–winter.

4. **B. perrinii** Lindley

ILLUSTRATIONS: Botanical Magazine, 3761 (1839); Cogniaux & Goossens, Dictionnaire Iconographique des Orchidées, Brassavola, t. 2 (1904).

HABIT: epiphytic or on rocks.

STEMS: to 20 cm × 5 mm, cylindric or slightly angular, branched, 3- or 4-noded, 1-leaved.

LEAVES: 10–25 cm × 5–80 mm, circular to triangular in cross-section, acute, deeply grooved, erect.

FLOWERING STEM: 2–8 cm, terminal, 1- or 2-flowered.

FLOWERS: 5–6 cm wide.

OVARY PLUS FLOWER-STALK: 5–10 cm.

SEPALS AND PETALS: 3.5–4.5 cm × 4–6 mm, narrowly lanceolate, acute, greenish yellow.

LIP: 3–4 × 2–2.5 cm, ovate, acute, base rolled around column, white with a central patch of greenish yellow and green veins.

COLUMN: 1 cm, partly exposed.

DISTRIBUTION: SE Brazil & Paraguay.

HARDINESS: G2.

FLOWERING: spring–summer.

5. **B. tuberculata** Hooker

SYNONYM: *B. fragrans* Lemaire.

ILLUSTRATIONS: Botanical Magazine 2878 (1829); Pabst & Dungs, Orchidaceae Brasilienses 1: 219 (1975); Bechtel et al,, Manual of cultivated orchid species, 172 (1981).

Like *B. perrinii* except as below.

FLOWERS: 1–12.

SEPALS: cream-coloured, marked with reddish purple spots and warts.

LIP: ovate, blunt, white with a central patch of yellow.

DISTRIBUTION: Brazil.

HARDINESS: G2.

FLOWERING: summer–autumn.

6. **B. flagellaris** Rodrigues

ILLUSTRATION: Bechtel et al., Manual of cultivated orchid species, 171 (1981).

Like *B. perrinii* except as below.

LEAVES: 35–50 cm × 5 mm.

FLOWERS: 5–15.
PETALS: 2–3 mm wide.
DISTRIBUTION: E Brazil.
FLOWERING: spring.

7. **B. nodosa** (Linnaeus) Lindley
ILLUSTRATIONS: Botanical Magazine, 3229 (1833); Williams et al., Orchids for everyone, 182 (1980); Bechtel et al., Manual of cultivated orchid species, 172 (1981); Orchid Review **92**: 343 (1984).
HABIT: epiphytic or on rocks.
STEMS: 8–16 cm, cylindric, slightly thicker above, 4- or 5-noded, 1-leaved.
LEAVES: 20–35 × 2.5–3.5 cm, linear, acute, fleshy and grooved.
FLOWERING STEM: 8–22 cm, terminal, bearing 1–several flowers.
FLOWERS: 7.5–8.5 cm wide, fragrant.
SEPALS AND PETALS: 6–9 cm × 5 mm, linear, acute, pale green to greenish yellow.
LIP: 3–5.5 × 2–4 cm, white; claw about one-third of total lip length, rolled around column; blade circular to ovate, acuminate.
COLUMN: *c.* 1 cm, concealed.
DISTRIBUTION: Mexico & West Indies south to Colombia, Venezuela & Surinam.
HARDINESS: G2.
FLOWERING: winter.

8. **B. subulifolia** Lindley
SYNONYM: *B. cordata* Lindley.
ILLUSTRATIONS: Botanical Magazine, 3782 (1840); Shuttleworth et al., Orchids, 68 (1970); Skelsey, Orchids, 45 (1979).
Like *B. nodosa* except as below.
STEMS: 2–8 cm, distinctly thicker above.
LEAVES: to 20 × 1.5 cm.
FLOWERS: 3–5, 5–7 cm wide, spotted with red.
COLUMN: about 5 mm.
DISTRIBUTION: Jamaica.
FLOWERING: autumn.

9. **B. acaulis** Lindley & Paxton
SYNONYM: *B. lineata* Hooker.
ILLUSTRATIONS: Botanical Magazine, 734 (1853); Shuttleworth et al., Orchids, 67 (1970); Graf, Tropica, 708 (1978).
Like *B. nodosa* except as below.
HABIT: pendent.
LEAVES: cylindric, grooved.
FLOWERING STEMS: lateral, 1- or 2-flowered.

FLOWERS: pale yellow to pale green.
DISTRIBUTION: Guatemala & Belize to Panama.
HARDINESS: G2.
FLOWERING: autumn.

56. SOPHRONITIS Lindley

A genus of about 8 species from Paraguay and E Brazil; often used to give bright red colours in intergeneric hybrids with *Cattleya*, *Laelia*, *Brassavola*, *Epidendrum*, etc. Cultivation as for *Cattleya* (p. 000), using tree-fern blocks or rafts in cool, shady conditions.

Literature: Fowlie, J.A., A contribution to a monographic revision of the genus Sophronitis Lindl., *Orchid Digest* **51**: 15–34 (1987).

HABIT: epiphytic or on rocks; rhizomes creeping, branched, bearing clusters of 1-leaved pseudobulbs.
LEAVES: ovate, elliptic or oblong, folded when young, leathery.
FLOWERS: terminal, solitary or in few-flowered racemes.
SEPALS AND PETALS: free, petals broader than sepals.
LIP: small, unlobed or 3-lobed, shortly fused to column at base and partly enclosing it.
COLUMN: 2-winged.
POLLINIA: 8.

Key to species

1. Leaves less than 3 cm wide; flowers 2–5, 2–3 cm wide; petals less than 1 cm wide **1. cernua**
2. Leaves 3 cm wide or more; flowers solitary, 6–7 cm wide; petals 2 cm wide or more **2. coccinea**

1. S. cernua Lindley

ILLUSTRATIONS: Botanical Magazine, 3677 (1839); Shuttleworth et al., Orchids, 69 (1970); Bechtel et al., Manual of cultivated orchid species, 274 (1981).
PSEUDOBULBS: 1–2 cm, ovoid to cylindric, compressed.
LEAVES: 2–2.8 × 1.5–2 cm, elliptic to oblong, slightly cordate at base, obtuse or abruptly pointed.
FLOWERING STEM: to 6 cm, bearing 2–5 flowers.
FLOWERS: 2–3 cm wide, reddish orange.
SEPALS AND PETALS: 1–1.5 cm × 5 mm; sepals ovate to lanceolate, acute; petals ovate to diamond-shaped, acute.
LIP: 1 cm × 8 mm, triangular to lanceolate, unlobed or shallowly 3-lobed, acute.
COLUMN: 5 mm, club-shaped with 2 small, purple-tipped wings.
DISTRIBUTION: E Brazil.
HARDINESS: G2.
FLOWERING: winter.

2. S. coccinea (Lindley) Reichenbach

SYNONYM: *S. grandiflora* Lindley.

ILLUSTRATIONS: Botanical Magazine, 3709 (1839); Williams et al., Orchids for everyone, 196 (1980); Bechtel et al., Manual of cultivated orchid species, 274 (1981).

PSEUDOBULBS: 1.5–3.5 cm, ovoid to fusiform.

LEAVES: 3–6 × 1–1.5 cm, ovate to elliptic, fleshy, sometimes glaucous.

FLOWERING STEM: to 8 cm, 1-flowered.

FLOWERS: 6–7 cm wide, yellow to red or pinkish red.

SEPALS: 1.5–2.5 × 1 cm, elliptic to oblong, acute.

PETALS: 2.5–3.5 × 2–3 cm, circular to broadly ovate, rounded.

LIP: 1.5–2 × 1.5–2 cm, 3-lobed; lateral lobes triangular, rolled around column; central lobe narrowly triangular.

COLUMN: 5–10 mm, club-shaped, partly concealed.

DISTRIBUTION: E Brazil.

HARDINESS: G2.

FLOWERING: winter.

57. SOPHRONITELLA Schlechter

A single species from Brazil, originally described in *Sophronitis*. Cultivation as for *Sophronitis*.

HABIT: epiphytic with creeping rhizomes.

PSEUDOBULBS: 1.5–3 cm × 3–10 mm, ovoid to fusiform, clustered, ridged, 1-leaved.

LEAF: 4–8 × 1 cm, linear, acute, channelled, folded when young, fleshy.

FLOWERING STEM: to 5 cm, 1- or 2-flowered.

FLOWERS: 2.5–4 cm wide, purple.

SEPALS AND PETALS: 1.7–2 cm × 3–5 mm, oblong to lanceolate, acute, free.

LIP: 1.7–2 cm × 5 mm, unlobed, lanceolate to ovate or diamond-shaped, abruptly acute, fused to column at base.

COLUMN: 5 mm, club-shaped.

POLLINIA: 8.

1. S. violacea (Lindley) Schlechter

SYNONYM: *Sophronitis violacea* Lindley.

ILLUSTRATIONS: Botanical Magazine, 6880 (1886); Ospina & Dressler, Orquideas de las Americas, t. 71 (1974); Bechtel et al., Manual of cultivated orchid species, 274 (1981).

DISTRIBUTION: E Brazil.

HARDINESS: G2.

FLOWERING: winter.

58. LEPTOTES Lindley

Three species from Brazil, Paraguay and Argentina. Cultivation requirements are similar to those for *Cattleya* (p. 167) though their dwarf habit is best accommodated on hanging rafts or baskets; the roots should be covered with moss.

HABIT: dwarf epiphytes with creeping rhizomes.

STEMS: short and cane-like, sheathed, 1-leaved, sometimes branched.

LEAVES: cylindric to linear, acute, fleshy and channelled, folded when young.

FLOWERS: few or solitary, in terminal racemes.

SEPALS AND PETALS: free, similar, or petals broader than sepals.

LIP: free, 3-lobed; lateral lobes small, sometimes enfolding but not concealing the short fleshy column.

POLLINIA: 6 (4 large and 2 small), waxy.

Key to species

1. Flowers erect, 4–5 cm wide; sepals similar to petals; lip with a dark purple patch at base **1. bicolor**

2. Flower pendent, 3–4 cm wide; sepals wider than petals; lip uniformly pink **2. unicolor**

1. L. **bicolor** Lindley

SYNONYM: *Tetramicra bicolor* (Lindley) Bentham.

ILLUSTRATIONS: Botanical Magazine, 3734 (1840); Shuttleworth et al., Orchids, 44 (1970); Pabst & Dungs, Orchidaceae Brasilienses 1: 222 (1975); Skelsey, Orchids, 118 (1979).

STEM: 1–4 cm, cylindric, fleshy.

LEAVES: 5–15 cm × 5–10 mm.

FLOWERING STEM: to 6 cm, bearing 2–4 flowers.

FLOWER-STALK PLUS OVARY: 4–5 cm.

FLOWER: 4–5 cm wide, erect, white to very pale pink, yellow or green.

SEPALS AND PETALS: 2–2.5 cm × 3–5 mm, linear to narrowly lanceolate, acute with curled tips.

LIP: 1.5–2 cm × 5–8 mm, 3-lobed; lateral lobes triangular to circular, white or pale green, curved around column; central lobe oblong to narrowly ovate, acute, pale to deep purple with darker veins.

COLUMN: 5 mm, greenish brown, exposed.

DISTRIBUTION: S Brazil & Paraguay.

HARDINESS: G2.

FLOWERING: winter–spring, autumn.

2. L. **unicolor** Rodrigues

ILLUSTRATIONS: Pabst & Dungs, Orchidaceae Brasilienses 1: 223 (1975); Northen, Miniature orchids, 100 (1980); Bechtel et al., Manual of cultivated orchid species, 221 (1981).

STEM: 1–3 cm, cylindric.

LEAVES: 2.5–6 cm × 4–8 mm, arched.
FLOWERING STEM: 1–3 cm, 2-flowered.
FLOWERS: 3–4 cm wide, pendent, white to pale pink or purple.
SEPALS: 2–2.5 cm × 5 mm, lanceolate, acute.
PETALS: 1.5–2 cm × 3 mm, strap-shaped, acute.
LIP: 1.5–2 × 1–1.3 cm, 3-lobed; lateral lobes triangular, obtuse, flat, not enclosing column; central lobe diamond-shaped, acute.
COLUMN: 5 mm, brown, exposed.
DISTRIBUTION: S Brazil & Argentina.
HARDINESS: G2.
FLOWERING: winter.

59. SCAPHYGLOTTIS Poeppig & Endlicher

A tropical American genus of about 40 species. Cultivation as for *Cattleya* (p. 167), though its straggling habit makes it more easily grown on rafts or bark suspended near the roof.

HABIT: epiphytic or on rocks.
PSEUDOBULBS: in superposed chains.
LEAVES: all apical, folded when young.
FLOWERS: terminal or at stem nodes, in racemes or clusters, white or pale shades of blue, green, yellow or purple, never bright orange or red.
COLUMN: base enlarged into a foot to which the lateral sepals are basally fused to form a mentum.
POLLINIA: 4 (in ours), waxy.

1. S. amethystina (Reichenbach) Schlechter

ILLUSTRATION: Bechtel et al., Manual of cultivated orchid species, 272 (1981).
LEAVES: 1 or 2, to 15 × 1.5 cm, apical, linear, notched at apex.
FLOWERS: 1–1.5 cm wide, in terminal clusters, white to pale purplish pink.
LIP: 3-lobed.
DISTRIBUTION: Guatemala to Panama.
HARDINESS: G2.
FLOWERING: spring.

60. NAGELIELLA Williams

A small genus of 2 species from Mexico and C America; long known as *Hartwegia* Lindley though this name had already been used for a genus in the Liliaceae. Cultivation requirements are similar to those for *Cattleya* (p. 167). Propagation is easily effected by division, though this should not happen frequently as the plants perform better when the pseudobulbs are densely clustered.

HABIT: epiphytic or terrestrial; rhizomes creeping.

STEMS: simple, 1-leaved, club-shaped.
LEAVES: fleshy, mottled with purple, folded when young.
FLOWERS: pinkish purple, terminal, in tall, slender, branched racemes or clusters.
SEPALS: pointing forwards; lateral sepals fused basally to column foot forming a mentum.
LIP: unlobed, bent downwards, basally fused to column foot, with a small apical pouch.
COLUMN: slender, slightly arched over upper part of lip.
POLLINIA: 4, waxy.

1. **N. purpurea** (Lindley) Williams
SYNONYM: *Hartwegia purpurea* Lindley.
ILLUSTRATIONS: Hamer, Orquideas de El Salvador **2**: 139 (1974); Orquidea (Mexico) **4**: 115 (1974); Bechtel et al., Manual of cultivated orchid species, 234 (1981); American Orchid Society Bulletin **55**: 252 (1986).
STEMS: 2–8 cm, narrowly fusiform to club-shaped, shallowly ridged.
LEAVES: 3–12 × 1–2.5 cm, elliptic to ovate with sheathing bases.
FLOWERING STEMS: 10–50 cm with small, chaffy bracts, gradually lengthening over a few months.
FLOWERS: several, 1–1.5 cm wide, opening in succession.
SEPALS: 7–10 × 3 mm, lanceolate to ovate, concave, acute or obtuse.
PETALS: shorter and narrower than sepals, linear to lanceolate.
LIP: 10 × 4 mm, basally fused to column foot; middle portion linear, S-shaped; apex a small triangular pouch.
COLUMN: *c*. 7 mm.
DISTRIBUTION: Mexico to Honduras.
HARDINESS: G2.
FLOWERING: summer.

61. **ERIA** Lindley

A very diverse genus of about 500 species from India, SE Asia and the Pacific islands. It is generally similar to *Dendrobium* but the flowers have 8 pollinia and are never spurred. Cultivation details are similar to those for the tropical species of *Dendrobium* (p. 234); the choice of pot, raft or bark will be governed by the stature and habit of the individual species. The resting period will depend on the fleshiness of the pseudobulb.

Literature: Seidenfaden, G., Orchid genera in Thailand, X, *Opera Botanica* **62**: 1–157 (1982).

HABIT: epiphytic or occasionally terrestrial, often densely hairy.
PSEUDOBULBS: ovoid, fusiform or cylindric, sometimes slender and cane-like, often densely sheathed.

LEAVES: 1–several, folded when young (rolled in Nos. 1 & 2).

FLOWERS: in terminal or axillary racemes.

SEPALS: upper free, laterals united at base with column foot forming a pronounced mentum.

PETALS: similar to sepals.

LIP: often 3-lobed, borne on column foot.

POLLINIA: 8, waxy, pear-shaped or ovoid.

Key to species

1. Leaves rolled when young

1.A. Stems short, concealed; flowers 10–30 per raceme, hairy outside **1. javanica**

1.B. Stems long, exposed; flowers 2–6 per raceme, not hairy outside **2. coronaria**

2. Flower-stalks not hairy; flowers in loose racemes **3. bractescens**

3. Stems pseudobulb-like, 5–15 × 1.5–3 cm; flowering stems erect **4. spicata**

4. Stems long and slender, to 30 × 1 cm; flowering stems horizontal to pendent **5. floribunda**

1. E. **javanica** (Swartz) Blume

SYNONYMS: *E. fragrans* Reichenbach; *E. stellata* Lindley.

ILLUSTRATIONS: Botanical Magazine, 3605 (1837); Williams et al., Orchids for everyone, 188 (1980); Bechtel et al., Manual of cultivated orchid species, 212 (1981); American Orchid Society Bulletin **55**: 463 (1986).

HABIT: epiphytic.

PSEUDOBULBS: 5–8 × 1–2 cm, narrowly ovoid, of 1 internode, sheathed, bearing 1 or 2 leaves.

LEAVES: 20–50 × 3.5–6 cm, lanceolate, acute, rolled when young.

FLOWERING STEM: 30–60 cm, erect to arched, terminal, bearing up to 30 flowers.

FLOWERS: *c.* 4 cm wide, fragrant, white or cream-coloured; flower-stalks and outsides of sepals hairy.

SEPALS: 2–2.5 cm, lanceolate to narrowly triangular, acute.

PETALS: smaller, slightly sickle-shaped.

LIP: 1–1.5 cm, oblong to lanceolate, 3- or 5-ridged, shallowly lobed, central lobe narrowly triangular, obtuse or acute; lateral lobes semi-ovate, erect.

COLUMN: *c.* 8 mm; foot *c.* 5 mm.

DISTRIBUTION: Himalaya, Thailand, Malaysia, Indonesia & Philippines.

HARDINESS: G2.

FLOWERING: spring–summer.

2. E. **coronaria** (Lindley) Reichenbach

SYNONYM: *Trichosma suavis* Lindley.

ILLUSTRATIONS: Orchid Review **82**: 283 (1974); Bechtel et al., Manual of cultivated orchid species, 212 (1981); Opera Botanica **62**: 41 & 151 (1982).

HABIT: terrestrial or epiphytic.

STEMS: 7.5–25 × 5 mm, sheathed at base only, bearing 2 leaves near the apex.

LEAVES: 10–25 × 3–5 cm, narrowly elliptic to ovate, leathery, acute, rolled when young.

FLOWERING STEM: to *c.* 15 cm, erect, robust, terminal, bearing 2–6 flowers.

FLOWERS: 2.5–4 cm wide, white or pale yellow with crimson markings on the lip, fragrant.

SEPALS: 1.5–2.2 cm × 5–8 mm, elliptic to ovate.

PETALS: *c.* 2 × 1 cm, oblong to elliptic, obtuse.

LIP: *c.* 2 × 1 cm, oblong to ovate, 3-lobed; lateral lobes oblong, rounded, erect; central lobe short, almost circular, with 5–7 wavy keels.

COLUMN: longer than column foot.

DISTRIBUTION: E Himalaya to Thailand & Malay peninsula.

HARDINESS: G2.

FLOWERING: autumn–winter.

3. **E. bractescens** Lindley

ILLUSTRATIONS: Botanical Magazine, 4163 (1845); Holttum, Flora of Malaya **1**: 386 (1953); Millar, Orchids of Papua New Guinea, 62 (1978).

PSEUDOBULBS: 6–15 cm, cylindric to fusiform, sheathed below, bearing 2 or 3 leaves.

LEAVES: 12–20 cm, oblong to narrowly elliptic, obtuse.

FLOWERING STEMS: usually 2 per pseudobulb, 15–20 cm, erect to horizontal, many-flowered.

BRACTS: conspicuous, to 2.5 cm, oblong, rounded, cream-coloured to pale green.

FLOWERS: *c.* 2.5 cm wide and long, white to pale green, sometimes flushed with pink.

SEPALS: narrowly triangular, acute; mentum pronounced.

PETALS: lying close under upper sepal.

LIP: oblong, rounded, shallowly 3-lobed; lateral lobes oblong, rounded, pale purple; central lobe almost circular, with 3 or 5 pink, wavy keels.

COLUMN: shorter than column foot.

DISTRIBUTION: E Himalaya, Andaman Islands, SE Asia, Indonesia & Philippines.

HARDINESS: G2.

FLOWERING: summer.

4. **E. spicata** (D. Don) Handel-Mazzetti

SYNONYM: *E. convallarioides* Lindley.

ILLUSTRATIONS: Shuttleworth et al., Orchids, 72 (1970);

Graf, Tropica, 732 (1978); Bechtel et al., Manual of cultivated orchid species, 213 (1981).

PSEUDOBULBS: 5–15 × 1.5–3 cm, densely sheathed, bearing up to 4 leaves.

LEAVES: narrowly elliptic, folded when young.

FLOWERING STEMS: shorter than leaves, densely many-flowered, erect.

FLOWERS: *c.* 1 cm wide, almost spherical, white to pale yellow.

DISTRIBUTION: E Himalaya & Burma.

HARDINESS: G2.

FLOWERING: spring.

5. **E. floribunda** Lindley

ILLUSTRATIONS: Edwards's Botanical Register **30**: t. 20 (1844); Orchid Review **76**(9): cover (1968).

Like *E. spicata* except as below.

STEMS: to 30 × 1 cm, cylindric, slender, bearing 3–7 leaves in upper half.

FLOWERING STEMS: several, *c.* 15 cm, axillary, horizontal to pendent, densely many-flowered.

FLOWERS: *c.* 1 cm wide white to pinkish red.

COLUMN: dark purple.

DISTRIBUTION: Himalaya to Vietnam, Malay peninsula, Sumatra & Borneo.

HARDINESS: G2.

FLOWERING: summer–autumn.

62. CRYPTOCHILUS Wallich

A genus of 2 species confined to the Himalaya. The affinities of *Cryptochilus* are uncertain and it was previously considered a member of the tribe *Vandae*; it is now thought to be allied to *Eria*. The flowers show adaptations to bird pollination. Cultivation conditions are similar to those for Himalayan species of *Dendrobium* (p. 234), allowing plants to become quite dry between waterings during the winter.

HABIT: epiphytic.

PSEUDOBULBS: cylindric to spherical, simple, crowded, developing from flowering stems.

LEAVES: 1 or 2 per pseudobulb, leathery, usually shortly stalked, folded when young, appearing after flowering.

FLOWERING STEMS: sheathed below, with many flowers in dense 2-rowed spikes.

FLOWERS: equal to or shorter than the persistent bracts.

SEPALS: fused into an urn-shaped to spherical tube with free, spreading, triangular points.

PETALS: linear to oblong, concealed in sepal-tube.

LIP: fused to column foot.

COLUMN: short, straight, smooth at apex.
POLLINIA: 8.

Key to species

1. Sepal-tube swollen at base, scarlet **1. sanguinea**
2. Sepal-tube not swollen at base, yellow **2. lutea**

1. C. sanguinea Wallich
ILLUSTRATIONS: Orchid Review **79**: 227 (1971); American Orchid Society Bulletin **46**: 244 (1977).
PSEUDOBULBS: to *c.* 7 cm, ovoid, slightly compressed.
LEAVES: 9–20 × 2–4 cm, oblong to lanceolate, acute.
FLOWERS: *c.* 2 cm long.
BRACTS: about equal to flowers in length.
SEPALS: united into a scarlet, urn-shaped tube, swollen at base; tips of lobes acute.
PETALS: small, oblong to obovate, obtuse, yellow, flushed with brown.
LIP: oblong, broader at tip, channelled, yellow.
COLUMN: about half as long as lip, straight.
POLLINIA: greenish blue.
DISTRIBUTION: Nepal & NE India (Sikkim, Meghalaya).
HARDINESS: G1.
FLOWERING: summer.

2. C. lutea Lindley
ILLUSTRATION: Annals of the Royal Botanic Garden Calcutta **8**: t. 221 (1898).
PSEUDOBULBS: cylindric.
LEAVES: 10–15 × 1.5–2.5 cm, narrowly elliptic, acute.
BRACTS: longer than flowers.
SEPAL-TUBE: *c.* 1 cm, spherical to cylindric, not swollen at base, yellow.
PETALS: diamond-shaped.
POLLINIA: yellow.
DISTRIBUTION: NE India (Sikkim, Naga hills), Bhutan.
HARDINESS: G1.
FLOWERING: summer.

63. MEIRACYLLIUM Reichenbach

A genus of 2 species from Mexico, Guatemala and El Salvador. Cultivation conditions are similar to those for *Cattleya* (p. 167), though night temperatures should be higher in winter.

Literature: Dressler, R.L., The relationships of Meiracyllium, *Brittonia* **12**: 222–5 (1960).

HABIT: creeping rhizomatous epiphytes.
STEMS: short, scaly, 1-leaved, slightly swollen, simple.
LEAVES: stalkless, thick and leathery, folded when young.

FLOWERS: solitary or several, borne at junction of leaf and stem.

SEPALS: all 3 similar, free except at the base where they are fused to the column foot, forming a short mentum.

PETALS: narrower than sepals, spreading.

LIP: simple, deeply concave or pouched.

POLLINIA: 8, in 2 groups of 4, narrowly club-shaped, waxy, unequal.

ROSTELLUM: prominent.

Key to species

1. One flower per stem; lip 1–1.3 cm — **1. wendlandii**
2. Several flowers per stem; lip 7–9 mm — **2. trinasutum**

1. M. wendlandii Reichenbach

SYNONYM: *M. gemma* Reichenbach.

ILLUSTRATIONS: Orchid Digest **41**: 69 (1977); Die Orchidee **30**: 76 (1979); Northen, Miniature orchids, C-57 (1980); American Orchid Society Bulletin **53**: 831 (1984).

STEMS: to 1 cm, ascending.

LEAVES: to 5 × 2.3 cm, oblong, obtuse to rounded.

FLOWERING STEMS: to *c.* 4 cm, 1-flowered.

FLOWERS: pinkish purple.

UPPER SEPAL: 1–1.7 cm × 5 mm, ovate, acute or slightly acuminate.

LATERAL SEPALS: similar but narrower and slightly asymmetric.

PETALS: 8–15 × 1–2.5 mm, oblong to lanceolate, acute, finely toothed.

LIP: 1–1.3 cm × 7 mm, ovate, concave; apex acuminate and curved downwards.

COLUMN: *c.* 6 mm, acute, slightly narrowed at base.

DISTRIBUTION: Mexico, Guatemala & El Salvador.

HARDINESS: G2.

FLOWERING: summer.

2. M. trinasutum Reichenbach

ILLUSTRATIONS: Orchid Digest **41**: 69 (1977); Northen, Miniature orchids, C-56 (1980); Bechtel et al., Manual of cultivated orchid species, 231 (1981).

LEAVES: 2.8–5 × 1.5–3.5 cm, elliptic to circular.

FLOWERS: several, pale pinkish purple, almost white below.

SEPALS: 8–11 × 3.5–5 mm, oblong to elliptic, acuminate.

PETALS: 7–10 × 3 mm, elliptic to spathulate, acute.

LIP: 7–9 × 4–5 mm, deeply pouched, acuminate.

COLUMN: broad at base, round-tipped.

DISTRIBUTION: Mexico & Guatemala.

HARDINESS: G2.

FLOWERING: summer.

64. PHYSOSIPHON Lindley

A genus of about 6 species from the New World tropics, of which only one is generally grown. Cultivation conditions are similar to those for *Masdevallia*.

HABIT: tufted epiphytes.
STEMS: slender, erect or ascending, sheathed, 1-leaved at apex.
LEAVES: tapering to the base, fleshy or leathery, folded when young.
FLOWERS: many, in long, narrow, erect or arching racemes, borne at the junction of leaf and stem.
SEPALS: equally fused for more than half their length, forming a 3-angled, inflated tube, slightly constricted at the mouth; free portions spreading.
PETALS AND LIP: very small, concealed in sepal-tube; lip 3-lobed or simple, fleshy, channelled.
COLUMN: small, 3-lobed.
POLLINIA: 2, ovoid, waxy.

1. **P. tubatus** (Loddiges) Reichenbach
SYNONYMS: *Stelis tubatus* Loddiges; *P. loddigesii* Lindley: *P. guatemalensis* Rolfe.
ILLUSTRATIONS: Botanical Magazine, 4869 (1855); Orchid Review **79**: 220 (1971); Bechtel et al., Manual of cultivated orchid species, 262 (1981).
STEMS: 1.5–12 cm.
LEAVES: 4–17 × 1.5–3 cm, elliptic to narrowly obovate.
FLOWERING STEM: 8–45 cm.
FLOWERS: 20–100, greenish yellow to brick-red.
SEPALS: 6–22 mm; free portion oblong to lanceolate or elliptic.
PETALS: 1.5–2.5 × 0.6–1 mm, obovate to spathulate, sometimes obscurely 3-lobed.
LIP: 2–3.2 × 1–2.5 mm, 3-lobed; lobes rounded.
COLUMN: 2–3.5 mm.
DISTRIBUTION: Mexico & Guatemala.
HARDINESS: G2.
FLOWERING: summer.
Very variable in size and colour of flowers.

65. MASDEVALLIA Ruiz & Pavon

About 300 species from Mexico, C and S America, mostly from the Andean cloud-forests, frequently grown for their large and brightly coloured sepals, whose tails often continue to grow after the flower has opened. They perform best if grown in epiphytic compost or on tree-fern bark in cool, moist conditions with plenty of ventilation. Having no pseudobulbs, plants must never be allowed to dry out; they should be shaded from direct sunlight. Well-established plants can be divided in early spring.

Literature: Woolward, F.H., *The genus Masdevallia*

(1890–6); Kraenzlin, F.L., Masdevallia, *Feddes Repertorium Beihefte* **34** (1925); Braas, L., The genus Masdevallia, *Orchid Review* 86: 30–43 (1978); Luer, C.A., *Thesaurus Masdevalliarum* 1 (1983), 2 (1986); Luer, C.A., A monograph of Masdevallia, *Systematic Botany* **16**: 1–62 (1986).

HABIT: epiphytic or on rocks.

STEMS: short and slender, erect, tufted, concealed in papery sheaths.

LEAVES: 1, apical, linear to lanceolate, somewhat fleshy, leathery, minutely 3-toothed at tip, folded when young.

FLOWERS: 1–several, in terminal racemes borne at the leaf-base.

SEPALS: fused below into a narrow or cup-shaped tube; lateral sepals usually further fused into a broad oblong blade; free portions of sepals spreading and usually long-tailed, separated by a sinus or notch.

PETALS AND LIP: small, often concealed in the sepal-tube; lip oblong, sometimes 3-lobed or fiddle-shaped.

COLUMN: toothed at apex; column foot forming a short mentum with base of lateral sepals.

POLLINIA: 2.

Key to species

1. Flowering stem sharply winged or angled — *Go to Subkey* 1
2. Two or more flowers open together on the same stem — *Go to Subkey* 2
3. Sepal-tube narrow and slightly curved — *Go to Subkey* 3
4. Sepal-tube broad and cup-shaped — *Go to Subkey* 4

Subkey 1. *Flowering stem sharply winged or angled.*

1. Lateral sepals fused in a blade less than 2 cm wide — **4. infracta**
2. Flowers all open together — **1. tovarensis**
3. Upper sepals more than 8 cm — **2. ephippium**
4. Upper sepal 6–8 cm — **3. maculata**

Subkey 2. *Flowering stem not winged or angled; two or more flowers open together on the same flowering stem.*

1. Flowering stem more than 20 cm; lateral sepals fused above the sepal-tube; petals not toothed

1.A. Sepal-tube *c.* 6 mm; sepals greenish yellow spotted with maroon below — **16. schlimii**

1.B. Sepal-tube *c.* 1.5 cm; sepals bright scarlet — **17. racemosa**

2. Flowering stem at most 12 cm; sepals *c.* 2 cm — **14. caloptera**
3. Flowering stem 15 cm or more; sepals *c.* 3 cm — **15. abbreviata**

Subkey 3. Flowering stem not winged or angled; only a single flower open at any time on each flowering stem; sepal-tube narrow and slightly curved.

1. Lateral sepals not drawn out into fine tails

1.A. Flowers held well above the leaves; tail of upper sepal erect or curved back **18. coccinea**

1.B. Flower held just above the leaves; tail of upper sepal drooping **19. militaris**

2. Flowering stem about equal to the leaves; lip fringed **20. rosea**

3. Free part of upper sepal (excluding the tail) 2 cm or more **21. veitchiana**

4. Free part of upper sepal (excluding tail) 1 cm or less

4.A. Sepals reddish orange with darker veins; petals pointed **22. amabilis**

4.B. Sepals pink with darker veins; petals with 3 teeth at the apex **23. barleana**

Subkey 4. Flowering stem not winged or angled; only a single flower open at any time on each flowering stem; sepal-tube broad and cup-shaped.

1. Free part of upper sepal (including tail) more than 7 cm

1.A. Leaves 20–30 × 5–7 cm; flowering stem 20–30 cm **5. macrura**

1.B. Leaves 15–20 × 2.5 cm; flowering stems 6–8 cm **6. velifera**

2. Apex of lip without papillae

2.A. Upper sepal white to greenish yellow; lateral sepals densely spotted with maroon **9. erinacea**

2.B. Upper sepal purple with a yellow margin; lateral sepals purple below, pink above, tapered to slender yellow tails **8. estradae**

2.C. All sepals dark purple, tapered into slender green and yellow tails **7. rolfeana**

3. Upper sepal abruptly contracted into a thick, fleshy tail; flowering stem equal to or only slightly shorter than the leaves **10. coriacea**

4. Flowers greenish yellow spotted with crimson **11. peristeria**

5. Flowers brownish purple below, green above, with crimson spots **12. civilis**

6. Flowers yellow, spotted and blotched with orange **13. corniculata**

1. **M. tovarensis** Reichenbach

ILLUSTRATIONS: Botanical Magazine, 5505 (1865); Skelsey, Orchids, 120 (1979); Bechtel et al., Manual of cultivated orchid species, 227 (1981).

HABIT: epiphytic.

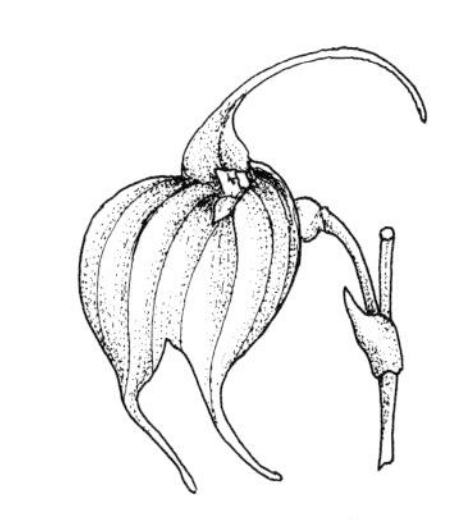
Masdevallia tovarensis – flower.

LEAVES: to 15 × 2 cm, lanceolate to obovate, obtuse or rounded.

FLOWERING STEM: to 20 cm, angled, bearing up to 4 white flowers, all open together.

SEPALS: tube *c.* 6 mm; upper sepal 4–5 cm, finely tapered; lateral sepals 3.5–4.5 cm, triangular to ovate, slender-tailed, fused for 2–2.5 cm.

PETALS: *c.* 6 × 1.5 mm, linear, acute, asymmetric.

LIP: *c.* 5 × 2 mm, oblong to spathulate, rounded or acute.

COLUMN: 4.5 mm, finely toothed.

DISTRIBUTION: Venezuela.

HARDINESS: G2.

FLOWERING: winter.

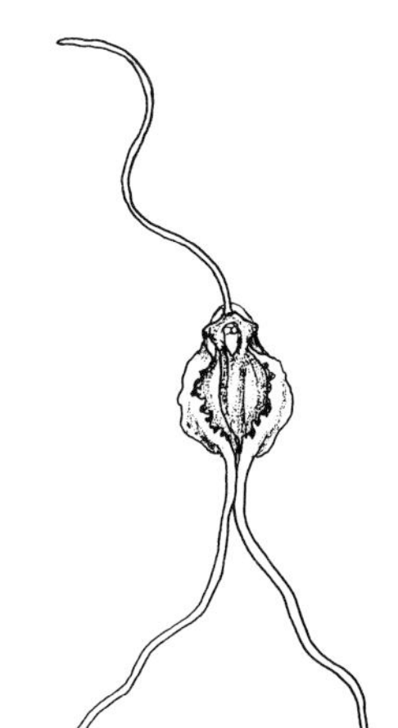
Masdevallia ephippium – flower.

2. M. ephippium Reichenbach

ILLUSTRATIONS: Botanical Magazine, 6208 (1876); Dunsterville & Garay, Venezuelan orchids illustrated **1**: 213 (1959); American Orchid Society Bulletin **45**: 104 (1976).

Like *M. tovarensis* except as below.

FLOWERING STEM: 20–30 cm, 3-angled, bearing several flowers in succession.

FLOWERS: purplish brown with yellow tails.

SEPALS: upper to 14 cm, blade small, circular, abruptly narrowed to a long tail; laterals fused into a deep cup; tails 7–10 cm.

DISTRIBUTION: Colombia & Ecuador.

FLOWERING: spring–summer.

3. M. maculata Klotzsch & Karsten

ILLUSTRATIONS: Orchid Digest **36**: 24 (1972); Bechtel et al., Manual of cultivated orchid species, 227 (1981).

Like *M. tovarensis* except as below.

FLOWERING STEM: to 25 cm, bearing several flowers in succession.

SEPALS: orange to purplish brown with yellow tails; sepal-tube *c.* 1.5 cm; upper sepal 6–8 cm, narrowly triangular; lateral sepals *c.* 8 cm, narrowly ovate, fused for *c.* 3 cm.

PETALS: *c.* 6 mm, oblong, asymmetric, each with a small point.

LIP: *c.* 6 mm, spathulate to fiddle-shaped, papillose.

COLUMN: *c.* 6 mm.

DISTRIBUTION: Venezuela, Colombia & Peru.

FLOWERING: summer–autumn.

4. M. infracta Lindley

ILLUSTRATIONS: Pabst & Dungs, Orchidaceae Brasilienses **1**: 225 (1975); Graf, Tropica, 737 (1978); Bechtel et al,, Manual of cultivated orchid species, 225 (1981).

HABIT: epiphytic.

LEAVES: 8–14 × 1.5–2.5 cm, lanceolate to oblong, obtuse.

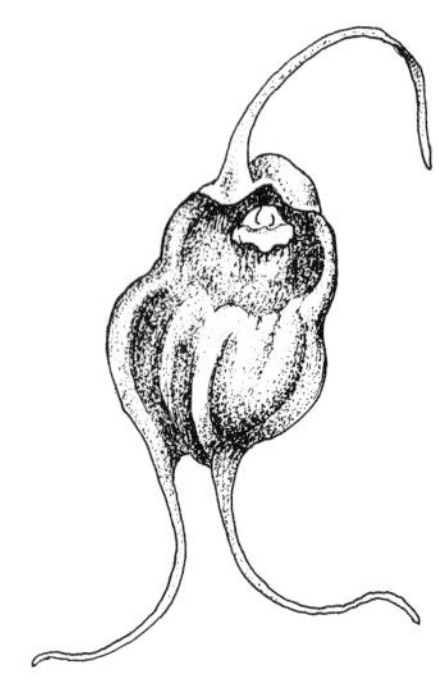

Masdevallia infracta – flower.

FLOWERING STEM: to 25 cm, angled, bearing several flowers in succession.

SEPALS: pinkish purple with greenish yellow tails; sepal-tube 1–1.5 cm; upper sepal 4–5 cm, triangular to ovate, slender-tailed; lateral sepals 4–5 cm, oblong to ovate, fused for about 2 cm, slender-tailed.

PETALS: 6–7 mm, linear, each with a small point, pink.

LIP: *c.* 6 mm, oblong, with a small point, pink and orange.

COLUMN: *c.* 6 mm.

DISTRIBUTION: Peru & Brazil.

HARDINESS: G2.

FLOWERING: summer.

5. M. macrura Reichenbach

ILLUSTRATIONS: Botanical Magazine, 7164 (1891); Die Orchidee **30**: 105 (1979); Bechtel et al., Manual of cultivated orchid species, 226 (1981).

HABIT: on rocks.

LEAVES: 20–30 × 5–7 cm, oblong to elliptic, obtuse, often notched.

FLOWERING STEM: 20–30 cm, 1-flowered.

SEPALS: red to brownish yellow with prominent veins and dark warts; tails long and slender, yellow; sepal-tube 1–1.5 cm, cup-shaped; upper sepal 10–15 cm, lanceolate to narrowly triangular, tapered; lateral sepals 9–13 cm, ovate to oblong, fused for 2–5 cm, tapered.

PETALS: 8–10 mm, lanceolate to rhombic, rounded, yellow with dark stippling.

LIP: 6–8 mm, oblong, rounded, papillose, yellow with thick red margins.

COLUMN: 6–8 mm, yellow.

DISTRIBUTION: Colombia & Ecuador.

HARDINESS: G2.

FLOWERING: winter–spring.

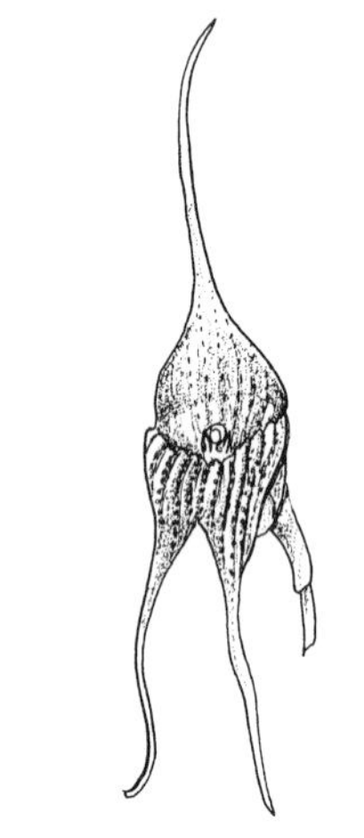

Masdevallia macrura – flower.

6. M. velifera Reichenbach

ILLUSTRATIONS: Gardeners' Chronicle **1**: 745 (1887); Journal of the Royal Horticultural Society 33: 393 (1908).

Like *M. macrura* except as below.

LEAVES: 15–20 × 2.5 cm, lanceolate, obtuse.

FLOWERING STEM: 6–8 cm.

SEPALS: tube *c.* 2 cm; upper sepal 7–8 cm, ovate, slender-tailed, brownish yellow spotted with brown; lateral sepals 7–8 cm, oblong, fused for *c.* 5 cm, shiny reddish brown with yellow tails.

PETALS: *c.* 1.2 cm, oblong, greenish yellow.

LIP: *c.* 1.2 cm, oblong, dark purple, fringed.

DISTRIBUTION: Colombia.

FLOWERING: winter.

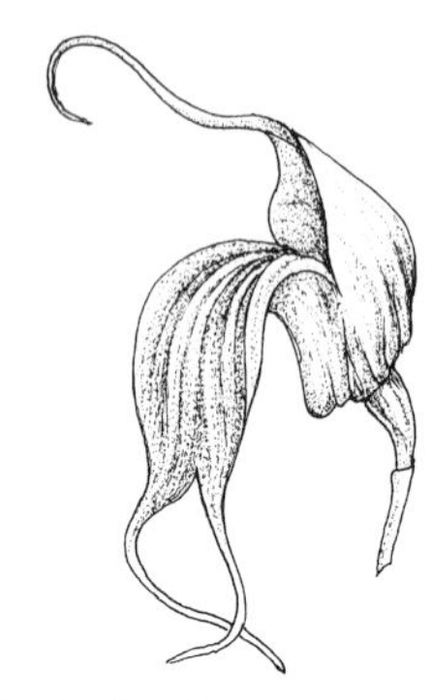

Masdevallia velifera – flower.

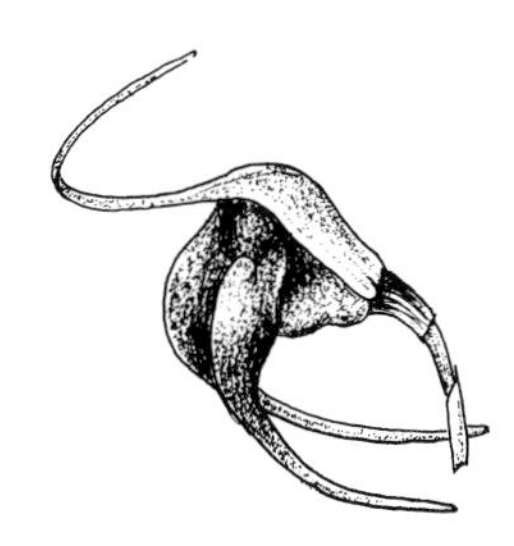

Masdevallia rolfeana – flower.

7. M. rolfeana Kränzlin

ILLUSTRATIONS: Woolward, The genus Masdevallia, 55 (1893); Orchid Digest **32**: 188 (1968).

HABIT: epiphytic.

LEAVES: 11–14 × 2 cm, obovate to elliptic, obtuse, keeled.

FLOWERING STEM: *c.* 8 cm, 1-flowered.

SEPALS: dark purple, tapered into slender green and yellow tails; tube to 1 cm, cup-shaped; upper sepal *c.* 4 cm, triangular to ovate; lateral sepals *c.* 6 cm, oblong to ovate, fused for about 2 cm.

PETALS: *c.* 6 mm, oblong to elliptic, obtuse, narrowed at base, purple.

LIP: 8–10 mm, oblong, apiculate, pink with darker spots.

COLUMN: *c.* 8 mm, finely toothed at apex.

DISTRIBUTION: Costa Rica.

HARDINESS: G2.

FLOWERING: spring–summer.

Masdevallia estradae – flower.

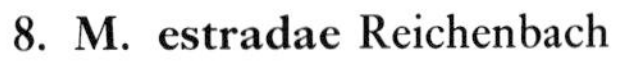

8. M. estradae Reichenbach

ILLUSTRATIONS: Botanical Magazine, 6171 (1875); Journal of the Royal Horticultural Society **26**: 146 (1901).

Like *M. rolfeana* except as below.

LEAVES: 5–8 × 2 cm.

FLOWERING STEM: *c.* 10 cm.

SEPALS: equally fused into a cup-shaped tube *c.* 6 mm long; upper sepal purple with a yellow margin, abruptly contracted into a slender, yellow tail; lateral sepals oblong, purple below, pink above, tapered to a slender yellow tail.

DISTRIBUTION: Colombia.

FLOWERING: spring.

9. M. erinacea Reichenbach

SYNONYM: *M. horrida* Teuscher & Garay.

ILLUSTRATIONS: American Orchid Society Bulletin **29**: 23 (1960) & **37**: 520 (1968); Die Orchidee 30: CXXV (1979); Northen, Miniature orchids, C-51 (1980).

Like *M. rolfeana* except as below.

HABIT: dwarf epiphyte.

LEAVES: 3–6 cm × 1–3 mm, linear, acute.

FLOWERING STEMS: 2.5–6 cm; bract *c.* 5 mm.

SEPALS: *c.* 1.5 cm, equally fused into a shallow cup *c.* 8 mm long, densely hairy on the veins; tails club-shaped, pendent; upper sepal white to greenish yellow; lateral sepals densely spotted with maroon.

DISTRIBUTION: Costa Rica & Ecuador.

FLOWERING: summer.

10. M. coriacea Lindley

ILLUSTRATIONS: Gardeners' Chronicle **21**: 95 (1897); Orquideologia **9**: 34 (1974); Bechtel et al., Manual of cultivated orchid species, 225 (1981).

HABIT: epiphytic or on rocks.

LEAVES: 12.5–20 × 1–2 cm, linear to lanceolate, obtuse, keeled.

FLOWERING STEM: 17–20 cm, 1 (rarely 2)-flowered; flowering stem, bract and ovary flecked with purple.

SEPALS: greenish white with lines of crimson spots; tube 1.5 cm, cup-shaped; upper sepal 3–4 cm, ovate, abruptly contracted into a broad, green, purple-flecked tail; lateral sepals *c.* 4 cm, narrowly triangular, fused for 2–2.5 cm, without distinct tails.

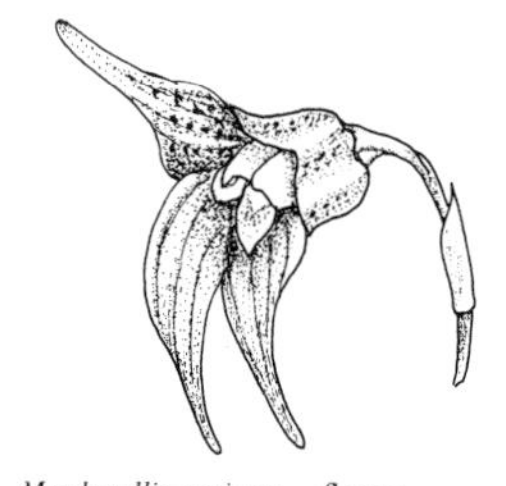

Masdevallia coriacea – flower.

PETALS: 1.2–1.5 cm, oblong to spathulate, obtuse, white with a central purple stripe.

LIP: *c.* 1.2 cm, oblong, obtuse, papillose at tip, purple and green.

COLUMN: *c.* 1 cm, finely toothed at tip, pale green.

DISTRIBUTION: Colombia.

HARDINESS: G2.

FLOWERING: spring–summer.

11. M. peristeria Reichenbach

ILLUSTRATIONS: Botanical Magazine, 6159 (1875); Orquideologia **10**: 159 (1975); Die Orchidee **27**(6): 20 (1976).

HABIT: epiphytic.

LEAVES: 10–15 × 2.5 cm, linear to lanceolate or oblong, obtuse.

FLOWERING STEM: 5–8 cm, 1-flowered.

SEPALS: 4.5–6 cm, triangular to narrowly ovate, finely tapered, greenish yellow spotted with crimson; tube 1.5–2 cm, cup-shaped; lateral sepals fused for *c.* 1.5 cm.

PETALS: 1–1.2 cm, lanceolate to spathulate, obtuse, green.

LIP: 1.2–1.5 cm, oblong, narrowed at base, rounded to truncate, papillose, purple.

COLUMN: *c.* 1.2 cm, finely toothed, greenish white.

DISTRIBUTION: Colombia.

HARDINESS: G2.

FLOWERING: spring–summer.

Masdevallia peristeria – flower.

12. M. civilis Reichenbach & Warscewicz

SYNONYM: *M. leontoglossa* misapplied.

ILLUSTRATION: Botanical Magazine, 5476 (1864); not Dunsterville & Garay, Venezuelan orchids illustrated **4**: 128 (1966).

Like *M. peristeria* except as below.

LEAVES: 15–25 × 1–1.5 cm, linear, acute.

FLOWERING STEM: 2.5–7 cm.

SEPALS: 3.5–4 cm, ovate, contracted into slender tails; outside of flower brownish purple below, green above, with crimson spots.

PETALS: white with a central purple stripe.

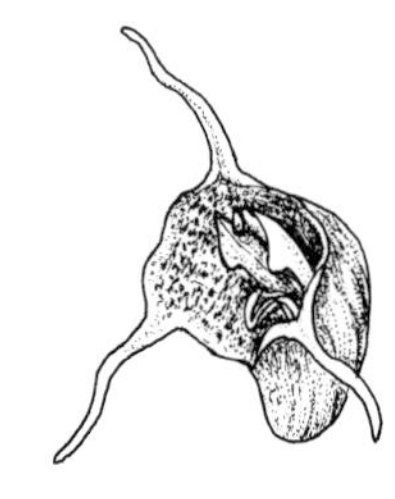

Masdevallia civilis – flower.

LIP: oblong with a rounded, papillose tip, purple.
DISTRIBUTION: Peru.

The name *M. leontoglossa* Reichenbach is frequently found in horticultural literature and catalogues. It is usually treated as a synonym of *M. civilis* though the leaf shape is quite distinct in Reichenbach's original description. Illustrations in the Botanical Magazine (7245) and Woolward (The genus Masdevallia, 20) show strongly pendent flowering stems. This is not referred to by Reichenbach.

13. M. corniculata Reichenbach

ILLUSTRATIONS: Botanical Magazine, 7476 (1896); Die Orchidee **30**: 105 (1979).

Like *M. peristeria* except as below.

LEAVES: 20–25 × *c.* 4 cm, lanceolate.
FLOWERING STEM: 7–10 cm, enclosed in an inflated sheath which also conceals the leaf-base.
FLOWERS: yellow, spotted and blotched with orange.
SEPALS: tube *c.* 2 cm, cup-shaped; lateral sepals fused for *c.* 3 cm.
DISTRIBUTION: Colombia.
FLOWERING: summer–autumn.

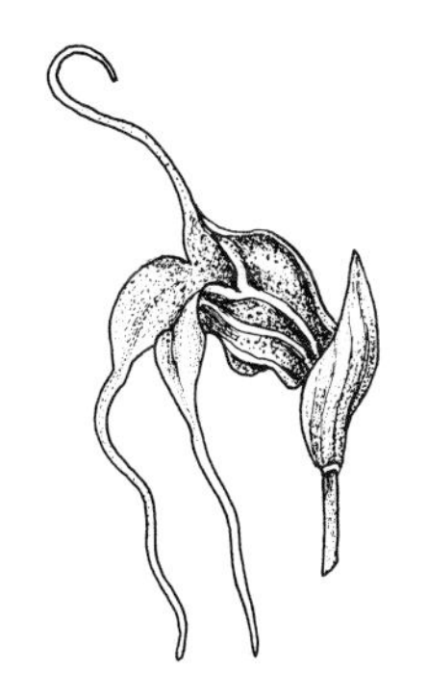

Masdevallia corniculata – flower.

14. M. caloptera Reichenbach

SYNONYM: *M. biflora* Regel.
ILLUSTRATIONS: Gartenflora **40**: 1341 (1891); Bechtel et al., Manual of cultivated orchid species, 224 (1981).
HABIT: epiphytic.
LEAVES: 6–8 × 1.4–2 cm, oblong to lanceolate, obtuse to rounded, older leaves tinged with red.
FLOWERING STEM: 8–12.5 cm, bearing 2 or more flowers, 2 or more open together.
SEPALS: *c.* 2 cm, equally fused for *c.* 6 mm, white with purple streaks, margins finely toothed; tails slender, yellow; tube sharply indented above mentum. Upper sepal almost circular; lateral sepals triangular to oblong.
PETALS: 4–5 mm, toothed, white with a central purple stripe.
LIP: *c.* 5 mm, oblong to shallowly 3-lobed, yellow and crimson.
COLUMN: *c.* 4 mm, green and purple.
DISTRIBUTION: Colombia, Ecuador & Peru.
HARDINESS: G2.
FLOWERING: summer.

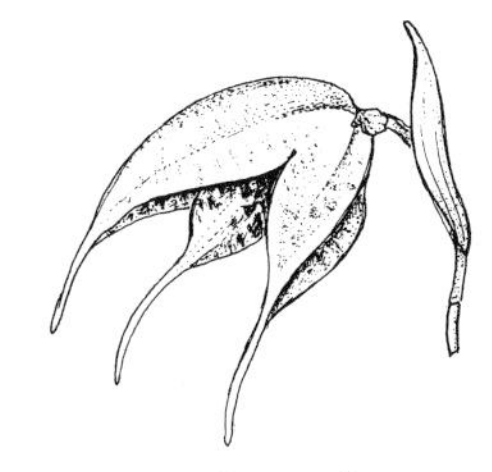

Masdevallia caloptera – flower.

15. M. abbreviata Reichenbach

SYNONYMS: *M. polysticta* Reichenbach; *M. melanopus* misapplied.
ILLUSTRATIONS: Botanical Magazine, 6258 (1876) & 6368 (1878).

Like *M. caloptera* except as below.

LEAVES: 12.5–15 cm.

Masdevallia abbreviata – flower.

FLOWERING STEM: 15–20 cm, bearing many flowers, several open together.
SEPALS: *c.* 3 cm.
LIP: yellow.
DISTRIBUTION: Peru.
FLOWERING: autumn–winter.
Very similar to *M. caloptera*. The name *M. melanopus*, which was given to a similar species with a single row of flowers all pointing in the same direction, has frequently been misapplied to *M. abbreviata*.

16. M. schlimii Lindley

Masdevallia schlimii – flower.

ILLUSTRATIONS: Botanical Magazine, 6740 (1884); Dunsterville & Garay, Venezuelan orchids illustrated **3**: 171 (1965); Bechtel et al., Manual of cultivated orchid species, 227 (1981); Orchid Review **92**: 40 (1984).
HABIT: epiphytic.
LEAVES: to 20 × 5 cm, obovate to elliptic, rounded or obtuse.
FLOWERING STEM: to 35 cm, bearing up to 8 flowers, several open together.
SEPALS: 5–6 cm, with long yellow tails; tube *c.* 6 mm, cup-shaped; upper sepal triangular to ovate, finely tapered, greenish yellow, spotted with maroon below; lateral sepals fused for 1.5–2 cm, oblong.
PETALS: *c.* 6 mm, linear, pale yellow.
LIP: *c.* 6 mm, oblong to fiddle-shaped, acute, pink and yellow with purple spots and blotches.
COLUMN: 6–8 mm.
DISTRIBUTION: Colombia & Venezuela.
HARDINESS: G2.
FLOWERING: summer.

17. M. racemosa Lindley

ILLUSTRATIONS: Die Orchidee **23**: 141 (1972) & **30**: 105 (1979); Orchid Review **86**: 35 (1978); Bechtel et al., Manual of cultivated orchid species, 227 (1981).
Like *M. schlimii* except as below.
HABIT: creeping terrestrial, rarely epiphytic.
LEAVES: 8–12.5 × 2 cm, oblong to lanceolate, obtuse.
SEPALS: bright scarlet; tube *c.* 1.5 cm, straight and narrow; upper sepal *c.* 1 cm, triangular, acuminate; lateral sepals 3–3.5 cm, elliptic to circular, shortly apiculate, fused for *c.* 2.5 cm.
DISTRIBUTION: Colombia.
FLOWERING: spring–autumn.

18. M. coccinea Lindley

SYNONYMS: *M. lindenii* André; *M. harryana* Reichenbach.
ILLUSTRATIONS: Botanical Magazine, 5990 (1872); Shu-

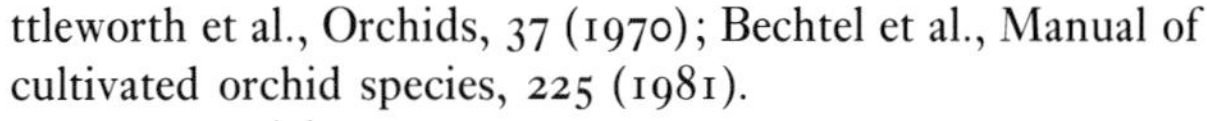
ttleworth et al., Orchids, 37 (1970); Bechtel et al., Manual of cultivated orchid species, 225 (1981).

HABIT: terrestrial

STEMS: *c.* 4 cm.

LEAVES: 15–20 × 2–3 cm, oblong to lanceolate, obtuse to rounded.

FLOWERING STEM: to 40 cm, 1-flowered, streaked with purple, flowers held well above leaves.

SEPALS: purple, crimson, scarlet, orange, yellow or white, base of tube white, greenish yellow or yellow; tube *c.* 2 cm, narrow and curved; upper sepal to 5 cm, narrowly triangular, finely tapered, tail erect or curved back; lateral sepals *c.* 6 × 2 cm, ovate to oblong, acuminate, fused for 2–3 cm.

PETALS: *c.* 1 cm, linear, notched, off-white.

LIP: *c.* 8 mm, oblong, shallowly fiddle-shaped, pinkish.

COLUMN: *c.* 6 mm.

DISTRIBUTION: Colombia & Peru.

HARDINESS: G2.

FLOWERING: spring.

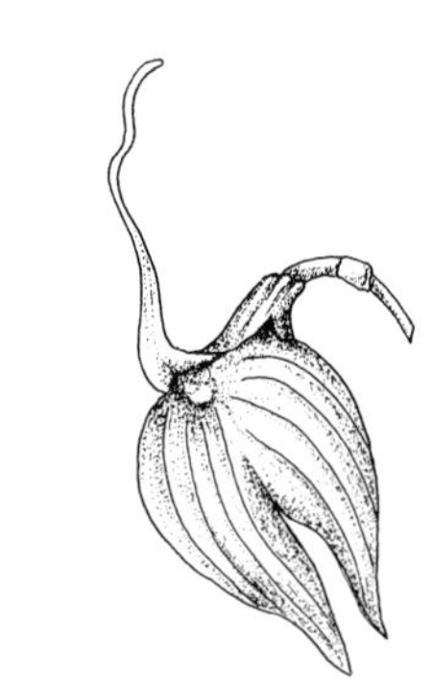
Masdevallia coccinea – flower.

19. M. militaris Reichenbach

SYNONYM: *M. ignea* Reichenbach.

ILLUSTRATION: Botanical Magazine, 5962 (1872).

Like *M. coccinea* except as below.

HABIT: on rocks.

STEMS: 5–8 cm.

FLOWERING STEM: to 30 cm; flowers held just above leaves.

SEPALS: usually reddish brown to orange; upper sepal tail drooping.

PETALS: acuminate.

DISTRIBUTION: Colombia & Venezuela.

FLOWERING: spring–summer.

Masdevallia militaris – flower.

20. M. rosea Lindley

ILLUSTRATIONS: Graf, Tropica, 736 (1978); Northen, Miniature orchids, C-52 (1980).

HABIT: epiphytic.

LEAVES: 12–20 × 2–3 cm, elliptic to obovate, acute.

FLOWERING STEM: 12–20 cm, 1-flowered, arched.

SEPALS: reddish purple and red; tube 2.5–3.5 cm, narrow, slightly curved; upper sepal 2.5–3.5 cm, tapering from the base into a long narrow tail; lateral sepals 4–5 cm, oblong to narrowly triangular, fused for *c.* 2 cm.

PETALS: 4–5 mm, oblong, off-white.

LIP: *c.* 5 mm, oblong, shallowly fiddle-shaped, pink and yellow at base; tip dark purplish brown, densely fringed.

DISTRIBUTION: Colombia & Ecuador.

HARDINESS: G2.

FLOWERING: spring–summer.

Masdevallia rosea – flower.

Masdevallia veitchiana – flower.

21. M. veitchiana Reichenbach

ILLUSTRATIONS: Botanical Magazine, 5739 (1868); Graf, Tropica, 736 (1978); Bechtel et al., Manual of cultivated orchid species, 228 (1981); American Orchid Society Bulletin **53**: 911 (1984).

HABIT: on rocks.

LEAVES: 15–25 × 2.5 cm, oblong to narrowly obovate, obtuse.

FLOWERING STEM: 30–45 cm, erect, 1-flowered.

SEPALS: bright orange inside, with short purplish hairs; yellowish with dark veins outside; tube *c.* 3 cm, cylindric; upper sepal 5–7 × 2 cm, ovate to triangular, finely tapered; lateral sepals 6–7.5 × 1.5–2.5 cm, triangular to ovate-oblong, fused for *c.* 3 cm, acuminate.

PETALS: 1–1.5 cm, oblong, each with a small point, off-white.

LIP: 1–1.5 cm, oblong to shallowly 3-lobed, purple and white.

DISTRIBUTION: Peru.

HARDINESS: G2.

FLOWERING: spring–summer.

Masdevallia amabilis – flower.

22. M. amabilis Reichenbach

ILLUSTRATIONS: Veitch, Manual of orchidaceous plants **5**: 24 (1889); American Orchid Society Bulletin **42**: 240 (1973).

LEAVES: 12–18 × 1.5–2 cm, oblong to lanceolate, obtuse to rounded.

FLOWERING STEM: 25–30 cm, erect, 1-flowered, tinged with pink.

SEPALS: tube 2–2.5 cm, narrow, slightly curved; upper sepal *c.* 4 cm, triangular to ovate, acuminate, long-tailed, reddish orange with darker veins; lateral sepals 4–4.5 cm, ovate to oblong, acuminate, fused for *c.* 1.5 cm, long-tailed, crimson.

PETALS: *c.* 6 mm, oblong, each with a small point, yellow.

LIP: *c.* 6 mm, spathulate; base oblong.

DISTRIBUTION: Peru.

HARDINESS: G2.

FLOWERING: winter.

23. M. barleana Reichenbach

ILLUSTRATIONS: American Orchid Society Bulletin **42**: 240 (1973); Bechtel et al., Manual of cultivated orchid species, 224 (1981).

Like *M. amabilis* except as below.

LEAVES: 8–12 × 2–2.5 cm.

FLOWERING STEM: 15–25 cm.

SEPALS: pink with darker veins; tube 1.5–2 cm; lateral sepals abruptly contracted into long tails.

PETALS: with 3 almost equal teeth at apex.

DISTRIBUTION: Peru.

FLOWERING: spring–summer.

66. DRACULA Luer

About 60 species, mostly from Colombia with a few representatives in C America Ecuador and Peru. The species in this genus were originally described in *Masdevallia* section *Saccilabiatae* Reichenbach, and more recently in section *Chimeroideae* of the same genus. They are, however, quite distinct in floral morphology and appearance. They should be grown in conditions similar to those described for *Masdevallia* (p. 208) though, being native to slightly lower altitudes, they prefer higher temperatures.

Literature: Hawley, R.M., Masdevallia chimaera and the marvellous monsters, *American Orchid Society Bulletin* **46**: 600–9 (1977); Luer, C.A., Dracula, a new genus in the Pleurothallininae (Orchidaceae), *Selbyana* **2**: 190–8 (1978); Kennedy, G., The genus Dracula, *Orchid Digest* **43**: 31–8 (1979); Pridgeon, A.M., Colombia, home of Dracula, *American Orchid Society. Bulletin* **53**: 908–917 (1984); Luer, C. A. & Escobar, R.R., *Thesaurus Dracularum: a monograph of the genus Dracula* (1988).

HABIT: epiphytic, terrestrial or on rocks.
STEMS: short, slender, erect, concealed in papery sheaths.
LEAVES: 1, apical, linear to lanceolate, thinly leathery, sharply keeled, minutely 3-toothed at tip, folded when young.
FLOWERS: 1 or a few in succession on erect or arched racemes borne at leaf-base.
SEPALS: fused below into a wide, shallow cup; lateral sepals usually further fused into a broad blade; free parts of sepals ovate, long-tailed.
PETALS: small, spathulate, split at tip into 2 bivalve-like flaps, often papillose.
LIP: hinged to column foot, clearly divided into a short claw and cup-shaped or pouched blade, 3-keeled below; marginally toothed.
COLUMN: short; column foot slender, fused to base of lateral sepals forming a mentum.
POLLINIA: 2.

Key to species

1. Lip blade broader than long, kidney-shaped; flowers produced singly, 1 per stem **1. bella**
2. Sepals hairless, very pale green with dense purplish black veins and tails; petals *c.* 6 mm; lip 1.5–2.5 cm **5. vampira**
3. Sepals hairy, unequally fused (the laterals joined for 1–1.5 cm above the sepal-tube); petals *c.* 8 mm, white with purplish markings; lip *c.* 2 cm **2. chimaera**
4. Sepals hairy, equally fused, tube *c.* 1.5 cm; petals 4–5 mm, yellow with purple spots; lip *c.* 6 mm **3. radiosa**

5. Sepals hairy, equally fused, tube *c.* 6 mm; petals *c.* 3 mm, pinkish; lip 1–1.5 cm **4. erythrochaete**

1. D. bella (Reichenbach) Luer

SYNONYM: *Masdevallia bella* Reichenbach.

ILLUSTRATIONS: Gardeners' Chronicle **16**: 237 (1881); Orchid Review **77**: 207 (1969); American Orchid Society Bulletin **46**: 601 (1977), **55**: 986 (1986).

HABIT: epiphytic or terrestrial.

LEAVES: 12–20 × 2.5 cm, oblong to lanceolate, obtuse.

FLOWERING STEMS: to 18 cm, pendent, purple to purplish green, 1-flowered.

FLOWERS: to *c.* 25 cm long.

SEPALS: hairless; tube 1–1.5 cm; upper sepal 10–12 × 2–2.5 cm, triangular to ovate, finely tapered, greenish yellow, densely spotted with purplish brown, tail long and slender; lateral sepals 8–10 × 1.5–2. cm, fused for *c.* 2.5 cm into an almost square blade; tails on outer corners, often crossed.

PETALS: 6–8 mm, divided into a fan-shaped blade and a claw; blade unequally 2-flapped, yellow with brown markings, with papillae and teeth on the flaps.

LIP-BLADE: 1.2 × 2 cm, kidney-shaped, concave, white with radiating lines; claw *c.* 4 mm, pink and fleshy.

COLUMN: *c.* 3 mm, stout, purplish brown.

DISTRIBUTION: Colombia.

HARDINESS: G2.

FLOWERING: winter–summer.

2. D. chimaera (Reichenbach) Luer

SYNONYMS: *Masdevallia chimaera* Reichenbach; *M. backhousiana* Reichenbach.

ILLUSTRATIONS: Shuttleworth et al., Orchids, 37 (1970); American Orchid Society Bulletin **46**: 600 (1977); Graf, Tropica, 738 (1978).

HABIT: terrestrial or on rocks, rarely epiphytic.

LEAVES: 15–25 × 3–5 cm, elliptic to obovate, obtuse.

FLOWERING STEMS: to 60 cm, erect or ascending, green to purplish green, bearing 3–8 flowers.

SEPALS: tube 1–1.5 cm; lateral sepals fused for *c.* 2.5 cm; sepal blades 5–6 cm, triangular to ovate, tapered, yellow or greenish yellow, densely spotted and blotched with brownish purple or black, hairy and warty; tails 8–20 cm slender, purple.

PETALS: *c.* 8 mm, spathulate, white with purple markings.

LIP: *c.* 2 cm, oblong to square or circular in surface view, deeply pouched, white to yellow or pinkish brown; claw pink and orange.

COLUMN: *c.* 1 cm, minutely toothed at apex; column foot pink.

DISTRIBUTION: Colombia.
HARDINESS: G2.
FLOWERING: winter.

3. D. radiosa (Reichenbach) Luer
SYNONYM: *Masdevallia radiosa* Reichenbach.
ILLUSTRATIONS: Orchid Review **38**: 247 (1930); Orquideologia **9**: 44 (1974) & **10**: 81 (1975); American Orchid Society Bulletin **46**: 604 (1977).
Like *D. chimaera* except as below.
FLOWERING STEMS: erect or pendent, bearing 2–4 flowers.
SEPALS: equally fused for *c.* 1.5 cm, brownish yellow with dark purplish brown papillae inside, pinkish or brownish yellow outside; free parts broadly ovate to rectangular, abruptly tapered; tails 7–10 cm, purplish brown.
PETALS: 4–5 mm, yellow with purple spots; flaps dissimilar, papillose.
LIP: *c.* 6 mm in diameter, pinkish white.
DISTRIBUTION: Colombia.
FLOWERING: spring–summer.

4. D. erythrochaete (Reichenbach) Luer
SYNONYMS: *Masdevallia erythrochaete* Reichenbach; *M. astuta* Reichenbach.
ILLUSTRATIONS: Orquideologia **9**: 43 (1974); Northen, Miniature orchids, C-30 (1980); Bechtel et al., Manual of cultivated orchid species, 225 (1981).
Like *D. chimaera* except as below.
HABIT: epiphytic.
LEAVES: 15–25 × 1–1.5 cm.
FLOWERING STEM: erect to pendent, bearing 1–3 flowers.
SEPALS: tube *c.* 6 mm; sepals *c.* 5 cm, ovate, abruptly tapered, pale yellow, spotted and veined with pinkish purple; tails purple.
PETALS: *c.* 3 mm, pinkish.
LIP: 1–1.5 cm, pale pinkish yellow.
DISTRIBUTION: Costa Rica to Colombia.
FLOWERING: autumn–winter.

5. D. vampira (Luer) Luer
SYNONYM: *Masdevallia vampira* Luer.
ILLUSTRATION: Orchid Digest **43**: 34 (1979).
Like *D. chimaera* except as below.
HABIT: epiphytic.
FLOWERING STEM: 20–40 cm or more, horizontal to pendent, bearing 5–7 flowers.
SEPALS: hairless, very pale green with dense purplish black veins and tails.

PETALS: *c.* 6 mm, white marked with purple.

LIP: 1.5–2.5 cm; blade concave, white with pink or yellow veins.

DISTRIBUTION: Ecuador.

FLOWERING: summer.

67. PORROGLOSSUM Schlechter

A genus of 8 or 9 montane species from Venezuela to Peru, originally included in *Masdevallia* but distinct in lip and column structure and in the sensitivity of the lip which folds up suddenly against the column if stimulated by touch or shaking. Only one species is generally grown. Cultivation should be in cool, moist, shady conditions similar to those for *Masdevallia* (p. 208).

Literature: Sweet, H.R., The genus Porroglossum, *American Orchid Society Bulletin* **41**: 513–24 (1972); Luer, C.A., Monograph of Porroglossum, *Systematic Botany* **24**: 25–89 (1987).

HABIT: similar to *Masdevallia.*

LEAVES: obovate, lanceolate or elliptic, long-stalked.

FLOWERS: solitary or a few open together or in succession, on a slender erect or arched stem.

SEPALS: equally fused, forming a cup-shaped tube, free parts spreading.

PETALS: small, meeting around column.

LIP: S-shaped in side profile, with a distinct claw, attached below apex of column foot, mobile and sensitive.

ROSTELLUM: upright.

ANTHER: erect, on column apex.

POLLINIA: 2, hard.

1. P. **echidnum** (Reichenbach) Garay

SYNONYMS: *Masdevallia echidna* Reichenbach; *M. muscosa* Reichenbach; *P. muscosum* (Reichenbach) Schlechter.

ILLUSTRATIONS: Botanical Magazine, 7664 (1899); Northen, Miniature orchids, 149 (1980); Bechtel et al., Manual of cultivated orchid species, 267 (1981).

LEAVES: to 10 cm, elliptic, obtuse; upper surface papillose.

FLOWERING STEM: to 15 cm, densely hairy, bearing 1 or 2 flowers.

SEPALS: tube 3–4 mm, free parts 2.5–3 cm × 6 mm, triangular to ovate; lateral sepals slightly asymmetric, yellow or yellowish brown.

PETALS: 5–8 × 1 mm, pinkish.

LIP: 5–8 × 4 mm, triangular to diamond-shaped, long-clawed, front margin hairy.

DISTRIBUTION: Venezuela, Colombia & Ecuador.

HARDINESS: G2.
FLOWERING: intermittent.

68. TRISETELLA Luer

A genus formerly known as *Triaristella*; cultivation as for *Masdevallia* (p. 208).

Like *Masdevallia*, distinguished as below.
UPPER SEPAL: small, almost free.
LATERAL SEPALS: large, almost totally fused, with tails attached to side of blade.
LIP: arrow-shaped.

1. T. reichenbachii (Brieger) Luer
SYNONYMS: *Masdevallia triaristella* Reichenbach; *Triaristella reichenbachii* Brieger.
ILLUSTRATIONS: Botanical Magazine, 6268 (1876); Bechtel et al., Manual of cultivated orchid species, 228 (1981).
LEAVES: 2.5–5 cm × 5–8 mm, cylindric, channelled, rounded, tinged with purple.
FLOWERING STEMS: 7.5–10 cm, erect, bearing 1–3 flowers in succession.
SEPALS: yellow and pinkish purple; upper *c.* 2 cm, triangular to ovate, long-tailed; lateral sepal blade 15 × 2–4 mm, tails 1–1.5 cm.
PETALS: *c.* 4 mm, linear, each with a small point.
LIP: *c.* 6 mm, pinkish brown.
DISTRIBUTION: Costa Rica, Panama & Colombia.
HARDINESS: G2.
FLOWERING: summer–autumn.

69. PLEUROTHALLIS R. Brown

A large genus of about 1000 species, widespread throughout the mountainous regions of the New World tropics and subtropics. Only a relatively small number is grown. Cultivation conditions are similar to those for *Masdevallia* (p. 208).

Literature: Luer, C.A., Monograph of Pleurothallis, *Systematic Botany* **20**: 1–101 (1986).

HABIT: epiphytic, rarely on rocks.
STEMS: clustered, slender, with a few overlapping sheaths, 1-leaved.
LEAVES: leathery, stalkless or shortly stalked, folded when young.
FLOWERING STEMS: 1-several, apical, erect or arched, bearing 1-many flowers in racemes.
FLOWERS: small, usually in 2 ranks, not widely open, often pendent.

SEPALS: upper free or very shortly fused to lateral sepals; lateral sepals partially to completely fused, concave, forming a small mentum with the column foot.
PETALS: usually much smaller than sepals.
LIP: small, loosely hinged to column foot.
POLLINIA: 2 (in our species).

Synopsis of characters

Stems. 7 cm or less: **1–4,8–10**; 8 cm or more: **1–3,5–10.** Pendent: **1–3.** Sheaths brown: **1–3,5,7–10**; sheaths grey: **6**; sheaths white: **1–4.**
Leaf. 7 cm or less: **1–4,8–10**; 8 cm or more: **1–3,5,7–10.** Less than 2 cm wide: **1–4,8–10.** Cone-shaped: **2.** Deeply grooved on midrib above: **1.**
Flowering stems. Solitary: **1–10**; 2 or more: **8–10.** 6 cm or less: **1–4,8–10**; shorter than leaf: **1–3,8–10.** Zig-zag: **1–4,8–10.**
Flowers. 3 or fewer: **1**; 25 or more: **6,8–10.** Yellow or greenish yellow: **1–4,6,8–10**; greenish brown: **7**; purple or purplish brown: **1–3,5**; pink: **1–3.** Translucent: **4,6,8–10.**
Hairy: **8–10.** With dangling marginal appendages: **8–10.**
Lateral sepals. Less than 1.5 cm: **1–4,6,8–10**; more than 2 cm: **5,7.** Almost totally fused: **4–6,8–10**; fused for two-thirds or more: **1–6,8**; fused for half or less; **7–10.**
Petals. 5 mm or less: **1–4,8–10**; 6–10 mm: **6,7**; *c.* 1.5 cm: **5.** Almost equal to sepals: **6**; distinctly shorter than sepals: **1–5,7,8–10.**
Lip. Warty: **1–3,5.** With erect basal margin: **6.**
Ovary. Spiny: **1.**

Key to species

1. Flowers 1–3 per flowering stem — **3. tribuloides**
2. Lateral sepals fused for half their length or less
2.A. Stems 30 cm or more, flowering stem solitary; sepals *c.* 3 cm — **7. grandis**
2.B. Stems to 1.5 cm; leaves to 4 cm x 8 mm — **10. schiedei**
2.C. Petals minute; sepals green outside, yellowish spotted with brown inside — **9. saurocephala**
2.D. Petals to 4 mm; sepals yellowish green, translucent — **8. gelida**
3. Flowering stem shorter than leaves, or, if longer, then lower part enclosed in a groove running about halfway up the leaf blade, or within the cone-shaped leaf
3.A. Leaf cone-shaped — **2. pectinata**
3.B. Leaf obovate — **1. immersa**
3.C. Stems 1 cm or less; leaf up to 1 cm wide — **4. grobyi**
3.D. Flowering stem arched; flowers purple; sepals 2.5–3 cm, twice as long as petals — **5. macrophylla**
3.E. Flowering stem erect; flowers yellowish green; sepals 5–15 mm, almost equal to petals — **6. quadrifida**

1. P. immersa Linden & Reichenbach

ILLUSTRATIONS: Botanical Magazine, 7189 (1891); Dunsterville & Garay, Venezuelan orchids illustrated 3: 254 (1965); Proceedings of 7th world orchid conference, 73 (1974).

STEMS: 2–7 cm, with brown sheaths.

LEAF: 7–20 × 2–4.5 cm, lanceolate to narrowly obovate, rounded to acute, with a central winged groove enclosing the lower part of flowering stem.

FLOWERING STEM: 17–40 cm, erect, zig-zag, sometimes arched above, bearing 12–20 bell-shaped flowers.

FLOWERS: purplish brown or bright greenish yellow.

SEPALS: stiff, finely hairy; upper 8–14 × 3–4.5 mm, lanceolate, obtuse; lateral sepals fused for *c.* two-thirds of their length into an elliptic blade, 8–13 × 5–7 mm.

PETALS: 3–4 × 2–3 mm, spathulate, fleshy with thin margins.

LIP: *c.* 4 × 2 mm, triangular, clawed, 2-ridged, warty at tip.

DISTRIBUTION: Mexico to Colombia & Venezuela.

HARDINESS: G2.

FLOWERING: winter.

2. P. pectinata Lindley

ILLUSTRATIONS: Schlechter, Die Orchideen, edn 2, 176 (1927); Pabst & Dungs, Orchidaceae Brasilienses 1: 226 & 345 (1975).

Like *P. immersa* except as below.

STEMS: 10–15 cm, pendent.

LEAF: 7–10 cm, broadly elliptic, cone-shaped by fusion of lower margins, enclosing the base of the flowering stem.

FLOWERING STEM: to 6 cm.

FLOWERS: *c.* 15 in 2 dense ranks, yellow to green.

DISTRIBUTION: S Brazil.

FLOWERING: spring–summer.

3. P. tribuloides Lindley

ILLUSTRATIONS: Reichenbach, Xenia Orchidacearum, t. 275 (1892).

Generally similar to *P. immersa*, differing as below.

STEM: less than 1 cm, with white sheaths.

LEAF: to 7 × 1.5 cm, obovate.

FLOWERING STEM: to 1 cm, bearing 1–3 pink to brownish red or purple flowers.

LATERAL SEPALS: fused from one-third of their length to completely so.

DISTRIBUTION: C America, West Indies.

4. P. grobyi Lindley

ILLUSTRATIONS: Botanical Magazine, 3682 (1839); Shuttleworth et al., Orchids, 38 (1970); Skelsey, Orchids, 133 (1979).

STEMS: 1 cm or less, with white sheaths.

LEAF: to 7 × 1.1 cm, lanceolate to narrowly obovate, obtuse or rounded.

FLOWERING STEM: to 15 cm, slender, erect, zig-zag, bearing 4–12 flowers in 2 opposite ranks.

FLOWERS: translucent, pale yellow to green with scattered purplish veins.

SEPALS: upper 4–10 × 1.5–3 mm, ovate, acuminate, rounded; lateral sepals fused into an oblong to elliptic, slightly bifid blade, 5–14 × 4–7 mm.

PETALS: 1.5–3 × 0.5–1 mm, lanceolate to spathulate.

LIP: 2–4 × 1 mm, tongue-shaped, arched, rounded.

DISTRIBUTION: Mexico & West Indies to Peru & Brazil.

HARDINESS: G2.

FLOWERING: summer.

5. **P. macrophylla** Humboldt, Bonpland & Kunth

SYNONYM: *P. roezlii* Reichenbach.

ILLUSTRATIONS: Cogniaux & Goossens, Dictionnaire Iconographique des Orchidées, Pleurothallis, t. 1 (1898); Journal of the Royal Horticultural Society **26**: 147 (1901); Gartenflora **50**: 272 (1901).

STEMS: 8–15 cm, with brown sheaths.

LEAF: 12–20 × 2.5–4 cm, oblong to lanceolate, acute, sometimes notched.

FLOWERING STEM: to 30 cm, arched to pendent, bearing 5–10 flowers; bracts 1–1.5 cm.

FLOWERS: pendent, dark purple.

SEPALS: upper 2.5–3 × 1–1.5 cm, elliptic, acute; lateral sepals fused into an elliptic, obtuse blade, 2.5–3 × 1.4–2 cm.

PETALS: *c.* 1.5 cm, lanceolate, acute, fleshy, prominently 3-veined.

LIP: *c.* 1.5 cm, spathulate, fleshy, 2-ridged, papillose above.

DISTRIBUTION: Colombia.

HARDINESS: G2.

FLOWERING: winter.

6. **P. quadrifida** (Llave & Lexarza) Lindley

SYNONYM: *P. ghiesbrechtiana* Richard & Galeotti.

ILLUSTRATIONS: Dunsterville & Garay, Venezuelan orchids illustrated **2**: 299 (1961); Graf, Tropica, 752 (1978); Bechtel et al., Manual of cultivated orchid species, 265 (1981).

STEMS: to 15 cm, with grey sheaths.

LEAF: to 17 × 3 cm, oblong to narrowly elliptic, rounded, often notched, sometimes grooved along the back of the midrib.

FLOWERING STEM: to 45 cm, usually longer than leaf, erect to spreading, bearing 10–30 flowers.

FLOWERS: pendent, translucent, pale yellow to yellowish green.

SEPALS: upper 7–10 × 1.5–2.5 mm, ovate, acuminate, concave,

3-veined; lateral sepals fused into an elliptic, rounded, minutely notched, 6-veined blade, 5–12 × 4–6 mm.

PETALS: 6–10 × 2–4 mm, lanceolate, acute, 1-veined.

LIP: 4–6 × 1.5–3 mm, oblong waisted; margin erect at base.

DISTRIBUTION: Mexico & West Indies south to Venezuela & Colombia.

HARDINESS: G2.

FLOWERING: winter.

7. P. grandis Rolfe

ILLUSTRATIONS: Botanical Magazine, 8853 (1920); Orchid Review **30**: 99 (1922).

STEMS: 30–45 cm, with pale brown sheaths.

LEAF: 18–35 × 9–18 cm, ovate to broadly elliptic, acute to obtuse.

FLOWERING STEM: to 50 cm, erect to spreading, loosely many-flowered.

FLOWERS: spreading to pendent, greenish brown, flushed and striped with brownish red.

SEPALS: upper 2.5–3 cm, linear to lanceolate, acute; lateral sepals *c.* 3 cm, lanceolate, obtuse, fused for about half their length.

PETALS: *c.* 1 cm, oblong, obtuse.

LIP: *c.* 2 cm, oblong to spathulate, with 2 fleshy incurved lobes at the tip.

DISTRIBUTION: Costa Rica.

HARDINESS: G2.

FLOWERING: summer–autumn.

8. P. gelida Lindley

ILLUSTRATIONS: Rickett Wild flowers of the United States **2**: 125 (1967); Luer, Native orchids of Florida, 185 (1972); American Orchid Society Bulletin **45**: 194 (1976).

STEMS: 5–35 cm, with brown sheaths.

LEAF: to 25 × 6 cm, oblong to elliptic, rounded, shortly stalked.

FLOWERING STEMS: 1-several, 5–30 cm, erect, bearing 5–25 flowers.

FLOWERS: pale yellowish green, translucent, softly hairy.

SEPALS: upper *c.* 8 × 3 mm, lanceolate to ovate, obtuse, concave; laterals *c.* 8 × 3 mm, lanceolate, acute, fused for about half their length, pouched below.

PETALS: *c.* 4 × 2 mm, oblong to spathulate, irregularly toothed at tip.

LIP: *c.* 2.5 mm, oblong, rounded or truncate.

DISTRIBUTION: S Mexico & Florida, south through C America & West Indies to Venezuela, Colombia, Ecuador & Peru.

HARDINESS: G2.

FLOWERING: intermittently.

9. P. saurocephala Loddiges

ILLUSTRATIONS: Botanical Magazine, 3030 (1830); Pabst & Dungs, Orchidaceae Brasilienses **1**: 351 (1975).

Like *P. gelida* except as below.

STEMS: to 15 cm, grooved.

LEAF: to 15 cm, elliptic.

FLOWERING STEMS: usually 2, equal to or slightly shorter than leaf; inflorescence dense, with fibrous chaffy bracts at the base.

SEPALS: *c.* 1 cm, green outside, yellow spotted with brown inside.

PETALS: minute.

DISTRIBUTION: S Brazil.

FLOWERING: spring–autumn.

10. P. schiedei Reichenbach

SYNONYM: *P. ornata* Reichenbach.

ILLUSTRATIONS: Hamer, Las Orquideas de El Salvador **2**: 237 & t. 27 (1974); Northen, Miniature orchids, C-77 (1980); American Orchid Society Bulletin **53**: 595, 597, 598 (1984).

Like *P. gelida* except as below.

STEMS: to 1.5 cm.

LEAF: to 4 cm × 8 mm, lanceolate, spotted on back.

FLOWERING STEMS: 1-several, longer than leaf; inflorescence loose, bearing 5–12 flowers.

SEPALS: *c.* 7.5 mm, broadly elliptic to spathulate, obtuse to rounded, greenish yellow with purple warts and dangling white marginal appendages.

DISTRIBUTION: Mexico.

FLOWERING: spring–summer.

70. RESTREPIA Humboldt, Bonpland & Kunth

A genus of about 40 species from Mexico through C America to Argentina. Cultivation conditions are similar to those for *Masdevallia* (p. 208).

Literature: Cokely, W.R., Growing Restrepias, *American Orchid Society Bulletin* **57**: 837–841 (1985).

HABIT: dwarf, tufted epiphytes, rarely on rocks, occasionally creeping.

STEMS: slender, 1-leaved at apex, concealed in flattened, papery, overlapping sheaths.

LEAVES: stiff and leathery, folded when young.

FLOWERS: solitary on slender stalks, borne at the junction of leaf and stem.

UPPER SEPAL AND PETALS: lanceolate or narrowly triangular, drawn out into long slender tails with club-shaped tips.

LATERAL SEPALS: almost totally fused into a broad, often bifid lower sepal.

LIP: oblong, often with fine lateral lobes near base; tip rounded, truncate or notched, papillose.

COLUMN: slender, arched, club-shaped, with small column foot.

POLLINIA: 4.

Key to species

1. Lower sepal marked with stripes or lines of spots separated by clear areas — **3. antennifera**
2. Leaves at least twice as long as wide — **1. guttulata**
3. Leaves less than twice as long as wide — **2. elegans**

1. **R. guttulata** Lindley

SYNONYM: *R. maculata* Lindley.

ILLUSTRATIONS: Dunsterville & Garay, Venezuelan orchids illustrated **4**: 263 (1966); American Orchid Society Bulletin **47**(7): cover (1978); Bechtel et al., Manual of cultivated orchid species, 269 (1981).

STEMS: 1–15 cm.

LEAVES: 5–8 × 2.5–3 cm, narrowly elliptic, obtuse, slightly keeled.

FLOWERING STEMS: to 7 cm, appearing behind the leaf through twisting of leaf-base.

UPPER SEPAL: to 2.5 cm × 3 mm, lanceolate, pinkish purple with pale margins, finely tapered into a slender tail with a club-shaped tip.

LOWER SEPAL: 2.5 × 1 cm, oblong to elliptic, slightly bifid, brownish green spotted with purple.

PETALS: 13 × 1 mm, narrowly triangular, pale pink with a darker central stripe.

LIP: 10 × 2 mm, oblong with 2 narrow basal lobes, similar in colour to the lower sepal, papillose.

COLUMN: *c.* 6 mm, club-shaped, pale with purple spots at base.

DISTRIBUTION: Venezuela, Colombia & Ecuador.

HARDINESS: G2.

FLOWERING: winter.

2. **R. elegans** Karsten

ILLUSTRATIONS: Botanical Magazine, 5966 (1872); Dunsterville & Garay, Venezuelan orchids illustrated **1**: 375 (1959); Bechtel et al., Manual of cultivated orchid species, 269 (1981).

STEMS: 4–6 cm, erect or ascending.

LEAVES: 4–6 × 3–4 cm, elliptic to circular, obtuse to rounded.

FLOWERING STEM: to *c.* 4 cm, slender.

UPPER SEPAL: 2.5 × 2–3 mm, lanceolate, attenuate, white with purple veins and tip.

LOWER SEPAL: 2.4 cm × 9 mm, oblong, obtuse, yellow to brown, spotted with purple.

PETALS: 14 × 1 mm, narrowly triangular, attenuate, pale with purple veins and tip.

LIP: 11 × 5 mm, 3-lobed; lateral lobes narrow and sickle-shaped, acute, central lobe fiddle-shaped, narrowing rapidly above the lateral lobes, then broadening above into an oblong notched blade.

COLUMN: *c.* 6 mm, pale green.

DISTRIBUTION: Venezuela.

HARDINESS: G2.

FLOWERING: winter–spring.

3. **R. antennifera** Humboldt, Bonpland Kunth

ILLUSTRATIONS: Dunsterville & Garay, Venezuelan orchids illustrated **4**: 259 (1966); Shuttleworth et al., Orchids, 39 (1970); American Orchid Society Bulletin **43**: 621 (1974); not Botanical Magazine, 6288 (1877).

STEMS: to *c.* 10 cm.

LEAVES: to 5.5 × 3 cm, elliptic to ovate, obtuse or rounded.

FLOWERING STEMS: *c.* 2.5 cm, slender.

UPPER SEPAL: 2.3 cm × 3 mm, lanceolate, attenuate, white with purple veins; tail green with pink tip.

LOWER SEPAL: 2.3 × 1.2 cm, oblong to ovate, acute, bifid, pale yellowish brown with purple lines or dense lines of spots.

PETALS: 1.5 cm × 1 mm, lanceolate to narrowly triangular, white with a purple mid-vein.

LIP: 10 × 3 mm, oblong to fiddle-shaped with very narrow, acute lateral lobes and a short, cylindric claw; tip truncate and papillose, pale brown, spotted with purple.

COLUMN: *c.* 5 mm, club-shaped, pale brown.

DISTRIBUTION: Venezuela, Colombia & Ecuador.

HARDINESS: G2.

FLOWERING: winter–spring.

71. CRYPTOPHORANTHUS Barbosa Rodrigues

About 25 species from the New World tropics. Cultivation conditions are similar to those described for *Masdevallia* (p. 208). If grown in pots, plants should be well shaded and frequently repotted to avoid the development of stale, soggy compost.

HABIT: tufted epiphytes, rarely terrestrial.

STEMS: erect or ascending, sheathed, 1-leaved at apex.

LEAVES: erect, usually hard and leathery, folded when young.

FLOWERS: 1 to many, clustered at junction of leaf and stem.

SEPALS: fused at base and tip, leaving 2 lateral openings or windows; lateral sepals forming a short mentum with column foot.

PETALS: very small.

LIP: very small, free, shallowly 3-lobed.

COLUMN: erect, club-shaped.
POLLINIA: 2, waxy.

Key to species

1. Leaves spotted with purple above **2. maculatus**
2. Leaves 1.5–3 cm broad; flowers purplish crimson **1. atropurpureus**
3. Leaves 3.5–5 cm broad; flowers white to pale yellow spotted with purplish crimson **3. dayanus**

1. C. atropurpureus (Lindley) Rolfe
SYNONYMS: *Pleurothallis atropurpurea* Lindley; *Masdevallia fenestrata* Hooker; not *C. fenestratus* Barbosa Rodrigues.
ILLUSTRATION: Botanical Magazine, 4164 (1845).
STEM: 3–5 cm.
LEAVES: 3–9 × 1.5–3 cm, oblong to elliptic, acute to obtuse, shortly stalked.
FLOWERS: solitary, purplish crimson.
SEPALS: *c.* 1.5 cm × 5 mm, acute, deflexed.
PETALS: *c.* 4 × 2 mm, oblong, obtuse to truncate.
LIP: *c.* 4.5 × 1 mm, spear-shaped.
COLUMN: *c.* 2 mm.
DISTRIBUTION: Cuba & Jamaica.
HARDINESS: G2.
FLOWERING: autumn.

2. C. maculatus Rolfe
Like *C. atropurpureus* except as below.
LEAVES: 4–5.5 × 1.8–3.2 cm, spotted with purple above.
FLOWERS: many, *c.* 5 mm long, yellow spotted with purple.
DISTRIBUTION: Brazil.
FLOWERING: summer.

3. C. dayanus Rolfe
ILLUSTRATIONS: Botanical Magazine, 8740 (1917); Ospina & Dressler, Orquideas de las Americas, f. 76 (1974).
Like *C. atropurpurea* except as below.
STEM: 10–15 cm.
LEAVES: 6–9 × 3.5–5.5 cm, elliptic to circular.
FLOWERS: solitary or few, 3.5–4 cm long, white to pale yellow, spotted with pink, purple or crimson.
DISTRIBUTION: Colombia.

72. OCTOMERIA R. Brown

A genus of 50–100 species mostly from Brazil with representatives in the West Indies and other parts of C and S America. Cultivation conditions are generally similar to those for *Masdevallia* (p. 208). The plants should never be allowed to dry out, though less water should be given in winter.

HABIT: creeping epiphytes, rarely on rocks.

STEMS: slender, often densely tufted, erect, sheathed below, bearing a single leaf; sheaths sometimes frayed into chaffy strands.

LEAVES: stalked or stalkless, folded when young, fleshy or leathery, flat to cylindric.

FLOWERS: 1–many, small, bell-shaped, in terminal clusters at the leaf-base.

SEPALS AND PETALS: free or slightly fused, lanceolate to narrowly ovate.

LIP: very small, simple or 3-lobed, sometimes shortly clawed, hinged to the base of the column.

COLUMN: very short, with a small column foot.

POLLINIA: 8.

Key to species

1. Leaves cylindric or semi-cylindric in section
1.A. Leaves 30–50 cm × 3–5 mm, cylindric **4. juncifolia**
1.B. Leaves 6–12 cm × 1.5–2.5 mm, semi-cylindric **5. gracilis**
2. Stems not densely tufted, 1–4 cm apart **3. graminifolia**
3. Flowers in a dense cluster; sepals and petals *c.* 6 × 2 mm **2. crassifolia**
4. Flowers loose; sepals 1–1.2 cm, petals 9–11 mm **1. grandiflora**

1. O. **grandiflora** Lindley

ILLUSTRATIONS: Dunsterville & Garay, Venezuelan orchids illustrated **1**: 258 (1959) & **6**: 286 (1976); Pabst & Dungs, Orchidaceae Brasilienses **1**: 228 & 365 (1975); Bechtel et al., Manual of cultivated orchid species, 235 (1981).

HABIT: densely tufted epiphyte.

STEMS: 6–20 cm, flattened above.

LEAVES: 10–20 cm × 7–17 mm, linear to narrowly lanceolate, acute, stiff and leathery.

FLOWERS: several, yellow and greenish yellow or white.

SEPALS: 1–1.2 cm × 3.5–5 mm, elliptic to ovate, obtuse.

PETALS: 9–11 × 2.5–3.5 mm, lanceolate to ovate, acute or obtuse.

LIP: 6–8 × 4–6 mm, almost oblong, truncate or notched at tip, with small lateral lobes near base, white to pale yellow with 2 purple ridges and spots near base.

COLUMN: 3–4 mm, curved, often with a pair of reddish spots near tip.

DISTRIBUTION: Venezuela, Trinidad, Surinam, Brazil, Bolivia, Paraguay.

HARDINESS: G2.

FLOWERING: autumn.

2. O. crassifolia Lindley

ILLUSTRATIONS: Martius, Flora Brasiliensis 3(4): t. 128 (1896); Pabst & Dungs, Orchidaceae Brasilienses **2**: 363 (1977).

Like *O. grandiflora* except as below.

STEMS: 15–20 cm.

LEAVES: 10–18 × 1.2–2 cm.

FLOWERS: densely clustered.

SEPALS AND PETALS: *c.* 6 × 2 mm.

LIP: 3.5–4 × 2.5–3 mm, oblong to elliptic, shallowly 3-lobed, acute to rounded at apex.

COLUMN: 1.5–2 mm.

DISTRIBUTION: Brazil, Uruguay & Paraguay.

HARDINESS: G2.

FLOWERING: winter–summer.

3. O. graminifolia R. Brown

ILLUSTRATION: Botanical Magazine, 2764 (1827).

HABIT: creeping epiphyte.

STEMS: 3–7 cm, 1–4 cm apart.

LEAVES: 6–11 cm × 5–10 mm, linear to narrowly lanceolate, acute.

FLOWERS: 1 or 2 on short individual stalks, pale yellow or greenish yellow, fragrant.

SEPALS: 6–8 × 2.5–3 mm, ovate to lanceolate, acute.

PETALS: 6–7.5 × 2–2.5 mm, lanceolate to elliptic, acute.

LIP: 4–6 × 2–3 mm shallowly 3-lobed, pale yellow with purple ridges; central lobe oblong to circular.

COLUMN: 2–2.5 mm, yellow.

DISTRIBUTION: West Indies & Brazil.

HARDINESS: G2.

FLOWERING: spring.

4. O. juncifolia Barbosa Rodrigues

ILLUSTRATIONS: Martius, Flora Brasiliensis 3(4): t. 132 (1896); Pabst & Dungs, Orchidaceae Brasilienses **1**: 228 & 372 (1975).

HABIT: tufted epiphyte.

STEMS: 20–35 cm, very slender.

LEAVES: 30–50 cm × 3–5 mm, very slender, cylindric with a shallow groove, arched.

FLOWERS: few (rarely numerous), yellow.

SEPALS AND PETALS: 8–10 × 2.5–3 mm, lanceolate to oblong, obtuse.

LIP: 4.5–5 × 3–3.5 mm, ovate to elliptic, centrally spotted with red.

COLUMN: 2–2.5 mm.

DISTRIBUTION: S Brazil.

HARDINESS: G2.
FLOWERING: summer–autumn.

5. O. gracilis Lindley
ILLUSTRATION: Pabst & Dungs, Orchidaceae Brasilienses 1: 372 (1975).
Like *O. juncifolia* except as below.
STEMS: 3–9 cm.
LEAVES: 6–12 × 1.5–2.5 mm, semi-cylindric, grooved.
FLOWERS: 1–8, pale yellow.
SEPALS AND PETALS: 3.5–4.5 × 1–1.5 mm, linear to oblong, obtuse.
LIP: 2.5–3 × 1.5 mm, circular to oblong, shallowly 3-lobed.
COLUMN: *c.* 1.5 mm.
DISTRIBUTION: S Brazil.
FLOWERING: spring.

73. DENDROBIUM Swartz

A genus of about 1000–1400 species, widely distributed in Asia, in particular the Himalaya, Burma, Thailand, Malaysia, Indonesia and New Guinea, and extending to Australia. Its supra-specific classification is rather uncertain, both taxonomically and nomenclaturally. The account below follows the sections (and their order) proposed by Schlechter (*Feddes Repertorium Beihefte* 1: 440–52, 1912), as modified by Schelpe and Schelpe & Stewart; the sections themselves, however, are referred to as groups, because of uncertainties with regard to their correct nomenclature. The sectional names used by Schlechter, Kränzlin, Schelpe and Schelpe & Stewart are cited as appropriate. In Schlechter (edited by Brieger, Maatsch & Senghas), *Die Orchideen*, edn 3, 1(11 & 12): 636–752 (1981), Brieger divides the genus, as recognised here, into numerous segregate genera, formed largely by the raising of the Schlechter sections to generic level. This produces many new names, which are not cited here as synonyms. There is no doubt that the older, inclusive concept of the genus is appropriate here; whether the new genera will be widely accepted remains in doubt. The distinction between stem and pseudobulb is very blurred in this genus. We have used the latter term only in those cases in which the stem is conspicuously swollen.

About 80 species are reputedly in general cultivation, together with numerous hybrids. Because of the large size of the genus, the methods of cultivation differ somewhat, depending on the origins of the species. The Indian and the Burmese species require heat and moisture when growing, exposure to air and light during autumn, and a cooler, drier, rest period during the winter. The species from Australia and New Guinea require similar treatment, but need more frequent watering and a slightly higher winter temperature. All the species benefit from appropriate shading, especially when the new growth appears. Epiphytic compost should be used, and the containers should be as small as possible. Many species do well on bark, slabs or tree fern. Propagation is mainly by division of old plants, by small pseudobulbs, and by plantlets produced on the stems.

Literature: Kränzlin, F., *Das Pflanzenreich* **45**: 25–313 (1910); Schelpe, E.A.C., Sections of cultivated Dendrobium *South African Orchid Journal* **16**: 36–59 (1985); Seidenfaden, G., Orchid genera in Thailand XII, Dendrobium Sw., *Opera Botanica* **83**: 1–295 (1985); Schelpe, S. & Stewart, J., *Dendrobiums: an introduction to the species in cultivation* (1990).

HABIT: sympodial epiphytes.
STEMS: thin or more frequently fleshy and pseudobulbous, bearing several leaves or leaf-scars.
LEAVES: variable, with or without a sheath enclosing the stem at the base, folded when young.
FLOWERS: 1–many in racemes borne on leafy or leafless stems.
SEPALS: upper free, laterals attached to the elongate column foot, forming a conspicuous, variously-shaped mentum which is sometimes spur-like.
PETALS: larger or smaller than the sepals.
LIP: entire or more frequently 3-lobed, attached to the apex and sometimes to the sides of the column foot, very variable in shape, usually with conspicuous longitudinal ridges.
COLUMN: short, though with an extended foot, frequently bearing 2 lateral horns.
POLLINIA: 4 in 2 pairs.

Key to species

1. Leaves without sheaths
1.A. Stems wiry, creeping or hanging, branched, each branch terminating in a single, thick, hard-pointed leaf **1. pugioniforme**
1.B. Lip hairy on the upper surface, often fringed *Go to Subkey* 1
1.C. Basal part of the lip erect, ridges conspicuous, close to or contacting the column *Go to Subkey* 2
1.D. Basal part of the lip diverging from the column, ridges usually shallow *Go to Subkey* 3
2. Stems and leaves strongly flattened, leaf-bases folded and overlapping **82. terminale**
3. Stems thickened at the base, the rest thin and normally stem-like **81. crumenatum**
4. Lip simple, unlobed, its margins attached to the column-foot for some distance *Go to Subkey* 4
5. Mentum double, consisting of 2 swellings at right angles to each other
5.A. Flowers 3–7 cm in diameter; petals broadly rounded,

occasionally with a small point; calluses on lip numerous and dense **64. bigibbum**

5.B. Flowers 6–10 cm in diameter; petals acuminate; calluses on lip small and obscure **65. phalaenopsis**

6. Leaf-sheaths hairy *Go to Subkey* 5

7. Flowers not resupinate; lip orange with purple or brown markings

7.A. Stems to 10 cm; lip lanceolate, acute, orange-red marked with purple lines and with ridges only near the base **22. seidenfadenii**

7.B. Stems to 25 cm; lip oblong-elliptic, brownish orange veined with brown and with 3 ridges extending over most of its length **23. unicum**

8. Lip hairy on the upper surface

8.A. Raceme with 5 or more flowers and an axis longer than the individual flower-stalks *Go to Subkey* 6

8.B. Lip fringed to the naked eye *Go to Subkey* 7

8.C. Stem with the internodes thickened upwards conspicuously, thus appearing beaded *Go to Subkey* 8

8.D. Flowers mostly yellow, without any mauve or purple except occasionally on the lip, usually without any white or pink

8.D.a. Racemes borne on leafy stems **30. luteolum**

8.D.b. Racemes borne on leafless stems **31. heterocarpum**

8.E. Sepals and petals either entirely white or white flushed pink or purple, without tips of contrasting colour *Go to Subkey* 9

8.F. Sepals and petals white with contrasting mauve, pink or purple tips *Go to Subkey* 10

9. Margin of lip fringed **63. delacourii**

10. Margin of lip not fringed *Go to Subkey* 11

Subkey 1. *Leaves without sheaths; lip hairy on the upper surface, often fringed.*

1. Stems creeping, bearing scattered pseudobulbs, each with 1 leaf **16. lindleyi**

2. Leaves and pseudobulbs white-hairy **17. senile**

3. Flowers entirely yellow

3.A. Pseudobulbs 4-angled; lip not fringed **19. densiflorum**

3.B. Pseudobulbs many-angled; lip fringed **18. chrysotoxum**

4. Pseudobulbs 4-angled; sepals and petals white flushed pink **21. farmeri**

5. Pseudobulbs not angled or with several slight ridges; sepals and petals white **20. thyrsiflorum**

Subkey 2. *Leaves without sheaths; lip not hairy on the upper surface, not fringed; basal part erect, with conspicuous ridges which are close to or contacting the column.*

1. Ovary and flower-stalk roughly hairy

1.A. Petals cream with purple spots and lines; lip with purple

lines on all lobes; leaves to 30 × 10 cm **7. macrophyllum**
1.B. Petals cream without spots or lines; lip with purple lines on the lateral lobes only; leaves 10–18 × 5–7 cm **8. forbesii**
2. Central lobe of lip obtuse and deeply notched; raceme hanging
2.A. Lip white flushed mauve, the lateral lobes also flushed with green **14. woodsii**
2.B. Lip entirely white **15. aberrans**
3. Lateral lobes of lip large, spreading to form a cup-like structure around the column; flowers 7–9 cm in diameter **9. spectabile**
4. Sepals unspotted
4.A. Flowers sweetly scented **13. engae**
4.B. Flowers not scented **11. rhodostictum**
5. Stems distinctly thickened upwards; central lobe of lip ovate **10. atroviolaceum**
6. Stem scarcely thickened upwards; central lobe of lip elongate-ovate **12. johnsoniae**

Subkey 3. Leaves without sheaths; lip not hairy on the upper surface, not fringed, basal part diverging from the column, ridges usually shallow.
1. Stems drooping, very slender and wiry in the lower half, then dilated and 4-angled above; sepals very acuminate **6. tetragonum**
2. Stem tapered evenly from base to apex; lateral lobes of lip acute **5. kingianum**
3. Central lobe of lip acute, upcurved; racemes 8–16 cm, with 4–20 flowers; leaves thin **4. falcorostrum**
4. Stems swollen at base, then tapering, then broadening again; central lobe of lip 3 times as broad as long **3. ruppianum**
5. Stems not swollen as under 4; central lobe of lip twice as broad as long **2. speciosum**

Subkey 4. Leaves with sheaths; lip simple, unlobed, its margins attached to the column-foot for some distance.
1. Plant crystalline-papillose on leaves, sheaths, ovaries and flower-stalks **58. cuthbertsonii**
2. Lip hollowed, its apical part fringed, upcurved
2.A. Warts absent from leaf-sheaths (rarely occasionally present at the very top): flowers scarlet or purplish **55. lawesii**
2.B. Stems erect; leaves 2.5–7 cm **56. subclausum**
2.C. Stems hanging; leaves to 2 cm **57. wentianum**
3. Leaves clustered at stem apex; ovary angled or winged; racemes with 2–4 flowers
3.A. Leaves linear; ovary with 5 obscure wings (3 close together along the top of the ovary)
3.A.a. Leaves terete, to 15 cm × 2.5 mm **61. hellwigianum**
3.A.b. Leaves flat, to 2.5 cm × 3–6 mm **62. violaceum**

3.B. Leaves oblong to oblong-lanceolate; ovary with 3 wings
3.B.a. Stems 1–25 cm x 3–15 mm; flowers purple, yellow or orange **59. vexillarius**
3.B.b. Stems usually smaller; flowers always bright orange **60. subacaule**
4. Sepals and petals white or rarely bluish; lip-apex pale purple; sepals 2–3 cm
4.A. Racemes with 15 or more flowers **51. amethystoglossum**
4.B. Racemes with up to 15 flowers **52. victoriae-reginae**
5. Flowers pink or reddish, rarely white; raceme conspicuously 1-sided **53. secundum**
6. Flowers orange; raceme not 1-sided **54. bullenianum**

Subkey 5. Leaves with sheaths; lip lobed, its margins free from the column-foot for most of its length; leaf-sheaths hairy.
1. Stems no more than 8 cm; mentum conical, blunt, not at all resembling a spur **80. bellatulum**
2. Leaves hairy; petals as broad as or a little broader than sepals
2.A. Lateral lobes of lip borne at its base; central lobe not fringed **78. draconis**
2.B. Lateral lobes of lip borne approximately halfway from base to apex; central lobe fringed **79. williamsonii**
3. Ovary 3-keeled; base of lip green or purple-streaked
3.A. Base of lip green **76. dearei**
3.B. Base of lip streaked with purple **77. sanderae**
4. Flower *c.* 10 cm in diameter; lip scarcely 3-lobed, notched **74. formosum**
5. Flower to 8 cm in diameter; lip distinctly 3-lobed, irregularly fringed at apex **75. infundibulum**

Subkey 6. Leaves with sheaths; lip 3-lobed, free from the column for most of its length; leaf-sheaths not hairy; lip hairy on the upper surface; raceme with 5 or more flowers and an axis longer than the individual flower-stalks.
1. Petals *c.* 3.5 cm broad, almost twice as broad as sepals; leaf-sheaths red-striped **41. pulchellum**
2. Lip with apical part and sides curving upwards, forming a pouch, margin entire **24. moschatum**
3. Stems hanging; flowering stem usually leafy; lip fringed to a depth of 4 mm **25. hookerianum**
4. Stem erect; flowering stem usually leafless; lip fringed to a depth of up to 2 mm **26. fimbriatum**

Subkey 7. Leaves with sheaths; lip 3-lobed, free from the column for most of its length, fringed; leaf-sheaths not hairy; lip hairy on the upper surface; raceme with fewer than 5 flowers and short axis.
1. Fringe of lip longer than the solid part; petals not broader than sepals **27. brymerianum**

2. Sepals and petals yellow
2.A. Lip densely hairy beneath **29. ochreatum**
2.B. Lip more or less hairless beneath **28. chrysanthum**
3. Sepals and petals 5–6 cm; lip mauve with 2 purple blotches, without any yellow **32. anosmum**
4. Stem 60–100 cm, hanging or arching; upper surface of lip minutely hairy **43. devonianum**
5. Stem to 15 cm, erect; upper surface of lip densely velvety **33. loddigesii**

Subkey 8. *Leaves with sheaths; lip 3-lobed, free from the column for most of its length; leaf-sheaths not hairy; lip hairy on the upper surface; raceme with fewer than 5 flowers and a short axis; stems with the internodes thickening conspicuously upwards, thus appearing beaded.*
1. Lip without any yellow; racemes usually borne on a leafy stem **34. moniliforme**
2. Leaves small, grass-like, 5–8 cm × 3–4 mm; centre of lip purple, outlined in pale yellow **44. falconeri**
3. Sepals up to 2.5 cm × 6–8 mm **35. findlayanum**
4. Flowers 5–8 cm in diameter **45. pendulum**
5. Flowers more than 8 cm in diameter **46. wardianum**

Subkey 9. *Leaves with sheaths; lip 3-lobed, free from the column for most of its length; leaf-sheaths not hairy; lip hairy on the upper surface; raceme with fewer than 5 flowers and a short axis; stem with unthickened internodes; sepals and petals entirely white or white flushed pink or purple, without tips of a contrasting colour.*
1. Flowers with the lip mostly yellow (sometimes very pale)
1.A. Petals broader than sepals; lip with a line of small, papilla-like hairs down the middle **39. aphyllum**
1.B. Sepals and petals equally broad; lip with most of the upper surface velvety **40. primulinum**
2. Flower 5 cm or more in diameter
2.A. Lip trumpet-shaped, its basal part more or less tubular, directed downwards, the apical part bell-shaped, directed forwards **36. lituiflorum**
2.B. Lip not trumpet-shaped, the whole directed obliquely forwards **42. nobile**
3. Free part of lip much smaller than mentum **50. aduncum**
4. Sepals white, lip with 2 red blotches **37. transparens**
5. Sepals purple, lip mauve with 2 deep purple blotches **38. parishii**

Subkey 10. *Leaves with sheaths; lip 3-lobed, free from the column for most of its length; leaf-sheaths not hairy; lip hairy on the upper surface; raceme with fewer than 5 flowers and a short axis; stem with unthickened internodes; sepals and petals white with mauve, pink or purple tips.*

1. Flowers more than 5 cm in diameter; lip without a yellow blotch **42. nobile**
2. Sepals and petals to 1.3 cm; lip not hairy beneath **47. crepidatum**
3. Lip densely velvety above and beneath; petals entire **48. gratiotissimum**
4. Lip densely velvety above, sparsely hairy beneath; petals finely toothed **49. amoenum**

Subkey 11. *Leaves with sheaths; lip 3-lobed, free from column for most of its length; leaf-sheaths not hairy; lip not hairy above.*

1. Petals one and a half to twice as long as the lateral sepals
1.A. Raceme with 5 or more flowers; petals conspicuously twisted
1.A.a. Flower spotted with dark red or brown; lip mauve at base **66. helix**
1.A.b. Flowers with dark red veins; lip white at base **67. tangerinum**
1.B. Central lobe of lip broader than long **68. bicaudatum**
1.C. Central lobe of lip longer than broad **69. stratiotes**
2. Sepals greenish white, petals red **70. taurinum**
3. Central lobe of lip conspicuously undulate-crisped, sepals and petals cream or yellowish (not dull yellow-brown or brown) **71. crispilinguum**
4. Sepals and petals very waved; racemes with many flowers **72. discolor**
5. Sepals and petals not waved; racemes with 4–12 flowers **73. mirbelianum**

GROUP A (Section *Rhizobium* Schlechter). Leaves without sheaths; stems hanging, wiry, branched, each branch terminating in a single, hard-pointed leaf.

1. D. **pugioniforme** Cunningham

ILLUSTRATIONS: Dockrill, Australian indigenous orchids, 369 (1969); Bechtel et al., Manual of cultivated orchid species, 201 (1981).

STEMS: thin, wiry, forming large hanging masses up to 2 m, each branch terminating in a single leaf.

LEAVES: 1–7 cm × 5–20 mm, flat, ovate-acuminate, very sharply pointed.

RACEME: with 1–3 flowers.

SEPALS & PETALS: 8–12 mm, petals slightly shorter, all spreading, pale green.

LIP: 3-lobed with oblong lateral lobes and a triangular, acute, deflexed central lobe which is notched or wavy on either side of the apex, very pale green with purple or red markings on the lateral lobes and the base of the central.

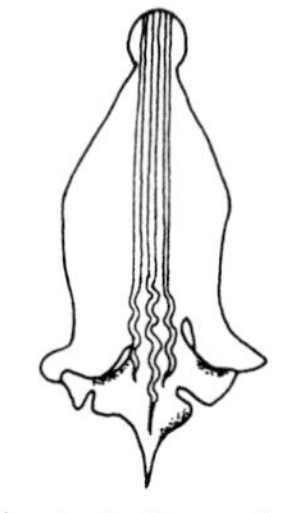

Dendrobium pugioniforme – lip.

DISTRIBUTION: Australia (New South Wales, Queensland).
HARDINESS: G2.

GROUP B (Subgenus *Dendrocoryne* Lindley: Section *Dendrocoryne* (Lindley) Schlechter). Leaves without sheaths; lip shorter than sepals and petals, its basal part diverging from the column, hairless on the upper surface.

2. **D. speciosum** J.E. Smith
ILLUSTRATIONS: Cogniaux & Goossens, Dictionnaire Iconographique des Orchidées, Dendrobium, t. 24 (1900): Dockrill, Australian indigenous orchids, 385 (1969); Bechtel et al., Manual of cultivated orchid species, 202 (1981); Schelpe & Stewart, Dendrobiums, 84 (1990).
STEMS: 8–100 cm.
LEAVES: 4–25 × 2–8 cm, usually ovate or oblong, thick and leathery.
RACEMES: 10–60 cm with many flowers.
FLOWERS: variable.
SEPALS AND PETALS: widely spreading or not; sepals 2–4 cm, petals slightly shorter and much narrower, all white, cream or yellow.
LIP: shorter than the petals, white, cream or yellow with red or purple spots, lobes obtuse, the central lobe *c.* twice as broad as long.
DISTRIBUTION: Australia (New South Wales, Queensland, Victoria).
HARDINESS: G1.
FLOWERING: spring.
A very variable species, divided by Dockrill into 5 varieties; it is uncertain which of these are in cultivation.

3. **D. ruppianum** Hawkes
SYNONYM: *D. speciosum* var. *fusiforme* Bailey.
ILLUSTRATION: Botanical Magazine n.s., 749 (1978).
Like *D. speciosum* except as below.
STEMS: swollen at the base and then tapering and then broadening.
LIP: central lobe 3 times broader than long.
DISTRIBUTION: Australia (Queensland), New Guinea.

4. **D. falcorostrum** Fitzgerald
ILLUSTRATIONS: Dockrill, Australian indigenous orchids, 407 & t. 11 (1969).
STEMS: 12–50 cm, erect, aggregated.
LEAVES: 2–4, 6–14 × 1.5–3 cm, thin, narrowly obovate or rarely narrowly ovate.

RACEMES: 8–16 cm with 4–20 flowers.

FLOWERS: 2.5–4 cm in diameter.

SEPALS AND PETALS: 3–4 cm, oblong, petals slightly shorter and narrower than sepals, all white.

LIP: shorter than the petals, white with yellow and purple markings, the central lobe acute, upcurved, the lateral lobes obtuse.

DISTRIBUTION: Australia (New South Wales, Queensland).

HARDINESS: G2.

FLOWERING: autumn.

5. D. kingianum Bidwill

ILLUSTRATIONS: Edwards's Botanical Register **31**: t. 61 (1845); Botanical Magazine, 4527 (1850); Dockrill, Australian indigenous orchids, 415 (1969); Bechtel et al., Manual of cultivated orchid species, 198 (1981); Schelpe & Stewart, Dendrobiums, 85 (1990).

STEM: 8–30 cm, broadest at the base, tapering upwards.

LEAVES: 3–10 × 1–2 cm, narrowly ovate to narrowly obovate.

RACEMES: with 2–10 flowers which are 1–2.5 cm in diameter.

SEPALS: 9–16 mm, triangular, the laterals erect and sickle-shaped.

PETALS: much narrower than sepals, all in various shades of pink or lilac, rarely white.

LIP: shorter than the petals to almost as long, pink with deep red stripes or blotches, the lateral lobes acute, the central transversely oblong, blunt or minutely apiculate.

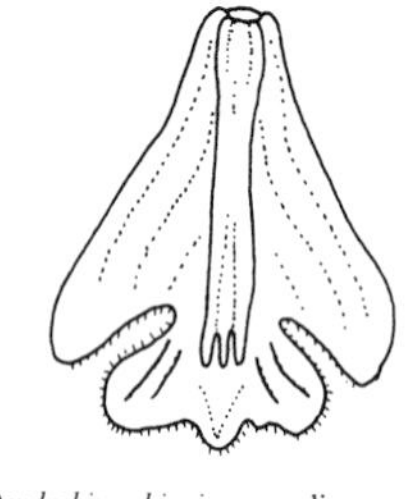

Dendrobium kingianum – lip.

DISTRIBUTION: Australia (New South Wales, Queensland).

HARDINESS: G1.

FLOWERING: spring.

6. D. tetragonum Cunningham

ILLUSTRATIONS: Botanical Magazine, 5956 (1872); Dockrill, Australian indigenous orchids, 421 (1969); Bechtel et al., Manual of cultivated orchid species, 202 (1981); Schelpe & Stewart, Dendrobiums, 86 (1990).

STEMS: hanging, aggregated, thin and wiry in the lower part, then dilated and 4-angled above.

LEAVES: 2–5 near the stem apex, 3–8 × 1.5–2.5 cm, ovate, acuminate.

RACEMES: short, with few flowers.

FLOWERS: 4–9 cm in diameter, very variable, very fragrant.

SEPALS: 2–5 cm, very narrowly lanceolate, very acuminate.

PETALS: erect, shorter and narrower than the sepals, all greenish yellow with irregular brown or purple blotches.

LIP: much shorter than the petals, 3-lobed, very variable in shape, yellow with dark red markings.

DISTRIBUTION: Australia (New South Wales, Queensland).

HARDINESS: G1.

FLOWERING: winter.

GROUP C (Section *Latouria* (Blume) Schlechter).

Leaves without sheaths, mostly borne at the stem apex; lip hairless, as long as the petals, the basal part erect, close to the column, basal callus conspicuous.

Literature: Cribb, P.I., Key to the species of Dendrobium section Latouria with setose ovaries, *Botanical Magazine* **183**(1): 28 (1980); Cribb, P.J., A preliminary key to Dendrobium section Latouria, *The Orchadian* **6**: 277–83 (1981); Cribb, P.J., A revision of Dendrobium section Latouria, *Kew Bulletin* **38**: 229–306 (1983).

7. D. macrophyllum Richard

ILLUSTRATIONS: Botanical Magazine, 5649 (1867); Illustration Horticole **35**: t. 57 (1888); American Orchid Society Bulletin **38**: 589 (1969); Schelpe & Stewart, Dendrobiums, 79 (1990).

STEMS: to 1 m × 2 cm, club-shaped.

LEAVES: 2–3, to 30 × 10 cm, elliptic to obovate.

RACEMES: to 30 cm with 8–10 (rarely more) flowers, erect.

SEPALS: *c.* 3–3.5 cm, triangular, pale yellowish green, crimson-spotted, bristly outside.

PETALS: shorter than the sepals, spathulate.

LIP: as long as petals, lateral lobes truncate, white, purple-veined inside, purple-spotted outside, the central lobe transversely elliptic, pale green with dark crimson spots outside; basal callus 1.5 cm, white, with 3 ridges.

DISTRIBUTION: Java to New Guinea & Polynesia.

HARDINESS: G2.

FLOWERING: summer–autumn.

8. D. forbesii Ridley

ILLUSTRATIONS: Botanical Magazine, 8141 (1907); Millar, Orchids of Papua New Guinea, 34 (1978); Kew Bulletin **38**: f. 1 & t. 11 (1983); Schelpe & Stewart, Dendrobiums, 82 (1990).

STEMS: 15–45 × 2.5 cm, club-shaped.

LEAVES: 2–3, 10–18 × 5–7 cm, elliptic.

RACEMES: to 17 cm, with 5 or more flowers, erect.

SEPALS: to 3.5 cm, triangular pale greenish yellow or cream, unspotted, bristly outside on the veins.

PETALS: white, longer than sepals.

LIP: pale greenish cream, unspotted, bases of the lateral lobes purple-veined, the central lobe transversely elliptic; callus obscurely 3-ridged.

DISTRIBUTION: New Guinea.

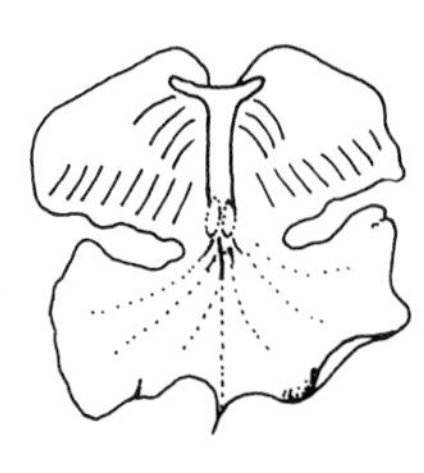

Dendrobium forbesii – lip.

HARDINESS: G2.
FLOWERING: summer–spring.

9. D. spectabile (Blume) Miquel
ILLUSTRATIONS: Botanical Magazine, 7747 (1900); Cogniaux & Goossens, Dictionnaire Iconographique des Orchidées, Dendrobium, t. 22 (1900); Orchid Review **74**: f. 59 (1966); Millar, Orchids of Papua New Guinea, (1978); Kew Bulletin **38**: t. 10 & 15 (1983).
STEMS: to 60 cm, club-shaped.
LEAVES: 2–5 near the apex of the stem, to 15 × 5 cm, oblong.
RACEME: to 30 cm, erect, with 5–25 slightly fragrant flowers.
SEPALS AND PETALS: *c.* 3.5 cm, narrowly triangular, pale yellow and marked with purple, margins wavy.
LIP: *c.* 5 cm, white, finely veined with purple, the lateral lobes large, forming a cup-shaped structure around the column, central lobe elongate, margins wavy.
DISTRIBUTION: New Guinea.
HARDINESS: G2.
FLOWERING: winter.

10. D. atroviolaceum Rolfe
ILLUSTRATIONS: Botanical Magazine, 7371 (1894); American Orchid Society Bulletin **38**: 587, 589 (1969); Millar, Orchids of Papua New Guinea, 40 (1978); Bechtel et al., Manual of cultivated orchid species, 193 (1981); Kew Bulletin **38**: f. 6, t. 10 & 15 (1983); Schelpe & Stewart, Dendrobiums, 78 (1990).
STEMS: 15–30 cm, thickened upwards.
LEAVES: 2–3, 7–15 × 2.5–5 cm, ovate-oblong.
RACEME: 10–20 cm, erect, bearing 3–8 fragrant flowers.
SEPALS & PETALS: *c.* 2.8 cm, sepals triangular, petals spathulate, slightly longer than sepals, all pale greenish white, purple-spotted outside.
LIP: pale green, the lateral lobes crimson-veined, margins violet, the central lobe ovate, purple-spotted outside.
DISTRIBUTION: New Guinea.
HARDINESS: G2.
FLOWERING: summer–spring.

11. D. rhodostictum Mueller & Kränzlin
ILLUSTRATIONS: Botanical Magazine, 7900 (1903); Kew Bulletin **38**: f. 8 & 10, t. 13 (1983); Kew Magazine **2**: pl. 41 (1985).

Like *D. atroviolaceum* except as below.
FLOWERS: unspotted.
LIP: with green veins and a few purple spots on the margins of the lateral lobes.

DISTRIBUTION: New Guinea.
FLOWERING: spring–summer.

12. D. johnsoniae Mueller
ILLUSTRATIONS: Journal of the Royal Horticultural Society **87**: f. 23 (1962); Botanical Magazine n.s., 560 (1969); Millar, Orchids of Papua New Guinea, 35 (1978); Kew Bulletin **38**: f. 7 & t. 12 (1983); Schelpe & Stewart, Dendrobiums, 82 (1990).

Also like *D. atroviolaceum* except as below.

STEMS: more slender.
LIP: with longer, ovate central lobe.
DISTRIBUTION: New Guinea. Solomon Islands.
FLOWERING: autumn–winter.

13. D. engae Reeve
ILLUSTRATIONS: The Orchadian **6**: 123–4 (1979); Kew Bulletin **38**: f.4 & t. 10 (1983); Die Orchidee **35**: centre-page pull-out (1984).

Like *D. atroviolaceum* except as below.

FLOWERS: sweetly scented.
SEPALS: without spots.
DISTRIBUTION: New Guinea.
FLOWERING: summer–spring.

14. D. woodsii Cribb
ILLUSTRATIONS: Botanical Magazine n.s., 727 (1977) — as D. fantasticum; The Orchadian **6**: 283 (1981); Kew Bulletin **38**: f. 4, t. 10 & 12 (1983).
STEMS: to 20 (rarely to 40) cm, slender at the base, thickened upwards.
LEAVES: 2 at the stem apex, *c.* 18 × 2–3.5 cm, elliptic to broadly lanceolate, gradually acuminate.
RACEMES: to 10 cm, hanging, with up to 10 fragrant flowers.
SEPALS AND PETALS: 1–1.2 cm, whitish, bluntly triangular, the petals truncate.
MENTUM: saccate.
LIP: white flushed with mauve, the lateral lobes flushed with green, the central lobe obtuse and deeply notched.
DISTRIBUTION: New Guinea.
HARDINESS: G2.
FLOWERING: autumn–winter.

15. D. aberrans Schlechter
ILLUSTRATION: Kew Bulletin **38**: f. 18 & t. 17 (1983).

Like *D. woodsii* except as below.

ALL PARTS: smaller.
FLOWERS: white without any other colour.

DISTRIBUTION: New Guinea.
FLOWERING: autumn–spring.

GROUP D (Section *Callista* (Loureiro) Schlechter).
Leaves without sheaths; lip 3-lobed, densely papillose-hairy on the upper surface.

Literature: Schelpe, E.A., Dendrobium section Callista, *Orchid Digest* **45**: 205–10 (1981).

16. D. lindleyi Steudel
SYNONYMS: *D. aggregatum* Roxburgh; *D. jenkinsii* misapplied.
ILLUSTRATIONS: Edwards's Botanical Register **20**: t. 1695 (1835); Botanical Magazine, 3643 (1838); Cogniaux & Goossens, Dictionnaire Iconographique des Orchidées, Dendrobium, t. 33 (1904); American Orchid Society Bulletin **55**: 381 (1986); Schelpe & Stewart, Dendrobiums, 49 (1990).
HABIT: rhizome creeping.
PSEUDOBULBS: aggregated, oblong-ovoid, angled, each bearing 1 leaf.
LEAVES: 5–8 × *c*. 2.5 cm, oblong.
RACEMES: from the upper axils of the pseudobulb, arching, with 5–15 flowers which are *c*. 3 cm in diameter.
SEPALS AND PETALS: *c*. 1.5 cm, petals broader than the sepals, at first pale, later deep yellow.
LIP: longer than the petals, broadly circular, concave, yellow which is deeper at the base, covered in short, white hairs.
DISTRIBUTION: Himalaya, Burma, SW China, Laos, Vietnam & Kampuchea.
HARDINESS: G2.
FLOWERING: spring.

17. D. senile Parish & Reichenbach
ILLUSTRATIONS: Botanical Magazine, 5520 (1865); Seidenfaden & Smitinand, Orchids of Thailand **2**(2): t. 7 (1960); Opera Botanica **83**: f. 20 & t. lvc (1985); Schelpe & Stewart, Dendrobiums, 78 (1990).
STEM: to 10 cm, club-shaped, bearing several leaves, covered, like the leaves, with long, white hairs.
LEAVES: 5–8 cm, lanceolate.
RACEMES: with 1–2 flowers on a short peduncle.
SEPALS AND PETALS: *c*. 2.5 cm, the petals broader than the sepals, all yellow.
LIP: ovate, as long as the petals, yellow, its surface densely downy.
DISTRIBUTION: Burma, Thailand & Laos.
HARDINESS: G2.
FLOWERING: spring–summer.

18. D. chrysotoxum Lindley

ILLUSTRATIONS: Botanical Magazine, 5053 (1858); Cogniaux & Goossens, Dictionnaire Iconographique des Orchidées, Dendrobium, t. 11, 11A (1898); Bechtel et al., Manual of cultivated orchid species, 195 (1981); Opera Botanica **83**: f. 6 & t. iia (1985); Schelpe & Stewart, Dendrobiums, 50 (1990).

PSEUDOBULBS: many-angled, bearing 2–5 leaves.

LEAVES: 3–10 cm, oblong or linear-oblong.

RACEMES: with 8–15 flowers, arching or hanging.

SEPALS AND PETALS: 2–2.2 cm, the petals a little broader than the sepals, all yellow.

LIP: as long as the petals, circular, the margin conspicuously fringed, orange-yellow, sometimes with a brown blotch at the base.

DISTRIBUTION: Himalaya, Burma, SW China (Yunnan), Laos & Thailand.

HARDINESS: G2.

FLOWERING: winter–spring.

Plants with the lip brown-blotched are sometimes called var. **suavissimum** (Reichenbach) Veitch (*D. suavissimum* Reichenbach).

19. D. densiflorum Lindley

ILLUSTRATIONS: Botanical Magazine, 3418 (1835); Cogniaux & Goossens, Dictionnaire Iconographique des Orchidées, Dendrobium, t. 14 (1898); Bechtel et al., Manual of cultivated orchid species, 195 (1981); Opera Botanica **83**: f. 8 (1985); Schelpe & Stewart, Dendrobiums, 52 (1990).

PSEUDOBULBS: 25–40 cm, 4-angled, bearing 3–5 leaves.

LEAVES: 7.5–15 × 2–2.5 cm, lanceolate or elliptic-lanceolate.

RACEMES: with many flowers, the scape arching over, the raceme completely pendent.

SEPALS AND PETALS: 2.5–2.8 cm, the petals somewhat broader than the sepals, all pale yellow.

LIP: as long as the petals, deep orange-yellow, margins somewhat finely toothed.

DISTRIBUTION: Himalaya, Burma, Vietnam.

HARDINESS: G2.

FLOWERING: spring.

Variants with white sepals and petals sometimes occur and have been cultivated; they are known as var. **schroederi** Anon. (lip orange-yellow) and var. **galliceanum** Linden (lip with white margin).

20. D. thyrsiflorum Reichenbach

ILLUSTRATIONS: The Garden **30**: 544 (1886); Cogniaux & Goossens, Dictionnaire Iconographique des Orchidées, Dendrobium, t. 18 (1899); Seidenfaden & Smitinand, Orchids of

Thailand 2 (2): t. 6 (1960); Orchid Review **92**: 242 (1984); Opera Botanica **83**: f. 9 (1985); Schelpe & Stewart, Dendrobiums, 54 (1990).

Like *D. densiflorum* except as below.

PSEUDOBULBS: not angled, or with many slight angles.

SEPALS AND PETALS: white.

LIP: deep yellow.

DISTRIBUTION: Burma, Thailand.

FLOWERING: spring–summer.

Often combined with *D. densiflorum*. The distinctions between the two are slight, and made more obscure by the existence of the varieties of *D. densiflorum* mentioned above. The precise distributions of both species are uncertain.

21. D. farmeri Paxton

ILLUSTRATIONS: Annals of the Royal Botanic Garden Calcutta 8: t. 80 (1898); Cogniaux & Goossens, Dictionnaire Iconographique des Orchidées, Dendrobium, t. 30 (1903); Bechtel et al., Manual of cultivated orchid species, 196 (1981); Opera Botanica **83**: f. 10 & t. iid (1985); Schelpe & Stewart, Dendrobiums, 53 (1990).

PSEUDOBULBS: 20–30 cm, 4-angled, bearing 3–5 leaves.

LEAVES: 8–18 × 3.5–5 cm, ovate-lanceolate.

RACEME: hanging, with arching scape, and with many flowers.

SEPALS AND PETALS: 2–2.5 cm, petals slightly broader than sepals, all white, often with pink tips.

LIP: as long as the petals, very densely papillose-hairy above, yellow with whitish margins and a pink apex.

DISTRIBUTION: Himalaya, Burma, Thailand, Vietnam & Malaysia.

HARDINESS: G2.

FLOWERING: spring.

Dendrobium farmeri – lip.

GROUP E (Section *Dendrobium*; Section *Eugenanthe* Schlechter). Leaves with sheaths; racemes usually from leafless stems; lip not or very obscurely 3-lobed, its sides free from the column foot, often curved upwards around the column, hairy (usually densely so) above.

22. D. seidenfadenii Senghas & Bockemühl

SYNONYM: *D. arachnites* Reichenbach not Thouars.

ILLUSTRATIONS: Gardeners' Chronicle **20**: 7 (1896); Botanisk Tidsskrift **65**: 335 (1970); Die Orchidee **29**: cii (1978); Opera Botanica **83**: f. 19 & t. ivb (1985).

STEM: to 10 cm, swollen; flowering stems leafless.

LEAVES: 4–6.5 cm, narrowly lanceolate.

RACEMES: with 2–3 flowers which are not resupinate.

FLOWERS: to 6 cm in diameter.

SEPALS AND PETALS: similar, 2–3 cm, orange-red.

LIP: as long as the petals, lanceolate, acute, orange marked with purple lines, and with ridges near the base only.

DISTRIBUTION: NE India, Burma, Thailand.

HARDINESS: G2.

FLOWERING: spring–summer.

23. D. unicum Seidenfaden

ILLUSTRATIONS: Botanical Magazine n,s., 616 (1972); Bechtel et al., Manual of cultivated orchid species, 193 (1981); Orchid Review **92**: 206 (1984); Opera Botanica **83**: f. 18 & t. iva (1985); Schelpe & Stewart, Dendrobiums, 44 (1990).

Very like *D. seidenfadenii* except as below.

STEMS: longer (to 25 cm).

LIP: oblong-elliptic, brownish orange, veined with brown, with 3 ridges extending for most of its length.

DISTRIBUTION: Laos, Thailand.

Much material in cultivation as *D. arachnites* could well be *D. unicum*.

24. D. moschatum Swartz

SYNONYMS: *D. calceolaria* Carey; *D. cupreum* Herbert.

ILLUSTRATIONS: Botanical Magazine, 3837 (1840); Annals of the Royal Botanic Garden Calcutta **8**: t. 84 (1898); Seidenfaden & Smitinand, Orchids of Thailand **2**(2): t. 8 (1960); Bechtel et al., Manual of cultivated orchid species, 200 (1981); Orchid Review **92**: 315 (1984); Opera Botanica **83**: f. 21 & t. ivd (1985); Schelpe & Stewart, Dendrobiums, 45 (1990).

STEMS: to 2 m, fleshy.

LEAVES: 10–15 cm, lanceolate, acuminate.

RACEMES: hanging, on leafless stems, with 10–15 flowers, axis elongate.

FLOWERS: 6–8 cm in diameter.

SEPALS AND PETALS: to 4 cm, pale orange, petals *c.* 2 cm broad, a little broader than the sepals.

LIP: shorter than the petals, concave, its margins erect, pale orange with 2 reddish brown blotches at the base.

DISTRIBUTION: Himalaya, Laos, Burma & Thailand.

HARDINESS: G2.

FLOWERING: spring–summer.

25. D. hookerianum Lindley

ILLUSTRATIONS: Botanical Magazine, 6013 (1873); Annals of the Royal Botanic Garden Calcutta **8**: t. 83 (1898).

STEMS: to 2.5 m, fleshy, pendent.

LEAVES: 5–15 × 2.5–3 cm, lanceolate or oblong-lanceolate, acute.

RACEMES: hanging, on leafy stems, with 5 or more flowers and an elongate axis.

FLOWERS: 7–10 cm in diameter.

SEPALS AND PETALS: to 4 cm, yellow, the petals finely toothed, broader than the sepals.

LIP: circular, yellow with 2 brown spots near the base, the margins fringed with the fringe *c.* 4 mm deep.

DISTRIBUTION: N India.

HARDINESS: G2.

FLOWERING: autumn.

26. D. fimbriatum Hooker

ILLUSTRATIONS: Cogniaux & Goossens, Dictionnaire Iconographique des Orchidées, Dendrobium, t. 9,9A (1898); Annals of the Royal Botanic Garden Calcutta **8**: t. 82 (1898); Seidenfaden & Smitinand, Orchids of Thailand **2**(2); t. 8 (1960); Bechtel et al., Manual of cultivated orchid species, 196, 197 (1981); American Orchid Society Bulletin **53**: 947, 948 (1984); Opera Botanica **83**: f. 16, 17 & t. iiid (1985); Schelpe & Stewart, Dendrobiums, 46 (1990).

STEMS: to 1.5 m, erect, fleshy.

LEAVES: 8–15 × 2–3 cm, oblong-lanceolate.

RACEMES: usually borne on leafless stems, with 5 or more flowers and an elongate axis.

FLOWERS: to 5.5 cm in diameter.

SEPALS AND PETALS: 2–3.5 cm, yellow-orange, the petals broader than the sepals.

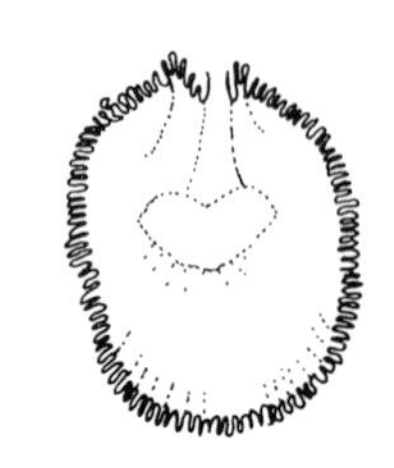

Dendrobium fimbriatum – lip.

LIP: orange, deeper orange or with 1 or 2 brown spots near the base, fringed, the fringe at most 2 mm deep.

DISTRIBUTION: Himalaya, Burma, Laos, Thailand, Vietnam & Malaysia.

HARDINESS: G2.

FLOWERING: spring.

Variable. Var. **oculatum** Hooker, with 1 or 2 brown spots on the lip, and var. **gibsoni** (Lindley) Gagnepain (*D. gibsoni* Lindley) with yellow, slightly smaller flowers, are both cultivated.

27. D. brymerianum Reichenbach

ILLUSTRATIONS: Botanical Magazine, 6383 (1878); Lindenia **4**: t. 183 (1888); Bechtel et al., Manual of cultivated orchid species, 195 (1981); Opera Botanica **83**: f. 15 & t. iiic (1985); Schelpe & Stewart, Dendrobiums, 51, 52 (1990).

STEMS: to 50 cm, fleshy.

LEAVES: 10–15 × 1.2–2 cm, lanceolate, acute.

RACEMES: with 1–2 flowers, axis very short, borne almost terminally on leafless stems.

FLOWERS: 5–6 cm in diameter.

SEPALS AND PETALS: similar in size, 2.5–3.5 cm, yellow.

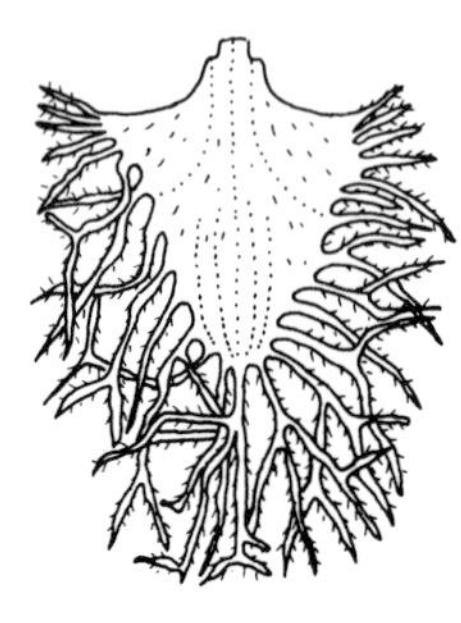

Dendrobium brymerianum – lip.

LIP: longer than the petals, mostly consisting of a fringe of branched threads, the central solid part small, all pale yellow.

DISTRIBUTION: Burma, Laos, Thailand.
HARDINESS: G2.
FLOWERING: winter–spring.

Var. **histrionicum** Reichenbach (*D. histrionicum* (Reichenbach) Schlechter) with shorter stems, smaller flowers and the lip fringe less conspicuous, is also perhaps in cultivation.

28. D. chrysanthum Lindley
ILLUSTRATIONS: Edwards's Botanical Register **15**: t. 1299 (1830); Cogniaux & Goossens, Dictionnaire Iconographique des Orchidées, Dendrobium, t. 2 (1897); Annals of the Royal Botanic Garden Calcutta **8**: t. 77 (1898); Bechtel et al., Manual of cultivated orchid species, 195 (1981); Opera Botanica **83**: f. 17, 27 & t. via (1985); Schelpe & Stewart, Dendrobiums, 40 (1990).
STEMS: to 1.5 m, fleshy.
LEAVES: 10–17 × 1–2.5 cm, lanceolate.
RACEMES: with 2–3 flowers on a very short axis, borne on leafless stems.
SEPALS AND PETALS: 2–3 × 1.6–2 cm, petals slightly broader than sepals, all yellow, the sepals with reddish, warty ridges on their backs.
LIP: circular, yellow with 2 reddish brown blotches, shorter than the petals, the margin fringed, more or less hairless and shiny beneath.
DISTRIBUTION: Himalaya, Burma, Thailand, Laos, Vietnam, China.
HARDINESS: G2.
FLOWERING: autumn.

29. D. ochreatum Lindley
SYNONYM: *D. cambridgeanum* Paxton.
ILLUSTRATIONS: Botanical Magazine, 4450 (1849); Cogniaux & Goossens, Dictionnaire Iconographique des Orchidées, Dendrobium, t. 16 (1899); Opera Botanica **83**: f. 17, 28 & t. vib (1985); Schelpe & Stewart, Dendrobiums, 41 (1990).

Like *D. chrysanthum* except as below.
STEMS: shorter.
PETALS: little broader than the sepals.
LIP: densely hairy beneath.
DISTRIBUTION: E Himalaya, Burma, Thailand, Laos.
HARDINESS: G2.
FLOWERING: spring.

30. D. luteolum Bateman
ILLUSTRATIONS: Botanical Magazine, 5441 (1864); Bateman, Second century of orchidaceous plants, t. 185 (1867).
STEMS: to 40 cm, fleshy.

LEAVES: 6–10 × *c.* 3 cm, oblong-ovate or oblong-lanceolate, obtuse.

RACEMES: with 2–3 flowers on a very short axis, borne on the leafy stems.

FLOWERS: 5–6 cm in diameter.

SEPALS AND PETALS: 2.5–2.8 cm, the petals a little broader than the sepals, all pale yellow.

LIP: oblong-ovate, almost as long as the petals, yellow with fine red lines towards the base; mentum rather conspicuous.

DISTRIBUTION: Burma.

HARDINESS: G2.

FLOWERING: winter–spring.

31. D. heterocarpum Lindley

SYNONYM: *D. aureum* Lindley.

ILLUSTRATIONS: Botanical Magazine, 4708 (1853); Annals of the Royal Botanic Garden Calcutta 8: t. 74 (1898); Cogniaux & Goossens, Dictionnaire Iconographique des Orchidées, Dendrobium, t. 10 (1898); Bechtel et al., Manual of cultivated orchid species, 198 (1981); Opera Botanica 83: f. 26 & t. vd (1985); Schelpe & Stewart, Dendrobiums, 40 (1990).

STEM: to 40 cm, erect, fleshy.

LEAVES: 10–20 × *c.* 2.5 cm, lanceolate or oblong-lanceolate, acute.

RACEMES: with 2–3 flowers on a very short axis, borne on the leafless stems.

FLOWERS: 5–7 cm in diameter.

SEPALS AND PETALS: 2.5–3.5 cm, petals broader than sepals, all pale yellow.

LIP: oblong-ovate, yellow with red-orange lines and veins.

DISTRIBUTION: India, Sri Lanka, Burma, Thailand, Malaysia, Indonesia & Philippines.

HARDINESS: G2.

FLOWERING: spring–summer.

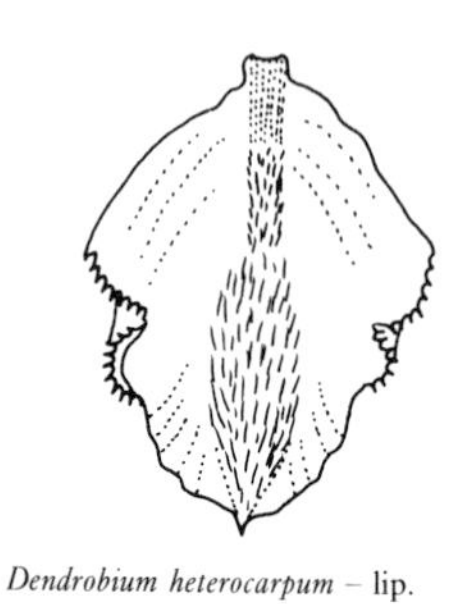

Dendrobium heterocarpum – lip.

32. D. anosmum Lindley

SYNONYMS: *D. macrophyllum* Lindley not Richard; *D. superbum* Reichenbach.

ILLUSTRATIONS: Cogniaux & Goossens, Dictionnaire Iconographique des Orchidées, Dendrobium, t. 20 (1899); Bechtel et al., Manual of cultivated orchid species, 193 (1981); Opera Botanica 83: f. 41, t. ixc (1985); Schelpe & Stewart, Dendrobiums, 32 (1990).

STEMS: to 1.2 m, fleshy, hanging.

LEAVES: 12–18 × *c.* 3 cm, oblong-lanceolate or ovate-oblong.

RACEMES: with usually 2 flowers, axis short or absent, borne on leafless stems.

FLOWERS: to 10 cm in diameter.

SEPALS AND PETALS: 5–6 cm, petals much broader than sepals, all pinkish to purple.
LIP: broadly ovate, mauve with 2 deep purple blotches, fringed.
DISTRIBUTION: Indo-China, Philippines, Borneo, Moluccas, New Guinea.
HARDINESS: G2.
FLOWERING: winter–spring.

33. D. **loddigesii** Rolfe
SYNONYM: *D. pulchellum* Loddiges not Lindley.
ILLUSTRATIONS: Botanical Magazine, 5037 (1858); Bechtel et al., Manual of cultivated orchid species, 199 (1981); Schelpe & Stewart, Dendrobiums, 37 (1990).
STEMS: 10–15 cm, thick.
LEAVES: 4–7 × 1.25–2 cm, oblong-lanceolate.
RACEMES: with 1–2 flowers, stalkless on leafless stems.
FLOWERS: *c.* 4 cm in diameter.
SEPALS AND PETALS: 2–2.5 cm, petals broader than sepals, all rose red.
LIP: as long as petals, circular, yellow with a purplish, conspicuously fringed margin.
DISTRIBUTION: SW China.
HARDINESS: G2.
FLOWERING: winter–spring.

34. D. **moniliforme** (Linnaeus) Swartz
SYNONYM: *D. monile* (Thunberg) Kränzlin.
ILLUSTRATIONS: Edwards's Botanical Register **16**: t. 1314 (1830); Botanical Magazine, 4153 (1845); Bechtel et al., Manual of cultivated orchid species, 200 (1981); Cribb & Bailes, Hardy orchids, f. 29A & t. 3 (1989); Schelpe & Stewart, Dendrobiums, 23 (1990).
STEM: to 30 cm, with the internodes thickened upwards, somewhat beaded.
LEAVES: to 7 cm, lanceolate, acute.
RACEMES: with 2–3 flowers on a short axis, borne on leafy stems.
FLOWERS: *c.* 4 cm in diameter.
SEPALS AND PETALS: 2–2.5 cm, the petals somewhat broader than the sepals, all white or white flushed with pink.
LIP: shorter than the petals, elliptic, acute, white with red spots at the base and a red apex.
DISTRIBUTION: Japan, Korea; ?China.
HARDINESS: G1.
FLOWERING: spring.

35. D. **findlayanum** Parish & Reichenbach
ILLUSTRATIONS: Botanical Magazine, 6438 (1879); The

Garden **49**: 496 (1896); Bechtel et al., Manual of cultivated orchid species, 197 (1981); Opera Botanica **83**: f. 33 & t. viic (1985); Schelpe & Stewart, Dendrobiums, 29 (1990).

STEM: to 30 cm, internodes thickened upwards, the stem appearing beaded.

LEAVES: 7–8 × 1.3–3 cm, lanceolate.

RACEMES: with 2 flowers on a short axis, usually borne on the leafless stems.

FLOWERS: 5–8 cm in diameter.

SEPALS AND PETALS: 2–2.5 cm, pale rose pink or mauve, rarely white suffused with these tones, the petals broader than the sepals.

LIP: ovate to almost circular, the centre deep yellow, the margin pale yellow or white, sometimes suffused reddish.

DISTRIBUTION: Burma, Thailand, Laos.

HARDINESS: G2.

FLOWERING: winter–spring.

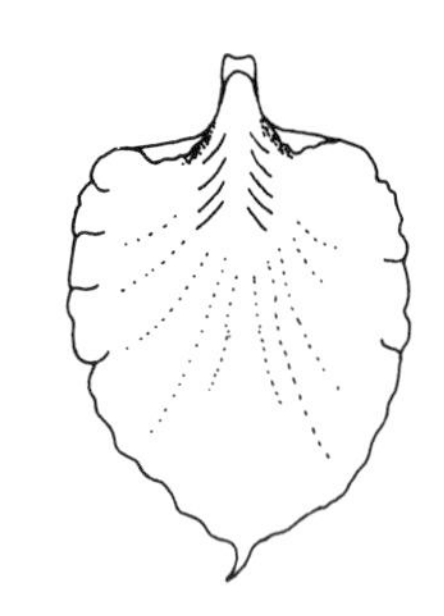

Dendrobium findlayanum – lip.

36. D. lituiflorum Lindley

ILLUSTRATIONS: Botanical Magazine, 6050 (1873); Bechtel et al., Manual of cultivated orchid species, 199 (1981); Opera Botanica **83**: f. 47 & t. xia (1985); Schelpe & Stewart, Dendrobiums, 36 (1990).

STEM: to 60 cm, fleshy.

LEAVES: 7.5–10 × 1.5–2 cm, lanceolate to linear-lanceolate, acute.

RACEMES: with 1–3 flowers on a short axis, borne on leafless stems.

FLOWERS: 5–7 cm in diameter.

SEPALS AND PETALS: 2.5–3.8 cm, petals broader than sepals, all uniformly purple.

LIP: as long as petals, trumpet-shaped, its basal part more or less tubular, directed downwards, the apical part bell-shaped, directed forwards, purple margined with white, the inner part of the white margin sometimes yellowish.

DISTRIBUTION: NE India (Manipur), Burma, SW China, Laos.

HARDINESS: G2.

FLOWERING: spring.

37. D. transparens Lindley

ILLUSTRATIONS: Botanical Magazine, 4663 (1852); Cogniaux & Goossens, Dictionnaire Iconographique des Orchidées, Dendrobium, t. 27 (1901); Bechtel et al., Manual of cultivated orchid species, 202 (1981); Schelpe & Stewart, Dendrobiums, 37 (1990).

STEMS: to 45 cm, rather thin but fleshy.

LEAVES: 8–10 × *c.* 1.2 cm, lanceolate or linear-lanceolate.

RACEMES: with 2–3 flowers and a very short axis, borne on leafless stems.

SEPALS AND PETALS: 1.8–2.5 cm, petals a little broader than the sepals, white or reddish.

LIP: ovate, shorter than petals, white with 2 red spots towards the base.

DISTRIBUTION: Himalaya.

HARDINESS: G2.

FLOWERING: spring–summer.

38. **D. parishii** Reichenbach

ILLUSTRATIONS: Botanical Magazine, 5488 (1865); Bechtel et al., Manual of cultivated orchid species, 200 (1981); Opera Botanica 83: f. 34 & t. viid (1985); Schelpe & Stewart, Dendrobiums, 29 (1990).

STEMS: to 30 cm, hanging or prostrate, fleshy.

LEAVES: to 10 cm, lanceolate to oblong-lanceolate.

RACEMES: usually with 2 flowers on a short axis, borne on leafless stems.

FLOWERS: *c.* 5 cm in diameter.

SEPALS AND PETALS: 2.2–3.5 cm, the petals broader than the sepals, all uniformly lilac-purple.

LIP: shorter than petals, obovate, mauve-purple marked with 2 large dark red-purple spots towards the base.

DISTRIBUTION: Burma, SW China, Thailand, Laos & Kampuchea.

HARDINESS: G2.

FLOWERING: spring–summer.

39. **D. aphyllum** (Roxburgh) Fischer

SYNONYMS: *Limodorum aphyllum* Roxburgh; *D. pierardii* Roxburgh.

ILLUSTRATIONS: Botanical Magazine, 2584 (1825); Annals of the Royal Botanic Garden Calcutta 8: t. 72 (1898); Cogniaux & Goossens, Dictionnaire Iconographique des Orchidées, Dendrobium, t. 26 (1901); Bechtel et al., Manual of cultivated orchid species, 193 (1981); American Orchid Society Bulletin 55: 380 (1986); Schelpe & Stewart, Dendrobiums, 31 (1990).

STEM: to 1 m, slender.

LEAVES: 6–12 × 2–3 cm, lanceolate to linear-lanceolate, acute.

RACEMES: with 2–3 flowers on a very short axis, borne on leafless stems.

FLOWERS: 4.5–5.5 cm in diameter.

SEPALS AND PETALS: 2.5–3 cm, petals almost 2 times broader than the sepals, all pale pinkish lilac.

LIP: broadly ovate, as long as the petals, pale yellow, with a line of hairs down the middle.

DISTRIBUTION: Himalaya, SW China, Burma, Thailand, Vietnam, Laos, Kampuchea & Malaysia.

HARDINESS: G2.

FLOWERING: spring.

40. D. primulinum Lindley

ILLUSTRATIONS: Lindenia **15**: t. 686 (1900); Annals of the Royal Botanic Garden Calcutta **9**: t. 98 (1906); Bechtel et al., Manual of cultivated orchid species, 201 (1981); Die Orchidee **32**: Orchideenbewertung, 9 (1982); Schelpe & Stewart, Dendrobiums, 30 (1990).

Like *D. aphyllum* except as below.

LEAVES: unequally 2-lobed at apex.

PETALS: little broader than the sepals.

LIP: upper surface velvety all over.

DISTRIBUTION: Himalaya, SW China, Burma, Thailand, Vietnam, Laos & Kampuchea.

HARDINESS: G2.

FLOWERING: spring.

41. D. pulchellum Lindley

SYNONYM: *D. dalhousieanum* Wallich.

ILLUSTRATIONS: Paxton's Magazine of Botany **11**: t. 145 (1844); Edwards's Botanical Register **32**: t. 10 (1846); Cogniaux & Goossens, Dictionnaire Iconographique des Orchidées, Dendrobium, t. 7 (1897); Orchid Review **92**: 242 (1984); Opera Botanica **83**: f. 37 & t. viiic (1985); Schelpe & Stewart, Dendrobiums, 45 (1990).

STEMS: to 1.2 m, thickened.

LEAF-SHEATHS: red-striped.

LEAVES: 10–15 × *c.* 3 cm, linear-oblong, 2-lobed at apex.

RACEMES: with 5–12 flowers on an elongate axis, borne on leafless stems.

FLOWERS: *c.* 7–8 cm in diameter.

SEPALS AND PETALS: petals *c.* 3.5 cm broad, almost twice as broad as the sepals, white.

LIP: shorter than the petals, slightly concave, white to pale yellow at the base, with 2 red-purple blotches.

DISTRIBUTION: Himalaya, Burma, Vietnam.

HARDINESS: G2.

FLOWERING: spring.

42. D. nobile Lindley

ILLUSTRATIONS: Cogniaux & Goossens, Dictionnaire Iconographique des Orchidées, Dendrobium, t. 1, A, B, C (1896–97); Annals of the Royal Botanic Garden Calcutta **8**; t. 71 (1898); Bechtel et al., Manual of cultivated orchid species, 200 (1981); Opera Botanica **38**: f. 48 & t. xib (1985); Schelpe & Stewart, Dendrobiums, 24, 25 (1990).

STEMS: to 1 m, fleshy.

LEAVES: 7–10 × 1–1.5 cm, lanceolate to ovate-lanceolate.

RACEMES: with 1–2 or rarely 3 flowers on a short axis, borne on leafless stems.

FLOWERS: 7–10 cm in diameter.

SEPALS AND PETALS: 3–5 cm, the petals broader than the sepals, all white, white tipped with purple or entirely mauve.

LIP: almost circular, white or yellowish white, variously marked with pink or mauve.

DISTRIBUTION: Himalaya, China, Taiwan.

HARDINESS: G2.

FLOWERING: spring–summer.

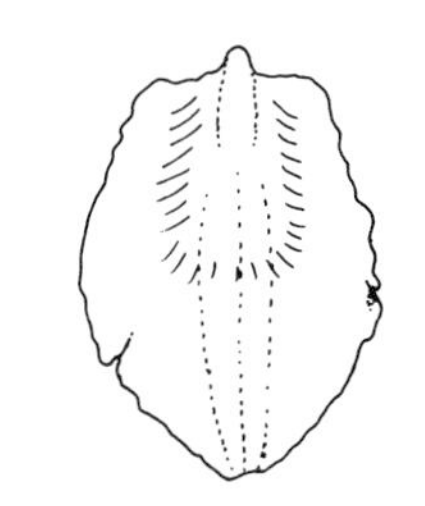

Dendrobium nobile – lip.

The precise coloration of the flowers is very variable, and several named varieties, based on this feature, have been in cultivation.

43. D. devonianum Paxton

ILLUSTRATIONS: Botanical Magazine, 4429 (1849); Cogniaux & Goossens, Dictionnaire Iconographique des Orchidées, Dendrobium, t. 23 (1900); Seidenfaden & Smitinand, Orchids of Thailand 2(2): t. 9 (1960); Opera Botanica **83**: f. 35 & t. viiia (1985); Schelpe & Stewart, Dendrobiums, 30 (1990).

STEMS: to 1 m, thin but fleshy, hanging.

LEAVES: 7–10 × *c.* 1.5 cm, lanceolate.

RACEMES: usually with 2 flowers on a short axis, borne on leafless stems.

FLOWERS: *c.* 5 cm in diameter.

SEPALS AND PETALS: 2–2.3 cm, petals broader than the sepals, white tipped with pinkish purple.

LIP: circular, as long as the petals, white with 2 large yellow blotches towards the base, the apex pinkish purple, margin conspicuously fringed.

DISTRIBUTION: Himalaya, SW China, Burma, Thailand & Vietnam.

HARDINESS: G2.

FLOWERING: summer.

44. D. falconeri Hooker

ILLUSTRATIONS: Botanical Magazine, 4944 (1856); Bateman, Second century of orchidaceous plants, t. 137 (1867); Opera Botanica **83**: f. 29 & t. vic (1985); Schelpe & Stewart, Dendrobiums, 26 (1990).

STEMS: to 1 m, thin, hanging, the internodes thickened upwards, appearing beaded.

LEAVES: grass-like, 5–8 cm × 3–4 mm.

FLOWERS: usually borne singly on the leafless stems, occasionally 2–3 together on a short axis, 5.5–7 cm in diameter.
SEPALS AND PETALS: 3–4.5 cm, the petals broader than the sepals, white tipped with purple.
LIP: ovate, almost as long as the petals, white, the apex purple and the base with a large purple blotch surrounded by an orange-yellow zone.
DISTRIBUTION: NE India, Burma, SW China, Taiwan.
HARDINESS: G2.
FLOWERING: spring–summer.

45. **D. pendulum** Roxburgh
SYNONYM: *D. crassinode* Benson & Reichenbach.
ILLUSTRATIONS: Cogniaux & Goossens, Dictionnaire Iconographique des Orchidées, Dendrobium, t. 19, 34 (1899); Bechtel et al., Manual of cultivated orchid species, 200 (1981); Opera Botanica **83**: f. 32 & t. viib (1985); Schelpe & Stewart, Dendrobiums, 28 (1990).
STEMS: to 30 cm, internodes thickened upwards, appearing beaded.
LEAVES: 10–12 × 1.5–2 cm, lanceolate.
RACEMES: with 2–3 flowers on a short axis, borne on the leafless stems.
FLOWERS: 5–8 cm in diameter.
SEPALS AND PETALS: 3–4 cm × 9–12 mm, the petals broader than the sepals, all white with red tips.
LIP: ovate to almost circular, the centre deep yellow, the margin white with a reddish apex.
DISTRIBUTION: Burma.
HARDINESS: G2.

46. **D. wardianum** Warner
ILLUSTRATIONS: Cogniaux & Goossens, Dictionnaire Iconographique des Orchidées, Dendrobium, t. 5, 5A (1897); Opera Botanica **83**: f. 30 & t. vid (1985); Schelpe & Stewart, Dendrobiums, 27 (1990).
Like *D. pendulum* except as below.
STEMS: with less swollen internodes.
FLOWERS: larger.
DISTRIBUTION: E Himalaya, Burma, SW China, Thailand.

47. **D. crepidatum** Lindley
ILLUSTRATIONS: Botanical Magazine, 4993, 5011 (1857); Cogniaux & Goossens, Dictionnaire Iconographique des Orchidées, Dendrobium, t. 40 (1906); Bechtel et al., Manual of cultivated orchid species, 195 (1981); Orchid Review **92**: 205 (1984); Opera Botanica **83**: f. 46 & t. xd (1985); Schelpe & Stewart, Dendrobiums, 35 (1990).

STEMS: to 45 cm, fleshy.

LEAVES: 5–10 × 1–1.25 cm, linear-lanceolate, acute.

RACEMES: with 1–3 flowers on a very short axis, borne on leafless stems.

FLOWERS: 2.5–4.5 cm in diameter.

SEPALS AND PETALS: 1–2 cm, the petals a little broader than the sepals, white with pink tips.

LIP: about as long as the petals, ovate, yellow at the base, then white, the margin pink, sparsely hairy above, hairless beneath.

DISTRIBUTION: Himalaya, Burma, Thailand & Laos.

HARDINESS: G2.

FLOWERING: spring.

48. D. **gratiotissimum** Reichenbach

ILLUSTRATIONS: Bechtel et al., Manual of cultivated orchid species, 197 (1981); Opera Botanica 83: f. 31 & t. viia (1985); Schelpe & Stewart, Dendrobiums, 28 (1990).

STEMS: to 1 m, fleshy, hanging.

LEAVES: 7–12 × 1–1.5 cm, lanceolate, sheaths reddish.

RACEMES: with 1–3 flowers on a very short axis, borne on leafless stems.

FLOWERS: 4–5 cm in diameter.

SEPALS AND PETALS: 1.8–3.5 cm, petals broader than sepals, all white tipped with pink.

LIP: broadly ovate, white, the base orange or yellow, the apex pink, densely velvety above and beneath.

DISTRIBUTION: NE India, Burma, SW China, Thailand, Laos.

HARDINESS: G2.

FLOWERING: spring.

49. D. **amoenum** Lindley

ILLUSTRATIONS: Botanical Magazine, 6199 (1875); Annals of the Royal Botanic Garden Calcutta 8: t. 69 (1898); Schelpe & Stewart, Dendrobiums, 36 (1990).

STEM: to 60 cm, slender.

LEAVES: to 10 cm, lanceolate or linear-lanceolate, acute.

RACEMES: with 2–3 flowers on a very short axis, borne on leafless stems.

FLOWERS: 3–4 cm in diameter.

SEPALS AND PETALS: 1.7–2.5 cm, the petals finely toothed, sometimes irregularly so, broader than the sepals, all white tipped with purple.

LIP: as long as the petals, white with a yellow base and purple apex, densely velvety above, sparsely hairy beneath.

DISTRIBUTION: Himalaya.

HARDINESS: G2.

FLOWERING: spring.

Sometimes considered to be the same as *D. aphyllum* (p. 256).

50. D. aduncum Wallich

ILLUSTRATIONS: Botanical Magazine, 6784 (1884); Annals of the Royal Botanic Garden Calcutta **8**: t. 67 (1898); Opera Botanica **83**: f. 51 & t. xid (1985); Schelpe & Stewart, Dendrobiums, 107 (1990).

STEMS: hanging or ascending, red-spotted.

LEAVES: to 8 cm, lanceolate or linear-lanceolate, somewhat red-lined beneath.

RACEMES: with 1–2 flowers on a short axis, borne on leafless stems.

FLOWERS: 2.5–3 cm in diameter.

SEPALS AND PETALS: 1.6–1.8 cm, petals scarcely broader than sepals, all pale pinkish mauve.

LIP: with the free part very small, mostly pouch-like with an oblong projection at the base, whitish, much shorter than the mentum.

DISTRIBUTION: India (Sikkim), Bhutan, Burma, W China, Thailand, Vietnam.

HARDINESS: G2.

FLOWERING: summer–autumn.

GROUP F (Section *Pedilonum* (Blume) Lindley).

Leaves with sheaths, evenly spread along the stems; racemes with many flowers; lip mostly simple, its margins attached to the column foot for some distance, flat, blunt; ovary not angled or winged.

51. D. amethystoglossum Reichenbach

ILLUSTRATIONS: Botanical Magazine, 5968 (1872); Davis & Steiner, Philippine orchids, 118 (1952); Schelpe & Stewart, Dendrobiums, 87 (1990).

STEMS: 60–90 cm.

LEAVES: 5–10 cm.

RACEMES: to 10 cm, drooping, borne on leafless stems and bearing 15–many flowers.

SEPALS AND PETALS: upper sepal *c.* 2 cm × 8 mm, lateral sepals to 3 cm, petals *c.* 2 cm, all white.

LIP: linear-spathulate, the basal part white with a backwardly pointing, fleshy, triangular ridge, the apical part amethyst-purple.

DISTRIBUTION: Philippines.

HARDINESS: G2.

FLOWERING: winter.

52. D. victoriae-reginae Loher

ILLUSTRATIONS: Cogniaux & Goossens, Dictionnaire Iconographique des Orchidées, Dendrobium, t. 21 (1899); Orchid Review **20**: f. 6 (1912); Botanical Magazine, 9071 (1925); Schelpe & Stewart, Dendrobiums, 88 (1990).

Like *D. amethystoglossum* except as below.

RACEME: with fewer flowers.
SEPALS AND PETALS: white in the lower part, violet-blue in the upper part.
LIP: orange-yellow at the base, violet-blue above.
DISTRIBUTION: Philippines.
FLOWERING: spring–summer.

53. D. **secundum** Lindley

ILLUSTRATIONS: Edwards's Botanical Register 15: t. 1291 (1829); Botanical Magazine, 4352 (1848); Cogniaux & Goossens, Dictionnaire Iconographique des Orchidées, Dendrobium, t. 35 (1905); Bechtel et al., Manual of cultivated orchid species, 201 (1981); Opera Botanica 83: f. 104 & t. viiia (1985); American Orchid Society Bulletin 55: 788 (1986): Schelpe & Stewart, Dendrobiums, 89 (1990).
STEMS: to 1 m.
LEAVES: *c.* 10 × 3 cm, broadly oblong, mostly in the upper two-thirds of the stem.
RACEMES: *c.* 12–15 cm, from the uppermost nodes, 1-sided.
FLOWERS: 1.5–1.8 cm in length.
SEPALS AND PETALS: upper sepal *c.* 7 × 4 mm, petals narrow, all pink or reddish or rarely white.
LIP: simple, pink or reddish.
DISTRIBUTION: Burma, Laos, Vietnam, Kampuchea, Malaysia, Indonesia & Philippines.
HARDINESS: G2.
FLOWERING: winter.

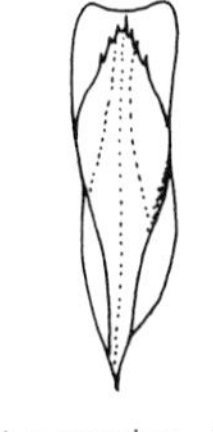
Dendrobium secundum – lip.

54. D. **bullenianum** Reichenbach

SYNONYM: *D. topaziacum* Ames.
ILLUSTRATION: Bechtel et al., Manual of cultivated orchid species, 194 (1981); Schelpe & Stewart, Dendrobiums, 91 (1990).
STEMS: to 60 cm, longitudinally grooved.
LEAVES: 7–14 × 1.5–2.8 cm. oblong.
RACEMES: short, borne on leafless stems, not 1-sided.
FLOWERS: *c.* 1.7 cm in length.
SEPALS AND PETALS: upper sepal lanceolate, *c.* 6 × 3 mm, lateral sepals obliquely triangular, *c.* 1.5 cm, petals *c.* 6 × 2 mm, oblong-spathulate, bluntly pointed, all orange.
LIP: *c.* 1.5 cm, narrowly elliptic or spathulate.
DISTRIBUTION: Philippines.
HARDINESS: G2.

GROUP G (Section *Calyptrochilus* Schlechter; Section *Glomerata* Kränzlin). Leaves with sheaths, uniformly spread along the fleshy stems; racemes short with few flowers; lip simple, its margins attached to the column foot for some distance, its apical part upcurved and fringed.

C. *Thunia bensoniae*

D. *Pleione forrestii*

E. *Masdevallia rosea*

F. *Dendrobium aphyllum*

G. *Calypso bulbosa*

H. *Dendrobium cuthbertsonii*

I. *Xylobium variegatum*

J. *Bifrenaria harrisoniae*

K. *Anguloa cliftonii*

L. *Zygopetalum crinitum*

M. *Maxillaria grandiflora*

N. *Cymbidium longifolium*

O. *Stanhopea grandiflora*

P. *Oncidium cheirophorum*

DISTRIBUTION: New Guinea.
HARDINESS: G1.

GROUP H (Section *Cuthbertsonia* Schlechter; Section *Leiotheca* Kränzlin in part). Small tufted plants with crystalline-papillose leaves, sheaths and flower-stalks; racemes each with 1 non-resupinate flower; lip simple, obtuse, boat-shaped, its margins attached to the column foot for some distance. Now generally included in Section *Oxyglossum* (see below).

58. D. cuthbertsonii Mueller

SYNONYM: *D. sophronites* Schlechter.

ILLUSTRATIONS: Journal of the Royal Horticultural Society **88**: f. 46 (1963); Orchid Review **76**: 248 (1968); The Orchadian **6**: 37,38 (1978); Millar, Orchids of Papua New Guinea, 14, 15 (1978); Kew Magazine **2**: 37 (1985); Notes from the Royal Botanic Garden Edinburgh **46**: f. 37 & t. 18 (1989); Schelpe & Stewart, Dendrobiums, 96 (1990).

STEMS: *c.* 1–2 cm (rarely to 5) × 4–7 mm, clustered, club-shaped or cylindric and narrowing abruptly upwards.

LEAVES: almost apical, 2–5, linear-lanceolate to almost ovate, dark green and finely crystalline-papillose above, often purplish beneath.

FLOWERS: 2.5–4 cm long, 1.3–3.5 cm in diameter, terminal, borne singly on leafless stems, the colour extremely variable, commonly scarlet but frequently purplish, mauve, orange or white, sometimes bicoloured.

LIP: boat-shaped and strongly concave, with darker marginal markings.

DISTRIBUTION: New Guinea.

HARDINESS: G1.

FLOWERING: summer–spring.

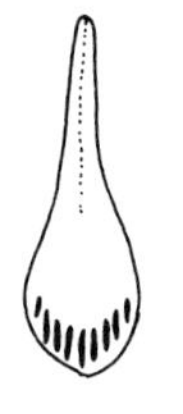

Dendrobium cuthbertsonii – lip.

GROUP I (Section *Oxyglossum* Schlechter). Small tufted plants; leaves few, borne towards the upper parts of the stems, with sheaths; racemes terminal, with 2–4 flowers; lip tightly attached to the column and its foot, simple, sharply pointed; ovary winged or angled.

Literature: Reeve, T.M. & Woods, P.J.B., A preliminary key to the species of Dendrobium section Oxyglossum, *The Orchadian* **6**: 195–208 (1980); Reeve, T.M. & Woods, P.J.B., A revision of Dendrobium Sect. Oxyglossum, *Notes from the Royal Botanic Garden Edinburgh* **46**: 161–305 (1989).

59. D. vexillarius J.J. Smith

ILLUSTRATIONS: Millar, Orchids of Papua New Guinea, 35,37 (1978), under various specific names; Van Royen, Orchids of the high mountains of New Guinea, 376–93

(1980); Australian Orchid Review **46**: 106 (1981); Kew Magazine **2**: 39 (1985); Notes from the Royal Botanic Garden Edinburgh **46**: f. 28–30 & t. 15, 16 (1989); Schelpe & Stewart, Dendrobiums, 98 (1990).

STEMS: 1–25 cm × 3–15 mm, clustered, cylindric, narrowing upwards.

LEAVES: 2–4, 2–10 cm (rarely to 16) × 3–18 mm, oblong to oblong-lanceolate, blunt or shortly pointed, green above, often purplish beneath.

RACEMES: stalkless from tips of leafy or leafless branches.

SEPALS AND PETALS: 2.5–5 × *c.* 3 cm, spreading, flat, variable in colour, purple, orange or yellow.

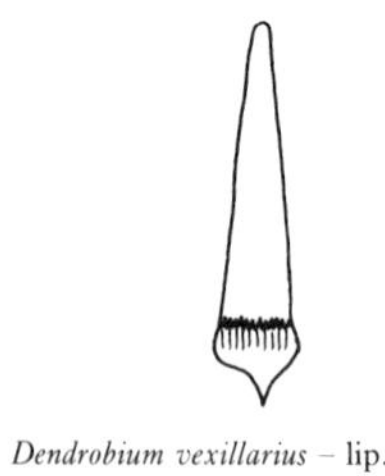

Dendrobium vexillarius – lip.

LIP: purplish, blackish or green, the sharp tip bright orange, reflexed.

OVARY: triangular in section.

DISTRIBUTION: New Guinea.

HARDINESS: G1.

FLOWERING: summer–spring.

A very variable plant both in habit and flower colour.

60. D. subacaule Lindley

SYNONYMS: *D. oreocharis* Schlechter; *D. tricostatum* Schlechter.

ILLUSTRATIONS: Orchid Digest **41**: 13 (1977); Northen, Miniature orchids, C-26 (1980); Notes from the Royal Botanic Garden Edinburgh **46**: f. 14 & t. 10 (1989).

Like *D. vexillarius* except as below.

HABIT: much smaller, stems and leaves to 1.5 cm.

FLOWERS: bright orange, 9–18 mm.

DISTRIBUTION: New Guinea, Moluccas, Solomon Islands.

61. D. hellwigianum Kränzlin

SYNONYMS: *D. cyananthum* Williams; *D. raphiotes* Schlechter.

ILLUSTRATIONS: Kew Magazine **2**: 38 (1985); Notes from the Royal Botanic Garden Edinburgh **46**: f. 25 & t. 14 (1989).

STEMS: 1–8 cm × 2–6 mm, clustered, cylindric.

LEAVES: 3–4, to 15 cm × 2.5 mm, round in section.

RACEMES: with 2–4 flowers, stalkless on leafless stems.

FLOWERS: pinkish to purple or bluish, lip-tip acute, bright orange, not reflexed.

OVARY: 5-winged (the upper 3 wings so close as to give a triangular appearance).

DISTRIBUTION: New Guinea.

HARDINESS: G1.

FLOWERING: summer–spring.

62. D. violaceum Kränzlin

ILLUSTRATIONS: Northen, Miniature orchids, C-21 (1980); Kew Magazine **2**: 40 (1985); Notes from the Royal Botanic

Garden Edinburgh **46**: f. 22, 23 & t. 12 (1989); Schelpe & Stewart, Dendrobiums, 97 (1990).

Like *D. hellwigianum* except as below.

LEAVES: not round in section, linear, to 25 cm × 3–6 mm.

FLOWERS: 3–4.5 cm, purplish, the lip-apex orange.

DISTRIBUTION: New Guinea.

GROUP J (Section *Stachyobium* Schlechter) Leaves with sheaths; racemes more or less terminal on the leafy stems; lip 3-lobed, fringed, its margin not attached to the column foot.

63. **D. delacourii** Guillaumin

SYNONYM: *D. ciliatum* Hooker not Persoon.

ILLUSTRATIONS: Botanical Magazine, 5430 (1864); Seidenfaden & Smitinand, Orchids of Thailand 2(2): 231 (1960); Bechtel et al., Manual of cultivated orchid species, 195 (1981); Opera Botanica **83**: f. 80 & t. xvic (1985); Schelpe & Stewart, Dendrobiums, 105 (1990).

STEMS: to 45 cm, though often shorter, erect.

LEAVES: 10–13 × 2–3 cm, oblong-elliptic, acute.

RACEMES: erect, with several flowers.

SEPALS AND PETALS: 1–1.2 cm, the petals somewhat narrower than the sepals, greenish yellow or yellow.

LIP: obovate, 3-lobed, as long as the petals, the central lobe fringed with spathulate divisions, all greenish yellow with red veins.

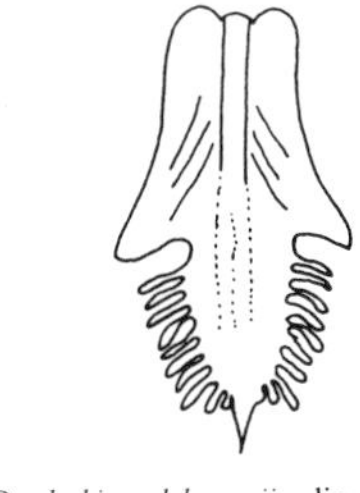

Dendrobium delacourii – lip.

DISTRIBUTION: Burma, Thailand, Laos, Vietnam Kampuchea.

HARDINESS: G2.

FLOWERING: winter.

GROUP K (Section *Phalaenanthe* Schlechter). Leaves with sheaths; racemes borne on leafless stems; petals much broader than sepals; mentum double, consisting of 2 swellings or humps at right angles to each other.

64. **D. bigibbum** Lindley

ILLUSTRATIONS: Botanical Magazine, 4898 (1856); Bateman, Second century of orchidaceous plants, t. 169 (1867); Lindenia **7**: t. 317 (1891); Bechtel et al., Manual of cultivated orchid species, 194 (1981); Schelpe & Stewart, Dendrobiums, 75, 76 (1990).

STEM: to 50 cm, fleshy, spindle-shaped or cylindric.

LEAVES: 5–12 × 2–3 cm, oblong-lanceolate or lanceolate.

RACEME: erect with many flowers which are 3–7 cm in diameter.

SEPALS AND PETALS: 2–3.5 cm, petals broader than sepals, broadly rounded, occasionally with a small point, all usually reddish purple, rarely white.

LIP: shorter than the petals, 3-lobed, with a blunt or notched central lobe which bears many small calluses in the middle.
MENTUM: bluntly conical.
DISTRIBUTION: Australia (N Queensland).
HARDINESS: G2.
Variable in flower colour and size. *D. superbiens* Reichenbach is the name applied to plants which are almost certainly natural hybrids between *D. bigibbum* and another Australian species, *D. discolor* (p. 269). These plants have been in cultivation (see Lindenia **7**: t. 294, 1891; Cogniaux & Goossens, Dictionnaire Iconographique des Orchidées, Dendrobium, t. 15, 1898) and differ from *D. bigibbum* in having the second hump of the mentum small and the petals relatively narrower (see Blake, S.T., Proceedings of the Royal Society of Queensland **74**: 29–44 1964).

65. D. phalaenopsis Fitzgerald
SYNONYM: *D. schroederanum* Gentil.
ILLUSTRATIONS: Cogniaux & Goossens, Dictionnaire Iconographique des Orchidées, Dendrobium, t. 4 (1897); Botanical Magazine, 6817 (1885); Lindenia **6**: t. 280 (1890); Bechtel et al., Manual of cultivated orchid species, 200 (1981); American Orchid Society Bulletin **55**: 1111 (1986); Schelpe & Stewart, Dendrobiums, 76 (1990).
Like *D. bigibbum* except as below.
HABIT: more robust, with stems to 60 cm.
LEAVES: to 15 cm.
FLOWERS: 6–10 cm in diameter.
SEPALS AND PETALS: 4–4.5 cm, petals much broader than sepals, acuminate, usually all red.
LIP: with a long, acute, central lobe which bears a few, small calluses, usually a deeper red than the petals and sepals.
DISTRIBUTION: Indonesia (Timor Laut).
HARDINESS: G2.
FLOWERING: winter–spring.

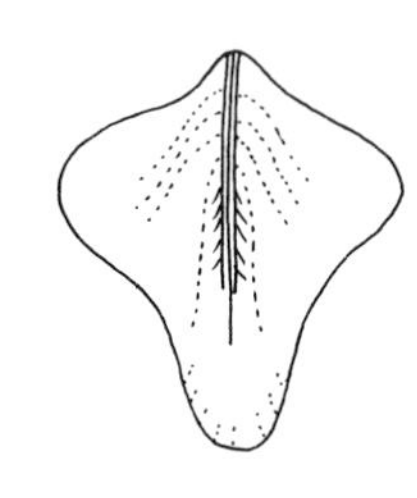
Dendrobium phalaenopsis – lip.

Like *D. bigibbum*, this species is very variable in flower colour (shades of red, purple and white), and numerous varieties (cultivars) have been described on the basis of this variation. The origin of the species has long been obscure, but S.T. Blake (reference under *D. bigibbum*) has shown conclusively that it is endemic to Timor Laut and does not occur as a native in Australia or New Guinea.

GROUP L (Section *Ceratobium* Schlechter). Leaves with sheaths; racemes from leafy or leafless stems; sepals and particularly petals often twisted, the petals usually very erect; lip with an entire central lobe.

Literature: Ossian, C.R., A review of the Antelope Dendrobiums (section Ceratobium), *American Orchid Society*

Bulletin **50**: 1213...1455 (1981); **51**: 23...362 (1982); Cribb, P.J., The Antelope Dendrobiums, *Kew Bulletin* **41**: 615–692 (1986).

66. D. helix Cribb

ILLUSTRATIONS: Orchid Review **88**: 144 (1980); The Orchadian **6**: 175 (1980); American Orchid Society Bulletin **51**: 255, 258 (1981); Orchid Review **92**: 289 (1984); Kew Bulletin **41**: f. 11 & t. 12 (1986); Schelpe & Stewart, Dendrobiums, 71 (1990).

STEMS: to 1 m, cane-like.

LEAVES: to 16 × 6.5 cm, elliptic or ovate-elliptic, obtuse and minutely 2-lobed at the apex.

RACEMES: erect with 5 or more flowers.

SEPALS: 2.6–3 cm × 8–12 mm, the upper spirally recurved.

PETALS: 3–4.5 cm, erect, spirally twisted 3 or 4 times; all yellow with red-brown blotches.

LIP: 3-lobed, shorter than the sepals, the central lobe recurved, base mauve, the rest yellow with red-brown spots.

DISTRIBUTION: New Guinea (New Britain).

HARDINESS: G2.

67. D. tangerinum Cribb

ILLUSTRATIONS: Orchid Review **88**: 145 (1980); The Orchadian **6**: 176 (1980); American Orchid Society Bulletin **50**: 1340, 1343 (1981); Orchid Review **92**: 242 (1984); Kew Bulletin **41**: f. 13 & t. 12 (1986); Schelpe & Stewart, Dendrobiums, 72 (1990).

STEMS: to 50 cm, erect, cane-like, clustered.

LEAVES: to 9 × 2.5 cm, oblong-elliptic.

RACEMES: spreading to erect, with *c.* 15 flowers.

UPPER SEPAL: 2.3–2.5 cm × 6–7 mm, recurved, margins wavy.

LATERAL SEPALS: 1.6–2.3 × 1–1.2 cm.

PETALS: 3–3.4 cm × 4–5 mm, erect, spirally twisted 2 or 3 times; all orange-yellow, with dark red venation.

LIP: 3-lobed, shorter than the sepals, orange-yellow with 3 lilac ridges and a white base.

DISTRIBUTION: New Guinea.

HARDINESS: G2.

68. D. bicaudatum Lindley

SYNONYM: *D. minax* Reichenbach.

ILLUSTRATIONS: Reichenbach, Xenia Orchidacearum, t. 145 (1868); Das Pflanzenreich **45**: 143 (1910); Kew Bulletin **41**: f. 2 & t. 6 (1986).

STEMS: to 30 cm.

LEAVES: to 10 × 2 cm, oblong, unequally 2-lobed at the apex.

RACEMES: with 2–3 flowers.

SEPALS: 2.5–2.8 cm.

PETALS: to 5 cm, not twisted or with 1 slight twist, all greenish red, intensely red-streaked.

LIP: 3-lobed, the lateral lobes broadly triangular, the central transversely oblong, broader than long, with a small point, greenish red with purple lines.

DISTRIBUTION: Indonesia (Amboina, Moluccas, Sulawesi).

HARDINESS: G2.

FLOWERING: summer.

69. D. stratiotes Reichenbach

ILLUSTRATIONS: Illustration Horticole **34**: t. 602 (1886); Bechtel et al., Manual of cultivated orchid species, 202 (1981); American Orchid Society Bulletin **50**: 1337, 1343 (1981), **55**: 1110 (1986); Kew Bulletin **41**: f. 1 & t. 6 (1986).

STEMS: to 60 cm or more.

LEAVES: 8–12 × *c.* 2 cm, oblong, obtuse.

RACEMES: with up to 5 flowers.

SEPALS AND PETALS: upper sepal, *c.* 3.5 cm, laterals *c.* 4.5 cm, petals to 6 cm, all white suffused with green.

LIP: with diamond-shaped lateral lobes and a central lobe which is broadly ovate or elliptic, longer than broad, acute, white with a greenish yellow base and red veins and spots.

DISTRIBUTION: Indonesia (Sunda Islands).

HARDINESS: G2.

FLOWERING: summer.

70. D. taurinum Lindley

ILLUSTRATIONS: Edwards's Botanical Register **29**: t. 28 (1843); Lindenia **13**: t. 621 (1897); American Orchid Society Bulletin **51**: 253, 258 (1982); Kew Bulletin **41**: f. 5 & t. 9 (1986); Schelpe & Stewart, Dendrobiums, 69 (1990).

STEMS: to 1.5 m, erect, spindle-shaped.

LEAVES: 10–15 × *c.* 6 cm, oblong or elliptic, rounded or slightly 2-lobed at the apex.

RACEMES: erect with many flowers.

SEPALS AND PETALS: upper sepal *c.* 3 cm, laterals *c.* 4 cm, reflexed; petals 3–4 cm, reflexed; sepals white suffused with green, petals brownish red to pink, slightly twisted.

LIP: oblong-elliptic, pale brownish red to pink, about as long as the petals, with large lateral lobes and a small, somewhat wavy central lobe.

DISTRIBUTION: Philippines.

HARDINESS: G2.

FLOWERING: autumn–winter.

71. D. crispilinguum Cribb

ILLUSTRATIONS: Orchid Review **88**: 146, 147 (1980); The

Orchadian **6**: 177 (1980); American Orchid Society Bulletin **51**: 32 (1982); Kew Bulletin **41**: f. 15 & t. 12 (1986).

STEMS: to 1 m or more, cane-like.

LEAVES: to 12 × 1.5–3.5 cm, lanceolate or ovate-lanceolate, acute.

RACEMES: with up to 20 flowers.

SEPALS AND PETALS: upper sepal recurved and somewhat twisted, 2.6–3 cm; lateral sepals reflexed, *c.* 2.6 cm; petals 2.5–3.3 cm, somewhat twisted; all cream or yellowish.

LIP: shorter than the petals with oblong lateral lobes and an ovate, very crisped central lobe, white suffused and veined red-purple.

DISTRIBUTION: New Guinea.

HARDINESS: G2.

72. D. discolor Lindley

SYNONYM: *D. undulatum* R. Brown not Persoon.

ILLUSTRATIONS: Edwards's Botanical Register **27**: t. 52 (1841); Dockrill, Australian indigenous orchids, 469 (1969); Bechtel et al., Manual of cultivated orchid species, 196 (1981); American Orchid Society Bulletin **51**: 256–9 (1982); Kew Bulletin **41**: t. 11 (1986); Schelpe & Stewart, Dendrobiums, 70 (1990).

STEMS: to 5 m, cane-like.

LEAVES: 5–15 × 2–5 cm, ovate or elliptic, obtuse or slightly 2-lobed.

RACEMES: erect with numerous flowers.

SEPALS AND PETALS: 2–5 cm, very wavy, somewhat twisted and reflexed, yellowish brown.

LIP: shorter than the petals, the central lobe ovate, bent downwards, not wavy, yellowish with 3 white or violet ridges.

DISTRIBUTION: Australia (Queensland), New Guinea.

HARDINESS: G1.

FLOWERING: winter–spring.

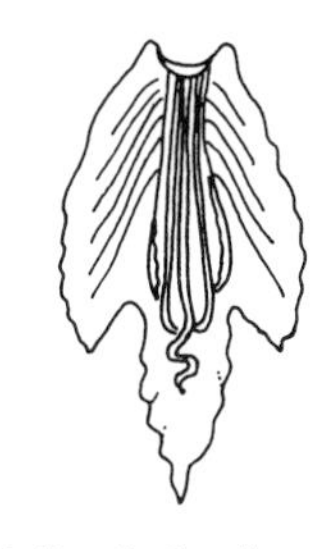

Dendrobium discolor – lip.

Variable in flower colour in the wild. See note under *D. bigibbum* (p. 266).

73. D. mirbelianum Gaudich

SYNONYM: *D. wilkianum* Rupp.

ILLUSTRATIONS: Dockrill, Australian indigenous orchids, 475 (1969); American Orchid Society Bulletin **51**: 32, 147, 152 (1982); Kew Bulletin **41**: f. 20 & t. 13 (1986); Schelpe & Stewart, Dendrobiums, 72 (1990).

STEMS: to 1 m, erect.

LEAVES: 8–15 × 1.5–4 cm, ovate, usually suffused with red or purple.

RACEMES: with 4–12 flowers.

SEPALS AND PETALS: 1.8–3 cm, petals slightly longer, not or little twisted, all pale to dark brown.

LIP: shorter than the petals, 3-lobed, the central lobe oblong-ovate, acute, yellowish green with deep red veins and 4 ridges.

DISTRIBUTION: Australia (Queensland); ?New Guinea.

HARDINESS: G2.

GROUP M (Section *Formosae*; Section *Oxygenianthe* Schlechter; Section *Nigro-hirsutae* Kränzlin). Leaves with sheaths which are conspicuously brown- or black-hairy; racemes borne on leafy stems; lip usually clearly 3-lobed, the lateral lobes usually arching upwards to the column, usually not hairy.

74. D. formosum Lindley

SYNONYM: *D. infundibulum* Reichenbach not Lindley.

ILLUSTRATIONS: Cogniaux & Goossens, Dictionnaire Iconographique des Orchidées Dendrobium, t. 8 (1897); Seidenfaden & Smitinand, Orchids of Thailand 2(2): t. 11 (1960); Orchid Digest **41**: 5 (1977); Bechtel et al., Manual of cultivated orchid species, 197 (1981); Opera Botanica **83**: f. 67 & t. xva (1985); Schelpe & Stewart, Dendrobiums, 55 (1990).

STEMS: to 45 cm, thick.

LEAVES: 10–12 × *c*. 3.5 cm, ovate-oblong, unequally 2-lobed at the apex.

RACEMES: with 2 or more flowers which are 9–11 cm in diameter.

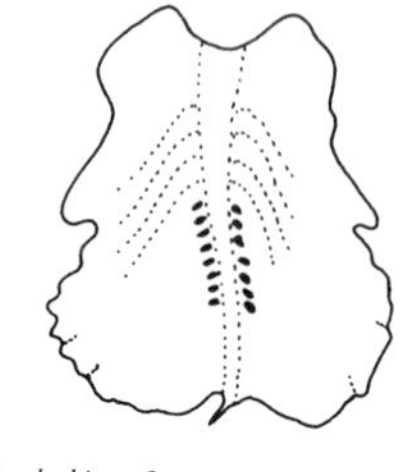

Dendrobium formosum – lip.

SEPALS AND PETALS: 4–5 cm, petals much broader than the sepals, all white.

LIP: longer than petals, scarcely 3-lobed, obovate, wavy, notched, white with yellow markings.

MENTUM: conical, acute.

DISTRIBUTION: Himalaya, Burma, Thailand.

HARDINESS: G2.

FLOWERING: winter–spring.

Variable in the size of the flowers and the precise coloration of the lip.

75. D. infundibulum Lindley

ILLUSTRATIONS: Botanical Magazine, 5446 (1864); Cogniaux & Goossens, Dictionnaire Iconographique des Orchidées, Dendrobium, t. 6 (1897); Orchid Digest **41**: 5 (1977); Bechtel et al., Manual of cultivated orchid species, 198 (1981); Orchid Review **92**: 28 (1984); Opera Botanica **83**: f. 72 & t. xva (1985); Schelpe & Stewart, Dendrobiums, 56 (1990).

Like *D. formosum* except as below.

FLOWERS: to 8 cm in diameter.

SEPALS AND PETALS: at most 4 cm.

LIP: distinctly 3-lobed, irregularly fringed.

DISTRIBUTION: Burma, Thailand.

FLOWERING: spring–summer.

Variable; var. **jamesianum** (Reichenbach) Veitch (*D. jamesianum* Reichenbach), which has the lip papillose and marked with orange-yellow, is often grown.

76. D. **dearei** Reichenbach

ILLUSTRATIONS: Cogniaux & Goossens, Dictionnaire Iconographique des Orchidées, Dendrobium, t. 36 (1905); Bechtel et al., Manual of cultivated orchid species, 195 (1981); Schelpe & Stewart, Dendrobiums, 62 (1990).

STEMS: to 1 m.

LEAVES: to 6 cm, ovate-oblong, unequally 2-lobed at the apex, hairless.

RACEME: with 3–4 flowers.

SEPALS AND PETALS: 2.5–3.5 cm, petals broader than sepals, all white.

LIP: 3-lobed, about as long as petals, the central lobe somewhat deflexed and wavy, white, green at the base.

MENTUM: conical, acute, *c.* 1.5 cm.

OVARY: 3-keeled.

DISTRIBUTION: Philippines.

HARDINESS: G2.

FLOWERING: spring–summer.

77. D. **sanderae** Rolfe

ILLUSTRATION: Botanical Magazine, 8351 (1910); Schelpe & Stewart, Dendrobiums, 63 (1990).

Like *D. dearei* except as below.

SEPALS AND PETALS: to 4.5 cm.

LIP: streaked with purple at the base.

DISTRIBUTION: Philippines.

FLOWERING: autumn–winter.

Apparently sometimes confused with *D. sanderianum* Rolfe, a totally different plant not often seen in cultivation, and with the artificial hybrid *D.* × *ainsworthii* Moore (*D. heterocarpum* × *nobile*), which is sometimes found in older horticultural literature as *D.* × *sanderae* (see Hunt & Summerhayes, Kew Bulletin **20**: 55, 1966).

78. D. **draconis** Reichenbach

SYNONYM: *D. eburneum* Reichenbach.

ILLUSTRATIONS: Bateman, Second century of orchidaceous plants, t. 166 (1867): Botanical Magazine, 5459 (1864); Orchid Digest **41**: 5 (1977); Opera Botanica 83: f. 66 & t. xivd (1985); Schelpe & Stewart, Dendrobiums, 57, 58 (1990).

STEMS: to 45 cm, fleshy.

LEAVES: 6–8 × 1–1.5 cm, lanceolate, apex unequally 2-lobed, with brown hairs on both surfaces.

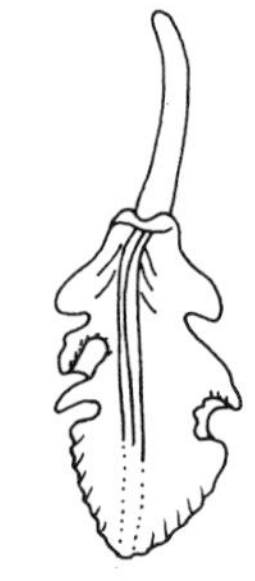

Dendrobium draconis – lip.

SEPALS AND PETALS: 3–4 cm, the petals broader than the sepals, all creamy white.

LIP: longer than the petals, with 2 triangular lateral lobes borne at the base, the central lobe oblong-obovate, acute, somewhat wavy, entire, white with red and yellow lines towards the base.

MENTUM: spur-like, 2–2.5 cm, narrow.

DISTRIBUTION: Burma, Thailand, Laos, Kampuchea & Vietnam.

HARDINESS: G2.

FLOWERING: spring–summer.

Cleistogamous variants of this species are known (see Pradhan, Orchid Digest **41**: 6, 1977).

79. **D. williamsonii** Day & Reichenbach

ILLUSTRATIONS: Annals of the Royal Botanic Garden Calcutta **5**: t. 9 (1895); Botanical Magazine, 7974 (1904); Bechtel et al., Manual of cultivated orchid species, 203 (1981); Opera Botanica **83**: f. 75, 77 & t.xvib (1985); Schelpe & Stewart, Dendrobiums, 59 (1990).

STEMS: to 25 cm.

LEAVES: 5–10 × 1.5–2.5 cm, linear-oblong, with brown hairs on both surfaces.

RACEMES: with 2 flowers.

SEPALS AND PETALS: 3–3.5 cm, equally broad, white inside, pale yellow outside.

LIP: longer than the petals with 2 rounded lateral lobes borne about half way along its length, the central lobe conspicuously fringed, creamy white to yellow with a large red-brown spot at the base.

MENTUM: spur-like, *c.* 2.5 cm.

DISTRIBUTION: NE India, Burma, W China, Vietnam.

HARDINESS: G1.

FLOWERING: spring.

80. **D. bellatulum** Rolfe

ILLUSTRATIONS: Botanical Magazine, 7985 (1904); Orchid Digest **41**: 5 (1977); Opera Botanica **83**: f. 60 & t xiiib (1985); Schelpe & Stewart, Dendrobiums, 61 (1990).

STEMS: short, usually less than 5 cm, clustered.

LEAVES: lanceolate or narrowly ovate, apex unequally 2-lobed, with dark brown hairs on both surfaces.

RACEMES: with 1–3 flowers.

SEPALS AND PETALS: 2–2.5 cm, equally broad, all white.

LIP: ovate, about as long as petals, with a notched, broadly oblong, deflexed central lobe and a thick, papillose ridge down most of its length, red at the base, the rest yellow.

MENTUM: broadly conical, blunt.

DISTRIBUTION: N India, SW China, Vietnam, Laos & Thailand.
HARDINESS: G1.
FLOWERING: spring.

GROUP N (Section *Rhopalanthe* Schlechter). Stems thickened only at the extreme base, tapering abruptly above this.

81. D. crumenatum Swartz

ILLUSTRATIONS: Botanical Magazine, 4013 (1843); Holttum, Orchids of Malaya, f. 3 (1953); Opera Botanica **83**: f. 138 (1985); American Orchid Society Bulletin **54**: 980 (1986); Schelpe & Stewart, Dendrobiums, 101 (1990).
STEMS: to 1 m, the lower 4–5 internodes swollen, the rest thin.
LEAVES: fleshy, obtuse or obscurely 2-lobed.
RACEME: leafless, with many fragrant flowers.
SEPALS AND PETALS: 2.5–3.5 cm, white or white suffused with pink.
LIP: as long as the petals, 3-lobed, the central lobe wavy and with a prominent crested callus, white with a yellow blotch.
DISTRIBUTION: Himalaya to Indonesia.
HARDINESS: G2.
FLOWERING: at any time.
Flowering in this sparingly cultivated species follows a sudden, though often short, drop in temperature. The flowers last for only 1 day.

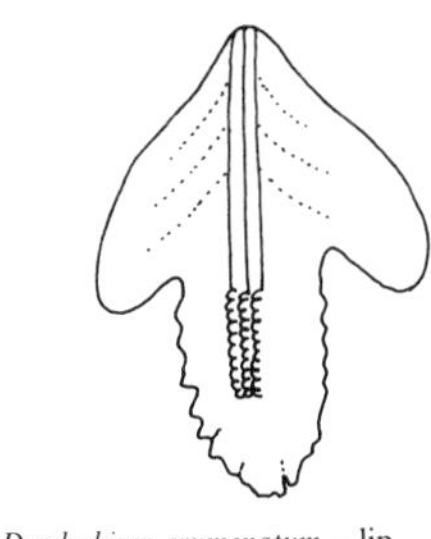

Dendrobium crumenatum – lip.

GROUP O (Section *Aporum* Schlechter).
Stems and leaves strongly laterally flattened, thick, leaves regularly alternating in 2 rows, jointed at the base; flowers produced singly from a cluster of chaffy bracts.

82. D. terminale Parish & Reichenbach

ILLUSTRATIONS: Annals of the Royal Botanic Garden Calcutta 8: t. 55 (1898); Opera Botanica **83**: f. 148 & t. xxiva (1985).
STEMS: 10–15 cm.
LEAVES: to 2 × 0.5 cm.
FLOWERS: produced subterminally.
SEPALS AND PETALS: upper sepal and petals *c.* 5 × 3 mm, lateral sepals *c.* 13 × 6 mm.
LIP: simple, notched at the apex, *c.* 10 × 5 mm.
DISTRIBUTION: Burma.
HARDINESS: G2.

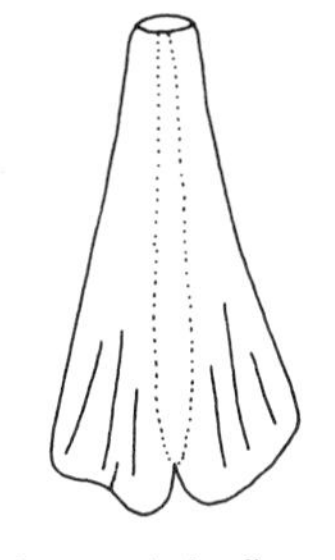

Dendrobium terminale – lip.

74. EPIGENEIUM Gagnepain

A genus of 35 species from Asia, formerly included in *Dendrobium* Swartz *Sarcopodium* Lindley or *Katherinea* Hawkes, of which 3 are occasionally grown. Cultivation in general as for the Indian and Burmese species of *Dendrobium* (p. 000), but these plants are better grown on rafts to accommodate their long rhizomes.

Literature: Summerhayes, V.S., Notes on Asiatic orchids II, *Kew Bulletin* 1957: 259–68; Seidenfaden, G., Orchid genera in Thailand IX, *Dansk Botanisk Arkiv* **34**: 68–82 (1980).

HABIT: epiphytes with long rhizomes.

PSEUDOBULBS: distant, simple, 4-angled, at least when dry, 2-leaved.

LEAVES: leathery, folded when young often unequally 2-lobed at the apex.

RACEMES: erect with 1-many flowers, terminal on the pseudobulbs.

SEPALS AND PETALS: similar, the lateral sepals attached to the short column foot, forming a small mentum.

LIP: 3-lobed, attached to the base of the column foot and somewhat movable on it.

COLUMN: short.

POLLINIA: 4, parallel.

Key to species

1. Racemes with 10–15 or more flowers; sepals and petals white with green or yellow tips and maroon bases **1. lyonii**
2. Flowers 8–9 cm in diameter; central lobe of lip diamond-shaped, tapered towards its base

2.A. Lip to 3.5 cm, brown, lateral lobes greenish suffused with brown **2. amplum**

2.B. Lip to 4.5 cm, purplish **3. coelogyne**

3. Flowers 3.5–5 cm in dimater, cetnral lobe of lip transversely oblong, not tapered towards the base **4. rotundatum**

1. **E. lyonii** (Ames) Summerhayes

SYNONYMS: *Dendrobium lyonii* Ames; *Sarcopodium lyonii* (Ames) Rolfe; *S. acuminatum* var. *lyonii* (Ames) Kränzlin; *Katherinea acuminata* var. *lyonii* (Ames) Hawkes.

ILLUSTRATIONS: Das Pflanzenreich **45**: 330 (1910); Davis & Steiner, Philippine orchids, 110 (1952).

PSEUDOBULBS: 4–4.5 cm.

LEAVES: 10–15 × 2–4 cm, oblong or elliptic.

RACEMES: with 10–15 (or more) flowers.

SEPALS AND PETALS: to 4 cm, triangular, white with green or yellow tips and maroon bases.

LIP: to 3 cm, central lobe triangular, acute, whitish, bases of lateral lobes dark red.

DISTRIBUTION: Philippines.

HARDINESS: G2.

2. E. amplum (Lindley) Summerhayes

SYNONYMS: *Dendrobium amplum* Lindley; *Sarcopodium amplum* (Lindley) Lindley; *Katherinea ampla* (Lindley) Hawkes.

ILLUSTRATIONS: Annals of the Royal Botanic Garden Calcutta 8: t. 89 (1898); Orchid Review 78: 117 (1970).

PSEUDOBULBS: to 5 cm, ovoid or spindle-shaped.

LEAVES: 1–15 × *c.* 5 cm, oblong or oblong-lanceolate, acute.

RACEME: with 1–2 flowers which are 8–9 cm in diameter.

SEPALS AND PETALS: sepals lanceolate, petals linear-lanceolate, all 3–4 cm, greenish suffused brownish purple and/or with brown spots.

LIP: to 3.5 cm, central lobe rhomboid, tapered to its base, brown, the lateral lobes coloured like the petals.

DISTRIBUTION: Nepal, N India.

HARDINESS: G2.

3. E. coelogyne (Reichenbach) Summerhayes

ILLUSTRATION: Bechtel et al., Manual of cultivated orchid species, 211 (1981).

Like *E. amplum* except as below.

FLOWERS: larger, more distinctly spotted and marked.

LIP: *c.* 4.5 cm, dark purple.

DISTRIBUTION: Burma.

HARDINESS: G2.

Not considered as a distinct species by Seidenfaden (reference above), who treats it as a synonym of *E. amplum*, but well known in cultivation.

4. E. rotundatum (Lindley) Summerhayes

SYNONYMS: *Sarcopodium rotundatum* Lindley; *Dendrobium rotundatum* (Lindley) Hooker; *Katherinea rotundata* (Lindley) Hawkes.

ILLUSTRATION: Annals of the Royal Botanic Garden Calcutta 8: t. 87 (1898).

PSEUDOBULBS: 3–4 cm, ovoid.

LEAVES: to 12 × 2.5 cm, lanceolate or oblong.

SEPALS AND PETALS: *c.* 2.5 cm, brown.

LIP: oblong *c.* 2.5 cm, the central lobe transversely oblong, not tapered to the base, yellowish and thickened in the middle.

DISTRIBUTION: N India, SW China.

HARDINESS: G2.

75. BULBOPHYLLUM Thouars

A genus of perhaps 1000 species from Asia, Australasia, Africa and tropical America. In this work *Cirrhopetalum*, which is often included in *Bulbophyllum* is treated as a separate genus. Being widely distributed, the species of *Bulbophyllum* show a range of cultural requirements. They can be grown in epiphytic compost in pots or on slabs of tree fern or bark. The deciduous species with hard pseudobulbs, mostly native to temperate regions, need a pronounced resting period once growth has ceased. The tropical species require a less marked rest and must not dry out completely. Propagation is by division.

Literature: Seidenfaden. G., Orchid genera in Thailand VIII: Bulbophyllum, *Dansk Botanisk Arkiv* 33(3), (1979).

HABIT: epiphytic.

PSEUDOBULBS: simple, distant or clustered, 1- or 2-leaved.

LEAVES: stalked or stalkless, folded when young, leathery.

FLOWERING STEMS: basal, erect or pendent (in our species), with distant, spreading, sheath-like scales.

FLOWERS: 1–many, in spikes, racemes or umbels.

SEPALS: usually almost equal; lateral sepals joined to column foot, forming a short mentum.

PETALS: usually shorter than sepals, sometimes much reduced.

LIP: thick, not obviously lobed, sometimes fringed, loosely hinged to column foot.

COLUMN: short, often with 2 or more apical arms; column foot long and narrow.

POLLINIA: 4.

Synopsis of characters

Pseudobulbs. 2-leaved: **1,2**. Cylindric: **3**; spherical and flattened: **3,4**; ovoid: **6,7**. Clustered: **3,4,6,7**.
Leaves. 10 cm or more: **1,2,5–7**. 4.5 cm wide or more: **5–7**.
Flowering stem. Pendent: **3,4**. Thickened or flattened: **1,2**.
Flowers. Solitary: **6,7**; more than 10: **1–4**. In umbels: **5–7**.
Sepals. More than 1.5 cm: **5–7**; less than 1.5 cm: **1–4**. Longer than petals: **1–7**; almost equal to petals: 1,2,6,7. Drawn out into long points: **5**. Upper sepal thickened: **1,2**.
Lip. Ovate to triangular: **5–7**. Fringed towards base: **1,2**; with apical thread-like hairs: **3,4**.

Key to species

1. Pseudobulbs 2-leaved; flowering stem flat or swollen above
1.A. Leaves to 2.3 cm wide; lip yellow **1. falcatum**
1.B. Leaves 2.5 cm or more wide; lip purple **2. leucorrhachis**
2. Flowers 10 or more per raceme; upper sepal 2 cm or more
2.A. Pseudobulbs clustered; flowering stems erect; lip-apex with a tuft of long, purplish hairs **3. barbigerum**
2.B. Pseudobulbs distant; flowering stems drooping; lip-apex without a tuft of hairs **4. careyanum**
3. Pseudobulbs 7 cm or more, cylindric or narrowly fusiform **5. ericssonii**

4. Flowering stem 10 cm or more, with a single flower; lip purplish with a bright orange circular callus **6. lobbii**
5. Flowering stem 1–2 cm, with 2 flowers; lip dark crimson, paler and spotted towards the margins **7. leopardinum**

1. **B. falcatum** (Lindley) Reichenbach

SYNONYM: *Megaclinium falcatum* Lindley.

ILLUSTRATIONS: Edwards's Botanical Register **12**: t. 989 (1826); Stewart & Campbell, Orchids of tropical Africa, 61 (1970); Bechtel et al., Manual of cultivated orchid species, 174 (1981); Fanfani, Macdonald encyclopaedia of orchids, t. 20 (1989).

PSEUDOBULBS: 2–6 cm, ovoid or narrowly ovoid, 4-angled, distant, 2-leaved.

LEAVES: to 16 cm × 6–23 mm, oblong to oblanceolate.

FLOWERING STEM: to 35 cm, erect, forming a flattened green or reddish blade to 1.4 cm wide.

FLOWERS: numerous in 2 ranks, 5–20 mm apart, *c.* 5 × 10 mm, green, reddish or purple.

SEPALS: upper to 7 mm, linear, tip wider and thicker; lateral sepals shorter than upper sepal, 2.5–3.5 mm wide, sickle-shaped.

PETALS: 1.7–3 mm, sickle-shaped, linear to oblong, obtuse and often thickened and yellow.

LIP: *c.* 2 mm, narrowly triangular, purplish.

DISTRIBUTION: W Africa to Uganda.

HARDINESS: G2.

FLOWERING: spring.

2. **B. leucorrhachis** (Rolfe) Schlechter

SYNONYM: *Megaclinium leucorrhachis* Rolfe.

ILLUSTRATION: Botanical Magazine, 7811 (1901).

Like *B. falcatum* except as below.

LEAVES: to 18 × 3 cm.

FLOWERING STEM: to 18 cm, swollen to *c.* 1.3 cm thick above; bracts triangular, reflexed.

FLOWERS: *c.* 8 mm, yellow.

LIP: small, tongue-shaped, fringed towards base.

DISTRIBUTION: W & C Africa.

3. **B. barbigerum** Lindley

ILLUSTRATIONS: Botanical Magazine, 5288 (1861); Schlechter, Die Orchideen, edn 2, 320 (1927); Orchid Review **76**: 318 (1968); Bechtel et al., Manual of cultivated orchid species, 173 (1981).

PSEUDOBULBS: to 3 × 1.5–2.5 cm, spherical, somewhat flattened, clustered, 1-leaved.

LEAVES: to 10 × 2.8 cm, broadly oblong, obtuse.

FLOWERING STEM: 10–20 cm, erect with *c.* 4 narrowly ovate-acuminate, sheath-like scales *c.* 9 mm long; floral bracts similar.

FLOWERS: 8–20 mm long, reddish.

SEPALS: 7–13 mm, greenish, narrowly lanceolate.

PETALS: minute.

LIP: *c.* 1 cm, yellowish, linear, with short marginal hairs; apex with a tuft of long, purplish, thread-like, apically swollen hairs; base hinged so that lip and hairs move in the slightest draught.

DISTRIBUTION: W & C Africa.

HARDINESS: G2.

FLOWERING: summer.

4. **B. careyanum** (Hooker) Sprengel

ILLUSTRATIONS: Botanical Magazine, 5316 (1862): Dansk Botanisk Arkiv 33: 146, 147 (1979); Pradhan, Indian orchids, t. 130 (1979); Bechtel et al., Manual of cultivated orchid species, 173 (1981).

Like *B. barbigerum* except as below.

PSEUDOBULBS: distant.

FLOWERING STEM: drooping.

FLOWERS: many, crowded, yellow.

LATERAL SEPALS: longer than upper sepal.

LIP: purple, downy, without long thread-like hairs.

DISTRIBUTION: E Himalaya, Burma & Thailand.

FLOWERING: winter.

5. **B. ericssonii** Kränzlin

ILLUSTRATION: Botanical Magazine, 8088 (1906).

PSEUDOBULBS: *c.* 9 × 1.5 cm, oblong-cylindric, distant, 1-leaved.

LEAF: to 20 × 9 cm, broadly elliptic to ovate; apex shortly pointed; base rounded or shortly drawn out.

FLOWERING STEM: 18–28 cm, erect, with 4 or 5 sheath-like scales *c.* 2 cm long.

FLOWERS: 4–7, in umbels.

SEPALS: to 11 × 1.5 cm, free, ovate, tapered to a long drawn out, often spirally curved point, greenish with reddish spots.

PETALS: to 4 × 1 cm, narrowly triangular, somewhat sickle-shaped, pale yellow, tapered to a long drawn out, often twisted, reddish point.

LIP: to 1.5 cm, triangular, strongly recurved, pale yellow with reddish apical markings.

DISTRIBUTION: Indonesia (Moluccas).

HARDINESS: G2.

FLOWERING: winter.

6. B. lobbii Lindley

ILLUSTRATIONS: Botanical Magazine, 4532 (1850); Schlechter, Die Orchideen, edn 2, 325 (1927); Shuttleworth et al., Orchids, 88 (1970); Bechtel et al., Manual of cultivated orchid species, 174 (1981).

PSEUDOBULBS: 3–5 × 2–3 cm, broadly ovoid *c.* 8 cm apart, 1-leaved.

LEAF: to 35 × 9 cm, elliptic to broadly elliptic.

FLOWERING STEM: 10 cm or more, 1-flowered.

FLOWERS: to 7.5 cm wide, pale to brownish yellow, flushed with red or brown.

SEPALS: upper to 5 × 1.7 cm, broadly lanceolate, acuminate, veins crimson-dotted; lateral sepals a little shorter and broader, downwardly curved, with purplish veins, apical half brownish.

PETALS: to 3.5 × 1 cm, lanceolate, horizontal, with *c.* 9 purplish veins.

LIP: *c.* 1.3 × 1 cm, narrowly ovate, curved, basal margins upcurved, purplish with a bright orange circular callus.

DISTRIBUTION: NE India, through SE Asia to the Philippines.

HARDINESS: G2.

FLOWERING: intermittently.

7. B. leopardinum (Wallich) Lindley

ILLUSTRATIONS: Botanical Magazine, 9631 (1941); Dansk Botanisk Arkiv 33: 29 (1979); Bechtel et al., Manual of cultivated orchid species, 174 (1981).

Like *B. lobbii* except as below.

PSEUDOBULBS: clustered.

FLOWERING STEM: 1–2 cm, with 1–3 flowers.

FLOWERS: 2–3 cm wide, pale yellowish green, tinged or spotted with pink.

PETALS: about two-thirds of the sepal length.

LIP: dark crimson, paler and spotted towards margins.

DISTRIBUTION: E Himalaya to Burma & N Thailand.

HARDINESS: G2.

FLOWERING: spring–summer.

76. CIRRHOPETALUM Lindley

A genus of about 40 species from SE Asia and India with a few representatives in Africa and the Pacific islands. Several wild species are morphologically intermediate between *Cirrhopetalum* and *Bulbophyllum* and the 2 genera are often united under the latter name. However, the characteristic *Cirrhopetalum* flower shape and arrangement is evident in most of the cultivated species. Cultivation conditions are similar to those for *Bulbophyllum* (p. 275).

Literature: Seidenfaden, G., Orchid genera in Thailand VIII: Bulbophyllum, *Dansk Botanisk Arkiv* **33**(3), (1979).

HABIT: epiphytic.
PSEUDOBULBS: simple, distant or rarely clustered, 1-leaved (in our species).
LEAVES: shortly stalked or stalkless, folded when young.
FLOWERING STEMS: erect or arching, basal, with a few evenly spaced scales.
FLOWERS: 1–many in umbels, often forming a regular arc or circle.
SEPALS AND PETALS: lateral sepals much longer than the upper sepal, usually lying side by side, often loosely united; sepals and petals often with marginal hairs or other appendages.
LIP: small, arched, narrow and fleshy, loosely hinged to column foot.
COLUMN: short, with a long, narrow foot.
POLLINIA: 4.

Synopsis of characters

Pseudobulbs. 3 cm or more and angled: **1–3,7,8,10,11.** Spherical in outline, 4-lobed: **6–8.** Brownish: **1,10,11.** Clustered: **10,11.**
Leaves. 7 cm or less: **2,3,6–9.** 4 cm wide or more: **1–3,7,8.**
Flowering stem. More than 30 cm: **2–5,10,11.** Pendent: **4–6.** Arched above: **10,11.** With large scales: **1,6.** With true leaf: **6.** Slender: **4,5.**
Flowers. Solitary: **7,8**; more than 10: **19–11**; many, 1.5 cm or less: **2–5,7–11**; many, 1.5 cm or more: **1,6–8.**
Bracts. Reaching base of sepals: **1,4,5**; not reaching base of sepals: **2–11.**
Upper sepal. Less than 1 cm: **2–5,9–11**; more than 2 cm; **1.** Finely tapered: **1,4–8**; abruptly tapered or with an apical hair or bristle: **2,3,9–11.** Almost circular: **2,3**; spathulate: **10,11.** With marginal hairs: **4–11**; with flat appendages: **6.**
Lateral sepals. 5 cm or less: **2–11**; 10 cm or more: **1,4–8.** Finely tapered: **1,4–8**; oblong or linear: **2,3,9–11.** Warty near base: **7,8.**
Petals. Finely tapered: **1,4,5**; oblong or linear **2,3,7,8.** Acute: **7,8,10,11.** With apical hair or bristle: **2,3,9.** With marginal hairs: **2–11**; with concentration or tuft of hairs at tip: **7,8**; with flat appendages: **6**; with irregular margin: **4,5.**
Lip. With fleshy auricles: **2,3,6.** 2-ridged: **7,8.**
Column. With 2 long projections: **1–3.**

Key to species

1. Margin of upper sepal entire, though an apical bristle may be present
1.A. Petal margins entire, not hairy **1. medusae**
1.B. Column with 2 long apical projections **2. umbellatum**

1.C. Column with 2 short, 2-toothed apical projections **3. picturatum**

2. Lateral sepals with drawn-out, tapered or thread-like tips

2.A. Petal margins evenly fringed or hairy

2.A.a. Flowers 8–10 per raceme; lateral sepals less than 20 cm **4. gracillimum**

2.A.b. Flowers 3–7 per raceme; lateral sepals 20–30 cm **5. longissimum**

3. Petals with 15 or fewer small, flat, marginal appendages **6. wendlandianum**

4. Petals with apical tufts of many hairs

4.A. Flowers 4–9 cm, yellow or pinkish yellow with purple veins **7. ornatissimum**

4.B. Flowers to 20 cm, greenish purple **8. putidum**

5. Lateral sepals about 10 times longer than broad **9. makoyanum**

6. Flowers 8–12, 2–2.5 cm **10. auratum**

7. Flowers 3–7, 4–5 cm **11. mastersianum**

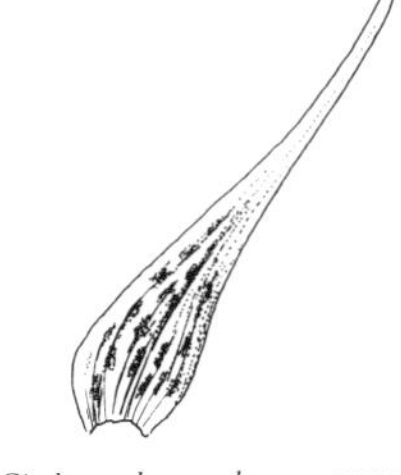

Cirrhopetalum medusae – upper sepal.

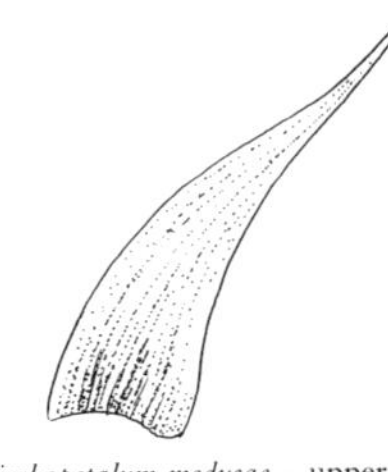

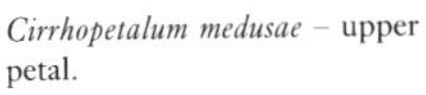

Cirrhopetalum medusae – upper petal.

1. C. **medusae** Lindley

ILLUSTRATIONS: Botanical Magazine, 4977 (1857); Shuttleworth et al., Orchids, 88 (1970); Skelsey, Orchids, 101 (1979).

PSEUDOBULBS: 2–5 × 1–2 cm, ovoid, angled, brownish, *c.* 3 cm apart.

LEAF: 10–20 × 2–4.5 cm, oblong to elliptic, rounded, sometimes notched.

FLOWERING STEM: 10–20 cm, erect, with large sheathing scales, bearing many small flowers in a dense umbel.

BRACTS: *c.* 2 cm.

FLOWERS: 7.5–17 cm long, white to pale yellow with pink spots.

SEPALS: upper 2.5–3 cm, linear, finely tapered; laterals to 14 cm, narrow and drawn out, acute.

PETALS: *c.* 5 mm, triangular, tapered, acute.

LIP: 1–2 mm.

COLUMN: with 2 long apical projections.

DISTRIBUTION: Malaysia, Indonesia, Borneo; ?Philippines.

HARDINESS: G2.

FLOWERING: winter.

2. C. **umbellatum** (Forster) Hooker & Arnott

SYNONYMS: *Bulbophyllum longiflorum* Thouars not Ridley; *C. thouarsii* Lindley; not *B. umbellatum* Lindley.

ILLUSTRATIONS: Botanical Magazine, 4237 (1846); Stewart & Campbell, Orchids of tropical Africa, 69 (1970); Millar, Orchids of Papua New Guinea, 50 (1978).

PSEUDOBULBS: 2.5–4.5 × 1–2 cm, conical to ovoid, angled, 2–4 cm apart.

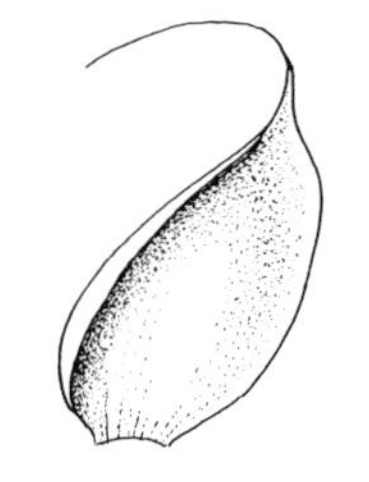

Cirrhopetalum umbellatum – upper sepal.

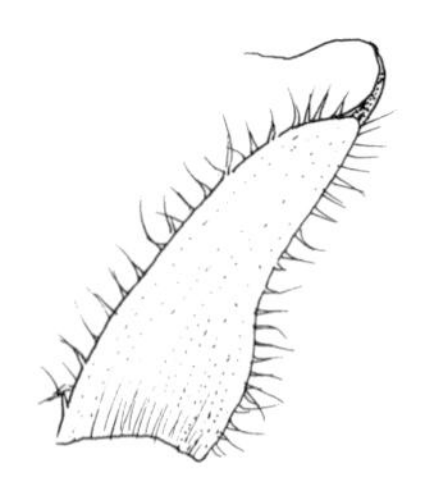

Cirrhopetalum umbellatum – upper petal.

LEAVES: 8–18 × 3–4 cm, elliptic to oblong, rounded, notched.
FLOWERING STEM: to 40 cm, erect or arched, with 3–7 flowers.
BRACTS: not reaching base of sepals.
FLOWERS: 4–5 cm long, pale pinkish yellow.
SEPALS: upper 7–10 mm, broadly ovate to circular, concave, purple-tipped, with a long apical bristle; laterals 2–4 cm × 4–7 mm, oblong.
PETALS: 4–10 mm, narrowly triangular, purple-tipped, with an apical bristle; margins hairy.
LIP: *c.* 7 mm, arched, purple.
COLUMN: with 2 long apical projections.
DISTRIBUTION: E Africa, Indian Ocean Islands, SE Asia, N Australia & Pacific islands.
HARDINESS: G2.
FLOWERING: summer.
If *Cirrhopetalum* is regarded as part of *Bulbophyllum*, this species must be called *B. longiflorum* Thouars and not *B. umbellatum*.

3. **C. picturatum** Loddiges
ILLUSTRATIONS: Botanical Magazine, 6802 (1885); Orchid Review **76**: 319 (1968); American Orchid Society Bulletin **42**(12): back cover (1973).
Like *C. umbellatum* except as below.
PETALS: broadly triangular.
LIP: with distinct fleshy auricles.
COLUMN: with short, broad, 2-toothed projections.
DISTRIBUTION: NE India, Burma. Thailand & Vietnam.
FLOWERING: spring–summer.

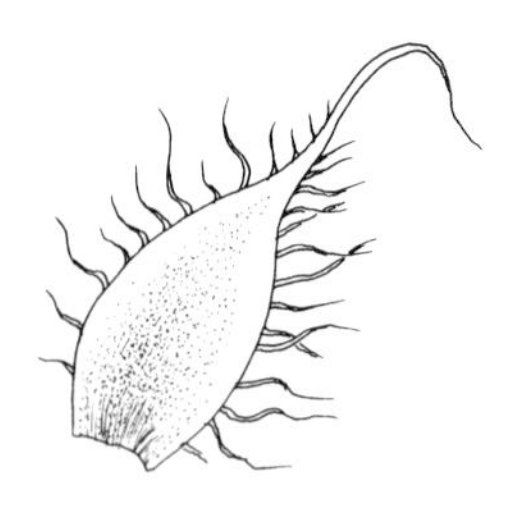

Cirrhopetalum gracillimum – upper sepal.

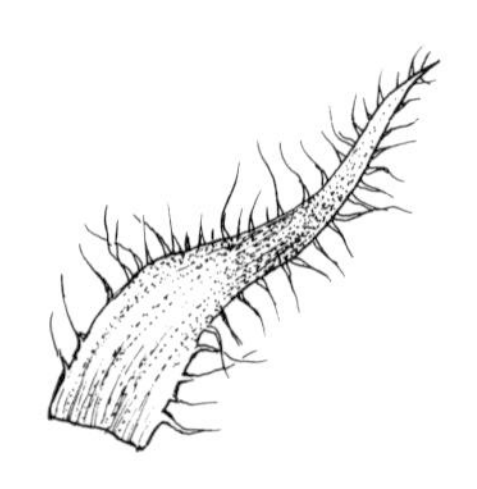

Cirrhopetalum gracillimum – upper petal.

4. **C. gracillimum** Rolfe
SYNONYM: *Bulbophyllum psittacoides* (Ridley) Smith.
ILLUSTRATIONS: Dansk Botanisk Arkiv **29**: 36 (1973); Bechtel et al., Manual of cultivated orchid species, 183 (1981).
PSEUDOBULBS: 1–2 cm × 5 mm, ovoid, *c.* 2 cm apart.
LEAF: to 14 × 2 cm, oblong, rounded.
FLOWERING STEM: to 35 cm, erect, very slender, bearing 8–10 flowers.
BRACTS: very small.
FLOWERS: 2–2.5 cm long, purplish red.
SEPALS: upper to 1 cm, ovate, concave, hairy on margin, finely tapered; laterals variable, 2 cm or more, finely tapered.
PETALS: 5–10 mm, narrowly triangular, finely tapered, hairy on margin.
LIP: 1.5–2.5 mm, pinkish purple with a paler tip.
DISTRIBUTION: Malay peninsula, Indonesia, New Guinea, Solomon & Fiji Islands.
HARDINESS: G2.
FLOWERING: summer–autumn.

5. C. longissimum Ridley

ILLUSTRATIONS: Botanical Magazine, 8366 (1911); Australian Orchid Review 33: 138 (1968); Bechtel et al., Manual of cultivated orchid species, 184 (1981).

Like *C. gracillimum* except as below.

PSEUDOBULBS: 3–7 cm apart.

FLOWERING STEM: arched to pendent, bearing 3–7 flowers.

BRACTS: 1–2 cm.

LATERAL SEPALS: 20–30 cm, finely tapered, pale pink with darker stripes.

DISTRIBUTION: Malay peninsula.

FLOWERING: winter.

6. C. wendlandianum Kränzlin

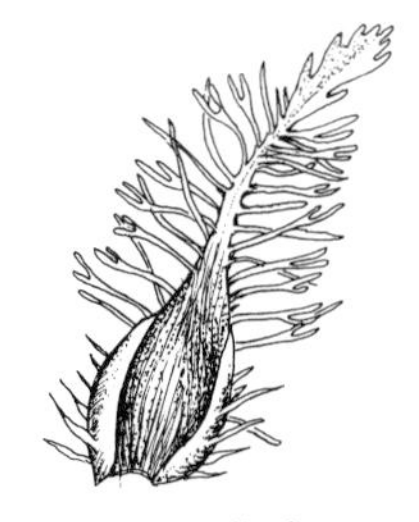

Cirrhopetalum wendlandianum – upper sepal.

SYNONYM: *C. collettii* Hemsley.

ILLUSTRATIONS: Botanical Magazine, 7198 (1891); Sheehan & Sheehan, Orchid genera illustrated, 55 (1979); Northen, Miniature orchids, t. 16 (1980).

PSEUDOBULBS: *c.* 2 × 2 cm, spherical but shallowly 4-lobed, 1.5–3 cm apart.

LEAF: 4–12 × 2–3 cm, elliptic, obtuse, sometimes notched.

FLOWERING STEM: 10–15 cm, arched to pendent, bearing *c.* 5 flowers; base sheathed, 1-leaved, later developing into a pseudobulb.

FLOWERS: 12–15 cm long.

BRACTS: short.

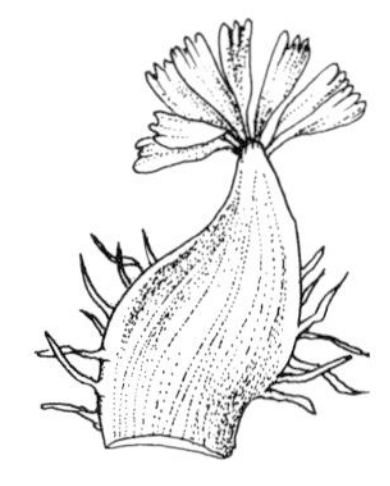

Cirrhopetalum wendlandianum – upper petal.

SEPALS: upper 1–2 cm, triangular, finely tapered, yellow with several small, flat, marginal appendages near the tip, margins hairy in lower half; laterals 8–12 cm, gradually tapered from the base, pink with darker red veins.

PETALS: to 1 cm, triangular; colour, marginal hairs and appendages as in upper sepal.

LIP: *c.* 5 mm with swollen, bristly, pinkish purple auricles.

DISTRIBUTION: Burma & Thailand.

HARDINESS: G2.

FLOWERING: spring–summer.

7. C. ornatissimum Reichenbach

ILLUSTRATIONS: Botanical Magazine, 7229 (1892); Dansk Botanisk Arkiv 29: 28 (1973); Bechtel et al., Manual of cultivated orchid species, 184 (1981).

PSEUDOBULBS: 2–5 × 1–2 cm, ovoid, angled, 2–5 cm apart.

LEAF: 7–16 × 3–6 cm, oblong to elliptic, obtuse or rounded.

FLOWERING STEM: 10–20 cm, erect, bearing 3–5 flowers.

FLOWERS: 4–9 cm long.

BRACTS: short.

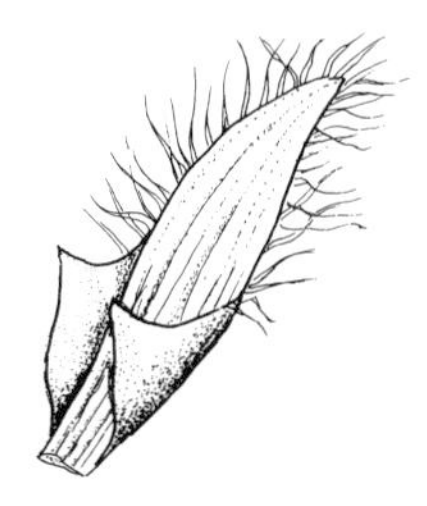

Cirrhopetalum ornatissimum – upper sepal.

SEPALS: upper 1–1.5 cm, ovate, tapered, hairy on margins, yellowish with purple veins; lateral sepals 5–8 cm, linear-

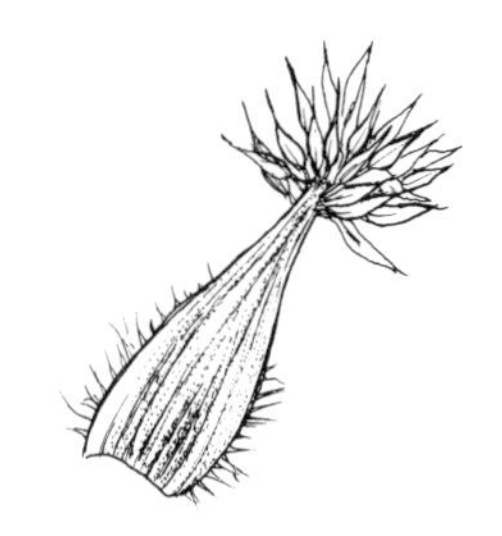

Cirrhopetalum ornatissimum – upper petal.

lanceolate, tapered, pinkish yellow with purple veins and spots.

PETALS: 1–1.5 cm, lanceolate to narrowly triangular, with a dense apical tuft of bulbous-based hairs, yellow with purple veins.

LIP: to 1 cm, 2-ridged, rounded, bristly, purple.

DISTRIBUTION: E Himalaya & NE India.

HARDINESS: G2.

FLOWERING: autumn–winter.

Cirrhopetalum putidum – upper sepal.

8. C. putidum Teijsmann & Binnendijk

SYNONYM: *C. fascinator* Rolfe.

ILLUSTRATIONS: Botanical Magazine, 8199 (1908); Kupper & Linsenmaier, Orchids, 55 (1961); American Orchid Society Bulletin **43**: 886, 888 (1974).

Like *C. ornatissimum* except as below.

PSEUDOBULBS: almost spherical, obscurely 4-lobed, smooth.

LEAVES: *c.* 5 × 1.5–3 cm, elliptic.

FLOWERING STEM: *c.* 10 cm, bearing 1–3 flowers.

FLOWERS: to 20 cm long, green and purple.

LATERAL SEPALS: to 18 cm, finely tapered, warty near base.

LIP: tapered, acute.

DISTRIBUTION: E Himalaya, NE India, Laos, Malaysia, Indonesia & Philippines.

FLOWERING: winter.

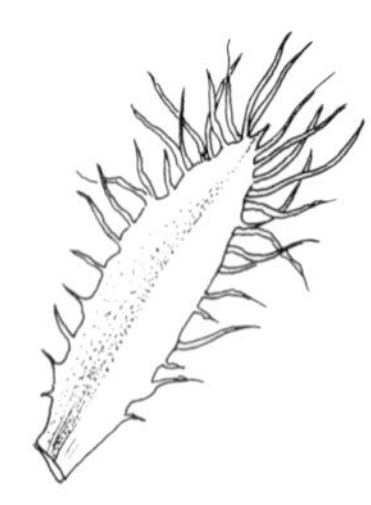

Cirrhopetalum putidum – upper petal.

9. C. makoyanum Reichenbach

ILLUSTRATIONS: Botanical Magazine, 7259 (1892); Shuttleworth et al., Orchids, 88 (1970); Graf, Tropica, 715 (1978).

PSEUDOBULBS: 1–2 cm × 7–10 mm, broadly ovoid, 1–4 cm apart.

LEAF: 5–10 × 1.5–3 cm, oblong to elliptic, rounded, notched.

FLOWERING STEM: 10–25 cm, erect, bearing 1–15 flowers.

FLOWERS: 3–4 cm long, mostly pale yellow with red spots.

SEPALS: upper *c.* 5 mm, triangular, acuminate, fringed with yellow hairs; laterals *c.* 3.5 cm, linear, parallel-sided.

PETALS: similar to upper sepal but narrower.

LIP: *c.* 2.5 mm, triangular.

DISTRIBUTION: Singapore, Borneo & Philippines.

HARDINESS: G2.

FLOWERING: winter.

10. C. auratum Lindley

SYNONYM: *Bulbophyllum campanulatum* Rolfe.

ILLUSTRATIONS: Botanical Magazine, 8281 (1909); Dansk Botanisk Arkiv **29**: 98 (1973).

PSEUDOBULBS: 2–3 × 1–1.5 cm, fusiform to ovoid, angled, greenish brown, 1–2 cm apart.

LEAF: 1–15 × 1.5–2.5 cm, oblong, obtuse, notched.
FLOWERING STEM: 10–15 cm, erect, arched above, bearing 8–12 flowers.
FLOWERS: 2–2.5 cm long, mostly pale yellow with reddish veins.
SEPALS: upper 6–7 mm, ovate-acuminate, with an apical hair, margins hairy; laterals 1.5–2 cm, oblong, sickle-shaped.
PETALS: *c.* 5 mm, triangular, margins hairy.
LIP: triangular, acute.
DISTRIBUTION: Malaysia, Sumatra & Borneo.
HARDINESS: G2.
FLOWERING: winter.

11. **C. mastersianum** Rolfe
ILLUSTRATIONS: Botanical Magazine, 8531 (1913); Dansk Botanisk Arkiv **29**: 84 (1973); Bechtel et al., Manual of cultivated orchid species, 184 (1981).
Like *C. auratum* except as below.
PSEUDOBULBS: ovoid, clustered.
FLOWERING STEM: 12–30 cm, bearing 6–8 flowers; flowers 4–5 cm long, yellow to orange.
LATERAL SEPALS: 3–4 × *c.* 1 cm.
DISTRIBUTION: Indonesia (Moluccas) & Borneo.

77. MALAXIS Swartz

A genus of over 200 species, widely distributed in temperate and tropical regions. Cultivation as for tropical, terrestrial species, with substitution of good leaf-mould (beech or oak) for crushed bark or peat.

HABIT: perennial herbs, growing in the soil, on rocks, or as epiphytes.
STEMS: fleshy, creeping, the tip leafy and erect, or crowded, or cylindric, or forming pseudobulbs.
LEAVES: few to many, usually deciduous on older stems, broad, often coloured, usually pleated, often unequal at the sheathing base.
INFLORESCENCE: terminal, erect; scape usually angled or shallowly winged, bracts deflexed.
FLOWERS: numerous, small, lip above the column (flowers non-resupinate).
SEPALS: broader than petals, spreading.
PETALS: linear, usually reflexed.
LIP: flat, entire, usually rounded or ovate, often with a hollow nectary at the base which is auriculate and clasps the column; apical margins often toothed or lobed.
COLUMN: short, winged.
POLLINIA: 4, in 2 pairs, waxy.

Key to species

1. Inflorescence dense, flowers crowded, their stalks less than 8 mm

1.A. Leaves purple; flowers at first yellow, later orange, ultimately purplish **2. discolor**

1.B. Leaves bronze-green with paler margins, spotted; flowers pink and cream **3. calophylla**

2. Inflorescence loose, flowers widely spaced, their stalks 8 mm or more **1. metallica**

1. **M. metallica** (Reichenbach) Kuntze

ILLUSTRATIONS: Botanical Magazine, 6668 (1883); La Belgique Horticole 34: plate opposite p. 281 (1884).

STEMS: indistinct, cylindric.

LEAVES: up to 6, to 7.5 × 3.8 cm, elliptic to ovate, acute, dark metallic red-purple, with 3–5 veins, margins wavy.

FLOWERS: *c.* 1.5 cm in diameter, purplish; stalk and ovary paler.

LIP: pale, apical margin irregularly toothed, auricles triangular.

DISTRIBUTION: Borneo.

HARDINESS: G2.

FLOWERING: late spring.

2. **M. discolor** (Lindley) Kuntze

ILLUSTRATIONS: Botanical Magazine, 5403 (1863); Jayaweera, Flora of Ceylon 2: 41 (1981).

STEMS: clustered, to *c.* 8 × 1 cm, oblong, leafless after flowering.

LEAVES: 3–6, to 12 × 4 cm, ovate, acuminate, rich purple, pleated, main veins 5–7, margins wavy, sometimes green.

FLOWERS: *c.* 3 mm in diameter, at first yellow, changing to orange, then purplish.

LIP: crescent- or kidney-shaped, margins not toothed, auricles shallow.

DISTRIBUTION: Sri Lanka.

HARDINESS: G2.

FLOWERING: summer.

3. **M. calophylla** (Reichenbach) Kuntze

SYNONYM: *Microstylis scottii* Ridley.

ILLUSTRATIONS: Botanical Magazine, 7268 (1892); Annals of the Royal Botanic Garden Calcutta 8: t. 20 (1898).

Like *M. discolor* except as below.

LEAVES: bronze-green, margins paler, with spots, the spots sometimes distributed to the middle of the blade.

FLOWERS: pink and cream.

LIP: cleft to form 2 teeth, auricles long, lanceolate.

DISTRIBUTION: Himalaya to Malaysia and Borneo.

FLOWERING: autumn.

78. LIPARIS Richard

A genus of over 200 species, distributed mostly in the tropics of the Old World. A few species also occur in temperate zones of the northern hemisphere. Many have little to recommend them as garden plants as their flowers are dull greenish or yellowish, but a few, not widespread in cultivation, are particularly attractive. *L. nervosa* should be grown in a compost recommended for terrestrial species whereas *L. viridiflora* should be grown in an epiphytic compost. *L. lilifolia* and *L. loeselii* are probably best grown under alpine-house or cold-frame conditions in a terrestrial compost containing one-third live Sphagnum moss.

Literature: Seidenfaden, G., Orchid genera in Thailand IV, *Dansk Botanisk Arkiv* **31**: 1–105 (1976).

HABIT: perennial herbs, terrestrial, growing on rocks or epiphytic; rhizome short.
STEMS: fleshy, thicker at base, sometimes pseudobulbous.
LEAVES: 1–7, membranous or leathery, jointed at the base or not, rolled or folded when young.
INFLORESCENCE: racemose with few to many flowers.
SEPALS AND PETALS: oblong, reflexed, the petals usually narrower and a little shorter than the sepals.
LIP: broadly oblong to rounded, sharply bent backwards at or about the middle.
COLUMN: long and slender, slightly winged.
POLLINIA: 4, in 2 pairs, waxy.

Key to species

1. Base of leaf jointed, not sheathing the stem; leaves leathery, oblong, lanceolate or oblanceolate **4. viridiflora**
2. Lip yellowish green, to 5 x 3 mm **3. loeselii**
3. Leaves 2–7; lip to 8 x 7 mm **1. nervosa**
4. Leaves 2; lip to 10 x 8 mm **2. lilifolia**

1. L. nervosa (Thunberg) Lindley

SYNONYM: *L. elata* Lindley.
ILLUSTRATIONS: Edwards's Botanical Register **14**: t. 1175 (1828); Maekawa, The wild orchids of Japan in colour, 325 (1971); Luer, Native orchids of Florida, 171 (1972); Dansk Botanisk Arkiv **31**: 33 (1976).
PSEUDOBULBS: 3–7 × *c.* 3 cm, conical.
LEAVES: 2–7, to 25 × 1.2 cm, elliptic to ovate or broadly lanceolate, pleated, acute or acuminate.
INFLORESCENCE: 15–50 cm, loosely flowered, scape with 5 wings or grooves.
FLOWERS: 10–30, purplish.
SEPALS AND PETALS: to 8 × 2–5 mm, petals narrower than sepals.

LIP: to 8 × 7 mm, broadly oblong to obovate, apex notched, base with 2 short protruberances.
DISTRIBUTION: tropics of America, Africa and Asia.
HARDINESS: G2.
FLOWERING: summer.

2. L. lilifolia (Linnaeus) Lindley
ILLUSTRATIONS: American Orchid Society Bulletin **43**: 199 (1974), **47**: 399 (1978); Luer, Native orchids of the United States and Canada, 309 (1975).
Like *L. nervosa* except as below.
PLANT: to 25 cm.
LEAVES: 2, deciduous.
LIP: purplish, *c.* 10 × 8 mm.
DISTRIBUTION: Eastern N America.
HARDINESS: H4.
FLOWERING: spring–summer.

3. L. loeselii (Linnaeus) Richard
ILLUSTRATIONS: Ross-Craig, Drawings of British plants, part **28**: t. 52 (1971); Danesch & Danesch, Orchideen Europas, Mitteleuropa, edn 3, 207, 211 (1972); Mossberg & Nilsson, Orchids of northern Europe, 63 (1979).
Like *L. nervosa* except as below.
PLANT: to 25 cm.
LEAVES: 2, deciduous.
LIP: yellowish green, *c.* 5 × 3 mm.
DISTRIBUTION: N & C Europe, NE USA.
HARDINESS: H1.
FLOWERING: spring–summer.

4. L. viridiflora Lindley
SYNONYM: *L. longipes* Lindley.
ILLUSTRATIONS: Annals of the Royal Botanic Garden Calcutta **8**: t. 37 (1898); Dansk Botanisk Arkiv **31**: 85 (1976); Lin, Native orchids of Taiwan **2**: 241 (1977).
PSEUDOBULBS: 3–9 × 1–1.5 cm, cylindric, swollen towards the base.
LEAVES: 2, to 28 × 2.5–3.5 cm, oblanceolate, apex shortly pointed.
INFLORESCENCE: to 27 cm, densely flowered; scape flattened or narrowly winged.
FLOWERS: numerous, greenish.
SEPALS AND PETALS: 2–3 × *c.* 1 mm, petals narrower than sepals.
LIP: to 3 × 2 mm, oblong, apex obtuse, yellowish orange.
DISTRIBUTION: India, Sri Lanka & SE Asia.
HARDINESS: G2.
FLOWERING: winter.

79. CALYPSO Salisbury

A genus with only 1 species, uncommon in cultivation. It succeeds best in a damp, semi-shaded position with plenty of leaf-mould.

HABIT: terrestrial, 5–20 cm; rootstock a corm, producing a solitary leaf which dies down in winter.
LEAF: 3–10 cm, broadly ovate, pleated, rolled when young.
SCAPE: bearing 1 terminal flower.
SEPALS AND PETALS: 1.5–2 cm, lanceolate, spreading and ascending, often twisted, purplish pink.
LIP: 1.5–2.5 cm, inflated and saccate, with 2 small horns at the base, whitish to pale pink, marked with purple and with white or yellow hairs towards the base.
POLLINIA: waxy, flat, 4 in 2 pairs fixed to a detachable viscidium.

1. **C. bulbosa** (Linnaeus) Oakes
ILLUSTRATIONS: Luer, Native orchids of the United States and Canada, t. 95, 96 (1975); Williams et al., Field guide to the orchids of Britain and Europe, 134 (1978).
DISTRIBUTION: N temperate regions.
HARDINESS: H2.
FLOWERING: late spring–summer.

80. APLECTRUM (Nuttall) Torrey

A genus with only 1 species which is rather difficult in cultivation. It grows best in a shady damp situation.

HABIT: terrestrial, to 55 cm; rootstock a corm, producing a solitary leaf which dies down in winter.
LEAF: 10–20 cm, broadly elliptic, pleated, with silvery veins, rolled when young.
SCAPE: with 6–15 flowers.
SEPALS AND PETALS: 1–1.4 cm, oblanceolate, brown or yellowish brown, tinged with purple or brown.
LIP: 1–1.2 cm, 3-lobed, obovate, whitish with purple marks.
POLLINIA: waxy, 4 in 2 pairs, fixed to a detachable viscidium.

1. **A. hyemale** (Willdenow) Nuttall
SYNONYM: *A. spicatum* (Walter) Britton et al.
ILLUSTRATIONS: Luer, Native orchids of the United States and Canada, t. 88 (1975).
DISTRIBUTION: NE USA, SE Canada.
HARDINESS: H2.
FLOWERING: late spring–early summer.

81. WARREA Lindley

A genus of 3 species from C and northern S America. They

grow best in a compost like that used for *Phaius* (p. 110). Propagation is by division.

HABIT: terrestrial.

PSEUDOBULBS: ovoid.

LEAVES: few, in 2 ranks, usually lanceolate, pleated, rolled when young.

RACEME: loose, with few to several flowers, produced from the base of an immature pseudobulb.

SEPALS AND PETALS: spreading or curved inwards, the lateral sepals joined to the column foot to form a mentum; petals similar to the upper sepal but smaller.

LIP: simple, usually with erect sides, disc bearing central ridges.

COLUMN: stout, somewhat swollen above.

POLLINIA: 4 in 2 pairs on a short, triangular stipe.

Key to species

1. Sepals and petals white or pale yellow **1. warreana**
2. Sepals and petals reddish brown **2. costaricensis**

1. W. warreana (Lindley) Schweinfurth

SYNONYMS: *W. tricolor* Lindley; *W. bidentata* Lindley.

ILLUSTRATIONS: Lasser, Flora de Venezuela 15(4): 31 (1970); Pabst & Dungs, Orchidaceae Brasilienses **2**: 230 (1977).

PLANT: to 1 m.

PSEUDOBULBS: 4–12 cm with 3–5 flowers.

LEAVES: 30–70 cm, oblong-lanceolate.

RACEME: erect, usually longer than the leaves, with 6–15 flowers.

FLOWERS: somewhat hanging, white or pale yellow.

SEPALS AND PETALS: nearly circular, upper sepal to 3.5 cm.

LIP: 2–3.5 cm, white or reddish, marked with yellow and reddish purple, wavy at the edges and often with an apical notch; disc with 3 ridges.

DISTRIBUTION: northwest S America.

HARDINESS: G2.

FLOWERING: summer.

2. W. costaricensis Schlechter

PLANT: to 75 cm.

LEAVES: to 70 cm, lanceolate.

RACEME: erect, about as long as leaves.

FLOWERS: reddish brown.

SEPALS AND PETALS: oblong-ovate, *c.* 3 cm.

LIP: to 3 cm, nearly circular, entire, paler than sepals and petals, with reddish brown markings, disc narrow.

DISTRIBUTION: Costa Rica, Panama.

HARDINESS: G2.

82. XYLOBIUM Lindley

A genus of 33 species from C and tropical S America and the West Indies. They should be grown in an epiphytic compost, and should be watered from the time that new leaves appear until November, then only occasionally. They should be freely ventilated during the summer.

HABIT: epiphytic or rarely terrestrial.

PSEUDOBULBS: ovoid or conical, more rarely slender and cylindric.

LEAVES: 1–3 per pseudobulb, stalked lanceolate to elliptic, pleated, with 3–6 conspicuous veins beneath, rolled when young.

RACEME: produced from the base of a mature pseudobulb.

BRACTS: linear, usually exceeding the ovary.

SEPALS AND PETALS: spreading, the lateral sepals broader than the upper and joined to the column foot to form a mentum; petals similar to the upper sepal but smaller.

LIP: curved upwards at the base, usually 3-lobed at the apex, the lateral lobes often standing more or less erect, with a central, usually elongate callus.

COLUMN: cylindric.

POLLINIA: 4 in 2 pairs, the pollinia of each pair unequal, borne on a short, triangular stipe.

Synopsis of characters

Pseudobulbs. Ovoid, conical or shortly cylindric, to 10 cm: **2–6,8–10**; cylindric and elongate, 15–27 cm: **1,7.**

Sepals and petals. Spotted or with other markings: **2,3,6,7**; lacking spots or other obvious markings: **1,4–6,8–10.**

Lip. With warts or papillae: **1,2,4–7**; without warts or papillae: **3,8–10.**

Key to species

1. Pseudobulbs with a single leaf

1.A. Pseudobulbs 15–20 cm, cylindric, elongate **1. pallidiflorum**

1.B. Flowers with spots

1.B.a. Leaves with 3 conpsicuous veins beneath; raceme hanging **3. colleyi**

1.B.b. Leaves with 5 or 6 conspicuous veins beneath; raceme more or less upright **2. leontoglossum**

1.C. Raceme hanging **4. palmifolium**

1.D. Raceme upright **5. bractescens**

2. Upper sepal to 1.5 cm

2.A. Racemes with 6–9 flowers **10. powellii**

2.B. Lip creamy brown with pale brown veins **9. hyacinthinum**

2.C. Lip white, sometimes with pink stripes **8. foveatum**
3. Pseudobulbs more than 12 cm, cylindric, elongate **7. elongatum**
4. Leaves with 3 conspicuous veins beneath **3. colleyi**
5. Leaves with 5 conspicuous veins beneath **6. variegatum**

1. X. **pallidiflorum** (Hooker) Nicholson

ILLUSTRATIONS: Dunsterville & Garay, Venezuelan orchids illustrated **1**: 439 (1959); American Orchid Society Bulletin **43**: 208 (1974).
HABIT: epiphytic or rarely terrestrial.
PSEUDOBULBS: 15–20 cm with 1 leaf.
LEAVES: to 40 cm.
RACEME: upright with 3–9 flowers.
FLOWERS: white, pale green or yellowish green.
UPPER SEPAL: *c.* 1.5 cm.
LIP: *c.* 5 cm, whitish grading to yellowish orange at the base, 3-lobed, the central lobe covered with white papillae or warts.
DISTRIBUTION: Northwest S America, West Indies.
HARDINESS: G2.

2. X. **leontoglossum** (Reichenbach) Rolfe

ILLUSTRATIONS: Botanical Magazine, 7085 (1889); American Orchid Society Bulletin **43**: 212 (1974); Dunsterville & Garay, Venezuelan orchids illustrated **6**: 443 (1976); Bechtel et al., Manual of cultivated orchid species, 280 (1981).
HABIT: epiphytic.
PSEUDOBULBS: 3.5–10 cm, ovoid to conical, with 1 leaf.
LEAVES: to 35 cm, with 5–6 veins beneath.
FLOWERS: pale to bright yellow with reddish brown or pink spots.
UPPER SEPAL: 1.8–2 cm.
LIP: 3-lobed, central lobe usually acute, densely covered with brown warts.
DISTRIBUTION: Northwest S America.
HARDINESS: G2.

3. X. **colleyi** (Lindley) Rolfe

SYNONYM: *X. brachystachyum* Kränzlin.
ILLUSTRATIONS: Dunsterville & Garay, Venezuelan orchids illustrated **3**: 331 (1965); Pabst & Dungs, Orchidaceae Brasilienses **2**: 22,279 (1977).
HABIT: epiphytic.
PSEUDOBULBS: to 4 cm, ovoid, with 1 or occasionally 2 leaves.
LEAVES: to 50 cm, with 3 veins beneath.
RACEME: hanging, with 3–6 flowers.
FLOWERS: brown or reddish brown, spotted with purple and smelling of cucumber.

UPPER SEPAL: 2–2.5 cm.

LIP: *c.* 2 cm, more or less oblong, smooth, dark purple to almost black, shining.

DISTRIBUTION: Venezuela, Guyana, Brazil & Trinidad.

HARDINESS: G2.

4. X. **palmifolium** (Swartz) Fawcett

SYNONYM: *X. decolor* (Lindley) Nicholson.

ILLUSTRATIONS: Botanical Magazine, 3981 (1843); Fawcett & Rendle, Flora of Jamaica 1: t. 23 (1910).

HABIT: epiphytic.

PSEUDOBULBS: to 7.5 cm, narrowly ovoid, with 1 leaf.

LEAVES: to 42 cm.

RACEME: somewhat hanging, loose, with few to many flowers.

FLOWERS: white or yellowish, scented (often unpleasantly so).

UPPER SEPAL: *c.* 2 cm.

LIP: *c.* 1.5 cm, 3-lobed with small lateral lobes, the central lobe rounded or notched, warty, with a callus with 4–5 thickened ridges.

DISTRIBUTION: Cuba, Dominican Republic, Trinidad & Jamaica.

HARDINESS: G2.

5. X. **bractescens** (Lindley) Kränzlin

PSEUDOBULBS: *c.* 3 cm, conical, with 1 leaf.

LEAVES: to 27 cm, 3-veined beneath.

RACEME: flexuous and arching, loose with 5–7 flowers.

FLOWERS: dull yellow.

UPPER SEPAL: *c.* 2 cm.

LIP: 3-lobed, reddish brown, recurved at the obtuse apex, bearing lines of warts, with the callus apically 3-lobed.

DISTRIBUTION: Ecuador, Peru, Brazil.

HARDINESS: G2.

6. X. **variegatum** (Ruiz & Pavon) Garay & Dunsterville

SYNONYMS: *Maxillaria squalens* Lindley: *X. squalens* (Lindley) Lindley; *X. scabrilingue* Schlechter.

ILLUSTRATIONS: Botanical Magazine, 2955 (1829); Dunsterville & Garay, Venezuelan orchids illustrated 2: 343 (1961); Bechtel et al., Manual of cultivated orchid species, 280 (1981).

HABIT: epiphytic or terrestrial.

PSEUDOBULBS: to 9 cm, ovoid to shortly cylindric, with 2–3 leaves.

LEAVES: to 70 cm, 5-veined beneath.

RACEME: with 7–many flowers.

FLOWERS: white, pale yellow, pale pink or brownish, often marked or tinged with purple.

UPPER SEPAL: *c.* 2.5 cm.

LIP: 1–2 cm, 3-lobed, the central lobe with rows of brown or purple papillae.

DISTRIBUTION: southern C America and northwest S America.

HARDINESS: G2.

7. X. elongatum (Lindley) Hemsley

ILLUSTRATION: American Orchid Society Bulletin **43**: 208 (1974).

HABIT: epiphytic or terrestrial.

PSEUDOBULBS: 15–27 cm, cylindric, elongate, with 2 leaves.

LEAVES: to 40 cm, with 3–5 veins beneath.

RACEME: arching or drooping, with 5–20 flowers.

FLOWERS: whitish, yellowish or pinkish with red, brown or violet markings.

UPPER SEPAL: 2–2.5 cm.

LIP: *c.* 2 cm, usually purplish brown, 3-lobed, central lobe covered with dense papillae.

DISTRIBUTION: C America.

HARDINESS: G2.

8. X. foveatum (Lindley) Nicholson

SYNONYM: *X. stachyobiorum* (Reichenbach) Hemsley.

ILLUSTRATIONS: American Orchid Society Bulletin **43**: 209 (1974); Dunsterville & Garay, Venezuelan orchids illustrated **6**: 441 (1976).

HABIT: epiphytic.

PSEUDOBULBS: to 10 cm, more or less ovoid, with 2–3 leaves.

LEAVES: to 40 cm, with 3 veins beneath.

RACEME: upright to arching, with 20–24 densely packed flowers.

FLOWERS: yellowish or sometimes white, fragrant.

UPPER SEPAL: 1–1.5 cm.

LATERAL SEPALS: each with a dorsal keel projecting at the apex.

LIP: to 2 cm, 3-lobed, white with pinkish stripes in the basal part.

DISTRIBUTION: C America, northern S America, Jamaica.

HARDINESS: G2.

9. X. hyacinthinum (Reichenbach) Schlechter

ILLUSTRATIONS: Dunsterville Garay, Venezuelan orchids illustrated **1**: 437 (1959); Lasser, Flora de Venezuela **15**(4): 333 (1970).

HABIT: epiphytic.

PSEUDOBULBS: to 6 cm, more or less ovoid, with 2–3 leaves.

LEAVES: 25–50 cm.

RACEME: upright with 10–20 flowers.

FLOWERS: greenish cream, smelling of hyacinth.
UPPER SEPAL: *c.* 1.3 cm.
LATERAL SEPALS: each with a dorsal keel projecting at the apex.
LIP: *c.* 1 cm, creamy brown with pale brown veins.
DISTRIBUTION: Venezuela.
HARDINESS: G2.

10. X. powellii Schlechter
ILLUSTRATION: Hawkes, Encyclopaedia of cultivated orchids, t. XIIB (1965).
HABIT: epiphytic.
PSEUDOBULBS: 4–5 cm, shortly cylindric, with 2 leaves.
LEAVES: to 20 cm.
RACEMES: upright with 6–9 flowers.
FLOWERS: yellow or reddish brown, sometimes tinged with pale green, faintly scented.
UPPER SEPAL: 1–1.5 cm.
LIP: *c.* 1.5 cm, 3-lobed, central lobe with 3 ridges, reddish brown.
DISTRIBUTION: Costa Rica, Panama.
FLOWERING: G2.

83. BIFRENARIA Lindley

A genus of 10 species from tropical S America, nearly all of which are in cultivation. It is closely related to *Maxillaria*, which differs in producing only 1 flower per inflorescence, and to *Lycaste*, which has larger, more persistent leaves. Cultivation as for *Lycaste* (p. 299).

HABIT: epiphytic.
PSEUDOBULBS: usually ovoid, often 4-angled.
LEAVES: 1 per pseudobulb, stalked, rather leathery, linear-lanceolate to oblong, rolled when young.
RACEMES: produced from the bases of the pseudobulbs.
SEPALS: spreading, the laterals fused to the column foot, forming a short mentum.
PETALS: similar to sepals or smaller, spreading or somewhat incurved.
LIP: joined to column foot, usually 3-lobed with the lateral lobes standing more or less erect, and with a central, usually elongate callus.
COLUMN: winged above.
POLLINIA: 4 in 2 pairs, each pair on a short, separate stipe attached to a broad, oblong, folded viscidium.

Synopsis of characters

Flower diameter (when opened out flat). 2–3 cm: **1–4**; 5–9 cm: **5–9**.

Sepals and petals. Basic colour orange or bright yellow: **1–3,5**; basic colour red or purple: **7,9**; basic colour white, pale yellow or green: **4–6,8**. Spotted: **1**.
Lip. Length 1.5 cm: **1,2,4**; length 1.8–3.2 cm; **3,8,9**; length 3.5–5 cm: **5–7**. Entire: **4,9**. Hairy: **5–7**.

Key to species

1. Flowers at most 3 cm in diameter when opened out flat
1.A. Sepals and petals greyish cream shaded with pink **4. racemosa**
1.B. Sepals and petals spotted **1. aurantiaca**
1.C. Lip with a purple patch; sepals and petals ovate, *c.* 10 mm wide **2. vitellina**
1.D. Lip without a purple patch; sepals and petals lanceolate, to 7 mm wide **3. aureo-fulva**
2. Sepals and petals basically red or purple
2.A. Flowers more than 5 cm in diameter when opened out flat; lip purple, 3-lobed **7. tyriantha**
2.B. Flowers to 5 cm in diameter when opened out flat; lip whitish or pink, entire **9. atropurpurea**
3. Lip not hairy **8. tetragona**
4. Sepals and petals basically green or yellowish green **6. inodora**
5. Sepals and petals basically white or pale yellow **5. harrisoniae**

1. **B. aurantiaca** Lindley

SYNONYM: *Rudolfiella aurantiaca* (Lindley) Schlechter.

ILLUSTRATIONS: Botanical Magazine, 3597 (1837); Pabst & Dungs, Orchidaceae Brasilienses **2**: 225, 281 (1977); Bechtel et al., Manual of cultivated orchid species, 271 (1981).

PSEUDOBULBS: more or less ovoid, 3–6 cm, somewhat 4-angled and often flattened, bright or yellowish green, often red-spotted.

LEAVES: 10–25 cm, often red-spotted beneath.

RACEMES: drooping or erect, with 4–18 flowers.

FLOWERS: *c.* 2.5 cm in diameter, orange or deep yellow with brown or purplish red spots.

LIP: 1–1.5cm, 3-lobed, the central lobe broadly triangular, often notched at apex, callus bright yellow.

DISTRIBUTION: Columbia, Venezuela, Guyana, Brazil & Trinidad.

HARDINESS: G2.

This species is sometimes treated as a member of *Rudolfiella* Hoehne, a genus of about 9 species, no other members of which are in cultivation.

2. **B. vitellina** (Lindley) Lindley

SYNONYM: *Stenocoryne vitellina* (Lindley) Kränzlin.

ILLUSTRATIONS: Edwards's Botanical Register **25**: t. 12 (1839); Pabst & Dungs, Orchidaceae Brasilienses **2**: 225, 281 (1977).
PSEUDOBULBS: ovoid, to 4 cm, bluntly angular.
LEAVES: to 30 cm.
RACEMES: drooping or erect, with 5–8 flowers.
FLOWERS: orange-yellow, *c.* 2.5 cm in diameter.
SEPALS AND PETALS: ovate, *c.* 10 mm wide.
LIP: 1.1–1.5 cm, 3-lobed, central lobe with a purple patch.
DISTRIBUTION: Brazil.
HARDINESS: G2.
FLOWERING: summer.

3. **B. aureo-fulva** (Hooker) Lindley
SYNONYM: *Stenocoryne aureo-fulva* (Hooker) Kränzlin.
ILLUSTRATION: Pabst & Dungs, Orchidaceae Brasilienses **2**: 225 (1977).
PSEUDOBULBS: ovoid, often wrinkled.
LEAVES: to 15 cm.
RACEMES: with 5–7 flowers.
FLOWERS: orange, 2–3 cm in diameter.
SEPALS AND PETALS: lanceolate, 5–7 mm wide.
LIP: *c.* 2 cm, 3-lobed, central lobe lanceolate.
DISTRIBUTION: Brazil.
HARDINESS: G2.
FLOWERING: autumn.
B. aureo-fulva and *B. vitellina* are sometimes included in *Stenocoryne* Lindley, a genus differing in the detailed structure of the pollinia.

4. **B. racemosa** (Hooker) Lindley
ILLUSTRATIONS: Loddiges' Botanical Cabinet **14**: t. 1318 (1828); Pabst & Dungs, Orchidaceae Brasilienses **2**: 225, 281 (1977).
PSEUDOBULBS: ovoid.
RACEMES: erect or drooping, with 4–10 flowers.
FLOWERS: greyish cream, shaded with pink, 2–3 cm in diameter.
LIP: 1.1–1.5 cm, entire, whitish, speckled with pink.
DISTRIBUTION: Brazil.
HARDINESS: G2.

5. **B. harrisoniae** (Hooker) Reichenbach
ILLUSTRATIONS: Pabst & Dungs, Orchidaceae Brasilienses **2**: 224, 280 (1977); Sheehan & Sheehan, Orchid genera illustrated, 45 (1979); Bechtel et al., Manual of cultivated orchid species, 170 (1981).

PSEUDOBULBS: broadly ovoid, 5–9 cm, weakly or strongly 4-angled.
LEAVES: to 30 cm.
RACEMES: usually erect with 1–3 flowers.
FLOWERS: fragrant, 6–7 cm in diameter.
SEPALS AND PETALS: whitish or pale yellowish, often tinged with red.
LIP: 4–5 cm, 3-lobed, purple to dark red, usually with darker veins, central lobe with a wavy margin, notched at apex, hairy; callus yellow, hairy.
DISTRIBUTION: Brazil.
HARDINESS: G2.
FLOWERING: spring–summer.
Several varieties have been described, differing in flower size and colour.

6. B. inodora Lindley
SYNONYM: *B. furstenbergiana* Schlechter.
ILLUSTRATIONS: American Orchid Society Bulletin **41**: 981 (1972); Pabst & Dungs, Orchidaceae Brasilienses **2**: 280 (1977).
Like *B. harrisoniae* except as below.
PSEUDOBULBS: usually with a dark band at the apex.
FLOWERS: a little larger, fragrant or not.
SEPALS AND PETALS: green or yellowish green, sometimes tinged with purple.
LIP: white, yellow or pale pink, often suffused with a darker colour; callus hairless.
DISTRIBUTION: Brazil.
HARDINESS: G2.
FLOWERING: early spring–summer.

7. B. tyrianthina (Loudon) Reichenbach
ILLUSTRATIONS: Botanical Magazine, 7461 (1896); Pabst & Dungs, Orchidaceae Brasilienses **2**: 224, 280 (1977).
Like *B. harrisoniae* except as below.
HABIT: more robust.
FLOWERS: to 9 cm in diameter, purple.
LIP: purple, whitish towards the base and densely hairy.
DISTRIBUTION: Brazil.

8. B. tetragona (Lindley) Schlechter
ILLUSTRATIONS: Edwards's Botanical Register **17**: t. 1428 (1831); Pabst & Dungs, Orchidaceae Brasilienses **2**: 224, 280 (1977).
PSEUDOBULBS: to 9 cm, ovoid, 4-angled.
LEAVES: to 45 cm.
RACEMES: with 3–5 flowers.
FLOWERS: fragrant, *c.* 5 cm in diameter.

SEPALS AND PETALS: basically greenish yellow, streaked with brownish purple.

LIP: *c.* 3 cm, about as wide as long, 3-lobed, whitish, marked with purple towards the base, callus large and conspicuous.

DISTRIBUTION: Brazil.

HARDINESS: G2.

FLOWERING: early summer.

9. **B. atropurpurea** (Loddiges) Lindley

ILLUSTRATIONS: Loddiges' Botanical Cabinet **19**: t. 1877 (1832); Pabst & Dungs, Orchidaceae Brasilienses **2**: 223, 280 (1977); Bechtel et al., Manual of cultivated orchid species, 170 (1981).

PSEUDOBULBS: to 8 cm, ovoid, 4-angled.

LEAVES: to 25 cm.

RACEME: usually borne horizontally, with 3–5 flowers.

FLOWERS: fragrant, *c.* 5 cm in diameter.

SEPALS AND PETALS: dark red flushed with yellow around the middle.

LIP: 1.8–2.2 cm, more or less entire, whitish or pink.

DISTRIBUTION: Brazil.

HARDINESS: G2.

FLOWERING: spring–summer.

84. **LYCASTE** Lindley

A genus of 45 species from C and tropical S America, of which fewer than half are in cultivation. They require an epiphytic compost and should be grown in a cool to intermediate house with a definite resting period in winter. Watering may begin again when new growth emerges. Spraying should be delayed until after the new leaves have fully developed, to avoid unsightly black blotches.

Literature: Fowlie, J.A., *The genus Lycaste* (1970).

HABIT: epiphytic or terrestrial.

PSEUDOBULBS: ovoid or ellipsoid, often laterally compressed, the base enclosed in papery or fibrous sheaths, of which the upper 1 or 2 are often leaf-bearing.

LEAVES: usually 1–3 per pseudobulb, stalked, lanceolate to oblong-elliptic, pleated, rolled when young.

FLOWERS: solitary from the bases of the pseudobulbs, scapes shorter than the leaves, with several papery bracts.

SEPALS: laterals broader than the upper, joined to the column foot to form a short mentum.

PETALS: more or less equal to the upper sepal or shorter.

LIP: 3-lobed, the base continuous with or jointed to the column foot; lateral lobes erect; disc with a thickened callus.

POLLINIA: 4 in 2 pairs, on a linear stipe.

Synopsis of characters

Pseudobulbs. With spines at the apex: **1–5,8,12,16,18**; without spines, or spines vestigial (less than 3 mm): **6,7,9–11,13–15, 17,19**.

Lip. Length 2.5–3 cm: **1,2,5,8,9,11,12,19**; length 3–4 cm: **4,6,10**; length 4–5.5 cm: **3,7,13–17**. Basic colour green: **18,19**; basic colour yellow or orange: **1–5,10,15,17**; basic colour cream or white: **5–12,14**; basic colour reddish, brownish or purple: **10,13,16,17**. Surface of central lobe hairless: **3,4,7–10,12–19**; surface of central lobe hairy: **1,2,5,6,11**. Callus hairless: **1,3,4,7–12,14–19**; callus hairy: **2,5,13**.

Key to species

1. Pseudobulbs armed with apical spines which are more than 3 mm — *Go to Subkey* 1
2. Pseudobulbs without apical spines, or spines vestigial, less than 3 mm — *Go to Subkey* 2

Subkey 1. *Pseudobulbs with apical spines which are more than 3 mm long.*

1. Lip green — **18. locusta**
2. Central lobe of lip entirely hairless

2.A. Petals more than 5 cm; lip reddish brown, callus notched at apex — **16. longipetala**

2.B. Lip 4–5 cm, deep yellow with red spots — **3. deppei**

2.C. Central lobe of lip narrowly ovate — **12. xytriophora**

2.D. Central lobe of lip more or less circular — **8. brevispatha**

3. Apex of callus a flap-like extension protruding over the base of the central lobe of the lip — **1. aromatica**
4. Central lobe of lip lanceolate — **5. crinita**
5. Callus hairy; lateral lobes of lip longer than central lobe — **2. cruenta**
6. Callus not hairy; lateral lobes of lip smaller than the central lobe — **4. macrobulbon**

Subkey 2. *Pseudobulbs without apical spines, or spines vestigial, less than 3 mm long.*

1. Flowers at least 12 cm in diameter

1.A. Lateral sepals *c.* 6.5 cm; sinuses between lateral and central lobes of the lip without a projection — **13. skinneri**

1.B. Lateral sepals to 9.5 cm; sinuses between lateral and central lobes of the lip each with a projection — **7. schilleriana**

2. Callus orange-red or reddish purple, if yellow then spotted

2.A. Pseudobulbs without any spines; callus with a broadened, truncate apex — **17. denningiana**

2.B. Pseudobulbs with vestigial spines; callus with a bluntly pointed apex — **10. macrophylla**

3. Lip at least 4.5 cm

3.A. Lip yellowish orange with the apex of the central lobe inrolled; upper sepal 5–6 cm **15. barringtoniae**
3.B. Lip creamy white or pale yellow, apex of the central lobe not inrolled; upper sepal 8–10 cm **14. ciliata**
4. Plant hanging; lip green **19. dyeriana**
5. Central lobe of lip narrowly oblong, densely hairy; upper sepal 6–6.5 cm **6. lasioglossa**
6. Lip whitish, heavily spotted with light pink, central lobe hairless **9. tricolor**
7. Lip creamy white flushed with brownish yellow, central lobe minutely downy **11. leucantha**

1. **L. aromatica** (Hooker) Lindley

ILLUSTRATIONS: Botanical Magazine, 9231 (1931); Fowlie, The genus Lycaste, 14 (1970); Fanfani, Macdonald encyclopaedia of orchids, t. 84 (1989).

PSEUDOBULBS: 7–10 cm, strongly ribbed, apex with spines.

LEAVES: 30–40 cm, developing after flowering.

SCAPES: 11–15 cm.

FLOWERS: 4–6 cm in diameter, smelling of cinnamon.

SEPALS: yellowish green, the upper 2.5–3 cm.

PETALS: orange-yellow, more or less equal to sepals.

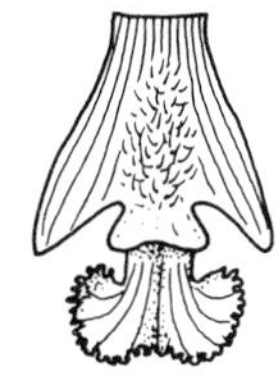

Lycaste aromatica – lip.

LIP: 2.5–3 cm, orange-yellow, spotted with dark orange, lateral lobes sickle-shaped, central lobe recurved with a finely toothed margin; callus very shortly hairy with a flap-like extension protruding over the base of the central lobe.

COLUMN: *c.* 2 cm, hairy below the stigma.

DISTRIBUTION: C America.

HARDINESS: G2.

FLOWERING: spring–summer.

2. **L. cruenta** (Lindley) Lindley.

ILLUSTRATIONS: Edwards's Botanical Register **28**: t. 13 (1842); Fowlie, The genus Lycaste, 26 (1970); Orchid Digest **42**: 209 (1978); Bechtel et al., Manual of cultivated orchid species, 223 (1981).

PSEUDOBULBS: to 10 cm, apex with spines.

LEAVES: 35–45 cm, deciduous at time of flowering.

SCAPES: *c.* 15 cm.

FLOWERS: 5–7 cm in diameter, smelling faintly of cinnamon.

SEPALS: greenish yellow, the upper 4–4.5 cm.

PETALS: yellowish orange with red spots towards the base, 3.5–3.8 cm, recurved at apices.

Lycaste cruenta – lip.

LIP: *c.* 2.5 cm, orange with red elongate spots towards the base and with a red triangular patch at the base, lateral lobes larger than central lobe, central lobe more or less circular, recurved, margin minutely scalloped, upper surface shortly hairy; callus ridged, hairy, red at the base.

COLUMN: 1.5 cm, red, hairy below stigma.

DISTRIBUTION: Northern C America.
HARDINESS: G2.
FLOWERING: spring–summer.

3. L. deppei (Loddiges) Lindley
ILLUSTRATIONS: Loddiges' Botanical Cabinet **17**: t. 1612 (1830); Fowlie, The genus Lycaste, 28 (1970); Bechtel et al., Manual of cultivated orchid species, 223 (1981).
PSEUDOBULBS: to 8 cm, apex with spines.
LEAVES: 30–50 cm.
SCAPES: 12–20 cm.
FLOWERS: to 10 cm in diameter.
SEPALS: green with reddish brown spots, the upper 4–7 cm.
PETALS: creamy white with tiny reddish brown spots, shorter than sepals, recurved at apices.
LIP: 4–5 cm, deep yellow with red spots and pale red stripes towards the base, central lobe recurved, margin minutely scalloped; callus cream to yellow with pale red markings, hairless.
COLUMN: *c.* 2.5 cm, whitish spotted with red, hairy.
DISTRIBUTION: C America.
HARDINESS: G2.
FLOWERING: spring–summer.

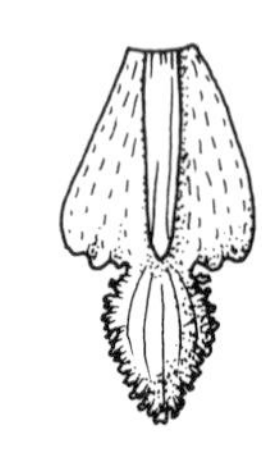
Lycaste deppei – lip.

4. L. macrobulbon (Hooker) Lindley
ILLUSTRATIONS: Botanical Magazine, 4228 (1846); Fowlie, The genus Lycaste, 32 (1970); Orchid Digest **42**: 209 (1978).
PSEUDOBULBS: 5–8 cm, apex with spines.
LEAVES: to 40 cm, developing after flowering.
SCAPES: to 25 cm.
FLOWERS: 6–7 cm in diameter, fragrant.
SEPALS: yellowish green, the upper 3.7–4 cm, gently recurved.
PETALS: yellow with orange spots, 3–3.3 cm, recurved at apices.
LIP: 3–3.3 cm, deep yellow spotted with reddish brown, especially on the central lobe which is recurved with a minutely toothed margin and hairy at the base; callus with raised margins.
COLUMN: *c.* 1.8 cm, pale yellow, spotted with red at base.
DISTRIBUTION: Colombia, Venezuela.
HARDINESS: G2.
FLOWERING: spring–summer.

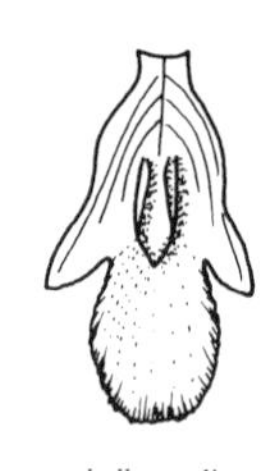
Lycaste macrobulbon – lip.

5. L. crinita Lindley
ILLUSTRATIONS: Fowlie, The genus Lycaste, 24 (1970); Orchid Digest **42**: 207 (1978).
PSEUDOBULBS: with apical spines.
LEAVES: to 35 cm.
SCAPES: 8–11 cm.

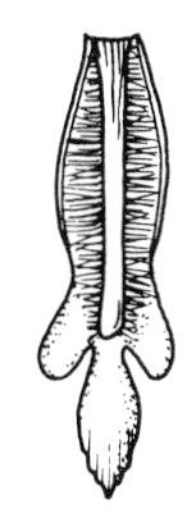
Lycaste crinita – lip.

SEPALS: light green, the upper 2.5–2.8 cm, somewhat recurved.
PETALS: pale yellow, *c.* 2 cm, recurved at apices.
LIP: *c.* 2.5 cm, pale yellow spotted with orange-red, central lobe lanceolate hairy; callus densely hairy on margins.
COLUMN: *c.* 1.5 cm, hairy at the base.
DISTRIBUTION: Mexico.
HARDINESS: G2.
FLOWERING: summer.

6. L. lasioglossa Reichenbach
ILLUSTRATIONS: Botanical Magazine, 6251 (1876); Fowlie, The genus Lycaste, 30 (1970); Bechtel et al., Manual of cultivated orchid species, 223 (1981).
PSEUDOBULBS: 5–10 cm, slightly ribbed, without apical spines.
LEAVES: 40–50 cm, present at flowering.
SCAPE: to 25 cm.
FLOWERS: *c.* 10 cm in diameter.
SEPALS: reddish brown, suffused with green outside and at apices, the upper 6–6.5 cm, the laterals with upswept apices.
LIP: 3–3.5 cm, pale yellow, suffused and spotted with red, central lobe recurved, narrowly oblong, densely hairy; callus cream, tongue-shaped, grooved.
COLUMN: 2.5–3 cm, hairy above the middle.
DISTRIBUTION: Guatemala.
HARDINESS: G2.
FLOWERING: winter–spring.

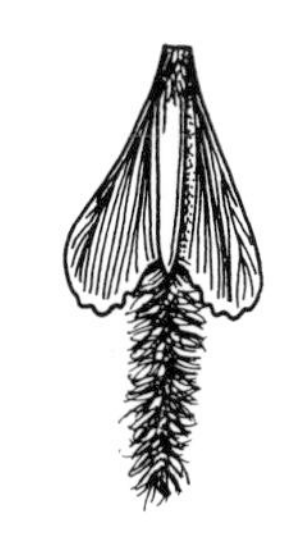
Lycaste lasioglossa – lip.

7. L. schilleriana Reichenbach
SYNONYMS: *L. hennisiana* Kränzlin; *L. longisepala* Schweinfurth.
ILLUSTRATIONS: Fieldiana **30**(1): 647 (1960); Fowlie, The genus Lycaste, 43 (1970).
PSEUDOBULBS: 7–10 cm, slightly ribbed, without apical spines.
LEAVES: to 40 cm.
SCAPES: 15–20 cm.
FLOWERS: 16–18 cm in diameter.
SEPALS: pale green suffused with brown towards the base, upper sepal 9–10 cm, gently recurved.
PETALS: 4–5 cm, white suffused with pale pink, recurved at apices.
LIP: 4.8–5.2 cm, white, suffused and spotted with pink, central lobe recurved, its margins finely scalloped, sinuses between lateral and central lobes with a tiny projection; callus yellowish, striped pinkish at the base.
COLUMN: *c.* 1.8 cm, shortly hairy.
DISTRIBUTION: Peru, Colombia & Surinam.
HARDINESS: G2.
FLOWERING: spring–summer.

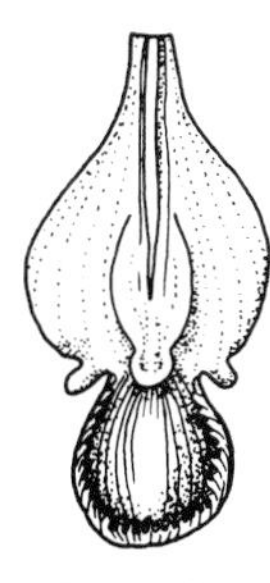
Lycaste schilleriana – lip.

8. L. brevispatha (Klotzsch) Lindley

SYNONYM: *L. candida* Reichenbach.

ILLUSTRATIONS: Reichenbach, Beiträge zu einer Orchideenkunde Central-Amerika's, t. 5 (1866); Fowlie, The genus Lycaste, 36 (1970); Bechtel et al., Manual of cultivated orchid species, 222 (1981).

PSEUDOBULBS: *c.* 6–7 cm, ribbed, apex with spines.

LEAVES: to 50 cm.

SCAPES: 6–10 cm.

FLOWERS: 4–5 cm in diameter, fragrant.

SEPALS: brownish green, sometimes spotted with pink, recurved at apices, the upper *c.* 2.5 cm.

PETALS: equal to sepals, paler, sometimes spotted or suffused with pink, apices recurved.

LIP: 2.5–3 cm, whitish with pink spots, concave towards the base, central lobe with wavy margins; callus tongue-shaped with 2 ridges, yellowish grading to pink at the base.

COLUMN: white with pink spots.

DISTRIBUTION: Southern C America.

HARDINESS: G2.

FLOWERING: winter–spring.

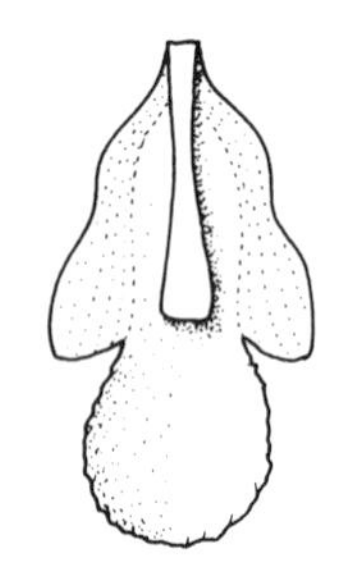

Lycaste brevispatha – lip.

9. L. tricolor (Klotzsch) Reichenbach

ILLUSTRATIONS: Fieldiana **26**(1): 555 (1952); Fowlie, The genus Lycaste, 39 (1970); Kew Magazine **1**: pl. 14 (1984).

PSEUDOBULBS: 5–9 cm, unarmed but with sharp apical ridges.

LEAVES: 20–40 cm, deciduous at flowering.

SCAPES: 8–12 cm.

FLOWERS: to 5.5 cm in diameter, never opening fully.

SEPALS: greenish brown, sometimes with pink spots, recurved, the upper 3–4.5 cm.

PETALS: shorter and paler than sepals, with pink spots.

LIP: 2.5–3 cm, whitish, heavily spotted with pink, central lobe with the apex strongly reflexed, margin finely wavy; callus cream.

COLUMN: 1.5–2 cm, sparsely hairy in front.

DISTRIBUTION: Southern C America.

HARDINESS: G2.

FLOWERING: spring–summer.

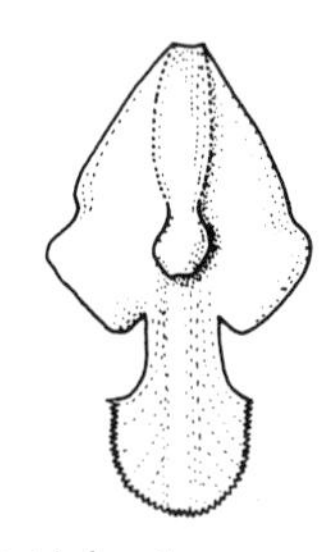

Lycaste tricolor – lip.

10. L. macrophylla (Poeppig & Endlicher) Lindley

SYNONYM: *L. dowiana* Endres & Reichenbach.

ILLUSTRATIONS: Fowlie, The genus Lycaste, 49, 53, 54 (1970); Pabst & Dungs, Orchidaceae Brasilienses **2**: 282 (1977).

PSEUDOBULBS: 8–12 cm, ribbed, apex with vestigial spines *c.* 0.3 mm.

LEAVES: to 65 cm.

SCAPES: 8–50 cm.

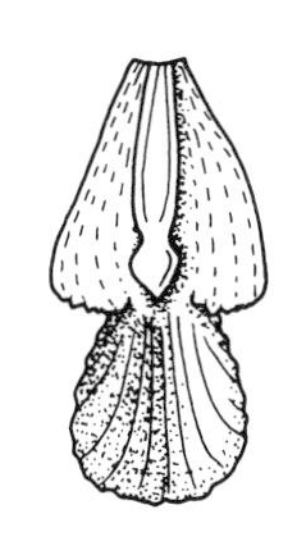

Lycaste macrophylla – lip.

FLOWERS: *c.* 9 cm in diameter, faintly scented or not.
SEPALS: brownish green to dark reddish, recurved at apices, the upper 3–5 cm.
PETALS: cream, suffused or not with yellow or pink, sometimes with pink spots, recurved at apices, shorter than sepals.
LIP: 3–4 cm, cream or yellowish, often spotted with pink to varying degrees, occasionally completely reddish; callus tongue-shaped, grooved, yellow with red spots or pale reddish purple.
COLUMN: *c.* 3 cm, cream, often with 2 purple spots at the base, hairy.
DISTRIBUTION: southern C America northwest S America.
HARDINESS: G2.
FLOWERING: late summer.
Related to *L. leucantha* and *L. xytriophora*. Several subspecies are recognised, some of which may be in cultivation.

11. **L. leucantha** (Klotzsch) Lindley
ILLUSTRATIONS: Paxton's flower garden **2**: 37–8 (1851–2); Fowlie, The genus Lycaste, 46 (1970); Bechtel et al., Manual of cultivated orchid species, 223 (1981).
PSEUDOBULBS: to 6 cm, without apical spines.
LEAVES: 40–50 cm.
SCAPE: 15–22 cm.
FLOWERS: *c.* 8 cm in diameter.
SEPALS: pale green, sometimes brownish green, recurved at apices, the upper *c.* 4 cm.
PETALS: creamy white, recurved towards apices 3.7–4 cm.
LIP: 2.7–3 cm, pale brownish yellow, central lobe ovate with a minutely wavy margin, minutely downy; callus brownish yellow with a shallow groove.
COLUMN: hairless.
DISTRIBUTION: southern C America.
HARDINESS: G2.
FLOWERING: winter.

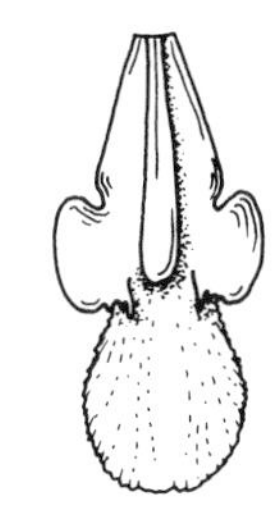

Lycaste leucantha – lip.

12. **L. xytriophora** Linden & Reichenbach
ILLUSTRATIONS: Reichenbach, Xenia Orchidacea **3**: t. 241 (1892); Fowlie, The genus Lycaste, 66 (1970); Bechtel et al., Manual of cultivated orchid species, 224 (1981).
PSEUDOBULBS: 7–8 cm, apex with spines.
LEAVES: 40–50 cm.
SCAPES: 4–13 cm.
FLOWERS: 7–8 cm in diameter.
SEPALS: pale green suffused with brown, the upper 3–4 cm, apices strongly reflexed.
PETALS: cream suffused with green, shorter than sepals, recurved towards apices.
LIP: 2.5–3 cm, creamy white suffused with yellow and with

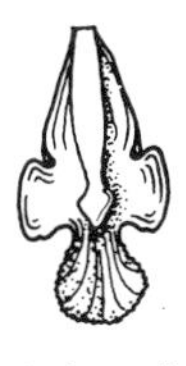

Lycaste xytriophora – lip.

mauve inside, central lobe narrowly ovate, sharply deflexed with minutely wavy margins; callus linear, expanded at apex into an ovate projection, deep yellow grading to red at the base.

COLUMN: 2.5–2.8 cm, white, hairy below stigma.

DISTRIBUTION: Panama, Ecuador.

HARDINESS: G2.

FLOWERING: summer.

13. L. skinneri (Lindley) Lindley

SYNONYM: *L. virginalis* (Scheidweiler) Linden.

ILLUSTRATIONS: Kupper & Linsenmaier, Orchids, 83 (1961); Fowlie, The genus Lycaste, 61 (1970); Bechtel et al., Manual of cultivated orchid species, 223 (1981).

PSEUDOBULBS: to 10 cm, slightly ridged, without apical spines.

LEAVES: 50–60 cm.

SCAPES: 15–30 cm.

FLOWERS: 12–15 cm in diameter.

SEPALS: cream, variously shaded with pink to lavender, recurved at apices the upper 7–8 cm.

PETALS: usually darker than sepals, reddish purple, recurved at apices, 4.5–7 cm.

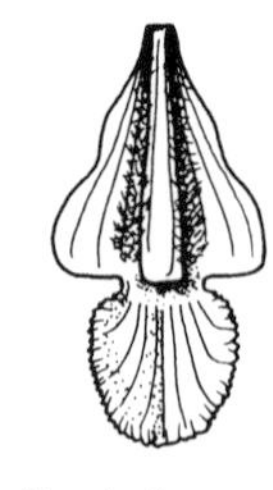

Lycaste skinneri – lip.

LIP: 4–5.5 cm, pinkish, unmarked or mottled with purple, central lobe ovate, reflexed; callus reddish, hairy.

COLUMN: 3–4 cm, white, spotted with red towards the base, hairy in front.

DISTRIBUTION: C. America.

HARDINESS: G2.

FLOWERING: winter–spring.

Probably the most commonly grown species. About 15 varieties, distinguished mainly by flower colour and size, have been in cultivation at various times, and some may still be available. *L. skinneri* has often been used in hybridisation.

14. L. ciliata (Ruiz & Pavon) Reichenbach

SYNONYM: *L. fimbriata* (Poeppig & Endlicher) Cogniaux.

ILLUSTRATIONS: Orchid Digest 38: 144 (1974); Botanical Magazine n.s., 767 (1978).

PSEUDOBULBS: to 12 cm, ridged, without apical spines.

LEAVES: 60–80 cm, present at flowering, then falling.

SCAPES: 8–24 cm.

FLOWERS: 9–10 cm in diameter, slightly fragrant, appearing not to open fully.

SEPALS: creamy white suffused with pale green, the upper 8–10 cm.

PETALS: usually paler than sepals, 4.5–5 cm.

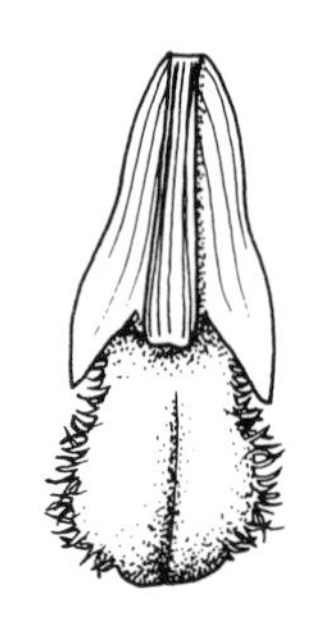

Lycaste ciliata – lip.

LIP: 4.5–5.5 cm, whitish or pale yellow with a slight green suffusion, central lobe oblong, recurved, lateral margins fringed; callus with 5–7 ribs, yellowish to pale orange, apex rounded or notched and projecting over the base of the central lobe.

COLUMN: white, hairy in front.

DISTRIBUTION: northern S America.

HARDINESS: G2.

FLOWERING: late winter–summer.

15. L. barringtoniae (Smith) Lindley

ILLUSTRATIONS: Fawcett & Rendle, Flora of Jamaica 1: t. 23 (1910); Fowlie, The genus Lycaste, 74 (1970).

PSEUDOBULBS: 6–8 cm, ribbed, without apical spines.

LEAVES: 35–50 cm, falling at or just before flowering.

SCAPES: 7–8 cm.

FLOWERS: 5–7 cm, fragrant, not opening fully.

SEPALS: pale green suffused with brown, the upper 5–6 cm.

PETALS: similar to sepals, but *c.* 4.5 cm.

LIP: 4–4.5 cm, yellow-orange, sometimes with red markings, central lobe ovate, fringed, apex inrolled to form a shallow pocket; callus yellow, with 5 ridges, the notched apex projecting forwards over the base of the central lobe.

COLUMN: 3–5 cm, white.

DISTRIBUTION: Cuba, Jamaica.

HARDINESS: G2.

FLOWERING: spring–summer.

Lycaste barringtoniae – lip.

16. L. longipetala (Ruiz & Pavon) Garay

SYNONYM: *L. gigantea* Lindley.

ILLUSTRATIONS: Botanical Magazine, 5616 (1866); Dunsterville & Garay, Venezuelan orchids illustrated 1: 203 (1959); Fowlie, The genus Lycaste; 83 (1970); Bechtel et al., Manual of cultivated orchid species, 223 (1981).

PSEUDOBULBS: 7–12 cm with shallow furrows and apical spines.

LEAVES: to 60 cm, present at flowering.

SCAPES: 50–60 cm.

FLOWERS: faintly fragrant, appearing to be only half open.

SEPALS: pale green suffused with brown, the upper 9.5–11 cm.

PETALS: similar in colour to the sepals, concave, *c.* 8 cm.

LIP: *c.* 6 cm, reddish brown, central lobe with a lighter orange border, recurved, with a deep central groove, apex folded upwards, lateral margins irregularly minutely toothed; callus greenish brown with 2 ridges which merge into a notched plate.

COLUMN: 2.5–3.5 cm, whitish, often orange at base, hairy below.

DISTRIBUTION: Northwest S America.

HARDINESS: G2.

FLOWERING: spring.

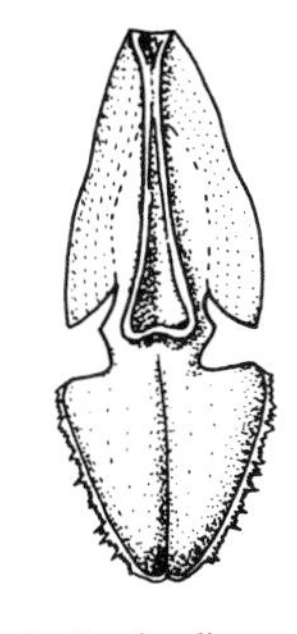

Lycaste longipetala – lip.

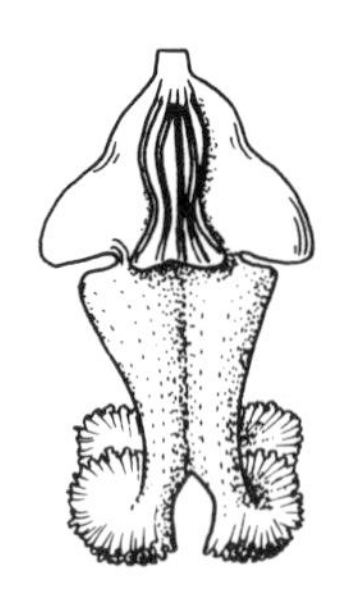
Lycaste denningiana – lip.

17. L. denningiana Reichenbach

SYNONYM: *L. cinnabarina* Rolfe.

ILLUSTRATIONS: Lindenia **9**: t. 394 (1893); Dunsterville & Garay, Venezuelan orchids illustrated **3**: 167 (1965); Fowlie, The genus Lycaste, 56 (1970).

PSEUDOBULBS: 7–10 cm, furrowed, without apical spines.

LEAVES: 45–70 cm.

SCAPES: to 16 cm.

FLOWERS: *c.* 10 cm in diameter.

SEPALS AND PETALS: cream, greenish or brownish yellow, the upper sepal 6–8 cm.

LIP: *c.* 4 cm, deep orange-red, central lobe ovate to circular, apex folded under and back, margin minutely toothed; callus orange-red, with 5 ridges and a broadened truncate apex.

COLUMN: 2–2.5 cm, white.

DISTRIBUTION: northwest S America.

HARDINESS: G2.

Fowlie suggests, on the basis of field evidence, that this species may be a hybrid between *L. longipetala* and *L. ciliata*.

18. L. locusta Reichenbach

ILLUSTRATIONS: Botanical Magazine, 8020 (1905); Fowlie, The genus Lycaste, 76 (1970); American Orchid Society Bulletin **55**: 352 (1986).

PSEUDOBULBS: 6–10 cm, ribbed, apex with spines.

LEAVES: 50–85 cm.

SCAPES: to 30 cm.

FLOWERS: *c.* 8 cm in diameter.

Lycaste locusta – lip.

SEPALS AND PETALS: pale greyish green, flushed or veined with darker green, sepals *c.* 5 cm, petals *c.* 4 cm with recurved apices.

LIP: 3.7–4 cm, dark green, lateral lobes lighter, central lobe oblong-ovate, recurved with a hairy border, apex notched; callus dark green, spathulate with 2 ridges.

COLUMN: white, hairy in front.

DISTRIBUTION: Peru.

HARDINESS: G2.

FLOWERING: spring–summer.

19. L. dyeriana Rolfe

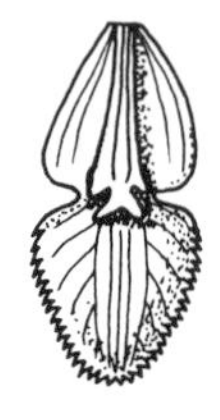
Lycaste dyeriana – lip.

ILLUSTRATION: Botanical Magazine, 8103 (1906).

HABIT: plant hanging.

PSEUDOBULBS: 4–7.5 cm, angular, without apical spines.

LEAVES: 18–30 cm.

SCAPES: 7–13 cm, hanging.

FLOWERS: 4–5 cm in diameter, fragrant, not opening fully.

SEPALS: pale green, the upper *c.* 5 cm.

PETALS: darker green than sepals.

LIP: *c.* 3 cm, green, central lobe ovate-elliptic, reflexed, margin fringed; callus green, grooved, 2-lobed on each side of the notched apex.

COLUMN: *c.* 1.7 cm.

DISTRIBUTION: Peru.

HARDINESS: G2.

FLOWERING: summer–autumn.

Fowlie considers that this species has a greater affinity with *Bifrenaria* (p. 295) than with *Lycaste*, a genus in which it is certainly anomalous. However, as the controversy is unresolved, it is retained in *Lycaste* here.

85. ANGULOA Ruiz & Pavon

A genus of 10 species from northern S America. Cultivation as for *Lycaste* (p. 299).

HABIT: terrestrial or occasionally epiphytic.

PSEUDOBULBS: oblong, becoming ridged with age.

LEAVES: usually 2–3 per pseudobulb, stalked, pleated, oblong, rolled when young.

SCAPES: often several from the base of the pseudobulb, 15–30 cm, clothed in leafy bracts, bearing 1 flower.

FLOWERS: fragrant.

SEPALS AND PETALS: fleshy, concave and incurving, almost concealing the lip.

LIP: 3-lobed, hinged to the column foot so that it rocks, the lateral lobes erect; callus linear-lanceolate with 2 apical lobes.

POLLINIA: 4 on a linear stipe which has a slightly expanded viscidium.

Synopsis of characters

Flower colour. Yellow: 3; yellow flushed red or green: **1,2,4**; whitish or pink: **5,6.**

Central lobe of lip. Surface hairless: **2,4–6**; surface hairy: **1,3.**

Key to species

1. Flowers whitish or pink

1.A. Central lobe of lip wider than callus **5. uniflora**

1.B. Central lobe of lip narrower than callus **6. virginalis**

2. Central lobe of lip with 2 tooth-like lobules, hairy

2.A. Flowers clear yellow **3. clowesii**

2.B. Flowers yellow flushed reddish and with red spots, or occasionally entirely red inside **1. ruckeri**

3. Central lobe of lip oblong, fringed at the apex, not or scarcely recurved **2. brevilabris**

4. Central lobe of lip more or less 3-lobed, not fringed, recurved **4. cliftonii**

1. A. ruckeri Lindley

ILLUSTRATIONS: Edwards's Botanical Register **32**: t. 41

(1846); Kupper & Linsenmaier, Orchids, 81 (1961); Orchid Review **92**: 253 (1984).

FLOWERS: dull yellowish flushed with red and with red spots or rarely entirely red inside.

LIP: lateral lobes more or less oblong, central lobe hairy with 2 tooth-like lobules.

DISTRIBUTION: Colombia.

HARDINESS: G2.

FLOWERING: spring–summer.

The albino variant (var. **albiflora** Anon.) with pure white flowers may be in cultivation. Var. **sanguinea** Lindley (Illustration: Botanical Magazine, 5384, 1863) has flowers entirely red inside.

2. A. **brevilabris** Rolfe

ILLUSTRATIONS: Botanical Magazine, 9381 (1935); Orchid Digest **42**: 140 (1978).

Like *A. ruckeri* except as below.

FLOWERS: smaller, flushed with dull reddish green.

LIP: lateral lobes very broad, central lobe oblong, hairless though fringed at the apex, and lacking lobules.

DISTRIBUTION: Colombia.

HARDINESS: G2.

FLOWERING: summer.

3. A. **clowesii** Lindley

ILLUSTRATIONS: Edwards's Botanical Register **30**: t. 63 (1844); Orchid Digest **40**: 132 (1976); Orchid Review **92**: 252 (1984).

FLOWERS: yellow.

LIP: whitish to orange-yellow, lateral lobes more or less oblong, central lobe hairy, ovate, reflexed, with 2 tooth-like lobules.

DISTRIBUTION: Colombia, Venezuela.

HARDINESS: G2.

FLOWERING: spring–summer.

Very similar to *A. ruckeri*; in the opinion of some authors the 2 should be treated as varieties of the 1 species.

4. A. **cliftonii** Rolfe

ILLUSTRATIONS: Botanical Magazine, 8700 (1917); Orchid Digest **40**: 132 (1976); Bechtel et al., Manual of cultivated orchid species, 167 (1981).

FLOWERS: yellow splashed and marked with reddish purple.

LIP: pouched at the base, lateral lobes rounded, central lobe with a basal claw, more or less 3-lobed, hairless.

DISTRIBUTION: Colombia.

HARDINESS: G2.

FLOWERING: spring–summer.

5. A. uniflora Ruiz & Pavon

ILLUSTRATIONS: Edwards's Botanical Register **30**: t. 60 (1844); Orchid Digest **40**: 132 (1976); Bechtel et al., Manual of cultivated orchid species, 168 (1981); Orchid Review **92**: 252, 253 (1984).

FLOWERS: whitish, sometimes flushed with green, spotted and tinged inside with pink.

LIP: yellowish, spotted with pink, lateral lobes rounded and inrolled, central lobe lanceolate, reflexed, wider than the callus.

DISTRIBUTION: Colombia, Venezuela, Peru.

HARDINESS: G2.

FLOWERING: spring– summer.

Several varieties, based on variation in flower colour, have been described.

6. A. virginalis Schlechter

ILLUSTRATIONS: Botanical Magazine, 4807 (1854); Orchid Digest **40**: 132 (1976); Sheehan & Sheehan, Orchid genera illustrated, 37 (1979).

Very like *A. uniflora*, except as below.

LIP: central lobe narrower than the callus.

DISTRIBUTION: Colombia.

Often considered to be the same as *A. uniflora*. Pink-flowered variants have been found in the wild.

86. PABSTIA Garay

A genus of 5 species from Brazil usually found under the name *Colax* Lindley. Cultivation as for *Zygopetalum* (p. 313), *Maxillaria* (p. 330) and *Lycaste* (p. 299).

HABIT: epiphytic.

PSEUDOBULBS: ovoid to oblong.

LEAVES: 2–4 per pseudobulb, lanceolate, pleated, rolled when young.

SCAPES: arising from the bases of the pseudobulbs, clothed in leafy bracts, with 1–4 flowers.

SEPALS AND PETALS: more or less equal or petals smaller.

LIP: 3-lobed.

COLUMN: without a foot, hairy in front.

POLLINIA: 4, on a short, linear or oblong stipe.

Key to species

1. Sepals white — **1. jugosa**
2. Sepals spotted — **2. placanthera**
3. Sepals unspotted — **3. viridis**

1. P. jugosa (Lindley) Garay

SYNONYM: *Colax jugosus* (Lindley) Lindley.

ILLUSTRATIONS: Botanical Magazine, 5661 (1867); Lindenia 8: t. 372 (1892); Orchid Digest **42**: 144 (1978); Bechtel et al., Manual of cultivated orchid species, 251 (1981).

PSEUDOBULBS: with 2 leaves.

LEAVES: 15–25 cm.

FLOWERS: 2–4 in each raceme, 5–7.5 cm in diameter, fragrant.

SEPALS: white, ovate-oblong.

PETALS: white spotted with brownish or reddish purple.

LIP: *c.* 2.5 cm, lateral lobes rounded and striped with bluish purple, central lobe rounded, striped and marked with bluish purple, velvety; callus with ridges.

DISTRIBUTION: Brazil.

HARDINESS: G2.

FLOWERING: winter–spring.

The following 2 varieties have been in cultivation, and may still be available:

Var. **rufina** (Reichenbach) Garay

SYNONYM: *Colax jugosus* var. *rufinus* Reichenbach.

FLOWERS: yellowish green spotted with dark purple and with brown on the lip.

Var. **viridis** (Godefroy-Leboeuf) Garay

SYNONYMS: *Colax jugosus* var. *punctatus* Reichenbach; *Colax puytdii* Linden & André.

ILLUSTRATION: Illustration Horticole **27**: t. 369 (1880).

FLOWERS: greenish yellow spotted with reddish purple.

2. P. placanthera (Hooker) Garay

SYNONYM: *Colax placanthera* (Hooker) Lindley.

ILLUSTRATION: Botanical Magazine, 3173 (1832).

PSEUDOBULBS: with 2–3 leaves.

FLOWERS: usually solitary.

PETALS: smaller than sepals, yellowish green with brownish purple spots.

LIP: lateral lobes pale green streaked with brown, central lobe with a whitish, purple-flushed central area, very shortly hairy; callus smooth.

DISTRIBUTION: Brazil.

HARDINESS: G2.

FLOWERING: summer.

3. P. viridis (Lindley) Garay

SYNONYM: *Colax viridis* (Lindley) Lindley.

ILLUSTRATIONS: Pabst & Dungs, Orchidaceae Brasilienses **2**: 226, 282 (1977).

LEAVES: 12–18 cm.

FLOWERS: 1–2 per scape.

SEPALS AND PETALS: directed forwards, green, petals with brownish spots which merge towards the centre.

LIP: with a pale purple, rhomboid central lobe.
DISTRIBUTION: Brazil.
HARDINESS: G2.
FLOWERING: summer.

87. ZYGOPETALUM Hooker

A genus of 20 species from C and tropical S America, of which 5 are found in cultivation. They require water throughout the year, though the amount should be reduced in winter. Propagation is usually by division of large plants. All are very prone to attack by thrips, scale and red spider mite.

HABIT: epiphytic or terrestrial.
PSEUDOBULBS: ovoid, sheathed in basal leaves, bearing 2–5 leaves.
LEAVES: lanceolate, pleated, with 5–9 conspicuous veins beneath, rolled when young.
RACEMES: produced from the bases of the pseudobulbs.
SEPALS AND PETALS: spreading, lateral sepals attached to the column foot to form a short mentum; petals similar to sepals or sometimes smaller.
LIP: 3-lobed, lateral lobes very small, erect or spreading, with a prominent callus, often ribbed.
POLLINIA: 4 in 2 pairs, stipe variable in length.

Synopsis of characters

Petals. 2–2.5 cm: 4,5; 3–4 cm: 1–3.
Lip. Hairy: 1–3; hairless: 1,4,5. Without spots, stripes or obvious veining: 5.
Column. Yellow or green: 1,3,4; white: 2; purple: 5.

Key to species

1. Lip neither spotted nor striped; callus shaped like a horse's hoof, more or less erect **5. maxillare**
2. Lip absolutely hairless
2.A. Leaves with 7–9 veins beneath; flowers *c.* 8 cm in diameter; lip 4–4.5 cm **1. mackayi**
2.B. Leaves with 5 veins beneath; flowers to 6 cm in diameter; lip 2.5–3 cm **4. brachypetalum**
3. Petals shorter than upper sepal; column hairless **1. mackayi**
4. Central lobe of lip more or less ovate **2. crinitum**
5. Central lobe of lip kidney-shaped **3. intermedium**

1. Z. mackayi Hooker

ILLUSTRATIONS: Martius, Flora Brasiliensis 3(5): t. 104 (1902); Pabst & Dungs, Orchidaceae Brasilienses 2: 227, 283 (1977); Gartenpraxis 1980: 272.
PSEUDOBULBS: 4–7 cm with 2–3 leaves.

LEAVES: 30–50 cm, with 7–9 veins beneath.

FLOWERING STEMS: to 75 cm, racemes usually with 5–7 flowers.

FLOWERS: fragrant, *c.* 8 cm in diameter.

SEPALS AND PETALS: greenish yellow with purplish or brownish blotches, sepals 4–4.5 cm, petals 3.5–4 cm.

LIP: 4–4.5 cm, central lobe ovate or fan-shaped, white with interrupted purple radiating lines, hairless or with tiny hairs, margins wavy, apex notched; callus horseshoe-shaped, ridged, toothed along the front margin.

COLUMN: *c.* 2 cm, hairless, yellowish green with reddish brown spots.

DISTRIBUTION: S & SC Brazil.

HARDINESS: G2.

FLOWERING: autumn–winter.

2. **Z. crinitum** Loddiges

ILLUSTRATIONS: Botanical Magazine, 3402 (1835); Orchid Digest **36**: 128 (1972); Pabst & Dungs, Orchidaceae Brasilienses **2**: 226, 282 (1977); Bechtel et al., Manual of cultivated orchid species, 280 (1981).

PSEUDOBULBS: 4–7 cm with 3–5 leaves.

LEAVES: 25–40 cm with 7–9 veins beneath.

FLOWERING STEMS: 30–45 cm, raceme with 4–7 flowers.

FLOWERS: fragrant, 7–8 cm in diameter.

SEPALS AND PETALS: *c.* 4 cm, yellowish green with purplish brown blotches or bars.

LIP: 3–4 cm, central lobe broadly obovate, velvety, white or cream with densely hairy bluish purple or reddish veins; callus yellow, more or less 2-lobed.

COLUMN: 1.6–2 cm, hairy in front, white with purple stripes.

DISTRIBUTION: S Brazil.

HARDINESS: G2.

FLOWERING: winter.

Several varieties have been described, based mainly on colour variations of the veins of the lip.

3. **Z. intermedium** Lindley

ILLUSTRATIONS: Pabst & Dungs, Orchidaceae Brasilienses **2**: 283 (1977); Bechtel et al., Manual of cultivated orchid species, 280 (1981).

PSEUDOBULBS: 4–8 cm with 3–5 leaves.

LEAVES: 25–50 cm with 7–9 veins beneath.

FLOWERING STEMS: 30–75 cm, raceme with 4–7 flowers.

FLOWERS: fragrant, 7–8 cm in diameter.

SEPALS AND PETALS: 3–4 cm, green or yellowish green with reddish brown or purplish markings.

LIP: 3–4 cm with a short claw, central lobe kidney-shaped,

whitish spotted with purple and with purple veins, downy, margin wavy, apex notched; callus horseshoe-shaped, ridged.
COLUMN: 1.3–1.5 cm, velvety in front, pale green with purple stripes.
DISTRIBUTION: S Brazil, Peru, Bolivia.
HARDINESS: G2.
FLOWERING: autumn–spring.

4. Z. **brachypetalum** Lindley
ILLUSTRATIONS: Orchid Digest **36**: 128 (1972); Pabst & Dungs, Orchidaceae Brasilienses **2**: 226 (1977).
PSEUDOBULBS: 4–6 cm, somewhat compressed, with 2–3 leaves.
LEAVES: 30–50 cm, with 5 veins beneath.
FLOWERING STEMS: 30–80 cm, raceme with 5–12 flowers.
FLOWERS: fragrant, 5–6 cm in diameter.
SEPALS AND PETALS: 2–3 cm, all green with brown blotches.
LIP: 2.5–3 cm, central lobe more or less obovate, white with bluish violet veins and spots, hairless, apex notched; callus white with purple stripes.
COLUMN: 1–1.2 cm, yellowish with purple spots and stripes.
DISTRIBUTION: S Brazil.
HARDINESS: G2.
FLOWERING: autumn–winter.

5. Z. **maxillare** Loddiges
ILLUSTRATIONS: Botanical Magazine, 3686 (1839); Orchid Digest **36**: 128 (1972); Pabst & Dungs, Orchidaceae Brasilienses **2**: 227, 283 (1977).
PSEUDOBULBS: 4–8 cm with 2–3 leaves, borne on creeping, branched rhizomes.
LEAVES: 20–40 cm with 5–7 veins beneath.
FLOWERING STEMS: 15–30 cm, racemes with 5–8 flowers, sometimes drooping.
FLOWERS: *c.* 5 cm in diameter.
SEPALS AND PETALS: 2–2.5 cm, all green with dark brown blotches or bars.
LIP: with the central lobe more or less circular, bluish purple, hairless; callus dark purple, more or less erect and shaped like a horse's hoof, ribbed.
COLUMN: *c.* 1 cm, hairless, bluish purple.
DISTRIBUTION: S Brazil, Paraguay, N Argentina.
HARDINESS: G2.
FLOWERING: at any time of year.
In the wild Z. *maxillare* grows only on tree ferns.

88. ACACALLIS Lindley

A genus of only 1 species, best grown in moderate shade in a basket using an epiphytic compost which is kept moist.

HABIT: climbing epiphyte with a creeping rhizome bearing pseudobulbs at intervals of 2–5 cm.

PSEUDOBULBS: ovoid, 2–5 cm, with 1 or 2 leaves.

LEAVES: 10–20 cm, narrowly elliptic, shortly stalked, rolled when young.

RACEMES: with 3–7 flowers, longer than leaves.

FLOWERS: 2.5–5 cm in diameter.

SEPALS AND PETALS: 3.5–4 cm, oval, bluish mauve shading to white outside, flushed with pink inside.

LIP: 2.5–3 cm, clawed, kidney-shaped with a wavy margin, yellowish brown with a reddish or purplish centre and paler veins; callus yellowish or pinkish, divided above into finger-like projections.

COLUMN: 3-angled, with 2 wings and a long slender foot.

POLLINIA: 4, in 2 pairs.

1. A. **cyanea** Lindley

SYNONYM: *Aganisia cyanea* (Lindley) Reichenbach).

ILLUSTRATIONS: Botanical Magazine, 8678 (1916); Dunsterville & Garay, Venezuelan orchids illustrated **2**: 33 (1961); Bechtel et al., Manual of cultivated orchid species, 161 (1981); American Orchid Society Bulletin **53**: 43 (1984).

DISTRIBUTION: Brazil, Venezuela & Colombia.

HARDINESS: G2.

89. PROMENAEA Lindley

A Brazilian genus containing 15 species. Cultivation as for *Zygopetalum* (p. 313). The pans or baskets are best hung near the glass.

HABIT: epiphytic.

PSEUDOBULBS: more or less ovoid, compressed.

LEAVES: 1–3 per pseudobulb, pleated, often rather glaucous, folded when young.

SCAPES: produced from the bases of the pseudobulbs.

SEPALS AND PETALS: spreading.

LIP: 3-lobed with usually narrow, erect lateral lobes; disc with a transverse callus.

COLUMN: with a short foot to which the lip is jointed.

POLLINIA: 4, sessile on a broad, rounded viscidium.

Key to species

1. Sepals and petals spotted or marked

1.A. Spots dark purple, coalescing into bars; callus of lip with a 2-lobed crest **4. stapelioides**

1.B. Spots reddish purple, distinct, scarcely coalescing into bars; callus with a 3-lobed crest **5. lentiginosa**

2. Sepals and petals pale green or pale brown **3. microptera**

3. Lip spotted with red towards the base **1. xanthina**

4. Lip whitish, spotted or barred with reddish purple **2. rollissonii**

1. P. xanthina (Lindley) Lindley

SYNONYM: *P. citrina* Don.

ILLUSTRATIONS: Botanical Magazine n.s., 697 (1975); Pabst & Dungs, Orchidaceae Brasilienses **2**: 214 (1977); American Orchid Society Bulletin **56**: 1163 (1987).

PSEUDOBULBS: to 2 cm.

LEAVES: to 7 cm.

SCAPE: usually with 1 flower, to 10 cm.

FLOWERS: *c.* 3.5 cm in diameter, fragrant.

SEPALS AND PETALS: bright yellow, petals slightly shorter than sepals.

LIP: bright yellow spotted with red on the lateral lobes and sometimes on the base of the central lobe which is deflexed and more or less circular; callus 3-lobed, the central lobe with a fleshy prominence and a toothed apex.

DISTRIBUTION: Brazil.

HARDINESS: G2.

FLOWERING: summer.

P. × crawshayana Anon. is a hybrid between *P. xanthina* and *P. stapelioides*. It has pale yellow flowers spotted with reddish brown, especially so on the petals and lip. The central lobe of the lip is ovate and with a small point at the apex. The callus has a tubercled crest.

2. P. rollissonii (Lindley) Lindley

ILLUSTRATIONS: Edwards's Botanical Register **24**: t. 40 (1838); Pabst & Dungs, Orchidaceae Brasilienses **2**: 276 (1977).

PSEUDOBULBS: to 2.5 cm.

LEAVES: 7.5–10 cm.

SCAPE: usually with 1 flower, arching or somewhat hanging.

FLOWERS: 3–5 cm in diameter.

SEPALS AND PETALS: pale yellow, petals shorter and broader.

LIP: whitish, spotted or barred with reddish purple, markings sometimes confined to the lateral lobes, central lobe slightly broader than long, and with a small point.

DISTRIBUTION: Brazil.

HARDINESS: G2.

FLOWERING: summer.

3. P. microptera Reichenbach

ILLUSTRATIONS: Botanical Magazine, 8631 (1915); Pabst & Dungs, Orchidaceae Brasilienses **2**: 276 (1977).

PSEUDOBULBS: 1.5–2.5 cm.

LEAVES: 5–9 cm.

SCAPES: with 1–3 flowers, 5–7 cm, arching.

FLOWERS: *c.* 5 cm in diameter.
SEPALS AND PETALS: pale green to pale brown.
LIP: white or yellowish white, central lobe elliptic-oblong with an acute, recurved apex, marked with purple; callus spotted with purple and with a tubercled crest.
DISTRIBUTION: Brazil.
HARDINESS: G2.
FLOWERING: summer.

4. **P. stapelioides** (Link & Otto) Lindley
ILLUSTRATIONS: Edwards's Botanical Register **25**: t. 17 (1839); Botanical Magazine, 3877 (1841); Pabst & Dungs, Orchidaceae Brasilienses **2**: 214, 276 (1977).
PSEUDOBULBS: to 2.5 cm.
LEAVES: 7.5–10 cm.
SCAPES: often with 2 flowers, usually longer than the leaves, hanging.
FLOWERS: 3–4 cm in diameter.
SEPALS AND PETALS: greenish yellow outside, deep green inside, barred and spotted with dark purple.
LIP: dark purple, central lobe ovate to circular, with a small point, somewhat concave, paler on the margin; callus with a 2-lobed crest and a tooth pointing to the base of the lip.
DISTRIBUTION: Brazil.
HARDINESS: G2.
FLOWERING: summer–autumn.

5. **P. lentiginosa** (Lindley) Lindley
Like *P. stapelioides* except as below.
FLOWERS: the spotting is reddish purple and the spots are more distinct, less coalesced into bars.
LIP: similar in colour to the petals; the transverse crest of the callus has a central prominence which is 3-toothed.
DISTRIBUTION: Brazil.
HARDINESS: G2.
FLOWERING: summer–autumn.

90. CHONDRORHYNCHA Lindley

A genus of about 7 species growing in C and northwestern S America, formerly included in *Warscewiczella* Reichenbach. Cultivation as for *Zygopetalum* (p. 000).

HABIT: epiphytic; pseudobulbs absent.
LEAVES: stalked, jointed to the sheaths, oblong-lanceolate to strap-shaped, folded when young.
SCAPE: with 1 flower.
SEPALS: almost equal, elliptic to lanceolate, joined to the column foot to form an acute mentum.
PETALS: obovate-elliptic.

LIP: not 3-lobed; callus in centre of disc, with 3 teeth.
POLLINIA: 4.

Key to species

1. Petal margins with fine teeth — **1. flaveola**
2. Petal margins with a long fringe — **2. chestertonii**

1. **C. flaveola** (Linden & Reichenbach) Garay

SYNONYM: *C. fimbriata* (Linden & Reichenbach) Reichenbach.
ILLUSTRATIONS: Saunders' Refugium Botanicum **2**: t. 107 (1872); Dunsterville & Garay, Venezuelan orchids illustrated **2**: 53 (1961); American Orchid Society Bulletin **44**: 1077 (1975).
LEAVES: to 45 cm, oblong, acute.
SCAPES: erect or arching.
SEPALS: *c.* 2.5 cm, elliptic, long-pointed, sometimes with wavy margins, pale yellow.
PETALS: *c.* 3 cm, margins finely toothed, pale yellow.
LIP: 3–3.5 cm, fan-shaped, lobed, pale yellow, marked and spotted with brown or red at the base, the margin wavy and fringed; callus hairless, surrounded by a lobed wall and numerous small calluses.
DISTRIBUTION: Colombia, Venezuela, Ecuador & Peru.
HARDINESS: G2.
FLOWERING: spring–autumn.

2. **C. chestertonii** Reichenbach

ILLUSTRATIONS: Lindenia **9**: t. 405 (1893); Kupper & Linsenmaier, Orchids, 87 (1961).
Very like *C. flaveola* except as below.
SEPALS AND PETALS: narrower and the petals have a long fringe on the margins.
LIP: callus with 3 teeth at the front.
DISTRIBUTION: Colombia.
FLOWERING: summer.

91. COCHLEANTHES Rafinesque

A genus of 14 species occurring in tropical America. Cultivation as for *Zygopetalum* (p. 313).

HABIT: epiphytic; pseudobulbs absent.
STEMS: short, covered by overlapping leaf-sheaths.
LEAVES: in 2 ranks, borne in a fan, membranous, jointed to basal sheaths, folded when young.
SCAPE: shorter than leaves, with 1 flower.
SEPALS AND PETALS: usually spreading, the lateral sepals attached to the column foot.
LIP: hinged to column foot, 3-lobed or not, often concave; callus fleshy, with ridges radiating from the base.

COLUMN: cylindric, with a short foot.
POLLINIA: 4, on a broad viscidium.

Key to species

1. Lip notched at apex **2. amazonica**
2. Callus yellow with ridges of unequal length **1. discolor**
3. Lip to 2.2 cm with an entire margin; callus oblong, with 5 ridges **3. wailesiana**
4. Lip more than 2.5 cm, with irregularly toothed or scalloped margins; callus semicircular, with more than 11 ridges **4. flabelliformis**

1. C. **discolor** (Lindley) Schultes & Garay

SYNONYMS: *Warscewiczella discolor* (Lindley) Reichenbach; *Zygopetalum discolor* (Lindley) Reichenbach.
ILLUSTRATIONS: Botanical Magazine, 4830 (1855); Orchid Digest 33: 225 (1969); Sheehan & Sheehan, Orchid genera illustrated, 65 (1979); Bechtel et al., Manual of cultivated orchid species, 185 (1981).
LEAVES: to 30 cm, oblanceolate to strap-shaped.
SCAPE: to 15 cm.
FLOWERS: 5–7.5 cm in diameter.
SEPALS AND PETALS: 3–3.5 cm, white, sepals greenish or yellowish towards apex, petals often tinged with lilac towards the apex.
LIP: more or less 3-lobed, 2.5–4 cm, deep purple, often with a whitish margin, central lobe rounded, entire; callus yellow, with ridges of unequal length.
DISTRIBUTION: Cuba, Honduras, Costa Rica, Panama & Venezuela.
HARDINESS: G2.
FLOWERING: spring–autumn.

2. C. **amazonica** (Reichenbach & Warscewicz) Schultes & Garay

SYNONYMS: *Warscewiczella amazonica* Reichenbach & Warscewicz; *Zygopetalum amazonicum* (Reichenbach & Warscewicz) Reichenbach.
ILLUSTRATION: Lindenia 8: t. 337 (1892).
LEAVES: to 30 cm.
SCAPE: to 11 cm.
SEPALS AND PETALS: 2.5–3.5 cm, white, reflexed.
LIP: obscurely 3-lobed, *c.* 5 cm, white, veined with purple or reddish purple in the lower part, central lobe rounded, notched at the apex; callus with a truncate apex with 3–7 teeth.
DISTRIBUTION: Peru, Brazil.
HARDINESS: G2.

3. C. wailesiana (Lindley) Schultes & Garay

SYNONYM: *Warscewiczella wailesiana* (Lindley) Morren.

ILLUSTRATIONS: Cogniaux & Goossens, Dictionnaire Iconographique des Orchidées, Warscewiczella, t. 2 (1897); Orchid Digest **33**: 225 (1969); Pabst & Dungs, Orchidaceae Brasilienses **2**: 232, 287 (1977).

LEAVES: 15–20 cm.

FLOWERS: *c.* 5 cm in diameter, scented.

SEPALS AND PETALS: white or cream.

LIP: to 2.2 cm, roundish, white, stained and striped with violet towards the centre, margin entire; callus oblong, white striped with purple, with 5 ridges.

DISTRIBUTION: Brazil.

HARDINESS: G2.

4. C. flabelliformis (Swartz) Schultes Garay

SYNONYMS: *Warscewiczella flabelliformis* (Swartz) Cogniaux; *W. cochlearis* (Lindley) Reichenbach.

ILLUSTRATIONS: Dunsterville & Garay, Venezuelan orchids illustrated **2**: 77, 241 (1961); Pabst & Dungs, Orchidaceae Brasilienses **2**: 232, 286 (1977); Bechtel et al., Manual of cultivated orchid species, 92 (1981).

LEAVES: to 30 cm, narrowly oblong to oblanceolate.

SCAPE: 5–15 cm.

FLOWERS: 5–6 cm in diameter, fragrant.

SEPALS AND PETALS: 2.5–3.5 cm, white or greenish white.

LIP: 2.5–3 cm, wider than long, broadly rounded, white marked with purple and with purple veins, margin irregularly toothed or scalloped; callus white or violet with purple veins, semicircular with more than 11 ridges.

DISTRIBUTION: West Indies, Venezuela, Brazil.

HARDINESS: G2.

FLOWERING: spring–autumn.

92. PESCATORIA Reichenbach

A genus of about 17 species in southern C America and western tropical S America. The name is often misspelled 'Pescatorea'. Cultivation as for *Zgyopetalum* (p. 313).

Literature: Fowlie, J.A., A key and annotated check list to the genus Pescatorea, *Orchid Digest* **32**: 86–91 (1968).

HABIT: epiphytic, without pseudobulbs.

LEAVES: lanceolate, pleated, arranged in 2 ranks, folded when young.

SCAPES: arising from the axils of basal bracts, each with 1 flower.

SEPALS AND PETALS: fleshy, rather concave, lateral sepals joined at the base and inserted on the column foot.

LIP: 3-lobed, contracted at the base into a claw which merges with the column foot; central lobe convex; callus semicircular, ribbed.

COLUMN: with a short basal foot.

POLLINIA: 4 in 2 pairs on a short, linear stipe.

Synopsis of characters

Sepals and petals. Striped: **5**.
Lip. Bearing warts and bristles: **5**; bearing warts: **1**; without warts or bristles: **2–4**.
Callus. Yellow marked with brown: **1,2**; reddish brown: **5**; purple or reddish purple: **3**; white, purple in front: **4**.

Key to species

1. Sepals and petals striped; lip bearing bristles — **5. lehmannii**
2. Lip purple
2.A. Callus purple or reddish purple — **3. dayana**
2.B. Callus white, purplish in front — **4. wallisii**
3. Callus purple or reddish purple — **3. dayana**
4. Flowers 6–8 cm in diameter; lip bearing warts — **1. cerina**
5. Flowers 5–6 cm in diameter; lip without warts — **2. lamellosa**

1. P. cerina (Lindley) Reichenbach

SYNONYM: *Zygopetalum cerinum* (Lindley) Reichenbach.

ILLUSTRATIONS: Botanical Magazine, 5598 (1866); Flore des Serres **17**: t. 1815 (1868); Orchid Digest **32**: 87 (1968); Bechtel et al., Manual of cultivated orchid species, 258 (1981).

LEAVES: 20–60 cm.

SCAPE: 4–10 cm.

FLOWERS: 6–8 cm in diameter.

SEPALS: white or cream, the laterals each with a greenish yellow blotch near the base.

PETALS: similar to sepals.

LIP: 2–3 cm, cream to yellow, central lobe oval, warty on the upper surface; callus semicircular, ribbed, yellowish orange striped with reddish brown.

COLUMN: white, often with a purple or reddish blotch near the base.

DISTRIBUTION: Costa Rica, Nicaragua, Panama.

HARDINESS: G2.

2. P. lamellosa Reichenbach

ILLUSTRATIONS: Botanical Magazine, 6240 (1876); Die Orchidee 33 (**4**): 142 (1982).

Like *P. cerina* except as below.

FLOWERS: slightly smaller (5–6 cm in diameter).

SEPALS AND PETALS: white to yellowish green.

LIP: pale yellow, the central lobe more or less round and without warts; callus large, yellowish orange striped with brown and sometimes with dark purple.
DISTRIBUTION: Colombia.
FLOWERING: summer.

3. **P. dayana** Reichenbach
LEAVES: 20–40 cm.
SCAPE: to 6 cm.
FLOWERS: 7–8 cm in diameter.
SEPALS AND PETALS: white with green tips.
LIP: white flushed with purple, with purple lines extending from the base of the purple or reddish purple callus.
COLUMN: yellow with a reddish band near the base.
DISTRIBUTION: Panama, Colombia.
HARDINESS: G2.
FLOWERING: autumn–winter.
Three varieties are cultivated; some authors recognise them as subspecies, but it appears that they are no more than horticulturally desirable colour forms

Var. **candidula** Reichenbach
ILLUSTRATION: Orchid Digest **32**: 87 (1968).
SEPALS AND PETALS: completely white.

Var. **rhodacra** Reichenbach
ILLUSTRATIONS: Botanical Magazine, 6214 (1876); Orchid Digest **32**: 87 (1968).
SEPALS AND PETALS: with reddish tips.

Var. **splendens** Reichenbach
SEPALS AND PETALS: with purple tips.
LIP: deep purple.

4. **P. wallisii** Reichenbach
ILLUSTRATIONS: Flore des Serres **18**: t. 1828 (1869); Brooklyn Botanic Garden Record **23**(2), Handbook on orchids, 40–1 (1967); Orchid Digest **32**: 87 (1968); Orchid Review **92**: 70 (1984).
FLOWERS: 7–8 cm in diameter.
SEPALS AND PETALS: white tipped with bluish violet.
LIP: deep violet, central lobe with a white margin and notched apex; callus white, ribbed, purplish in front.
COLUMN: white, pale violet below, hairy in front.
DISTRIBUTION: Ecuador.
HARDINESS: G2.

5. **P. lehmannii** Reichenbach
ILLUSTRATIONS: Cogniaux & Goossens, Dictionnaire Iconographique des Orchidées, Pescatorea, t. 2 (1898); Orchid Digest **32**: 90 (1968); Die Orchidee **33** (4): 140 (1982).

LEAVES: 30–45 cm.
SCAPE: to 15 cm.
FLOWERS: 6–9 cm in diameter.
SEPALS AND PETALS: white, yellowish towards the tips, striped longitudinally with reddish purple.
LIP: purple, central lobe oblong, apex notched, covered with warts and bristle-like whitish or purplish papillae; callus reddish brown, ribbed.
COLUMN: purple.
DISTRIBUTION: Colombia, Ecuador.
HARDINESS: G2.
FLOWERING: spring–autumn.

93. BOLLEA Reichenbach

A genus of about 6 species from western tropical S America. Cultivation as for *Zygopetalum* (p. 313).

HABIT: epiphytic, without pseudobulbs.
LEAVES: several, lanceolate, in 2 ranks, with prominent veins beneath, folded when young.
INFLORESCENCE: lateral; scapes shorter than the leaves, each with 1 flower.
SEPALS AND PETALS: free, spreading, somewhat concave, the lateral sepals joined to the base of column which is produced into a short foot.
LIP: with a narrow claw, movably attached to the column base to form a short mentum, the limb broad, entire; callus large, ribbed, partly concealing the column.
COLUMN: short and wide with a short foot, deeply hollowed in front, hooded.
POLLINIA: 4, waxy.

Key to species

1. Lip violet **1. coelestis**
2. Flowers to 8 cm in diameter; lip-callus yellow **2. lalindei**
3. Flowers *c.* 10 cm in diameter; lip-callus pinkish **3. patinii**

1. B. coelestis (Reichenbach) Reichenbach

SYNONYM: *Zygopetalum coeleste* Reichenbach.
ILLUSTRATION: Botanical Magazine, 6458 (1879); Ospina & Dressler, Orquideas de las Americas, t. 101 (1974).
LEAVES: 30–60 cm.
FLOWERS: 7–10 cm in diameter, fragrant.
SEPALS AND PETALS: pale blue to deep violet blue, darker above the middle, yellowish at the tips, margin wavy.
LIP: heart-shaped at the base, with recurved tip and margin, dark violet, yellowish towards the base; callus semicircular, pale brownish or yellow, with many ribs.
DISTRIBUTION: Colombia.

HARDINESS: G2.
FLOWERING: summer.

2. B. lalindei (Reichenbach) Reichenbach
SYNONYM: *Zygopetalum lalindei* Reichenbach.
ILLUSTRATION: Botanical Magazine, 6331 (1877).
FLOWERS: 6–8 cm in diameter, fragrant.
SEPALS AND PETALS: pale pink or purple, darker above the middle, upper sepal yellowish green at the tip, lateral sepals with yellowish lower margin; petals often with white margins.
LIP: ovate, narrowly heart-shaped at base, with recurved tip and margin, bright yellow; callus semicircular, yellow, with many ribs.
DISTRIBUTION: Colombia.
HARDINESS: G2.
FLOWERING: summer–autumn.

3. B. patinii Reichenbach.
Like *B. lalindei* except as below.
FLOWERS: larger (*c*. 10 cm in diameter).
PETALS: without white margins.
LIP: with a flesh-coloured callus.
DISTRIBUTION: Colombia.
FLOWERING: summer.

94. HUNTLEYA Lindley

A genus from C and northern S America. The number of species is doubtful, some authors maintaining that there are about 10, while others recognise only 4. Only 1 is generally grown. Cultivation as for *Zygopetalum* (p. 313).

HABIT: epiphytic, without pseudobulbs.
LEAVES: narrowly lanceolate, arranged in a fan, folded when young.
SCAPES: arising from the leaf axils, shorter than the leaves, each with 1 flower.
SEPALS AND PETALS: more or less similar, spreading, glossy, the lateral sepals attached to the short column foot.
LIP: 3-lobed with small lateral lobes, attached to the column foot by a long, narrow claw, central lobe curving downwards; callus semicircular, densely fringed.
COLUMN: with broad, apical, lateral wings.
POLLINIA: 4 in 2 pairs in which the members are unequal, on a short, reversed-triangular stipe.

1. H. meleagris Lindley
SYNONYM: *H. burtii* (Endres & Reichenbach) Pfitzer.
ILLUSTRATIONS: Dunsterville & Garay, Venezuelan orchids illustrated 3: 147 (1965); Hawkes, Encyclopaedia of cul-

tivated orchids, 249 (1965); Bechtel et al., Manual of cultivated orchid species, 217 (1981).

LEAVES: 25–40 cm.

FLOWERS: 7.5–10 cm in diameter, fragrant.

SEPALS AND PETALS: lanceolate to ovate, pointed, whitish at the base shading into yellow, marked and flushed with reddish brown, usually with a median yellow cross-bar, and often appearing chequered.

LIP: 2.5–3.5 cm, white at the base, shading upwards to reddish brown, with a small point; callus white or yellowish.

COLUMN: 1.5–2 cm, creamy green, often striped with reddish brown, with a semicircular, down-pointing wing on each side and a hood covering the anther cap.

DISTRIBUTION: C America, Northwestern S America, Trinidad.

HARDINESS: G2.

FLOWERING: spring–autumn.

95. **ERIOPSIS** Lindley

A genus of 6 species from southern C America and tropical S America. They should be grown in an epiphytic compost, with frequent watering when in growth. Light shading should be provided when young growth is present, but should be removed when the growth is mature; propagation by division.

HABIT: epiphytic or terrestrial.

PSEUDOBULBS: elongate-cylindric or ovoid-oblong.

LEAVES: 2–4 per pseudobulb, pleated, strongly veined, rolled when young.

RACEME: produced from the base of a pseudobulb, with many flowers.

SEPALS AND PETALS: more or less equal, sepals free or the laterals joined to the column foot to form a short mentum.

LIP: 3-lobed, lateral lobes erect, larger than the central lobe; callus composed of 2 or more fleshy plates.

COLUMN: curved, slightly swollen above.

POLLINIA: 2 or 4, stipe very short, almost non-existent.

Key to species

1. Callus of 2 erect fleshy plates with 2 separate lumps in front of each plate **1. biloba**
2. Callus of 2 erect, plate-like horns only **2. sceptrum**

1. E. **biloba** Lindley

SYNONYMS: *E. rutidobulbon* Hooker; *E. schomburgkii* (Reichenbach) Reichenbach; *E. wercklei* Schlechter.

ILLUSTRATIONS: Botanical Magazine n.s., 611 (1972); Pabst & Dungs, Orchidaceae Brasilienses 2: 214, 276 (1977).

HABIT: epiphytic or terrestrial.

PSEUDOBULBS: ranging from elongate-cylindric (to 45 × 3 cm) to ovoid-oblong (to 14 × 8 cm), surface smooth or finely wrinkled, greenish, brown or dark purplish brown, with 2–4 leaves at the apex.

LEAVES: 20–50 cm, narrowly lanceolate to narrowly ovate.

SCAPE: 30–110 cm, erect to arching, raceme with 20–35 flowers.

FLOWERS: 2.5–5 cm in diameter.

SEPALS AND PETALS: 1.2–2.5 cm, yellow to orange-yellow or brownish yellow, reddish purple or brownish on margins or flushed with red to varying degrees.

LIP: 1.2–2.5 cm, lateral lobes yellowish with reddish purple spots, central lobe white with reddish purple spots, notched at the apex; callus of 2 erect fleshy plates with 2 separate fleshy lumps in front of each.

COLUMN: *c.* 1 cm, greenish yellow.

DISTRIBUTION: Tropical America from Costa Rica to Peru & Bolivia.

HARDINESS: G2.

FLOWERING: summer–autumn.

Extremely variable in flower colour, lip shape and colour of pseudobulbs.

2. **E. sceptrum** Reichenbach & Warscewicz

SYNONYMS: *E. sprucei* Reichenbach; *E. helenae* Kränzlin.

ILLUSTRATIONS: Botanical Magazine, 8462 (1912); Pabst & Dungs, Orchidaceae Brasilienses **2**: 214, 277 (1977).

Like *E. biloba* except as below.

HABIT: epiphytic, but generally smaller.

FLOWERS: *c.* 3.5 cm in diameter

SEPALS AND PETALS: 1.3–1.6 cm.

LIP: *c.* 1 cm; the callus consisting of 2 erect plate-like horns which are cream in colour.

DISTRIBUTION: Venezuela, Peru, Brazil.

FLOWERING: summer.

96. GALEOTTIA Richard

A genus of 9 species, most of which have at some time been included in *Zygopetalum* or *Batemannia* and which will be found in much recent literature under the incorrect name *Mendoncella* Hawkes. They are native to C and tropical S America, and are best grown in small pans, using an epiphytic compost with very free drainage to allow liberal watering during the growing season.

Literature: Garay, L.A., El complejo Zygopetalum, *Orquideologia* 8: 15–34 (1973).

HABIT: epiphytic or terrestrial.

LEAVES: 2–3 per pseudobulb, pleated, rolled when young.

RACEME: with few flowers, arising laterally from the base of the new growth.

SEPALS AND PETALS: spreading, lateral sepals more or less swollen at the base and joined to the column foot.

LIP: 3-lobed; callus formed from a series of longitudinal ribs or plates.

COLUMN: with distinct wings.

POLLINIA: 4 in 2 pairs on a short, fleshy stipe.

Key to species

1. Flower *c.* 5 cm in diameter; central lobe of lip with entire margins; wings of column entire **3. burkei**
2. Central lobe of lip hairy or warty between the veins; callus violet-streaked **1. fimbriata**
3. Central lobe of lip hairless or sparsely hairy; callus orange **2. grandiflora**

1. **G. fimbriata** Linden & Reichenbach

SYNONYM: *Mendoncella fimbriata* (Linden & Reichenbach) Garay.

ILLUSTRATIONS: Gardeners' Chronicle 1856: 660; American Orchid Society Bulletin **48**: 1223 (1979), **50**: 37 (1981).

HABIT: epiphytic.

PSEUDOBULBS: 5–6.5 cm, ovoid, with 2 leaves.

LEAVES: 15–30 cm.

RACEME: with 2–5 flowers.

FLOWERS: *c.* 8 cm in diameter.

SEPALS AND PETALS: greenish or brownish yellow striped with reddish brown, acuminate.

LIP: white, striped longitudinally with violet, lateral lobes erect, fringed, central lobe ovate, recurved, fringed, acuminate, hairy or warty between the veins; callus violet-streaked, formed of a number of plates arranged like a fan.

COLUMN: whitish yellow, wings toothed.

DISTRIBUTION: Colombia, Venezuela.

HARDINESS: G2.

2. **G. grandiflora** Richard.

SYNONYM: *Mendoncella grandiflora* (Richard) Hawkes.

ILLUSTRATIONS: Botanical Magazine, 5567 (1866); Annals of the Missouri Botanical Garden **36**: 82 (1949).

Like *M. fimbriata* except as below.

FLOWERS: 7–8 cm in diameter.

LIP: central lobe hairless or sparsely hairy and less sharply pointed at the apex; callus orange.

DISTRIBUTION: C America, Colombia.

HARDINESS: G2.

FLOWERING: spring–summer.

3. G. **burkei** (Reichenbach) Dressler & Christensen

SYNONYM: *Zygopetalum burkei* Reichenbach; *Mendoncella burkei* (Riechenbach) Garay.

ILLUSTRATIONS: Dunsterville & Garay, Venezuelan orchids illustrated 3: 333 (1965); American Orchid Society Bulletin 48: 1223 (1979).

HABIT: terrestrial.

PSEUDOBULBS: 5–10 cm, narrowly oblong, with 2–3 leaves.

LEAVES: 20–45 cm.

RACEME: with 5–6 flowers.

FLOWERS: *c.* 5 cm in diameter.

SEPALS AND PETALS: light green or light brown thickly striped or spotted with brown.

LIP: white, lateral lobes with entire margins, central lobe with recurved sides, margin entire; callus reddish purple, ribbed.

COLUMN: yellow or greenish, streaked in front with purple, wings entire.

DISTRIBUTION: Guyana, Venezuela.

HARDINESS: G2.

97. SCUTICARIA Lindley

A genus of 5 species from tropical S America. They are best grown in epiphytic compost on rafts or blocks of tree fern suspended near the glass. They require ample moisture when growing, but should be kept dry when growth is completed, but not so much as to allow the foliage to shrivel.

HABIT: epiphytic.

PSEUDOBULBS: small, somewhat stem-like, borne on a stout, branching rhizome.

LEAVES: narrowly cylindric with a longitudinal groove, continuous with the pseudobulbs each of which bears 1 leaf.

RACEME: with 1–several flowers, arising laterally from a pseudobulb.

SEPALS AND PETALS: spreading, similar, the lateral sepals inserted on the column foot to form a mentum.

LIP: 3-lobed, joined to the column foot, lateral lobes erect, central lobe concave.

POLLINIA: 2.

Key to species:

1. Leaves 60–150 cm; raceme with 1–3 flowers; central lobe of lip hairless — **1. steelei**
2. Leaves hanging or somewhat drooping — **2. hadwenii**
3. Leaves erect — **3. strictrifolia**

1. S. **steelei** (Hooker) Lindley

ILLUSTRATIONS: Pabst & Dungs, Orchidaceae Brasilienses **2**:

240, 296 (1977); American Orchid Society Bulletin **49**: 31 (1980).

LEAVES: hanging, 60–150 cm.

RACEME: usually hanging, with 1–3 flowers.

FLOWERS: 6–10 cm in diameter, fragrant.

SEPALS AND PETALS: yellow, blotched with reddish brown.

LIP: 3–4 cm, whitish to pale yellow marked with brownish pink especially on the lateral lobes, central lobe 2-lobed at apex, hairless; callus yellowish orange with 3–5 apical teeth, hairy.

COLUMN: whitish spotted with dark pink.

DISTRIBUTION: Brazil, Guyana, Colombia & Venezuela.

HARDINESS: G2.

FLOWERING: throughout the year.

2. S. **hadwenii** (Lindley) Hooker

ILLUSTRATIONS: Botanical Magazine, 4629 (1852); Pabst & Dungs, Orchidaceae Brasilienses **2**: 238 (1977).

LEAVES: hanging or sometimes more or less erect, 30–45 cm.

RACEME: with 1 flower, hanging or arching.

FLOWERS: 6–10 cm in diameter, fragrant.

SEPALS AND PETALS: yellowish green blotched with reddish brown.

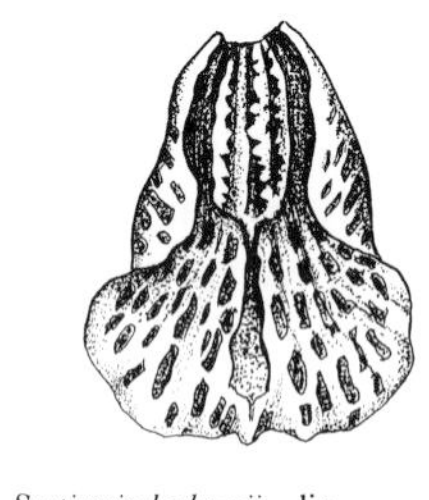

Scuticaria hadwenii – lip.

LIP: pale yellow irregularly spotted with pinkish red, central lobe notched at apex, hairy; callus with 3 apical teeth.

COLUMN: whitish yellow flushed and spotted with red.

DISTRIBUTION: Brazil, Guyana.

HARDINESS: G2.

FLOWERING: spring–autumn.

3. S. **strictifolia** Hoehne

ILLUSTRATION: Orchid Digest **41**: 235 (1977).

Very like *S. hadwenii*, except as below.

LEAVES: erect.

DISTRIBUTION: Brazil.

Plants named as *S. hadwenii* growing in bright light may have the ability to produce erect leaves, and so this may not be a distinct species.

98. MAXILLARIA Ruiz & Pavon

A genus of about 250 species from tropical and subtropical America. They should be grown in epiphytic compost; those with long rhizomes are best accommodated on rafts or blocks of tree fern. All should be kept moist throughout the year and should be shaded from bright sunlight. Maxillarias are prone to black spot on the foliage; this can be controlled by adequate ventilation and heating.

Literature: Ashwood, J., Maxillaria, a forgotten genus *American Orchid Society Bulletin* **54**: 25–30 (1985).

HABIT: epiphytic or terrestrial, with long or short rhizomes clothed in often papery sheaths.

PSEUDOBULBS: usually laterally compressed and with 1–2 leaves.

LEAVES: thin to leathery, with a prominent central vein, folded when young.

SCAPES: produced in the axils of sheaths on the rhizome, solitary or in bundles, each bearing clasping, usually overlapping bracts and 1 flower.

SEPALS: free, the laterals usually oblique and joined to the column foot to form a mentum.

PETALS: similar but smaller, usually joined to the upper sepal.

LIP: hinged to or continuous with the column foot, often erect and parallel to the column, entire or 3-lobed, usually shorter than the sepals.

COLUMN: erect, somewhat curved, without wings.

POLLINIA: 4 in 2 pairs, often 1 large and 1 small in each pair.

Synopsis of characters

Pseudobulbs. Absent: **12**. Bearing 1 leaf: **1–11,13–19, 22–28**; bearing 2 leaves: **2,3,16,17,20–25**. Spaced at intervals along the rhizome: **2,6,8,10,11,16–20,22,23,24**; clustered: **1–7,9,13–17, 21–27**.

Leaves. Linear: **10,11,17–20,27**; relatively broader: **1–9,12–17, 21–26,28**.

Sepals. Linear: **1–6**; relatively broader: **7–28**. Length to 3 cm: **8,10,12,16–28**; length 3–8 cm: **1,3–7,9,13–17,20,21,23–25**; length 10–15 cm: **2**.

Sepals and petals. Striped: **13**.

Central lobe of lip. Basic colour white, cream, yellow or orange: **2–25,27,28**; basic colour red to maroon: **1,10,16–19,23–27**. Spotted: **5,7,10,12,14,18,19,23–25**; not spotted: **1–4,6–9,11, 13–28**.

Entire: **1,9,10,15–19**; 3-lobed: **2–9,11–14,18–28**.

Key to species

1. Sepals linear to narrowly strap-shaped, always 3 cm or more — *Go to Subkey* 1
2. Sepals entirely white or creamy, never marked with another colour — *Go to Subkey* 2
3. Leaves linear, grass-like — *Go to Subkey* 3
4. Lateral sepals directed upwards towards the upper sepal — *Go to Subkey* 4
5. Lateral sepals directed downwards or spreading horizontally — *Go to Subkey* 5

Subkey 1. *Sepals linear to narrowly strap-shaped, always* 3 *cm or more.*

1. Sepals maroon, yellow at base, apex maroon-orange **1. nigrescens**
2. Sepals at least 10 cm, twisted **2. fractiflexa**
3. Sepals white on the inner surface **7. venusta**
4. Central lobe of lip white, recurved **3. setigera**
5. Central lobe of lip obtuse at apex, margins wavy **4. luteo-alba**
6. Lip with some purplish spots; pseudobulbs 2.5–5 cm **5. lepidota**
7. Lip without spots; pseudobulbs 5–8 cm **6. ochroleuca**

Subkey 2. Sepals ovate to oblong or broadly lanceolate, if narrower then less than 3 cm long, entirely white or creamy, never marked with another colour.

1. Flowers *c.* 2 cm in diameter; sepals 1.8–2.5 cm

1.A. Pseudobulbs borne at intervals along the rhizome; lip *c.* 1.2 cm with a downy central lobe **8. alba**

1.B. Pseudobulbs clustered; lip 3.5–5 cm, central lobe hairless **27. densa**

2. Sepals lanceolate **7. venusta**
3. Sepals ovate-oblong to triangular-ovate **9. grandiflora**

Subkey 3. Sepals ovate to oblong or broadly lanceolate, if narrower then less than 3 cm long, variously coloured, if white or cream, then marked with a contrasting colour; leaves linear, grass-like.

1. Pseudobulbs with 2 leaves **20. chrysantha**
2. Lip with reddish brown or purple spots **10. tenuifolia**
3. Pseudobulbs clustered; lip less than 6 mm **27. densa**
4. Apex of central lobe of lip white **11. sanguinea**
5. Lip deep red with yellow apex **18. variabilis**
6. Lip basically entirely yellow, though with purple spots **19. curtipes**

Subkey 4. Sepals ovate to oblong or broadly lanceolate, if narrower then less than 3 cm long, variously coloured, if white or cream, then marked with a contrasting colour; lateral sepals directed upwards, towards the upper sepal; leaves not grass-like.

1. Plant hanging, leaves in a fan; pseudobulbs absent; sepals and petals pale greenish yellow **12. valenzuelana**
2. Sepals and petals bright pink to orange-red

2.A. Pseudobulbs 1–2 cm, borne at intervals along the rhizome **28. sophronitis**

2.B. Pseudobulbs 2.5–4 cm, clustered **26. coccinea**

3. Sepals and petals white with purple blotches at the base **15. sanderiana**
4. Sepals with longitudinal stripes **13. striata**
5. Sepals with spots towards the apex **14. fucata**

Subkey 5. Sepals ovate to oblong or broadly lanceolate, if narrower then less than 3 cm long, variously coloured, if white or cream then marked with a contrasting colour; lateral sepals directed downwards or spreading horizontally; leaves not grass-like.

1. Lip entire
1.A. Scapes up to 4 cm
1.A.a. Lip deep red with yellow apex **18. variabilis**
1.A.b. Lip basically entirely yellow, though with purple spots **19. curtipes**
1.B. Sepals and petals reddish yellow **16. elatior**
1.C. Sepals and petals white, yellow towards the tips, flushed and spotted with purple, or deep blue **17. elegantula**
2. Pseudobulbs with 2 leaves
2.A. Central lobe of lip acute
2.A.a. Sepals and petals not spotted **22. marginata**
2.A.b. Lip not hinged to column-foot **23. picta**
2.A.c. Sepals 2.5–5 cm **24. cucullata**
2.A.d. Sepals 1–2.5 cm **25. rufescens**
2.B. Flowers 2–3 cm in diameter **21. porphyrostele**
2.C. Flowers 4–4.5 cm in diameter **20. chrysantha**
3. Callus downy
3.A. Lip not hinged to column-foot **23. picta**
3.B. Sepals 2.5–5 cm **24. cucullata**
3.C. Sepals 1.5–2.5 cm **25. rufescens**
4. Central lobe of lip pale yellow with a dark purple margin **22. marginata**
5. Pseudobulbs clustered; scapes 4–6 cm **26. coccinea**
6. Pseudobulbs 1–2 cm; leaves 1–8 cm; central lobe of lip warty **28. sophronitis**
7. Pseudobulbs 1.5–6 cm; leaves 5–25 cm; central lobe of lip smooth
7.A. Lip deep red with yellow apex **18. variabilis**
7.B. Lip basically yellow, though with purple spots **19. curtipes**

1. **M. nigrescens** Lindley

ILLUSTRATIONS: Cogniaux & Goossens, Dictionnaire Iconographique des Orchidées, Maxillaria, t. 3 (1899); Dunsterville & Garay, Venezuelan orchids illustrated 3: 189 (1965); American Orchid Society Bulletin 43: 963 (1974).

HABIT: epiphytic.

PSEUDOBULBS: to 9 cm, clustered, ovoid, compressed, each with 1 leaf.

LEAVES: strap-shaped, 20–35 cm.

SCAPES: usually erect, to 12 cm.

FLOWERS: *c.* 10 cm in diameter.

SEPALS: narrowly strap-shaped, 4.5–6 cm, maroon, yellow at the base and maroon-orange at the apex.

PETALS: similar to sepals, but only 3.8–5.5 cm and usually darker.
LIP: entire, oblong, obtuse, 1.7–2 cm, deep maroon, flushed with brownish purple.
DISTRIBUTION: Colombia, Venezuela.
HARDINESS: G2.
FLOWERING: summer–autumn.

2. M. **fractiflexa** Reichenbach
ILLUSTRATION: Gardeners' Chronicle **31**: 359 (1902).
LEAVES: strap-shaped, to 50 cm.
SCAPES: erect.
FLOWERS: *c.* 15 cm in diameter.
SEPALS AND PETALS: linear, 10–15 cm, curved and twisted, yellowish tinged with brown and often purplish at the base and apex.
LIP: 3-lobed, white, part of the rounded lateral lobes and the central lobe red; callus white.
DISTRIBUTION: Ecuador.
FLOWERING: G2.

3. M. **setigera** Lindley
ILLUSTRATIONS: Botanical Magazine, 4434 (1849); Dunsterville & Garay, Venezuelan orchids illustrated 3: 192 (1965); American Orchid Society Bulletin **43**: 963 (1974).
HABIT: epiphytic.
PSEUDOBULBS: 3–6 cm, clustered, spherical, somewhat compressed, each with 1–2 leaves.
LEAVES: narrow, 12–30 cm.
SCAPES: 7–30 cm.
FLOWERS: fragrant, 10 cm or more across.
SEPALS: narrowly strap-shaped, to 6.5 cm, white at base, greenish yellow towards apex.
PETALS: similar to sepals, incurved, to 5 cm, with recurved margins.
LIP: 3-lobed, to 2.5 cm, lateral lobes white striped with purple, central lobe oblong, reflexed, white, margin toothed; callus yellow, hairy.
DISTRIBUTION: Colombia, Venezuela, Guyana & Brazil.
HARDINESS: G2.
FLOWERING: autumn.

4. M. **luteo-alba** Lindley
ILLUSTRATIONS: Cogniaux & Goossens, Dictionnaire Iconographique des Orchidées, Maxillaria, t. 1 (1899); American Orchid Society Bulletin **43**: 962 (1974); Bechtel et al., Manual of cultivated orchid species, 229 (1981).
HABIT: epiphytic.

PSEUDOBULBS: 3–5 cm, clustered, oblong-ovoid, compressed, each with 1 leaf.
LEAVES: linear-lanceolate, 25–45 cm.
SCAPES: erect, 9–15 cm.
FLOWERS: very fragrant, 10 cm or more across.
SEPALS: narrowly strap-shaped, 4–7.5 cm, whitish, or yellow at the base, becoming brownish orange towards the apex.
PETALS: shorter, similar in colour although sometimes with purplish streaks.
LIP: 3-lobed, to 2.5 cm, downy inside, cream or yellow, lateral lobes purple-streaked, central lobe ovate, obtuse with a white wavy margin; callus with 2 lateral ridges, downy towards base.
DISTRIBUTION: Panama, Colombia, Venezuela & Ecuador.
HARDINESS: G2.
FLOWERING: summer.

5. **M. lepidota** Lindley
ILLUSTRATIONS: Dunsterville & Garay, Venezuelan orchids illustrated **6**: 253 (1976); Bechtel et al., Manual of cultivated orchid species, 229 (1981).
HABIT; epiphytic.
PSEUDOBULBS: 2.5–5 cm, clustered, oblong, each with 1 leaf.
LEAVES: linear-lanceolate, 15–35 cm.
SCAPES: to 15 cm, often with the flowers hanging.
SEPALS: linear, 4–6 cm, yellow with red at the base and often brown at the tip.
PETALS: similar to sepals but shorter.
LIP: 3-lobed, *c.* 2 cm, deep cream, spotted with dark purple or maroon towards the apex, reflexed, mealy on the upper surface; callus grooved, hairy.
DISTRIBUTION: Colombia, Ecuador & Venezuela.
HARDINESS: G2.

6. **M. ochroleuca** Lindley
ILLUSTRATIONS: Pabst & Dungs, Orchidaceae Brasilienses **2**: 235, 291 (1977).
HABIT: epiphytic.
PSEUDOBULBS: 5–8 cm, compressed, each with 1 leaf.
LEAVES: 25–40 cm.
SCAPES: erect, 5–10 cm, many borne together in a bundle.
FLOWERS: fragrant.
SEPALS: 3–3.5 cm, linear, slenderly pointed, pale yellow.
PETALS: similar to sepals but shorter.
LIP: 3-lobed, pale yellow, central lobe oblong, obtuse, downy, deep yellow at apex.
DISTRIBUTION: Brazil.
HARDINESS: G2.

7. M. venusta Linden & Reichenbach

ILLUSTRATIONS: Botanical Magazine, 5296 (1862); Cogniaux & Goossens, Dictionnaire Iconographique des Orchidées, Maxillaria, t 6 (1902).

HABIT: epiphytic.

PSEUDOBULBS: 4–7.5 cm, clustered, ovoid, much compressed, each with 1 leaf.

LEAVES: strap-shaped, 30–38 cm.

SCAPES: 10–25 cm.

FLOWERS: very fragrant, 12–15 cm in diameter, hanging.

SEPALS: lanceolate, to 6.5 cm, white, usually with recurved margins.

PETALS: similar to sepals, to 4.5 cm.

LIP: 3-lobed, to 2 cm, lateral lobes oblong, pale yellow with red margins, central lobe ovate, obtuse, reflexed, pale yellow with red margin and sometimes with red spots, mealy; callus oblong, downy.

DISTRIBUTION: Colombia, Venezuela.

HARDINESS: G2.

FLOWERING: summer–autumn.

8. M. alba (Hooker) Lindley

ILLUSTRATIONS: Hoehne, Flora Brasilica 12 (7): t. 91 (1953); Dunsterville & Garay, Venezuelan orchids illustrated **2**: 197 (1961).

HABIT: epiphytic.

PSEUDOBULBS: 2.5–5 cm, borne at short intervals along the rhizome, ovoid to narrowly ovoid, strongly compressed, each with 1 leaf.

LEAVES: strap-shaped, 25–40 cm.

SCAPES: 2.5–4 cm.

FLOWERS: slightly fragrant, *c.* 2 cm across.

SEPALS: broadly lanceolate, 1.8–2.5 cm, creamy white.

PETALS: similar to sepals, 1.6–1.8 cm.

LIP: 3-lobed towards the apex, hinged to the column foot, *c.* 1.2 cm, pale to bright yellow with an ovate or elliptic central lobe which is downy; callus strap-shaped, downy.

DISTRIBUTION: most of C America & northern S America.

HARDINESS: G2.

FLOWERING: spring.

9. M. grandiflora (Humboldt, Bonpland & Kunth) Lindley

ILLUSTRATIONS: Dunsterville & Garay, Venezuelan orchids illustrated **5**: 185 (1966); Die Orchidee **23**: 92 (1972); American Orchid Society Bulletin **54**: 30 (1984); Orchid Review **92**: 43 (1984).

HABIT: epiphytic.

PSEUDOBULBS: 3.5–7 cm, clustered, ovoid, compressed, each with 1 leaf.

LEAVES: strap-shaped, 25–50 cm.

SCAPES: erect but bent at apex, 10–25 cm.

FLOWERS: slightly scented, *c.* 10 cm across, drooping.

SEPALS: ovate-oblong to triangular-ovate, 3.5–5.5 cm, white, sometimes with margins rolled under; lateral sepals with reflexed tips.

PETALS: shorter and narrower than sepals, white, sometimes with pink streaks on the basal half, tips reflexed.

LIP: entire or very slightly 3-lobed, 2.5–3 cm, lateral lobes (if present) deep pink or purple, streaked paler, central lobe white or yellowish sometimes with a purple margin, wavy, reflexed, mealy; callus mealy.

DISTRIBUTION: Northwestern S America.

HARDINESS: G2.

FLOWERING: spring–summer.

10. **M. tenuifolia** Lindley

ILLUSTRATIONS: Shuttleworth et al., Orchids, 110 (1970); Die Orchidee **23**: 92 (1972); Bechtel et al., Manual of cultivated orchid species, 230 (1981); American Orchid Society Bulletin **54**: 25 (1984).

HABIT: terrestrial or epiphytic.

PSEUDOBULBS: 2–6 cm, borne at intervals of 2–4 cm along the rhizome, ovoid, somewhat compressed, each with 1 leaf.

LEAVES: linear, 12–50 cm.

SCAPES: 2–6 cm.

FLOWERS: coconut-scented, 4–5 cm in diameter.

SEPALS: narrowly elliptic, 1.7–2.8 cm, dark red, heavily mottled with red or yellow.

PETALS: similar to sepals, 1.5–2.4 cm, directed forwards.

LIP: 1.5–2.5 cm, entire, oblong, reflexed, dark red at base, yellow or whitish towards apex, spotted with purple or reddish brown; callus downy, dark purple.

DISTRIBUTION: Mexico to Costa Rica.

HARDINESS: G2.

FLOWERING: summer–autumn.

11. **M. sanguinea** Rolfe

ILLUSTRATION: Hartmann, Introduction to the cultivation of orchids, 89 (1971).

PSEUDOBULBS: to 3 cm, borne at intervals along the rhizome, ovoid, compressed, each with 1 leaf.

LEAVES: linear, to 40 cm.

SCAPES: to 3 cm.

SEPALS: elliptic, to 2.5 cm, dark red or brownish, mottled with red or yellow.

PETALS: similar to sepals, to 2 cm.

LIP: scarcely 3-lobed, to 1.3 cm, elliptic, dark red, pale yellow at base, white at apex; callus dark red.

DISTRIBUTION: Costa Rica; ?Colombia.
HARDINESS: G2.
FLOWERING: summer.

12. M. valenzuelana (Richard) Nash
ILLUSTRATIONS: Shuttleworth et al., Orchids, 111 (1970); Bechtel et al, Manual of cultivated orchid species, 231 (1981).
HABIT: hanging epiphyte without pseudobulbs.
LEAVES: growing in a fan, 8–20 cm, lanceolate.
SCAPES: to 5 cm, solitary or in bundles in the leaf axils.
SEPALS: ovate, 1–1.4 cm, pale greenish yellow.
PETALS: similar to sepals, 8–10 mm.
LIP: scarcely 3-lobed, *c.* 1 cm, pale brownish yellow with purple spots; callus in 3 parts, glandular-downy.
DISTRIBUTION: C America, northern S America, West Indies.
HARDINESS: G2.
FLOWERING: summer.

13. M. striata Rolfe
ILLUSTRATIONS: Cogniaux & Goossens, Dictionnaire Iconographique des Orchidées, Maxillaria, t. 4 (1899); Lindenia 9: t. 398 (1883–84).
PSEUDOBULBS: 4.5–8 cm, clustered, ovoid, compressed, each with 1 leaf.
LEAVES: broadly elliptic, to 25 cm.
SCAPES: more or less erect or arching, to 30 cm.
FLOWERS: *c.* 12 cm in diameter.
SEPALS: greenish yellow with reddish purple stripes, 4.5–7 cm, upper sepal oblong-lanceolate, lateral sepals triangular-lanceolate, forming a mentum with the column foot.
PETALS: similar in colour to sepals, shorter, lanceolate, often with wavy margins.
LIP: erect, recurved, slightly 3-lobed towards apex, lateral lobes more or less obovate, white striped with reddish purple, central lobe ovate-lanceolate, white; callus linear-oblong with many grooves.
DISTRIBUTION: Peru, Ecuador.
HARDINESS: G2.
FLOWERING: summer–autumn.

14. M. fucata Reichenbach
ILLUSTRATIONS: Botanical Magazine, 9376 (1934); Shuttleworth et al., Orchids, 109 (1970).
PSEUDOBULBS: 3.5–6 cm, clustered, narrowly ovoid, compressed, each with 1 leaf.
LEAVES: narrowly oblong, 30–40 cm.
SCAPES: erect, usually numerous, to 22 cm.
FLOWERS: *c.* 6 cm across, fragrant.

SEPALS: 3–3.5 cm, ovate, reflexed at tips, white at the base, purplish in the middle and yellow at the apex, with reddish brown spots towards apex.

PETALS: *c.* 2.5 cm, lanceolate-ovate, similar in colour to sepals but sometimes without spots.

LIP: 3-lobed, reflexed, lateral lobes yellowish, striped and edged with brown, central lobe ovate-oblong, notched at apex, yellowish, striped, spotted or suffused with brownish red; callus yellow, strap-shaped, grooved and hairless.

DISTRIBUTION: ?Ecuador.

HARDINESS: G2.

FLOWERING: late summer–autumn.

The name is often given erroneously as *M. fuscata* or *M. furcata*.

15. **M. sanderiana** Reichenbach

ILLUSTRATIONS: Kupper & Linsenmaier, Orchids, t. 42 (1961); Bechtel et al., Manual of cultivated orchid species, 230 (1981); Orchid Review **92**: 43 (1984).

HABIT: terrestrial or epiphytic.

PSEUDOBULBS: 2.5–5 cm, clustered, ovoid, compressed, each with 1 leaf.

LEAVES: broadly oblong, 15–40 cm.

SCAPES: ascending or horizontal, to 25 cm.

FLOWERS: 10–15 cm across, fragrant.

SEPALS: white, blotched at base with reddish purple, upper sepal oblong, to 7.5 cm, lateral sepals triangular-ovate, slightly longer than upper sepal, forming a mentum with the column foot.

PETALS: shorter than upper sepal, lanceolate-triangular, reflexed at tips, similar in colour to sepals.

LIP: entire, erect, recurved, elliptic with a crisped margin, white or yellow, streaked and stained with red or purple; callus tongue-shaped, yellow, hairless.

DISTRIBUTION: Ecuador, Peru.

HARDINESS: G2.

FLOWERING: summer–early autumn.

16. **M. elatior** Reichenbach

ILLUSTRATION: Botanical Magazine, 9206 (1930).

HABIT: epiphytic.

PSEUDOBULBS: 4.5–9 cm, ovoid, compressed, scattered along the rhizome or often solitary at the end of the rhizome, each with 1–2 leaves.

LEAVES: 12–40 cm, lanceolate.

SCAPES: erect, 5–7.5 cm.

SEPALS: 2.2–3.2 cm, oblong-elliptic, reddish yellow, often spotted or striped with red, upper sepal concave, lateral sepals spreading or recurved.

PETALS: 1.7–2.5 cm, dull red.

LIP: entire, to 2 cm, ovate-oblong, recurved at apex, dark red; callus *c.* 1 cm, blackish, hairless.

DISTRIBUTION: Mexico, Guatemala, Honduras & Costa Rica.

HARDINESS: G2.

FLOWERING: spring–summer.

17. M. elegantula Rolfe

Like *M. elatior* except as below.

PSEUDOBULBS: each with 1 leaf

SCAPES: to 25 cm.

SEPALS AND PETALS: white, yellow towards the tips, flushed and spotted with brownish purple or deep blue, margins revolute.

LIP: to 1.7 cm, elliptic, apex recurved and minutely toothed and crisped, usually purplish red on margin.

DISTRIBUTION: Ecuador, Peru.

HARDINESS: G2.

18. M. variabilis Lindley

ILLUSTRATION: Die Orchidee **23**: 92 (1972); American Orchid Society Bulletin **54**: 26 (1984).

HABIT: terrestrial or epiphytic.

PSEUDOBULBS: 1.5–6 cm, narrowly ellipsoid to ovoid, borne at 2–3 cm intervals along the rhizome, each with 1 leaf.

LEAVES: 5–25 cm, very narrowly strap-shaped.

SCAPES: 2–2.5 cm.

FLOWERS: 2–2.5 cm in diameter.

SEPALS: 6–20 mm, ovate, white, greenish yellow, orange, deep red or purplish, often marked with red.

PETALS: 5–14 mm, oblong-lanceolate, similar in colour to sepals.

LIP: entire or slightly 3-lobed near apex, up to 1.2 cm, hinged to column foot, erect, usually deep red and often with yellow blotch at apex; callus strap-shaped, deep red, hairless.

DISTRIBUTION: C America, West Indies, Guyana.

HARDINESS: G2.

19. M. curtipes Hooker

SYNONYM: *M. houtteana* Reichenbach.

ILLUSTRATION: Cogniaux & Goossens, Dictionnaire Iconographique des Orchidées, Maxillaria, t. 2 (1899); Orchid Review **92**: 265 (1984).

Like *M. variabilis* except as below.

SCAPES: 2–4 cm.

SEPALS AND PETALS: yellowish, overlaid with deep red inside and somewhat blotched towards the base, margins orange.

LIP: entire, oblong to broadly elliptic, reflexed, deep yellow, spotted and striped with reddish purple; callus warty.

DISTRIBUTION: Mexico, Guatemala, El Salvador & Costa Rica.

HARDINESS: G2.

FLOWERING: spring.

20. M. chrysantha Rodriguez

ILLUSTRATIONS: Botanical Magazine, 8979 (1923); Hoehne, Flora Brasilica **12**(7): t. 127 (1953).

HABIT: epiphytic.

PSEUDOBULBS: 1.5–2.5 cm, ovoid-conical, slightly compressed, produced at intervals of 2.5–4 cm along the rhizome, each with 2 leaves.

LEAVES: 20–35 cm, linear.

SCAPES: 7–10 cm.

FLOWERS: 4–4.5 cm in diameter.

SEPALS: 2.8–3 cm, oblong, yellow with reddish or mauve margins.

PETALS: 2.2–2.4 cm, similar in colour to sepals.

LIP: 3-lobed, recurved at apex, more or less circular, yellowish with violet blotches on the lateral lobes, margin of central lobe wavy; callus strap-shaped, downy.

DISTRIBUTION: Surinam, Brazil.

HARDINESS: G2.

21. M. porphyrostele Reichenbach

ILLUSTRATIONS: Botanical Magazine, 6477 (1880); Cogniaux & Goossens, Dictionnaire Iconographique des Orchidées, Maxillaria, t. 10 (1904); Pabst & Dungs, Orchidaceae Brasilienses **2**: 234, 290 (1977).

PSEUDOBULBS: 2.5–3 cm, clustered, ovoid, compressed, each with 2 leaves.

LEAVES: 12–20 cm, lanceolate.

SCAPES: to 7.5 cm.

FLOWERS: *c.* 3 cm across, slightly fragrant.

SEPALS: oblong, incurving, pale yellow.

PETALS: shorter than sepals, incurving, pale yellow with purple stripes at the base.

LIP: 3-lobed, lateral lobes yellow striped with purple, central lobe oblong-circular, yellow; callus yellow marked with purple.

DISTRIBUTION: Brazil.

HARDINESS: G2.

FLOWERING: winter–spring.

22. M. marginata (Lindley) Fenzl

ILLUSTRATIONS: Hoehne, Flora Brasilica **12**(7): t. 141 (1953); Pabst & Dungs, Orchidaceae Brasilienses **2**: 290 (1977).

PSEUDOBULBS: 4–6 cm, each with 1 or 2 leaves.

LEAVES: 12–30 cm, lanceolate.

SCAPES: 7.5–15 cm, erect or arching.

SEPALS: 2–2.5 cm, narrowly oblong, yellow with a dark red margin and a dorsal red line.

PETALS: 1.7–2.2 cm, similar in colour to sepals.

LIP: 3-lobed, lateral lobes pale yellow streaked with reddish purple, central lobe oblong, reflexed, acute, pale yellow with a scalloped purple margin; callus strap-shaped, hairless.

DISTRIBUTION: Colombia, Ecuador, Brazil.

HARDINESS: G2.

FLOWERING: summer–autumn.

23. M. picta Hooker

ILLUSTRATIONS: Botanical Magazine, 3154 (1832); Bechtel et al., Manual of cultivated orchid species, 230 (1981).

HABIT: epiphytic.

PSEUDOBULBS: 4–7 cm, ovoid, somewhat compressed, clustered or borne at intervals along the rhizome, each with 1 or 2 leaves.

LEAVES: to 30 cm, narrowly oblong.

SCAPES: 12–16 cm.

FLOWERS: fragrant.

SEPALS: *c.* 3 cm, oblong, incurved, whitish to yellow with purple or dark brown spots usually on the dorsal surface.

PETALS: 2.5–2.8 cm, similar to sepals but with a red basal streak, very incurved.

LIP: 3-lobed, lateral lobes white with purple marks, central lobe oblong, reflexed, acute, whitish with purple spots; callus oblong, downy.

DISTRIBUTION: Colombia, Brazil.

HARDINESS: G2.

FLOWERING: spring–summer.

24. M. cucullata Lindley

SYNONYMS: *M. meleagris* Lindley; *M. praestans* Reichenbach.

ILLUSTRATION: Bechtel et al., Manual of cultivated orchid species, 229 (1981).

Like *M. picta* except as below.

SEPALS: 2.5–5 cm, usually deep red, occasionally yellow, brownish, pinkish or almost black, variously striped and spotted with dark red.

PETALS: shorter, similar in colour, with the tips meeting above the column.

LIP: 3-lobed, hinged to column foot, yellow or whitish with reddish brown or purple spots, or occasionally wholly dark red, central lobe finely warty.

DISTRIBUTION: Mexico to Panama.

FLOWERING: summer–autumn.

Very variable in the size of the plant and the flowers and in the width of the sepals and petals.

25. M. rufescens Lindley

ILLUSTRATIONS: Dunsterville & Garay, Venezuelan orchids illustrated **2**: 219 (1961); American Orchid Society Bulletin **43**: 966 (1974).

Like *M. picta* except as below.

SCAPES: 1–5 cm.

SEPALS: 1–2.5 cm, greenish white, pinkish or yellow, usually flushed with brownish pink or maroon.

PETALS: very slightly shorter, similar in colour, sometimes with deep red splashes on veins.

LIP: 3-lobed, hinged to column foot, lateral lobes pointed, somewhat sickle-shaped, central lobe oblong, slightly reflexed, usually pale or deep yellow, or reddish, spotted with maroon or brownish-red; callus brownish yellow or deep red, warty and velvety.

DISTRIBUTION: tropical America.

FLOWERING: autumn.

An extremely variable species.

26. M. coccinea (Jacquin) Hodge

SYNONYM: *Ornithidium coccineum* (Jacquin) R. Brown.

ILLUSTRATIONS: Loddiges' Botanical Cabinet **4**: t. 301 (1819); Hooker, Exotic Flora **1**: t. 38 (1823).

HABIT: epiphytic.

PSEUDOBULBS: 2.5–4 cm, ovoid, more or less compressed, clustered, each with 1 leaf.

LEAVES: 10–35 cm, lanceolate.

SCAPES: 4–6 cm, borne in bundles.

FLOWERS: bright pink to red.

SEPALS: 1.1–1.2 cm, concave, ovate to narrowly ovate.

PETALS: 7–8 mm, similar.

LIP: 3-lobed, 6–8 mm, the central lobe curved downwards, narrowly ovate, apex acute; callus fleshy, transverse.

DISTRIBUTION: West Indies, Colombia, Venezuela.

HARDINESS: G2.

FLOWERING: summer.

27. M. densa Lindley

SYNONYM: *Ornithidium densum* (Lindley) Reichenbach.

ILLUSTRATIONS: Hamer, Orquideas de El Salvador **2**: 111 (1974); Bechtel et al., Manual of cultivated orchid species, 229 (1981).

Like *M. coccinea* except as below.

PSEUDOBULBS: often larger.

LEAVES: linear.

FLOWERS: slightly smaller, variable in colour, which ranges from greenish white to yellow, purplish, maroon or brownish red.

LIP: 3.5–5 mm, central lobe more or less circular to broadly ovate, obtuse at the apex and with a small point; callus plate-like, obtuse.

DISTRIBUTION: C America from Mexico to Honduras.

HARDINESS: G2.

28. M. sophronitis (Reichenbach) Garay

SYNONYM: *Ornithidium sophronitis* Reichenbach.

ILLUSTRATIONS: Dunsterville & Garay, Venezuelan orchids illustrated **1**: 237 (1959).

HABIT: epiphytic.

PSEUDOBULBS: 1–2 cm, borne at 2.5–6 cm intervals along the creeping rhizome, slightly compressed, each with 1 leaf.

LEAVES: 1–8 cm, oblong, thick, apex often asymmetric.

FLOWERS: borne on scapes *c.* 1 cm.

SEPALS AND PETALS: ovate, bright orange-red; sepals 9–13 mm, petals smaller.

LIP: 3-lobed, *c.* 7 mm, yellow to orange-yellow, central lobe oblong, the upper surface warty and margin minutely toothed; callus smooth, flat.

DISTRIBUTION: Venezuela.

HARDINESS: G2.

99. MORMOLYCA Fenzl

A genus of 5 species native to C and tropical S America. Its name is often misspelled 'Mormolyce'. Cultivation as for *Maxillaria* (p. 330).

HABIT: epiphytic with creeping rhizomes.

PSEUDOBULBS: with 1–4 leaves, subtended by several bracts.

LEAVES: folded when young.

FLOWERS: solitary, arising from an axil of a pseudobulb bract.

SEPALS: spreading, more or less equal, free.

PETALS: similar to sepals but smaller.

LIP: simple or 3-lobed.

COLUMN: without a foot.

POLLINIA: 4 in 2 pairs, waxy.

1. M. ringens (Lindley) Schlechter

ILLUSTRATIONS: Fieldiana **26**: 591 (1952); Canadian Journal of Botany **37**: 487 (1959); Bechtel et al., Manual of cultivated orchid species, 234 (1981).

PSEUDOBULBS: 2–5 cm, clustered or arising at intervals of 1–2 cm along the rhizome, rounded to ovoid, compressed, each bearing 1 leaf.

LEAVES: 10–35 cm.

SEPALS: 1.5–2 cm, yellowish green to light lavender, striped longitudinally with violet or dark red.

PETALS: similar in colour.

LIP: *c.* 2 cm, pale purple to dark red, almost erect, jointed with the base of the column, central lobe broadly rounded to oblong-elliptic, curved downwards, hairless or minutely warty towards the margin; callus triangular, apex entire or 3-toothed.

DISTRIBUTION: C America from Mexico to Costa Rica.

HARDINESS: G2.

FLOWERING: summer–autumn.

The flowers have a characteristic shape resulting from the upper sepal and petals being directed upwards and the lateral sepals and lip downwards.

100. TRIGONIDIUM Lindley

A genus of 14 species from C and tropical S America. Cultivation as for *Maxillaria* (p. 330).

HABIT: epiphytic, with creeping rhizomes.

PSEUDOBULBS: with 1–5 leaves, ribbed, subtended by several fibrous bracts.

LEAVES: strap-shaped, leathery.

FLOWERS: usually solitary, arising from the axils of the pseudobulb bracts.

SEPALS: forming a basal tube, recurved in the upper half.

PETALS: much smaller than sepals.

LIP: 3-lobed, shorter than petals, almost erect, lateral lobes erect and parallel to column, central lobe fleshy; callus usually strap-shaped.

POLLINIA: 4 on a short, laterally crescent-shaped stipe.

1. T. egertonianum Lindley

SYNONYM: *T. seemanni* Reichenbach.

ILLUSTRATIONS: Hawkes, Encyclopaedia of cultivated orchids, 475 (1965); Dunsterville & Garay, Venezuelan orchids illustrated 3: 321 (1965).

PSEUDOBULBS: 4–9 cm, more or less spherical to ellipsoid, compressed, each with 2 leaves.

LEAVES: 20–60 cm.

SCAPES: 15–35 cm.

SEPALS AND PETALS: yellowish green or pinkish with reddish brown or purple veins, upper sepal 2.5–4.5 cm, lateral sepals 2.5–4 cm, petals 1–2 cm with a purple or brown thickening just below the apex.

LIP: 5–10 mm, green or yellowish brown striped with dark purple or brownish red, central lobe reflexed, warty.

DISTRIBUTION: C America, Colombia, Venezuela.

HARDINESS: G2.

FLOWERING: spring.

101. DICHAEA Lindley

A genus of 40 species occurring in tropical America and the

West Indies. Only 4 species are generally cultivated. They are best grown on a block of tree fern or bark. Cultivation otherwise as for *Angraecum* (p. 394).

HABIT: epiphytes without pseudobulbs.
LEAVES: in 2 ranks, folded when young, the sheaths concealing the stem.
FLOWERS: solitary from the leaf axils.
SEPALS AND PETALS: free, similar, though petals usually smaller.
LIP: joined to the base of the column, clawed, usually 3-lobed, the lateral lobes often recurved, usually without a callus.
POLLINIA: 4, ovoid.

Key to species

1. Leaf-blade not jointed to the sheath, leaves persistent; lip with the narrow lateral lobes pointing backwards; capsule spiny
1.A. Sepals orange, lanceolate, warty on the outer surface **1. muricata**
1.B. Sepals pale yellow to pinkish, ovate, not warty **2. pendula**
2. Lip 4.5–6 mm, with a callus; leaves green **3. graminoides**
3. Lip 6–7 mm, without a callus; leaves glaucous **4. glauca**

1. **D. muricata** (Swartz) Lindley
ILLUSTRATIONS: Schultes, Native orchids of Trinidad and Tobago, 259 (1960); Lasser, Flora de Venezuela 15(5): 461 (1970).
STEMS: hanging, 30–50 cm.
LEAVES: 1.5–2 cm, ovate-oblong, acuminate, persistent, not jointed to the sheaths.
SEPALS: orange, lanceolate, 7–10 mm, warty outside.
PETALS: purplish-spotted, 5–8 mm, warty on the outer surfaces.
LIP: 5–8 mm, pale purple with 2 recurved, narrow, lateral lobes.
CAPSULE: 1–2 cm, densely spiny.
DISTRIBUTION: Mexico to Brazil, West Indies.
HARDINESS: G2.
FLOWERING: winter.

2. **D. pendula** (Aublet) Cogniaux
ILLUSTRATION: Lasser, Flora de Venezuela 15(5): 466 (1970).
STEMS: hanging.
LEAVES: 1–2.5 cm, elliptic-oblong to lanceolate, each with a small point, persistent, not jointed to the sheaths.
SEPALS: 8–9 mm, pale yellow to pinkish, ovate.
PETALS: *c.* 8 mm, pinkish-buff, sometimes with bluish purple marks near the tips, obovate.
LIP: 5–9 mm, white or pale purple with bluish purple blotches and bands, with 2 recurved, narrow side lobes.

CAPSULE: 1–1.5 cm, spiny.

DISTRIBUTION: Southern C America to northern S America, West Indies.

HARDINESS: G2.

FLOWERING: summer–autumn.

3. **D. graminoides** (Swartz) Lindley

STEM: erect, 10–30 cm, strongly flattened.

LEAVES: 3–4 cm, linear, jointed to the sheaths, eventually falling.

SEPALS AND PETALS: 5–8 mm, whitish, ovate to lanceolate.

LIP: 4.5–6 mm, whitish, heart-shaped at the base, shallowly 3-lobed, central lobe almost circular, with a small point; callus present, linear.

CAPSULE: 1.5–2 cm, not spiny.

DISTRIBUTION: C and tropical S America, West Indies.

HARDINESS: G2.

4. **D. glauca** (Swartz) Lindley

STEMS: ascending or hanging, flattened.

LEAVES: 3.5–7 cm, linear-oblong, glaucous, jointed to the sheaths, eventually falling.

FLOWERS: fragrant.

SEPALS AND PETALS: 7–8 mm, generally white but sometimes tinged and spotted with lilac and yellow.

LIP: 6–7 mm, whitish with a dark red spot at the base, claw broad, lateral lobes very short or absent, central lobe broadly ovate to kidney-shaped.

CAPSULE: 1–2 cm, not spiny.

DISTRIBUTION: Northern C America, Cuba, Hispaniola & Jamaica.

HARDINESS: G2.

FLOWERING: summer.

102. ORNITHOCEPHALUS Hooker

A genus of about 50 species from the American tropics. They are best grown in small shallow pans suspended near the glass, using an epiphytic compost. They require light but must be shaded from the direct rays of the sun. Moisture must be provided throughout the year, particularly during the summer.

HABIT: epiphytic, without pseudobulbs.

LEAVES: fleshy, leathery, in fan-shaped clusters, densely overlapping and concealing the stem, jointed to sheaths, the part below the joint often thickened.

RACEMES: erect or hanging, with several–many flowers, arising from the leaf axils.

FLOWERS: not resupinate.

SEPALS: usually equal, free.

PETALS: equal to or larger than sepals.

LIP: joined to the base of the column, simple or 3-lobed; callus present.

ROSTELLUM: long-beaked.

ANTHER: with a long terminal appendage.

POLLINIA: 4 on a slender stipe.

Key to species

1. Leaves 7.5–15 cm; racemes 22–25 cm **3. grandiflorus**
2. Sepals 2–2.5 mm; lip 4–5 mm, central lobe obtuse **1. iridifolius**
3. Sepals 3–4 mm; lip 5–6 mm, central lobe acute **2. gladiatus**

1. **O. iridifolius** Reichenbach

LEAVES: 2.5–7.5 cm, linear-lanceolate.

RACEMES: 4–8 cm, loose, with many flowers.

FLOWERS: white.

SEPALS: 2–2.5 mm, almost circular to broadly elliptic, with a keel projecting beyond the tip.

PETALS: 3–3.5 mm, fan-shaped.

LIP: 4–5 mm, deeply 3-lobed, lateral lobes almost circular, central lobe triangular-ovate.

DISTRIBUTION: Mexico, Guatemala.

HARDINESS: G2.

FLOWERING: summer.

2. **O. gladiatus** Hooker

ILLUSTRATIONS: Hooker, Exotic flora **2**: t. 127 (1824); Howard, Flora of the Lesser Antilles, 214 (1974).

LEAVES: 2.5–4 cm, lanceolate.

RACEME: to 5 cm, loose, with few flowers.

FLOWERS: greenish yellow to greenish white.

SEPALS: 3–4 mm, each with a keel projecting beyond the tip.

PETALS: 3–4 mm.

LIP: 5–6 mm, 3-lobed, central lobe linear-oblong with recurved margins, acute.

DISTRIBUTION: C America, northwest S America, West Indies.

HARDINESS: G2.

FLOWERING: spring–summer.

3. **O. grandiflorus** Lindley

SYNONYM: *Dipteranthus grandiflorus* (Lindley) Pabst.

ILLUSTRATIONS: Pabst & Dungs, Orchidaceae Brasilienses **2**: 263, 322 (1977).

LEAVES: 7.5–15 cm, narrowly oblong.

RACEMES: 22–25 cm with many flowers.

SEPALS: concave, white with a greenish mark at the base, petals similar but a little larger.

LIP: 3-lobed, white, greenish or yellowish at the base, concave, margins finely crisped.

DISTRIBUTION: Brazil.

HARDINESS: G2.

FLOWERING: summer.

103. SARCOCHILUS R. Brown

A genus of about 12 species, mainly native to Australia. Though good drainage is essential to their culture, they should never be allowed to dry out. Treatment varies somewhat, depending on the origin of the species. Some are best grown in pots in an epiphytic compost, others on rafts or tree fern, close to the glass.

HABIT: epiphytic, pseudobulbs absent.

STEMS: covered with the remains of old sheathing leaf-bases.

LEAVES: in 2 ranks, folded when young.

RACEMES: lateral with 3–25 fragrant flowers.

SEPALS AND PETALS: similar, lateral sepals joined to the column foot.

LIP: hinged to column foot, shallowly pouched, 3-lobed, central lobe much smaller than lateral lobes, with a forward-pointing spur; surface of lip with a large callus, and some calluses present on the inside walls of the spur.

POLLINIA: 4, in 2 pairs.

Key to species

1. Sepals and petals white, pink or purple, without spots at the base; stems 5–12 cm, unbranched

1.A. Racemes hanging; flowers 2–3.5 mm in diameter **1. falcatus**

1.B. Racemes erect; flowers 4–6 mm in diameter **2. ceciliae**

2. Callus at opening of spur 2-lobed, with additional small calluses on the inside walls of the spur **3. hartmannii**

3. Callus at opening of spur not 2-lobed, with a smaller, additional callus behind **4. fitzgeraldii**

1. S. falcatus R. Brown

ILLUSTRATIONS: Nicholls, Orchids of Australia, t. 452, 453 (1969); Dockrill, Australian indigenous orchids, t. 25 (1969).

STEMS: 5–7.5 cm.

LEAVES: narrowly oblong, sickle-shaped, 5–14 cm, channelled.

RACEMES: hanging, with 3–12 flowers.

FLOWERS: 2–3.5 cm in diameter.

SEPALS AND PETALS: oblong, 8–16 mm, white, sometimes with a purplish midrib on the back.

LIP: 3.5–6 mm, lateral lobes ovate-oblong, flushed with orange and with red stripes, central lobe broad, about a quarter of the length of the lateral lobes; spur fleshy, sometimes marked with purple; callus at opening of spur 2-lobed, spotted, some calluses on inside wall of spur.

DISTRIBUTION: Australia.

HARDINESS: G2.

FLOWERING: autumn–winter.

2. **S. ceciliae** Mueller

ILLUSTRATIONS: Nicholls, Orchids of Australia, t. 456, 457 (1969).

Like *S. falcata* except as below.

STEMS: to 12 cm.

LEAVES: often brown-spotted.

RACEMES: erect with 6–12 flowers which are 4–6 mm in diameter, facing upwards, pale pink, mauve or purple.

LIP: densely hairy, callus at opening of spur yellow, pronounced calluses present on inside walls of the spur.

DISTRIBUTION: Australia.

3. **S. hartmannii** Mueller

ILLUSTRATIONS: Nicholls, Orchids of Australia, t. 455 (1969); American Orchid Society Bulletin **47**: 1122 (1978); Bechtel et al., Manual of cultivated orchid species, 271 (1981).

STEMS: to 1 m, usually branched.

LEAVES: linear-oblong, 5–15 cm, channelled.

RACEMES: arching or erect with 5–25 flowers.

FLOWERS: 1.5–3.5 cm in diameter.

SEPALS AND PETALS: white spotted with red at the base (rarely completely white or red).

LIP: lateral lobes spotted and streaked with red, central lobe yellowish, spur white flushed with yellow; callus at opening of spur 2-lobed, small calluses on inside walls of spur.

DISTRIBUTION: Australia.

HARDINESS: G2.

4. **S. fitzgeraldii** Mueller

ILLUSTRATIONS: Botanical Magazine n.s., 201 (1953); Nicholls, Orchids of Australia, t. **454** (1969): Bechtel et al., Manual of cultivated orchid species, 271 (1981).

Like *S. hartmannii* except as below.

LEAVES: sickle-shaped.

RACEMES: hanging, with up to 16 flowers, scape reddish

FLOWERS: 2.5–3.5 cm in diameter.

LIP: *c.* 5 mm, callus at opening of spur yellowish with a smaller one behind.

DISTRIBUTION: Australia.

FLOWERING: winter–spring.

104. CHILOSCHISTA Lindley

A genus of 3 species found in tropical Asia, of which only 1 is generally cultivated. Cultivation as for *Sarcochilus* (p. 349).

Like *Sarcochilus* except that the plants are leafless when flowering.

PETALS: larger than the sepals.

POLLINIA: 4, in 2 pairs.

1. C. **lunifera** (Reichenbach) J.J. Smith

SYNONYM: *Sarcochilus luniferus* (Reichenbach) Hooker.

ILLUSTRATION: Botanical Magazine, 7044 (1899).

RACEMES: drooping, with many flowers, scapes purple-spotted, shortly hairy.

FLOWERS: 1–1.5 cm in diameter.

SEPALS AND PETALS: yellow spotted with red or purple.

LIP: white, lateral lobes erect, central lobe minute, recurved, with a forward-pointing spur; disc warty and hairy with 2 thick ridges.

ANTHER CAP: with 3 spurs.

DISTRIBUTION: N India (Sikkim), Nepal & Burma.

HARDINESS: G2.

105. PTEROCERAS Hasskarl

Thirty species found in NE India, SE Asia and Malaysia. Cultivation as for *Sarcochilus* (p. 349).

Literature: Senghas, K., Die Gatting Pteroceras, *Die Orchidee* **40**: 2–6 (1989).

Like *Sarcochilus* except as below.

LIP: movably attached to the column foot, the spur of the lip lacking calluses on its inside walls.

POLLINIA: 4 in 2 pairs.

1. P. **pallidum** (Blume) Holttum

SYNONYMS: *Sarcochilus pallidus* (Blume) Reichenbach; *S. unguiculatus* Lindley.

ILLUSTRATIONS: Lindenia **16**: t. 756 (1901).

LEAVES: oblong, fleshy, up to 20 cm.

RACEME: usually hanging, axis somewhat thickened.

FLOWERS: to 5 cm across, fragrant.

SEPALS AND PETALS: white to pale yellowish.

LIP: *c.* 6 mm, lateral lobes white streaked with red or purple, central lobe and spur yellowish, spotted with red or purple.

DISTRIBUTION: Malay peninsula, Java, Sumatra, Borneo & Philippines.

HARDINESS: G2.

106. AERIDES Loureiro

A genus of some 40 species from tropical Asia. Cultivation as for *Vanda* (p. 377).

Literature: Seidenfaden, G., Contributions to the orchid flora of Thailand V, *Botanisk Tidsskrift* **68**: 68–80 (1973); Wood, J. & Kennedy, G.C., Some showy members of the genus Aerides, *Orchid Digest* **41**: 205–8 (1977); Christensen, E., The taxonomy of Aerides and related genera, *Proceedings of the 12th World Orchid Congress*, 35–40 (1987); Seidenfaden, G., Orchid Genera in Thailand, *Opera Botanica* **95**: 242–251 (1988).

HABIT: epiphytic, pseudobulbs absent.

STEM: producing long aerial roots which either hang or attach themselves to a support.

LEAVES: usually strap-shaped, thick, in 2 ranks, their bases sheathing the stem, folded when young, occasionally terete.

FLOWERS: usually fragrant, in hanging, many-flowered racemes arising from the leaf axils.

SEPALS AND PETALS: spreading, the 2 lateral sepals attached to the column foot.

LIP: simple or 3-lobed, continuous with the column foot, prolonged at the base into a hollow spur.

COLUMN: short, usually with a long foot.

POLLINIA: 2 unequally divided, on a single slender stipe.

Synopsis of characters

Leaves. Terete: **1,2**.

Sepals and petals. Basic colour olive green: **11–13**; white, pale green or yellowish: **1–10**; pinkish or purplish: **6–8,11–13**. With defined spots at the tips: **3–10**. With wavy margins: **2**.

Lip. Simple: **6–8,11–13**.

Margin of lip, or if lip 3-lobed, of central lobe. Entire: **1,3–8,13**; toothed or fringed: **2–5,9–12**.

Key to species

1. Leaves terete

1.A. Leaves 7.5–10 cm; margins of sepals and petals not wavy; lateral lobes of lip with violet markings **1. cylindrica**

1.B. Leaves 10–20 cm; margins of sepals and petals wavy; lateral lobes of lip without violet markings **2. vandara**

2. Basic colour of sepals and petals white, pale greenish or yellowish

2.A. Spur curved upwards towards the rest of the flower

2.A.a. Sepals and petals creamy white tipped with purple or pink **3. odorata**

2.A.b. Sepals and petals greenish at the base **4. quinquevulnera**

2.A.c. Sepals and petals greenish throughout **5. lawrenceae**

2.B. Lip simple with an entire margin **6. multiflora**

2.C. Sepals and petals white, tipped and sometimes dotted with pink **7. falcata**

2.D. Sepals and petals yellow-buff, tipped with reddish purple or brown **8. houlletiana**

3. Lip simple and fringed, or 3-lobed, the central lobe with a toothed margin

3.A. Sepals and petals white flushed with pinkish purple **9. crispa**

3.B. Sepals and petals olive green marked with reddish brown **10. flabellata**

4. Flowers 2–3 cm in diameter; lip simple, rounded or slightly notched at apex **6. multiflora**

5. Lip simple, with a pointed tip **12. rosea**

6. Central lobe of lip sharply curved upwards **13. jarckiana**

7. Central lobe of lip not sharply curved upwards **11. crassifolia**

1. **A. cylindrica** Lindley

SYNONYM: *Papilionanthe subulata* (Koenig) Garay.

ILLUSTRATION: Orchid Digest **41**: 204 (1977).

LEAVES: terete, 7.5–10 cm.

RACEME: with 1 or 2 flowers.

FLOWERS: *c.* 2.5 cm in diameter.

SEPALS AND PETALS: white, sometimes flushed with pale pink, ovate-oblong.

LIP: lateral lobes white with violet lines, central lobe white or yellowish with violet spots and 3 broad yellow ridges; spur funnel-shaped, pinkish.

DISTRIBUTION: S India, Sri Lanka.

HARDINESS: G2.

2. **A. vandara** Reichenbach

SYNONYMS: *A. cylindrica* Hooker not Lindley; *Papilionanthe vandarum* (Reichenbach) Garay.

ILLUSTRATIONS: Botanical Magazine, 4982 (1857); Orchid Digest **41**: 204 (1977); Bechtel et al., Manual of cultivated orchid species, 164 (1981).

LEAVES: terete, 10–20 cm.

RACEME: with 1–3 flowers.

FLOWERS: *c.* 5 cm in diameter.

SEPALS AND PETALS: white, somewhat reflexed, with wavy margins, petals twisted.

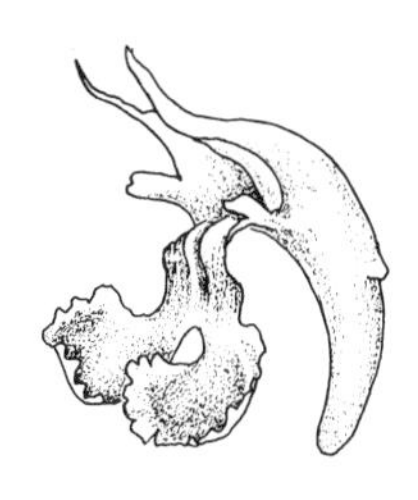

Aerides vandara – lip.

LIP: white flushed with yellow in the middle, central lobe clawed, expanding into 2 fringed lobes; spur curved, 2–2.5 cm.

DISTRIBUTION: NE India, Burma.

HARDINESS: G2.

FLOWERING: spring.

3. A. odorata Loureiro

SYNONYMS: *A. suavissima* Lindley; *A. cornuta* Roxburgh; *A. ballantiana* Reichenbach.

ILLUSTRATIONS: Botanical Magazine, 4139 (1845); Botanisk Tidsskrift **68**: 72 (1973); Sheehan & Sheehan, Orchid genera illustrated, 31 (1979); Orchid Review **92**: 53, 54 (1984).

STEMS: to 50 cm, in old plants to 150 cm.

LEAVES: 20–25 cm.

RACEMES: dense, with many flowers.

FLOWERS: 2–3 cm in diameter.

SEPALS AND PETALS: creamy white tipped with pink or purple.

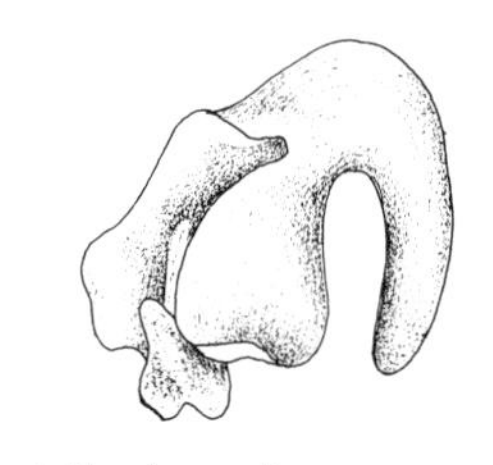

Aerides odorata – lip.

LIP: lateral lobes white or occasionally yellowish, flushed and spotted with pink or purple, especially at the base, central lobe ovate to narrowly oblong, white with a broad central purplish band, margin entire or toothed; spur curved upwards towards the flower, tipped with green or yellow.

DISTRIBUTION: SE Asia, from the Himalaya to Indonesia.

HARDINESS: G2.

FLOWERING: summer–autumn.

4. A. quinquevulnera Lindley.

ILLUSTRATION: Orchid Digest **43**: 126 (1979).

Like *A. odorata* except as below.

SEPALS AND PETALS: greenish at the base.

LIP: central lobe deep pinkish purple, with a revolute margin which is always toothed.

DISTRIBUTION: Philippines.

HARDINESS: G2.

FLOWERING: summer–autumn.

Often regarded as a variety of *A. odorata*.

5. A. lawrenceae Reichenbach

ILLUSTRATIONS: Lindenia **9**: t. 401 (1893); Bechtel et al., Manual of cultivated orchid species, 164 (1981).

Like *A. odorata* except as below.

SEPALS AND PETALS: greenish white becoming yellowish.

LIP: lateral lobes white with toothed margins, central lobe purple, or purple-tipped with lines of purple spots, margin toothed.

DISTRIBUTION: Philippines.

Possibly also merely a variety of *A. odorata*.

6. A. multiflora Roxburgh

SYNONYM: *A. affine* Lindley.

ILLUSTRATIONS: Botanical Magazine, 4049 (1844); Cogniaux & Goossens, Dictionnaire Iconographique des Orchidées, Aerides, t. 2 (1898); Gartenpraxis 1980(11): 512.

LEAVES: 18–25 cm.

Aerides multiflora – lip.

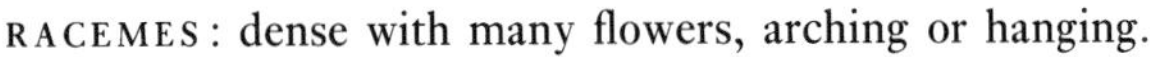

RACEMES: dense with many flowers, arching or hanging.
FLOWERS: 2–3 cm in diameter.
SEPALS AND PETALS: white, spotted, flushed and tipped with purplish pink, or occasionally completely pink.
LIP: pale purplish pink with a deeper central area, simple, oval, with the sides recurved and the apex rounded or slightly notched; spur pointing forwards.
DISTRIBUTION: N India, Bhutan, SE Asia.
HARDINESS: G2.
FLOWERING: summer–autumn.

7. A. falcata Lindley
SYNONYM: *A. larpentae* Reichenbach.
ILLUSTRATIONS: Botanisk Tidsskrift **68**: 74 (1973); Orchid Digest **41**: 204 (1977).
STEMS: to 1.8 m.
LEAVES: 15–35 cm.
RACEME: with many flowers, hanging.
SEPALS AND PETALS: white tipped and sometimes dotted with pink.
LIP: lateral lobes white or mauve, sickle-shaped, central lobe pink or purple with a toothed margin; spur short, greenish, pointing forwards.
DISTRIBUTION: NE India, SE Asia.
HARDINESS: G2.
FLOWERING: spring–summer.

Aerides falcata – lip.

8. A. houlletiana Reichenbach
SYNONYM: *A. falcata* var. *houlletiana* (Reichenbach) Veitch.
ILLUSTRATIONS: Cogniaux & Goossens, Dictionnaire Iconographique des Orchidées, Aerides, t. 3 (1899); Orchid Digest **41**: 206 (1977).

Like *A. falcata* except as below.

SEPALS AND PETALS: yellow-buff tipped with reddish purple or brown, upper sepal and petals with finely toothed margins.
LIP: lateral lobes yellowish, marked with pinkish purple, central lobe pinkish purple, toothed and notched at the apex.
DISTRIBUTION: E Thailand, Vietnam, Laos & Kampuchea.
FLOWERING: spring.

9. A. crispa Lindley
SYNONYM: *A. brookei* Lindley.
ILLUSTRATIONS: Orchid Digest **43**: 126 (1979).
STEMS: to 1.5 m, often purplish.
LEAVES: 15–25 cm.
RACEME: with many flowers, arching or hanging.
FLOWERS: *c.* 5 cm in diameter.
SEPALS AND PETALS: white flushed with pinkish purple.
LIP: lateral lobes white striped with pinkish purple, central lobe

Aerides crispa – lip.

broadly ovate, deep pinkish purple, margin toothed; spur compressed, slightly curved.

DISTRIBUTION: Himalaya, India, Burma.

HARDINESS: G2.

FLOWERING: summer.

10. **A. flabellata** Downie

ILLUSTRATIONS: Botanisk Tidsskrift **68**: 77 (1973); Orchid Digest **41**: 206 (1977).

Like *A. crispa* except as below.

STEMS: much shorter.

RACEME: loose.

SEPALS AND PETALS: olive green marked with reddish brown.

LIP: simple, clawed, white blotched with red or purple, margin fringed; spur curved upwards.

DISTRIBUTION: Burma, Laos, Thailand.

A distinct species which may be more closely related to *Vanda* (p. 377).

11. **A. crassifolia** Parish & Reichenbach

SYNONYM: *A. expansa* Reichenbach.

ILLUSTRATIONS: Cogniaux & Goossens, Dictionnaire Iconographique des Orchidées, Aerides, t. 1 (1897); Botanisk Tidsskrift **68**: 76 (1973); Orchid Digest **41**: 204 (1977).

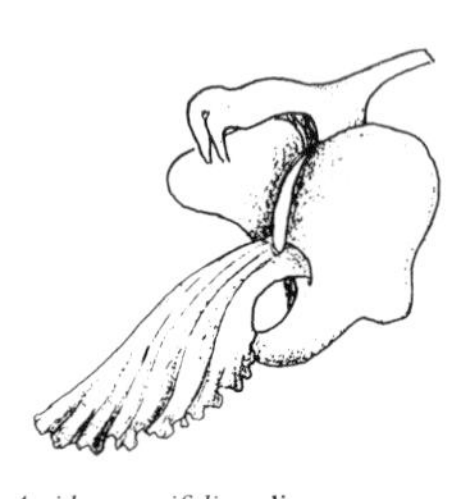

Aerides crassifolia – lip.

LEAVES: to 20 cm.

RACEMES: with 10–50 flowers.

FLOWERS: 3–5 cm in diameter, fragrant.

SEPALS AND PETALS: pinkish purple, deeper towards the apex and white towards the base.

LIP: lateral lobes crescent-shaped, pinkish purple, central lobe broadly ovate, often notched at the apex, darker in colour than sepals and petals; spur with a greenish tip.

DISTRIBUTION: Burma, Laos, Thailand.

HARDINESS: G2.

FLOWERING: spring–summer.

12. **A. rosea** Lindley & Paxton

SYNONYM: *A. fieldingii* Williams.

ILLUSTRATIONS: La Belgique Horticole **26**: facing p. 283 (1876); Botanisk Tidsskrift 68: 78 (1973); Orchid Digest **41**: 206 (1977); Bechtel et al., Manual of cultivated orchid species, 163 (1981).

Like *A. crassifolia* except as below.

RACEMES: branched.

LIP: simple with a pointed tip, spur white.

DISTRIBUTION: N India, N Thailand, Laos.

HARDINESS: G2.

FLOWERING: spring–summer.

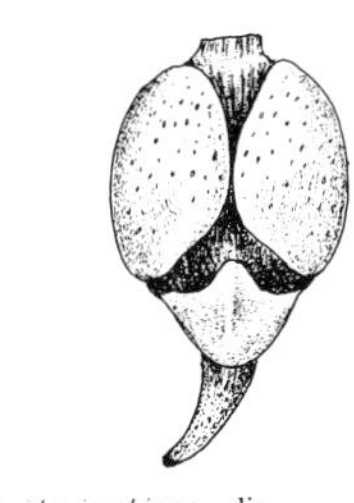

Aerides jarckiana – lip.

13. A. jarckiana Schlechter

ILLUSTRATIONS: Botanical Magazine, 9274 (1932); Orchid Digest **43**: 126 (1979); Bechtel et al., Manual of cultivated orchid species, 163 (1981).

Like *A. crassifolia* except as below.

LIP: lateral lobes very small, rounded, united nearly to their apex with the central lobe which is sharply curved upwards.

DISTRIBUTION: Philippines.

FLOWERING: spring.

107. SEDIREA Garay & Sweet

A genus of only 1 species, growing in Japan and Korea. Cultivation as for *Vanda* (p. 377).

HABIT: epiphytic, pseudobulbs absent.

STEM: short, *c.* 10 cm.

LEAVES: in 2 ranks, linear-oblong, thick.

RACEME: axillary, loose, drooping, to 15 cm with few flowers.

FLOWERS: *c.* 3 cm in diameter, fragrant.

SEPALS AND PETALS: white, cream or greenish, lateral sepals barred with purplish brown.

LIP: hinged to the column foot, 3-lobed, lateral lobes triangular, central lobe obovate, recurved, scalloped, white blotched with reddish purple; spur funnel-shaped, obtuse, pointing forwards.

COLUMN: *c.* 8 mm, curved, with a short foot.

POLLINIA: 2, ovoid, each with a groove, on a narrowly triangular stipe.

1. S. japonica (Linden & Reichenbach) Garay & Sweet

SYNONYM: *Aerides japonica* Linden & Reichenbach.

ILLUSTRATION: Orchid Digest **41**: 207 (1977); American Orchid Society Bulletin **55**: 701 (1986).

DISTRIBUTION: Japan, Korea.

HARDINESS: G1.

FLOWERING: summer.

108. RHYNCHOSTYLIS Blume

A genus of about 15 species native to tropical Asia often found under the name *Anota* Schlechter. Cultivation as for *Gastrochilus* (p. 390).

HABIT: epiphytic.

STEMS: short, thick.

LEAVES: fleshy, strap-shaped or linear, in 2 ranks, 2-lobed at the apex.

RACEMES: lateral, erect or drooping, many-flowered.

SEPALS AND PETALS: spreading, petals narrower than sepals.

LIP: entire, spurred.

COLUMN: short, dilated at base with a short foot; rostellum beaked.

POLLINIA: 2.

Key to species

1. Apex of lip 3-lobed

1.A. Lip pinkish purple, white towards the middle and base and with 2 ridges near the base **3. gigantea**

1.B. Lip completely purple, with 5 ridges near the base **4. violacea**

2. Racemes erect; sepals and petals tipped with blue **1. coelestis**

3. Racemes drooping; sepals and petals with purplish or pinkish markings **2. retusa**

1. **R. coelestis** (Reichenbach) Veitch

ILLUSTRATIONS: Lindenia **7**: t. 300 (1891); Brooklyn Botanic Garden Record 23 (2), Handbook on orchids, 40–1 (1967).

RACEMES: erect.

FLOWERS: *c.* 2 cm in diameter, fragrant.

SEPALS AND PETALS: white tipped with blue.

LIP: obovate-oblong, blue or violet-blue, white at the base; spur slightly curved, compressed.

DISTRIBUTION: Thailand.

HARDINESS: G2.

FLOWERING: summer.

2. **R. retusa** (Linnaeus) Blume

ILLUSTRATIONS: Botanical Magazine, 4108 (1844); Orchid Digest **39**: 8 (1975); Bechtel et al., Manual of cultivated orchid species, 269 (1981).

RACEMES: drooping.

FLOWERS: *c.* 2 cm in diameter, fragrant.

SEPALS AND PETALS: white or pinkish, marked with pinkish or purple, lateral sepals broader than upper sepal and petals.

LIP: variable in shape, usually lanceolate, tip rounded or sometimes notched, mauve; spur truncate-conical, compressed.

DISTRIBUTION: from India and Sri Lanka to the Philippines.

HARDINESS: G2.

FLOWERING: summer.

'Alba' has completely white flowers and 'Holfordiana' has the sepals and petals spotted with crimson and a crimson lip.

3. **R. gigantea** (Lindley) Ridley

SYNONYMS: *R. densiflora* (Lindley) Williams; *Anota densiflora* (Lindley) Schlechter.

ILLUSTRATIONS: Orchid Digest **43**: 124 (1979); Bechtel et al., Manual of cultivated orchid species, 269 (1981).

RACEMES: drooping.
FLOWERS: 2.5–3 cm in diameter, fragrant.
SEPALS AND PETALS: white spotted with pinkish purple or pale purple.
LIP: oblong, 3-lobed at apex, pale purple or pinkish purple, white towards the middle and base and with 2 small ridges near the base; spur very short, obtuse, compressed.
DISTRIBUTION: Burma, Thailand, Laos.
HARDINESS: G2.
FLOWERING: autumn–winter.

Var. **harrisoniana** (Hooker) Holttum has pure white flowers and longer racemes; var. *petotiana* Anon. is probably the same.

4. **R. violacea** (Lindley) Reichenbach
SYNONYMS: *Anota violacea* (Lindley) Schlechter; *Saccolabium violaceum* Lindley.
ILLUSTRATION: Edwards's Botanical Register **33**: t. 30 (1847).
RACEMES: arching or drooping.
FLOWERS: 2–2.5 cm in diameter, fragrant.
SEPALS AND PETALS: white spotted with pale violet.
LIP: oblong, 3-lobed at apex, purple, with 5 ridges near the base; spur greenish, compressed.
DISTRIBUTION: Philippines.
HARDINESS: G2.
FLOWERING: winter–spring.

109. DORITIS Lindley

A genus of 2 species occurring in tropical Asia. Cultivation as for *Phalaenopsis* (p. 360).

Literature: Holttum, R.E., Cultivated species of the orchid genus Doritis Lindl., *Kew Bulletin* **19**: 207–12 (1965).

Like *Phalaenopsis* except that the claw of the lip bears 2 linear appendages and there are 4 pollinia.

1. **D. pulcherrima** Lindley
SYNONYMS: *Phalaenopsis pulcherrima* (Lindley) J.J. Smith; *P. esmeralda* Reichenbach; *P. buyssoniana* Reichenbach).
ILLUSTRATIONS: Skelsey, Orchids, 112 (1978); Sheehan & Sheehan, Orchid genera illustrated, 84 (1979).
HABIT: epiphytic.
LEAVES: to 20 cm, in 2 ranks, oblong, bright green, often purplish beneath.
FLOWERS: 2–2.5 cm wide, 20–30 in erect racemes which are 30–75 cm long.
SEPALS AND PETALS: 1–1.2 cm, pale pink to dark purple, usually somewhat reflexed.

LIP: 3-lobed with a long claw bearing a linear appendage on each side, appendages pale pink sometimes flushed with yellow, or with purple spots; lateral lobes erect, rounded, purple or striped with orange-red; central lobe oblong, acute, with a central ridge, purple or reddish.

DISTRIBUTION: Burma, Kampuchea, Thailand, Malay peninsula.

HARDINESS: G2.

FLOWERING: summer–autumn.

The flowers are variable in size and colour and are produced over a period of 2–3 months. Plants which are larger in all their parts and with the sepals and petals not reflexed are usually sold under the name *D. pulcherrima* var. *buyssoniana*, and are tetraploid. A white-flowered variant, 'Alba', is commercially available.

110. PHALAENOPSIS Blume

A genus of 40–55 species occurring in the Himalaya, SE Asia and N Australia. Their classification is based to a considerable extent on the complex lip-structure, and the identification of the species can be difficult. They are best grown in baskets in an epiphytic compost. They lack pseudobulbs and therefore they should not be allowed to dry out and the atmosphere should always be kept moist. Shading is essential in spring and early summer to prevent leaf scorch. Flowering can be encouraged by a drop in temperature.

Literature: Sweet, H.R., *The genus Phalaenopsis* (1980).

HABIT: epiphytic, without pseudobulbs.

STEM: short.

LEAVES: in 2 ranks, usually persistent, fleshy, rarely stalked.

RACEME OR PANICLE: lateral, with 1–many flowers and with a straight or zig-zag axis.

SEPALS: almost equal, free, usually spreading.

PETALS: similar to or larger than sepals.

LIP: relatively small, often more brightly coloured than the sepals and petals, 3-lobed, continuous with the column foot or inserted at a right angle, lateral lobes erect with cushion-like swellings in the middle, central lobe fleshy, usually firmly attached to the lateral lobes; between the lateral lobes and the base of the central lobe are various calluses and other ornamented processes.

COLUMN: often constricted in the middle, usually wingless, produced into a short foot at the base.

POLLINIA: 2, more or less spherical.

Key to species

1. Apex of lip with 2 tendril- or horn-like appendages *Go to Subkey 1*

2. Leaves mottled or spotted above *Go to Subkey* 2
3. Sepals and petals unmarked, or with dots at the base only *Go to Subkey* 3
4. Sepals and petals generally barred, blotched or spotted *Go to Subkey* 4
5. Sepals and petals longitudinally striped at the base and apex *Go to Subkey* 5

Subkey 1. *Apex of lip with 2 tendril- or horn-like appendages.*
1. Leaves not purplish beneath, even when young
1.A. Scape purplish; callus horseshoe-shaped **4. sanderiana**
1.B. Scape green; callus almost square **1. amabilis**
2. Leaves spotted above
2.A. Callus irregularly toothed, more or less square **6. × leucorrhoda**
2.B. Petals pink **5. schilleriana**
2.C. Petals white **7. stuartiana**
3. Lip white; callus notched at apex **2. aphrodite**
4. Lip pinkish purple; callus not notched **3. × intermedia**

Subkey 2. *Apex of lip without tendril- or horn-like appendages; leaves mottled or spotted above.*
1. Callus at junction of the lobes of the lip peltate
1.A. Inflorescence purplish; lip uniformly deep pink or purplish **12. equestris**
1.B. Inflorescence green; lip with white lateral lobes **13. lindenii**
2. Sepals and petals yellowish green blotched with brown
2.A. Sepals 1.4–1.8 cm; lip yellow **14. fuscata**
2.B. Sepals 1.5 cm or more **15. kunstleri**
3. Sepals 1.5 cm or more **16. cochlearis**
4. Sepals up to 1.5 cm **17. celebensis**

Subkey 3. *Apex of lip without tendril- or horn-like appendages; leaves not mottled or spotted; sepals and petals unmarked, or with dots at the base only.*
1. Callus with 4 bristle-like appendages originating at the base **8. parishii**
2. Inflorescence with a flattened axis **21. violacea**
3. Inflorescence with a terete axis **25. fimbriata**

Subkey 4. *Apex of lip without tendril- or horn-like appendages; leaves not mottled or spotted; sepals and petals generally barred, blotched or spotted.*
1. Central lobe of lip anchor-shaped
1.A. Scape not winged above **11. mannii**
1.B. Lip *c.* 8 mm **9. cornu-cervi**
1.C. Lip *c.* 1.5 cm **10 pantherina**
2. Petals broadly elliptic, up to twice as long as wide

2.A. Lip 2 cm or more **18. amboinensis**
2.B. Callus between the lateral lobes of the lip with a pair of toothed appendages **19. gigantea**
2.C. Callus between the lateral lobes of the lip with a lanceolate, backwardly-pointing appendage and a 2-toothed forwardly-pointing appendage **20 javanica**
3. Callus with elongate, backwardly-pointing warts **22. fasciata**
4. Lip more than 1.7 cm; petals 2 cm or more
4.A. Sepals and petals marked with spots and circles **27. hieroglyphica**
4.B. Sepals and petals with transverse bars in reddish brown or purple **26. lueddemanniana**
5. Central lobe of lip hairless
5.A. Lip with the callus between the lateral lobes 2-toothed; central lobe of lip red to deep purple **30. maculata**
5.B. Lip with the callus between the lateral lobes with a pair of forked appendages; central lobe of lip white, purple towards the apex **31. modesta**
6. Callus between the lateral lobes of the lip with a pair of forked appendages **29. pallens**
7. Callus between the lateral lobes of the lip with a series of 2-lobed appendages **28. mariae**

Subkey 5. Apex of lip without tendril- or horn-like appendages; leaves not mottled or spotted; sepals and petals longitudinally striped at base and apex.
1. Sepals and petals whitish to pale yellow with transverse reddish brown or purplish bars **23. sumatrana**
2. Sepals and petals greenish yellow towards apex, with longitudinal brown or purplish stripes merging into blotches near the base **24. corningiana**

1. **P. amabilis** (Linnaeus) Blume

ILLUSTRATIONS: Kupper & Linsenmaier, Orchids, 111 (1961); Orchid Digest **38**: 192 (1974); Bechtel et al., Manual of cultivated orchid species, 259, 362 (1981); Orchid Digest **46**: 29 (1982).

LEAVES: elliptic to obovate, 15–50 cm, green above and beneath.

INFLORESCENCE: simple or branched, often 1 m or more, loose with few to many flowers.

FLOWERS: fragrant, 6–10 cm in diameter.

SEPALS: 3–4 cm, elliptic-ovate, white, often pinkish on the back.

PETALS: to 4.5 cm, more or less circular, similar in colour to the sepals.

LIP: white, to 2.3 cm, lateral lobes oblanceolate with yellow margins and basal red markings, central lobe somewhat cross-shaped with triangular side-arms and 2 terminal, tendril-like

appendages with yellow tips; callus more or less square, yellow, spotted with red.
DISTRIBUTION: E Indies, Australia (NE Queensland).
HARDINESS: G2.
FLOWERING: autumn–early spring.
There are 3 varieties in cultivation:

Var. **aurea** Rolfe.
LIP: most of the lateral lobes, the margin of the central lobe, and the terminal appendages yellow.
DISTRIBUTION: Borneo.

Var. **moluccana** Schlechter
LIP: central lobe linear-oblong with small lateral lobes.
DISTRIBUTION: Indonesia (Moluccas, Sulawesi).

Var. **papuana** Schlechter
LIP: central lobe narrowly triangular.
DISTRIBUTION: New Guinea, Australia (NE Queensland).

2. **P. aphrodite** Reichenbach
SYNONYMS: *P. amabilis* var. *formosa* Shimadzu; *P. amabilis* var. *dayana* Warner & Williams.
ILLUSTRATIONS: Orchid Digest **38**: 192 (1974); American Orchid Society Bulletin **46**: 217 (1977); Bechtel et al., Manual of cultivated orchid species, 362 (1981).
LEAVES: broadly elliptic, 20–38 cm, green above, purplish beneath especially when young.
INFLORESCENCE: simple or branched, arching or hanging, longer than the leaves, purplish.
FLOWERS: 7 cm or more in diameter.
SEPALS: to 4 cm, elliptic, white, sometimes with tiny pink dots.
PETALS: similar to the sepals but more or less circular.
LIP: white, *c.* 3 cm, lateral lobes ovate with yellow margins and dark pink dots and streaks towards the base, central lobe triangular with a dark pink flush at the base and yellow on the side-arms, apex with 2 tendril-like appendages; callus with notched apex, each side with tooth-like projections, deep yellow with deep pink dots.
DISTRIBUTION: Philippines, Taiwan.
HARDINESS: G2.
FLOWERING: all the year.
According to Wallbrunn (American Orchid Society Bulletin **40**: 228, 1971) it is possible that *P. amabilis*, *P. aphrodite* and *P. sanderiana* are all one species.

3. **P. × intermedia** Lindley
ILLUSTRATIONS: Die Orchidee **24**: 144 (1973); Rittershausen & Rittershausen, Orchids in colour, 111 (1979); Bechtel et al., Manual of cultivated orchid species, 260, 362 (1981).

LEAVES: elliptic, to 30 cm, green above, brownish purple beneath.

INFLORESCENCE: to 60 cm, simple or branched, with many flowers.

SEPALS: 2.5–4 cm, ovate-elliptic, white with pink suffusion.

PETALS: 2–3.5 cm, broadly elliptic, white with pink dots at the base.

LIP: pinkish purple, lateral lobes ovate, purplish spotted with deep pink, central lobe obovate, tapered towards the tip, deep pinkish purple, apex notched or with 2 horn-like or tendril-like appendages; callus more or less square, yellow spotted with red.

DISTRIBUTION: Philippines, Borneo.

FLOWERING: spring.

A hybrid between *P. aphrodite* and *P. equestris.*

4. **P. sanderiana** Reichenbach

ILLUSTRATIONS: Orchid Digest **38**: 221 (1974); Rittershausen & Rittershausen, Orchids in colour, 121 (1979); Sweet, The genus Phalaenopsis, 41 (1980); Bechtel et al., Manual of cultivated orchid species, 362 (1981).

LEAVES: 15–35 cm, elliptic, dark green above, often marked with silvery grey beneath.

INFLORESCENCE: to 80 cm but usually less, purplish, simple or branched.

FLOWERS: 6–7.5 cm in diameter.

SEPALS: to 3.5 cm, ovate-elliptic, pink, sometimes with white mottling or completely white.

PETALS: similar to sepals but broader.

LIP: *c.* 3 cm, lateral lobes ovate, white spotted with pink, central lobe triangular with narrowly triangular side-arms, white streaked with brown or purple and sometimes yellow, apex with 2 tendril-like appendages; callus horseshoe-shaped, white or yellow with brown, red or purple spots.

DISTRIBUTION: Philippines.

HARDINESS: G2.

FLOWERING: spring.

Very variable in flower colour and in the markings on the lip. On this basis several varieties have been recognised but it is generally accepted that they have no botanical standing.

5. **P. schilleriana** Reichenbach

ILLUSTRATIONS: Flore des Serres **5**: t. 1559 (1862–65); Orchid Digest **38**: 221 (1974); Hunt & Grierson, Country Life book of orchids, 61 (1978); Bechtel et al., Manual of cultivated orchid species, 261, 362 (1981).

LEAVES: to 45 cm, elliptic, dark green spotted with silvery grey above, purple beneath.

INFLORESCENCE: to 90 cm and often bearing over 200 flowers, branched, hanging or arched, purplish.

FLOWERS: 5–7.5 cm in diameter.

SEPALS: ovate-elliptic, to 3.5 cm, pink fading at margins to white, lateral sepals with deep pink spots towards the base.

PETALS: more or less circular, to 4 cm, similar in colour to the sepals.

LIP: 2–2.3 cm, white to deep pink, lateral lobes elliptic, often broadly so, with reddish brown spots, central lobe ovate to rounded with 2 curved, horn-like appendages of varying size at the apex; callus more or less square, yellow.

DISTRIBUTION: Philippines.

HARDINESS: G2.

FLOWERING: winter–spring, sometimes also late summer.

6. P. × leucorrhoda Reichenbach

ILLUSTRATIONS: Cogniaux & Goossens, Dictionnaire Iconographique des Orchidées, Phalaenopsis hybrides, t. 1 (1902); Rittershausen & Rittershausen, Orchids in colour, 113 (1979).

LEAVES: green above, spotted with silvery grey, purplish beneath.

INFLORESCENCE: usually hanging, to 70 cm, with many flowers, purplish.

SEPALS: 2.2–3.5 cm, elliptic to ovate.

PETALS: 2.5–4 cm, kidney-shaped to almost circular; sepals and petals white, often flushed with pink, or deep pink with white margins, lateral sepals sometimes with purple dots at the base.

LIP: 2.2–3 cm, lateral lobes spathulate, white or yellow with reddish purple spots, central lobe white with yellow and purple markings, sometimes entirely purple, variable in shape, tapering towards the apex which has 2 anchor-shaped or tendril-like appendages; callus irregularly toothed, deep yellow, occasionally paler, densely spotted with dark red.

DISTRIBUTION: Philippines.

HARDINESS: G2.

A hybrid between *P. aphrodite* and *P. schilleriana*.

7. P. stuartiana Reichenbach

SYNONYMS: *P. schilleriana* var. *alba* Roebelen; *P. schilleriana* var. *vestalis* Reichenbach.

ILLUSTRATIONS: Orchid Digest 38: 221 (1974); Der Palmengarten 1979(4): 187 (1979); Bechtel et al., Manual of cultivated orchid species, 261, 362 (1981).

LEAVES: elliptic-oblong, 15–35 cm, green blotched with grey above, purplish beneath.

INFLORESCENCE: to 80 cm, with many flowers, branched, hanging.

FLOWERS: 5–6 cm in diameter, slightly fragrant.

SEPALS: to 3.5 cm, upper sepal white, lateral sepals white with yellow on basal half, and with brownish red dots.

PETALS: to 3.3 cm, somewhat square to circular, white, often with purplish brown dots at the base.

LIP: *c.* 2.5 cm, lateral lobes obovate, white at apex, yellow at base with reddish brown spots and with 2 white, horn-like appendages; callus almost square with an apical projection on each side, orange, spotted.

DISTRIBUTION: Philippines.

HARDINESS: G2.

FLOWERING: winter.

In var. **punctatissima** Reichenbach the dots on the flowers are pale purple.

8. **P. parishii** Reichenbach

ILLUSTRATION: Botanical Magazine, 5815 (1870).

LEAVES: to 12 cm, elliptic to obovate.

INFLORESCENCE: erect or arching, to 15 cm, with few or several flowers.

FLOWERS: opening simultaneously.

SEPALS: white, the upper 6–8 mm, elliptic to almost circular, the laterals obovate to almost circular, 7–10 mm.

PETALS: white, elliptic to obovate, 6–7 mm.

LIP: 10–15 mm, the short claw fused to the column foot, lateral lobes very small, triangular, yellow or whitish spotted with purple or brown, central lobe mobile, purple, triangular; callus semicircular, fringed at margin; at the junction of the lobes there is a plate-like projection ending in 4 bristle-like appendages.

COLUMN: white with purple spots.

DISTRIBUTION: E Himalaya, India.

HARDINESS: G2.

Var. **lobbii** Reichenbach

SYNONYM: *P. lobbii* (Reichenbach) Sweet.

ILLUSTRATIONS: Orchid Digest **37**: 168 (1973): American Orchid Society Bulletin **50**: 33–4 (1981).

LIP: white with a pair of vertical brown stripes on each side, and the callus margin entire or only slightly toothed.

9. **P. cornu-cervi** (Breda) Blume & Reichenbach

ILLUSTRATIONS: Botanical Magazine, 5570 (1866); Rittershausen & Rittershausen, Orchids in colour, 118 (1979); American Orchid Society Bulletin **50**: 32 (1981); Bechtel et al., Manual of cultivated orchid species, 362 (1981).

LEAVES: to 25 cm, oblong.

INFLORESCENCE: simple or branched, 10–40 cm, axis flattened.

FLOWERS: 3–5 cm in diameter.

SEPALS: 1.8–2.3 cm, elliptic, keeled on the back, upper sepal with recurved margins, yellow or greenish with reddish brown blotches, lateral sepals often with blotches only on the upper half.

PETALS: 7–18 mm, lanceolate, similar in colour to upper sepal.

LIP: *c.* 8 mm, lateral lobes almost square, with a callus below the truncate apex, whitish with reddish brown stripes, lateral lobes running into base of central lobe to form a flat, semicircular swelling, in front of which the central lobe is constricted and then expanded into a whitish, anchor-shaped portion with recurved lobes; callus complex with 3 appendages in series, the first upcurved, yellow, the second forked, white, and the third lanceolate, purple; the second and third project over the central lobe.

DISTRIBUTION: SE Asia.

HARDINESS: G2.

FLOWERING: summer.

10. **P. pantherina** Reichenbach

ILLUSTRATIONS: American Orchid Society Bulletin **38**: 510 (1969); Sweet, The genus Phalaenopsis, 61 (1980).

Like *P. cornu-cervi* except as below.

LIP: larger (*c.* 1.5 cm), differently shaped: below the swelling at the base of the central and lateral lobes, the central lobe has an isthmus which expands into a somewhat anchor-shaped portion with lobes at right angles; the callus appendages are complex.

DISTRIBUTION: Borneo.

11. **P. mannii** Reichenbach

SYNONYM: *P. boxallii* Reichenbach.

ILLUSTRATIONS: Orchid Digest **40**: 207 (1976); Bechtel et al., Manual of cultivated orchid species, 260, 362 (1981); American Orchid Society Bulletin **50**: 30–2 (1981).

LEAVES: to 35 cm, oblong.

INFLORESCENCE: usually simple, hanging, with many flowers.

FLOWERS: *c.* 5 cm in diameter, opening in succession.

SEPALS: obovate-lanceolate, yellow or green with brownish blotches, keeled on the back towards the apex, upper sepal 2–2.4 cm, margins rolled under, lateral sepals 2.2–2.5 cm.

PETALS: lanceolate, margins rolled under, 1.7–2 cm, similar in colour to sepals.

LIP: white and purple, 9–11 mm, lateral lobes almost square with a fleshy callus, base of lateral lobes running into the central lobe to form a semicircular swelling from which the variable central lobe expands from a basal isthmus; central lobe anchor-shaped, the lateral lobes toothed, upper surface warty and with an apical, often hairy callus; the basal callus appendages are complex.

DISTRIBUTION: Himalaya, Vietnam.

HARDINESS: G2.

FLOWERING: summer.

12. P. equestris (Schauer) Reichenbach

SYNONYM: *P. rosea* Lindley.

ILLUSTRATIONS: Flore des Serres **6**: t. 1645 (1865–67); Rittershausen & Rittershausen, Orchids in colour, 118 (1979); Die Orchidee **31**: 225 (1980); Bechtel et al., Manual of cultivated orchid species, 259, 362 (1981).

LEAVES: to 20 cm, oblong.

INFLORESCENCE: simple or branched, almost erect to arching, purplish, with many flowers.

FLOWERS: 2.5–4 cm in diameter.

SEPALS: oblong-elliptic, white to pink, the upper 1–1.7 cm, the laterals 1–1.5 cm.

PETALS: elliptic, 8–15 mm, similar in colour to sepals.

LIP: deep pink or purplish, 1–1.4 cm, lateral lobes oblong, often with dark streaks, central lobe ovate with apex pointed, more or less concave in the middle; callus peltate, 6–8-sided, yellow with red spots.

DISTRIBUTION: Philippines, Taiwan.

HARDINESS: G2.

FLOWERING: spring–autumn.

13. P. lindenii Loher

ILLUSTRATIONS: Orchid Digest **37**: 108 (1973); Bechtel et al., Manual of cultivated orchid species, 260 (1981); Orchid Digest **41**: 6 (1982).

Like *P equestris* except as below.

LEAVES: green mottled silvery white.

INFLORESCENCE: simple, usually green.

LIP: 1.2–1.4 cm, lateral lobes obovate to strap-shaped, white with tiny orange or reddish dots at the base and 3 purplish lines above, central lobe more or less circular, with a small point, slightly concave in the middle, purplish pink towards the apex with 5–7 darker lines.

DISTRIBUTION: Philippines.

HARDINESS: G2.

Perhaps a hybrid between *P. equestris* and *P. schilleriana*.

14. P. fuscata Reichenbach

ILLUSTRATION: Sweet, The genus Phalaenopsis, 64 (1980).

LEAVES: to 30 cm, oblong.

INFLORESCENCE: to 30 cm, simple or branched, with few flowers.

SEPALS AND PETALS: yellowish green blotched with brown, margins rolled under, sepals 1.4–1.8 cm, petals 1.2–1.5 cm.

LIP: yellowish with brown markings, 1.1–1.4 cm, lateral lobes almost square, twisted so that they touch, central lobe ovate to elliptic with a central ridge; callus at junction of lobes with 2 forked appendages.

DISTRIBUTION: Malay Peninsula.

HARDINESS: G2.

15. P. kunstleri Hooker

ILLUSTRATIONS: Botanical Magazine, 7885 (1903); Orchid Digest **40**: 209 (1976).

Like *P. fuscata* except as below.

FLOWERS: larger, *c.* 5 cm in diameter.

SEPALS AND PETALS: yellow at base and apex, brown in the central part, sepals 1.8–2 cm, petals 1.5 cm.

LIP: yellow, 1–1.2 cm, lateral lobes almost square, central lobe almost circular with 1 or 2 brown stripes on each side of the central ridge; callus at junction of lobes with 2 forked appendages.

DISTRIBUTION: Malaysia.

HARDINESS: G2.

Considered to be the same as *R. fuscata* by some authorities.

16. P. cochlearis Holttum

ILLUSTRATION: Orchid Digest **40**: 209 (1976).

Like *P. fuscata* except as below.

INFLORESCENCE: to 50 cm, branched.

SEPALS AND PETALS: white to pale yellow with 2 pale brown bars at the base, upper sepal 1.5–1.8 cm, lateral sepals 1.4–1.5 cm, petals 1.2–1.4 cm.

LIP: white or yellow, 9–12 mm, lateral lobes narrowly oblong, notched at apex with reddish streaks at the base, central lobe circular, concave, sometimes notched at apex, with reddish brown stripes and 5 shallow ridges; callus with a pair of 2-lobed appendages.

DISTRIBUTION: Sarawak.

17. P. celebensis Sweet

ILLUSTRATION: Sweet, The genus Phalaenopsis, 66 (1980).

Like *P. fuscata* except as below.

LEAVES: mottled with silvery white.

SEPALS AND PETALS: white with lilac stripes at the base, sepals to 1.4 cm, petals to 9 mm, slightly wider than sepals.

LIP: lateral lobes deep yellow, rounded, with a central fleshy ridge, central lobe up to 9 mm, concave, elliptic-ovate, with a small point at apex; callus at junction of lobes more or less triangular.

DISTRIBUTION: Indonesia (Sulawesi).

HARDINESS: G2.

18. P. amboinensis J.J. Smith

ILLUSTRATION: Orchid Digest **36**: 88 (1972).

LEAVES: to 25 cm, elliptic to obovate.

INFLORESCENCE: arching, to 45 cm, with few flowers; flowers opening in succession.

SEPALS AND PETALS: white or yellowish with greenish tips and brownish red bars, elliptic to ovate, sepals 2–3 cm, petals shorter.

LIP: white or yellowish, 2–2.2 cm, lateral lobes oblong, each with an orange spot and a callus, central lobe ovate, the central ridge with a toothed edge; callus with a pair of 2-lobed appendages.

DISTRIBUTION: Indonesia (Amboina).

HARDINESS: G2.

FLOWERING: summer.

19. P. gigantea J.J. Smith

ILLUSTRATIONS: Orchid Digest **36**: 38 (1972); Bechtel et al., Manual of cultivated orchid species, 259, 362 (1981); Orchid Digest **41**: 29 (1982).

Like *P. amboinensis* except as below.

LEAVES: to 90 cm, oblong-ovate.

FLOWERS: *c.* 5 cm in diameter, slightly fragrant, opening simultaneously, numerous, in hanging, dense inflorescences.

SEPALS AND PETALS: greenish or yellow, whitish at the base, blotched with reddish brown or purple.

LIP: *c.* 1.5 cm, lateral lobes triangular, curved, orange towards the apex, central lobe ovate with some tiny teeth on the lateral margins and an ovoid callus at the apex, striped and blotched with red or purple; callus between lateral lobes with a pair of 2-toothed, deep yellow appendages.

DISTRIBUTION: Borneo.

20. P. javanica J.J. Smith

SYNONYM: *P. latisepala* Reichenbach.

ILLUSTRATION: Orchid Digest **43**: 57 (1979).

Like *P. amboinensis* except as below.

LEAVES: to 22 cm, elliptic.

FLOWERS: *c.* 3 cm in diameter.

SEPALS AND PETALS: broadly elliptic, white to yellow with brownish purple spots arranged in longitudinal lines.

LIP: distinctly clawed 1–1.8 cm, lateral lobes linear-oblong, grooved in the middle, yellowish, apex with 2 teeth, central lobe elliptic, very fleshy, apex with a few hairs, purple, whitish towards the base; callus between lobes consisting of a lanceolate, backwardly pointing appendage and a 2-toothed forwardly pointing appendage.

DISTRIBUTION: Java.

HARDINESS: G2.

21. P. violacea Witte

ILLUSTRATIONS: Orchid Digest **36**: 12 (1972); Die Orchidee **30**: 122–3 (1979); Bechtel et al., Manual of cultivated orchid species, 362 (1981).

LEAVES: 20–25 cm, elliptic to obovate.

INFLORESCENCE: more or less erect or arching, simple, with a few, distant flowers; axis flattened.

FLOWERS: 5–7.5 cm in diameter.

SEPALS AND PETALS: bright purple with pale greenish tips; sepals elliptic, 2–3.5 cm, keeled on the back towards the apex, petals elliptic, 2–3 cm.

LIP: reddish purple, 1.8–2.8 cm, lateral lobes linear-oblong, marked with yellow, central lobe ovate, abruptly pointed with a central ridge running into an apical callus; callus between lateral lobes with warts and complex appendages, yellow.

DISTRIBUTION: Malay peninsula, Borneo, Sumatra.

HARDINESS: G2.

FLOWERING: summer.

In var. **alba** Teijsmann & Binnedijk (Illustration: American Orchid Society Bulletin **34**: 206, 1965) the sepals and petals are white with greenish tips.

22. P. fasciata Reichenbach

ILLUSTRATIONS: Orchid Digest **40**: 207 (1976); Bechtel et al., Manual of cultivated orchid species, 259 (1981).

LEAVES: 14–20 cm, elliptic.

INFLORESCENCE: more or less erect or arching, branched, longer than leaves.

FLOWERS: *c.* 4 cm in diameter.

SEPALS AND PETALS: yellow to yellowish green with transverse red-brown bars; sepals ovate-elliptic, 2–3 cm, petals ovate to ovate-elliptic, 2.2–2.6 cm.

LIP: 2–2.7 cm, lateral lobes strap-shaped, erect, yellow, central lobe pale purple, oblong-ovate, convex with a central ridge running into an apical callus; callus between lateral lobes orange, with elongate, backwardly pointing warts and a forwardly pointing forked appendage.

DISTRIBUTION: Phillippines.

HARDINESS: G2.

In cultivation many of the plants offered under the name *P. lueddemanniana* var. *ochracea* are *P. fasciata*.

23. P. sumatrana Korthals & Reichenbach

SYNONYMS: *P. zebrina* Witte; *P. zebrina* Teijsmann & Binnedijk.

ILLUSTRATIONS: American Orchid Society Bulletin **37**: 1093 (1968); Orchid Digest **36**: 25 (1972); Bechtel et al., Manual of cultivated orchid species, 362 (1981).

LEAVES: 15–30 cm, oblong to obovate.

INFLORESCENCE: erect or slightly arching, usually simple.

FLOWERS: *c.* 5 cm in diameter.

SEPALS AND PETALS: oblong-lanceolate, whitish to pale yellow with transverse, reddish brown or purplish bars; sepals 2–4 cm, petals 2.2–3.5 cm.

LIP: to 2.5 cm, lateral lobes linear-oblong, apex 2-toothed with a third tooth on the inside, cream with yellow or brown margin and often orange spots, central lobe oblong-elliptic, very thick especially towards the hairy apex, convex with a central ridge, white with reddish or purple stripes on each side of the ridge; callus between the lateral lobes a complex series of forked plates.

DISTRIBUTION: Thailand, Malay peninsula, Java, Sumatra & Borneo.

HARDINESS: G2.

24. P. corningiana Reichenbach

ILLUSTRATION: Orchid Digest **40**: 105 (1976).

Like *P sumatrana* except as below.

SEPALS AND PETALS: greenish yellow towards the apex with longitudinal brown or purplish stripes merging into blotches towards the base, upper sepal obovate to oblanceolate, 3–4 cm, lateral sepals ovate, 2.5–3 cm, petals lanceolate 2.5–3.5 cm.

LIP: *c.* 2 cm, lateral lobes strap-shaped, whitish with a central orange-yellow callus, central lobe deep purplish pink, narrowly elliptic, convex with a central ridge running into a hairy apical callus; callus between lateral lobes yellow, forked with a central groove; below the callus is an appendage with finger-like projections.

DISTRIBUTION: Malaysia, Sarawak.

HARDINESS: G2.

Possibly the same as *P. sumatrana*.

25. P. fimbriata J.J. Smith

ILLUSTRATIONS: Sweet, The genus Phalaenopsis, 97 (1980); Bechtel et al., Manual of cultivated orchid species, 259, 362 (1981).

LEAVES: 14–23 cm, oblong-elliptic.

INFLORESCENCE: to 30 cm, usually hanging, loose, with many flowers.

FLOWERS: opening simultaneously.

SEPALS AND PETALS: white, shaded green towards the tips and sometimes with fine transverse reddish purple lines, sepals 1.5–2 cm, petals 1.5–1.8 cm, all elliptic.

LIP: 1.4–2 cm, lateral lobes oblong, white flushed with reddish purple, convex with a central toothed ridge running into an apical callus with white hairs, lateral margins of central lobe expanded and toothed towards the apex; callus between the lateral lobes with 3 overlapping plates, each with 2 teeth.

DISTRIBUTION: Java, Sumatra.
HARDINESS: G2.

26. P. lueddemanniana Reichenbach
ILLUSTRATIONS: Orchid Digest **40**: 207 (1976); Bechtel et al., Manual of cultivated orchid species, 362 (1981).
LEAVES: 20–30 cm, oblong-elliptic.
INFLORESCENCE: erect or hanging, simple or branched, longer than the leaves.
FLOWERS: 5–6 cm in diameter.
SEPALS AND PETALS: white with transverse brownish purple bars, sepals oblong-elliptic, 2–3 cm, petals slightly smaller.
LIP: pink to purple, 1.8–2.2 cm, lateral lobes oblong, central lobe oblong to ovate with a central ridge running into an apical callus with white hairs; callus between lateral lobes warty and with a forked appendage.
DISTRIBUTION: Philippines.
HARDINESS: G2.
FLOWERING: summer.
Very variable in the size and colour of the flowers.

Var. **delicata** Reichenbach
ILLUSTRATION: Flore des Serres **6**: t. 1636 (1865–67).
SEPALS AND PETALS: white, with narrow brown bars on the apical part and purple bars on the basal part.
LIP: purple.

Var. **ochracea** Reichenbach
ILLUSTRATION: Orchid Digest **40**: 207 (1976).
SEPALS AND PETALS: yellowish with yellowish brown bars and a basal pale purple flush.
This is often confused with *P. fasciata* (p. 371) and many plants cultivated under this name are, in fact, *P. fasciata*. According to Wallbrunn (American Orchid Society Bulletin **40**: 225, 1971), it may be better to consider *P. fasciata*, *P. hieroglyphica* and perhaps *P. pallens* as subspecies of a variable *P. lueddemanniana*.

27. P. hieroglyphica (Reichenbach) Sweet
SYNONYM: *P. lueddemanniana* var. *hieroglyphica* Reichenbach.
ILLUSTRATIONS: Orchid Digest **40**: 207 (1976); Bechtel et al., Manual of cultivated orchid species, 260, 362 (1981); Orchid Review **92**: 70 (1984).
LEAVES: to 30 cm, strap-shaped.
INFLORESCENCE: more or less erect or arching, simple or branched, with many flowers.
SEPALS AND PETALS: ovate-elliptic, white, greenish or sometimes yellowish with brownish spots and circles; sepals keeled on the back towards the apex, the upper 2.3–3.8 cm, the laterals 2.4–4.1 cm, petals 2.2–3.3 cm.
LIP: white or yellowish, 2–2.5 cm, lateral lobes oblong, apex

notched, central lobe wedge-shaped with a central ridge running into an apical hairy callus; callus between lateral lobes with elongate, forwardly pointing warts and 2 forked appendages.

DISTRIBUTION: Philippines.

HARDINESS: G2.

FLOWERING: summer.

28. P. mariae Warner & Williams

ILLUSTRATIONS: Botanical Magazine, 6964 (1887); Orchid Digest **35**: 308 (1971); Bechtel et al., Manual of cultivated orchid species, 260 (1981).

LEAVES: 15–30 cm, strap-shaped.

INFLORESCENCE: hanging, simple or branched.

FLOWERS: 2.5–5 cm in diameter.

SEPALS AND PETALS: oblong-elliptic, white or cream with transverse brownish red bars and blotches, sometimes with purplish spots at the base; sepals 1.6–2.2 cm, petals 1.5–1.7 cm.

LIP: pale mauve or purple, lateral lobes broadly strap-shaped, toothed at the apex, central lobe ovate, the margins expanded and toothed towards apex, with a central ridge at the base and a hairy callus at the apex; callus between lateral lobes with a series of 2-lobed appendages.

DISTRIBUTION: Philippines.

HARDINESS: G2.

FLOWERING: summer.

29. P. pallens (Lindley) Reichenbach

SYNONYM: *P. foerstermanii* Reichenbach.

ILLUSTRATIONS: Orchid Digest **40**: 207 (1976); Sweet, The genus Phalaenopsis, 109 (1980); Bechtel et al., Manual of cultivated orchid species, 261, 362 (1981).

Like *P. mariae* except as below.

LEAVES: elliptic to obovate, 12–18 cm.

INFLORESCENCE: erect or arching, simple, with 1 or a few flowers.

FLOWERS: *c.* 5 cm in diameter.

SEPALS AND PETALS: oblong-elliptic, yellowish green with transverse brown bars, lines and spots; sepals with a keel on the back, the upper 1.2–2.3 cm, the laterals 1.5–2.2 cm, petals 1.1–2 cm.

LIP: 1.3–1.7 cm, lateral lobes oblong, yellowish, central lobe white, narrowly ovate, margins usually expanded and toothed towards the apex, with a central ridge at the base and hairy callus towards the apex; callus between lateral lobes with a pair of forked appendages.

DISTRIBUTION: Philippines.

Var. **denticulata** (Reichenbach) Sweet has 2 or 3 reddish purple lines on either side of the central ridge of the central lobe of the lip.

30. P. maculata Reichenbach

ILLUSTRATIONS: Orchid Digest **35**: 308 (1971), **41**: 29 (1982).

LEAVES: 15–20 cm, strap-shaped.

INFLORESCENCE: more or less erect to arching, usually simple, with few flowers.

FLOWERS: less than 2.5 cm in diameter.

SEPALS AND PETALS: oblong-lanceolate, cream to greenish white with a few purplish brown or red-brown blotches; upper sepal 1.3–1.8 cm, lateral sepals 1.3–1.5 cm, petals *c.* 1.2 cm.

LIP: *c.* 1 cm, lateral lobes diamond-shaped, the apical part with a horseshoe-shaped callus, central lobe red to deep purple, elliptic with faint longitudinal grooves, very thick and fleshy. Callus between side lobes 2-toothed; at the base of central lobe is a shallow, 2-toothed process.

DISTRIBUTION: Malay peninsula, Sarawak.

HARDINESS: G2.

31. P. modesta J.J. Smith

SYNONYM: *P. psilantha* Schlechter.

ILLUSTRATION: Orchid Digest **43**: 212 (1979).

Like *P. maculata* except as below.

LEAVES: obovate, 11–15 cm.

INFLORESCENCE: arched, usually simple, with few flowers.

SEPALS AND PETALS: elliptic, white or pale pink, with fine transverse purple stripes towards the base, sepals 1.2–1.5 cm, petals 1.1–1.5 cm.

LIP: 1.2–1.4 cm, lateral lobes oblong, narrowing towards the notched apex, yellow callus at middle, central lobe oblong, white, purple towards the apex, convex, with a basal ridge and an apical, usually hairless callus; callus between the lateral lobes with a pair of forked appendages.

DISTRIBUTION: Borneo, Sulawesi.

HARDINESS: G2.

111. PARAPHALAENOPSIS Hawkes

A genus of 3 species from W. Borneo (Kalimantan), all of which are in cultivation. Though the genus has been crossed with several other (e.g. *Vanda*, *Renanthera*), attempts to hybridise it with *Phalaenopsis* have so far failed. Cultivation as for *Phalaenopsis* (p. 360).

Literature: Sweet, H.R., The genus *Phalaenopsis*, 118–23 (1980).

Like *Phalaenopsis* except as below.

LEAVES: terete.

RACEMES: often more than 1 raceme from the same point on the stem.

Key to species

1. Sepals and petals greenish yellow to yellowish brown **1. denevei**
2. Flowers 3–4 cm in diameter; central lobe of lip with 2 divergent apical lobes like a snake's tongue **2. serpentilingua**
3. Flowers 5–8 cm in dimaeter; central lobe of lip with 2 short divergent lobes **3. laycockii**

1. P. denevei (J.J. Smith) Hawkes

SYNONYM: *Phalaenopsis denevei* J.J. Smith.

ILLUSTRATION: Orchid Digest **37**: 12–13 (1973).

LEAVES: erect or hanging, to 70 cm.

RACEMES: usually upright with 3–15 flowers, to 15 cm.

FLOWERS: *c.* 5 cm in diameter, fragrant.

SEPALS AND PETALS: pale greenish yellow to yellowish brown, with paler, wavy margins, *c.* 2.5 cm.

LIP: lateral lobes triangular, yellow at base, purple above, with white margin, central lobe narrowly oblong, somewhat expanded and notched at apex, whitish, reddish or purplish towards apex; callus more or less square with a slightly toothed margin, yellow with red lines.

DISTRIBUTION: W Borneo.

HARDINESS: G2.

FLOWERING: spring–summer.

2. P. serpentilingua (J.J. Smith) Hawkes

SYNONYM: *Phalaenopsis serpentilingua* J.J. Smith.

ILLUSTRATION: Orchid Digest **37**: 12 (1973).

LEAVES: as for *P. denevei*.

RACEMES: usually upright, to 35 cm, with 7–many flowers.

FLOWERS: 3–4 cm in diameter, fragrant.

SEPALS AND PETALS: white or cream, sometimes tinged with violet.

LIP: lateral lobes pale orange with reddish markings, central lobe narrowly oblong with 2 divergent apical lobes like a snake's tongue, white or pale yellow with transverse yellow and purple marks; callus with a toothed margin, yellow with red marks.

DISTRIBUTION: W Borneo.

HARDINESS: G2.

3. P. laycockii (Henderson) Hawkes

SYNONYM: *Phalaenopsis laycockii* Henderson.

ILLUSTRATION: Orchid Digest **37**: 12–13 (1973).

Like *P. serpentilingua* except as below.

RACEMES: with 9–15 flowers which are 5–8 cm in diameter.

SEPALS AND PETALS: 3.5–4.5 cm, whitish flushed with pink or pale purple.

LIP: lateral lobes white flushed with dark purple or red, central lobe narrowly oblong, margins reflexed with 2 short, divergent apical lobes, callus margins untoothed.

DISTRIBUTION: W Borneo.

HARDINESS: G2.

112. VANDA R. Brown

A genus of 50 or more species from the Himalaya and SE Asia. Its classification is confused, as are its relations with other genera. Species No. 1 is often referred to *Euanthe* Schlechter (recognised by its flat flowers with unequal sepals and the particular form of the spurless lip); Nos. 2 & 3 have been separated off into the genus *Papilionanthe* (a genus proposed by Schlechter in 1915, but hardly taken up until recently—see Garay, Botanical Museum Leaflets, Harvard University **23**: 369–72, (1974), recognised by the terete leaves and the extension of the short column into a conspicuous foot); No. 4 is referred by Garay to the genus *Holcoglossum* Schlechter (recognised by its narrow, cylindric spur, grooved leaves, footless, winged column and linear stipe). However, for the present purpose it seems sensible to maintain a wide circumscription of *Vanda*, citing synonyms as appropriate.

The number of species in cultivation is uncertain; 16 are included here, though only 8 are treated fully. A large number of hybrids is also available. Plants should be given as much light as possible, with only light shading in very sunny weather. They require an epiphytic compost and can be successfully grown in pots or on lengths of tree fern or bark. Propagation is by means of pieces of stem cut back to a well-developed aerial root.

Literature: Christensen, E.A., An infrageneric classification of Holcoglossum with a key to the genera of the Vanda-Aerides alliance, *Notes from the Royal Botanic Garden Edinburgh* **44**: 249–256 (1987).

HABIT: monopodial epiphytes with long or short stems, sometimes scrambling.

LEAVES: in 2 ranks, often fleshy, flat or terete, often lobed or toothed at the apex, folded when young.

RACEMES: axillary with few to many flowers.

SEPALS AND PETALS: similar or the lateral sepals larger than the rest, all more than 1.5 cm, usually spreading.

LIP: often complex, 3-lobed, the lateral lobes large or small, the central lobe often itself notched or lobed, bearing calluses of various forms, generally with 2 humped ridges at the entrance to the spur; spur usually present, conical or more rarely cylindric.

COLUMN: short and thick with or without a foot.

POLLINIA: 2, usually borne on a short, broad stipe and with a large viscidium.

Synopsis of characters

Leaves. Terete: **2,3**; V-shaped in section: **4,5**; flat: **1,6–16**.
Sepals and petals. Spreading: **1–13**; curved forwards: **14–16**.
Lip. Central lobe with 2 divergent lobes at apex: **10–16**; with a downwardly projecting horn at apex: **14**. Shorter than sepals: **1,6**; as long as or longer than sepals: **2–5,7–16**.
Spur. Absent: **1,14–16**; cylindric, *c.* 2.5 cm: **4**; conical and much shorter: **2,3,5–16**.

Key to species

1. Leaves terete or very narrowly V-shaped in section — *Go to Subkey 1*
2. Sepals and petals curved forwards; racemes with 2–5 flowers — *Go to Subkey 2*
3. Flowers flat with broad, overlapping sepals and petals; base of the lip hollowed out, but spur absent — **1. sanderiana**
4. Central lobe of lip expanded into a kidney-shaped apical part from an oblong base — *Go to Subkey 3*
5. Central lobe of lip oblong, sometimes waisted but then the apical part rounded, not kidney-shaped — *Go to Subkey 4*

Subkey 1. Leaves terete or narrowly V-shaped in section.

1. Leaves terete, not grooved

1.A. Flowers 8–10 cm in diameter; petals and upper sepals violet or pink — **2. teres**

1.B. Flowers to 5 cm in diameter; sepals and petals all white flushed with pink — **3. hookeriana**

2. Spur cylindric, acute, *c.* 2.5 cm; lateral sepals sickle-shaped, larger than upper — **4. kimballiana**

3. Spur shorter, blunt, conical; lateral sepals similar to the upper — **5. amesiana**

Subkey 2. Leaves flat; sepals and petals curved forwards; racemes with 2–5 flowers.

1. Sepals and petals at most 1.8 cm; flowers hanging or drooping — **15. alpina**
2. Central lobe of lip oblong, 2-lobed at the apex, bearing a downwardly-pointing projection between the lobes, all bright yellow to green with brown stripes — **14. cristata**
3. Central lobe of lip ovate, concave, without lobes or a projection, red-striped — **16. pumila**

Subkey 3. Leaves flat; sepals and petals spreading; central lobe of lip expanded into a kidney-shaped apical part from an oblong base.

1. Sepals and petals greenish to creamy white; lip white with a yellow patch at the base — **10. denisoniana**

2. Sepals and petals pale violet; lip dark violet **11. caerulescens**

3. Sepals and petals olive-green with brownish spots; lip violet-pink, white at the base **12. bensonii**

4. Sepals and petals violet, yellow or white; lip of the same colour **13. insignis**

Subkey 4. Leaves flat; sepals and petals spreading; central lobe of lip oblong, if waisted, then apical part not kidney-shaped.

1. Flower 7–10 cm in diameter; lip considerably smaller than the sepals **6. caerulea**

2. Sepals and petals wavy, tessellated yellow and brown **9. tesselata**

3. Sepals and petals wavy, yellow spotted with dark brown (white on their backs) **7. tricolor**

4. Sepals and petals white or flushed with red **8. lamellata**

1. **V. sanderiana** Reichenbach

SYNONYM: *Euanthe sanderiana* (Reichenbach) Schlechter.

ILLUSTRATIONS: Botanical Magazine, 6983 (1888); Lindenia **12**: t. 547 (1896); Cogniaux & Goossens, Dictionnaire Iconographique des Orchidées, Vanda, t. 12 (1900); Orchid Digest **43**: 84 (1979); American Orchid Society Bulletin **57**: 269 (1988).

STEM: to 60 cm.

LEAVES: to 45 × 5 cm, flat, oblong-linear, sharply truncate and 3-toothed at the apex.

RACEME: to 30 cm with 5–10 flowers.

FLOWERS: 9–10 cm in diameter, flat.

SEPALS AND PETALS: overlapping, broadly elliptic; upper sepal to 5.5 cm, violet-blue like the petals, which are *c.* 4 cm; lateral sepals to 6 cm, yellow or greenish with red-brown marbling.

LIP: shorter than the petals, with a lower, hemispherical, saccate part bearing 2 semicircular lateral lobes, the central lobe broadly kidney-shaped with 3 low ridges.

DISTRIBUTION: Philippines.

HARDINESS: G2.

FLOWERING: autumn–winter.

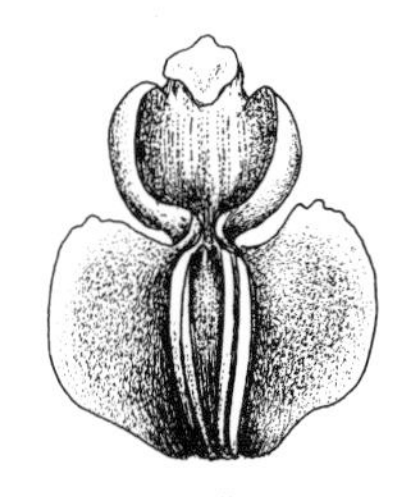

Vanda sanderiana – lip.

2. **V. teres** (Roxburgh) Lindley

SYNONYM: *Papilionanthe teres* (Roxburgh) Schlechter.

ILLUSTRATIONS: Botanical Magazine, 4114 (1844); Orchid Digest **43**: 84 (1979); Bechtel et al., Manual of cultivated orchid species, 279 (1981).

STEM: to 2 m.

LEAVES: cylindric, terete, upright, rather blunt.

RACEMES: to 30 cm with 3–5 flowers.

FLOWERS: 8–10 cm in diameter.

SEPALS AND PETALS: elliptic, lateral sepals white, petals and upper sepal violet or pink.

Vanda teres – lip.

LIP: as long as sepals, lateral lobes large, arching over the column, reddish striped with yellow at the base, central lobe obovate, deeply 2-lobed, violet-pink to yellow with red spots towards the base; spur short, conical.
DISTRIBUTION: E Himalaya to Thailand & Laos.
HARDINESS: G2.
FLOWERING: spring.

3. V. hookeriana Reichenbach
SYNONYM: *Papilionanthe hookeriana* (Reichenbach) Schlechter.
ILLUSTRATIONS: The Garden **23**: 10 (1883); Illustration Horticole **30**: t. 484 (1883).
Like *V. teres* except as below.
FLOWERS: to 5 cm in diameter.
SEPALS AND PETALS: white, flushed and spotted with red.
LIP: with spreading lateral lobes and a fan-shaped, white and red-spotted, shallowly 3-lobed central lobe.
DISTRIBUTION: Malaysia, Sumatra, Borneo.
FLOWERING: autumn.

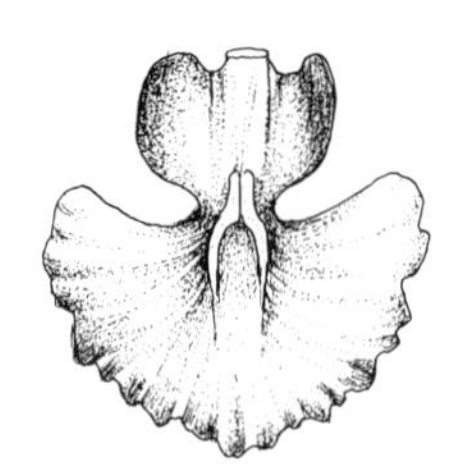

Vanda hookeriana – lip.

4. V. kimballiana Reichenbach
SYNONYM: *Holcoglossum kimballianum* (Reichenbach) Garay.
ILLUSTRATIONS: Lindenia **5**: t. 204 (1889); Botanical Magazine, 7112 (1890); Cogniaux & Goossens, Dictionnaire Iconographique des Orchidées, Vanda, t. 7 (1898).
STEM: elongate.
LEAVES: to 20 cm, fleshy, narrowly V-shaped in section, acute.
SEPALS AND PETALS: *c.* 2.5 cm, oblong-elliptic, white, wavy, the lateral sepals longer, sickle-shaped.
LIP: about as long as the sepals with small, acuminate, yellow, red-spotted lateral lobes and a large, nearly circular, toothed and notched, rose-red central lobe; spur *c.* 2.5 cm, cylindric, curving downwards, acute.
DISTRIBUTION: Burma.
HARDINESS: G2.
FLOWERING: autumn.

Vanda kimballiana – lip.

5. V. amesiana Reichenbach
SYNONYM: *Holcoglossum amesianum* (Reichenbach) Christensen.
ILLUSTRATIONS: Botanical Magazine, 7139 (1890); Cogniaux & Goossens, Dictionnaire Iconographique des Orchidées, Vanda, t 1 (1897); Orchid Digest **43**: 44 (1979).
STEM: short.
LEAVES: to 20 cm, fleshy, narrowly V-shaped in section, acute.
RACEME: to 50 cm, with 15–30 flowers, axis red-spotted.
SEPALS AND PETALS: *c.* 2 cm, broadly oblong, white.
LIP: slightly longer than sepals, pink with white margin and stripes, lateral lobes erect, short, central lobe broadly ovate, wavy; spur conical, very short and blunt.

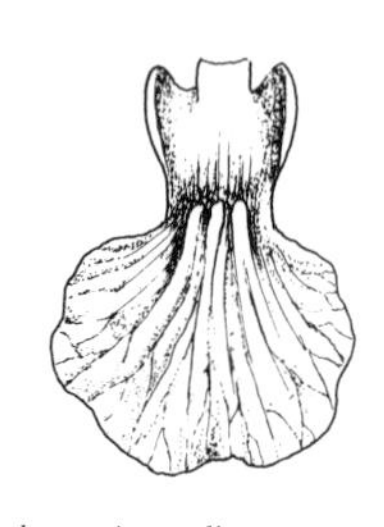

Vanda amesiana – lip.

DISTRIBUTION: Kampuchea, Burma SW China.
HARDINESS: G1.
FLOWERING: winter

6. V. caerulea Lindley
ILLUSTRATIONS: Schlechter, Die Orchideen, edn 2, t. 13 (1927); Orchid Digest **43**: 44, 49 (1979); Bechtel et al., Manual of cultivated orchid species, 278 (1981); American Orchid Society Bulletin **54** 324 (1985).
STEMS: short, densely leafy.
LEAVES: to 25 × 2.5, oblong, blunt, flat.
RACEMES: to 45 cm with 7–15 flowers.
FLOWERS: 7–10 cm in diameter.
SEPALS AND PETALS: oblong-obovate, clear blue, often with darker tessellation, petals often with twisted claws.
LIP: much shorter than the sepals, dark violet-blue with small whitish lateral lobes and an oblong, convex, 2–3-ridged central lobe; spur conical, short, blunt.
DISTRIBUTION: E Himalaya, Burma, Thailand.
HARDINESS: G2.
FLOWERING: winter.

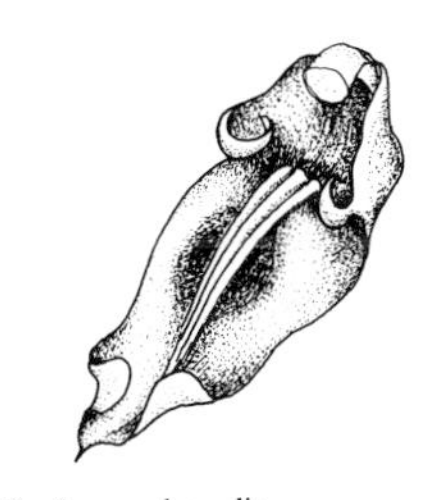
Vanda caerulea – lip.

7. V. tricolor Lindley
SYNONYM: *V. suavis* Reichenbach.
ILLUSTRATIONS: Botanical Magazine, 4432 (1894); Bechtel et al., Manual of cultivated orchid species, 279 (1981).
STEMS: to 1 m.
LEAVES: to 45 × 5 cm, oblong.
RACEMES: erect with 8–10 flowers.
FLOWERS: *c.* 5 cm in diameter, fragrant.
SEPALS AND PETALS: broadly ovate with broad, claw-like bases, wavy, yellow spotted with dark brown, their backs white.
LIP: as long as the sepals, lateral lobes small, erect, white, central lobe broadly oblong, waisted at the middle, the basal part broader than the apical, violet-red with purple stripes; spur short, broad, laterally compressed.
DISTRIBUTION: Java.
HARDINESS: G2.
FLOWERING: winter.

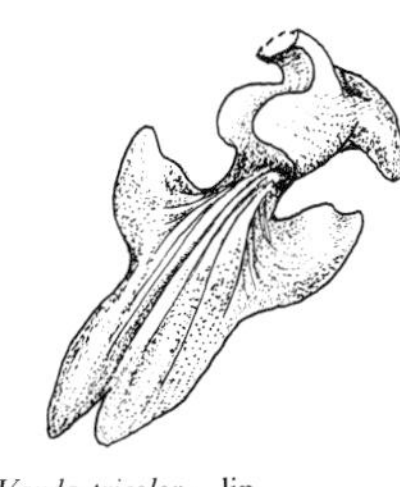
Vanda tricolor – lip.

8. V. lamellata Lindley
SYNONYM: *V. boxallii* Reichenbach.
ILLUSTRATION: Cogniaux & Goossens, Dictionnaire Iconographique des Orchidées, Vanda, t. 9 (1898).
Like *V. tricolor* except as below.
FLOWERS: smaller, non-fragrant.
SEPALS AND PETALS: *c.* 2 cm, whitish (flushed red in 'Boxallii').

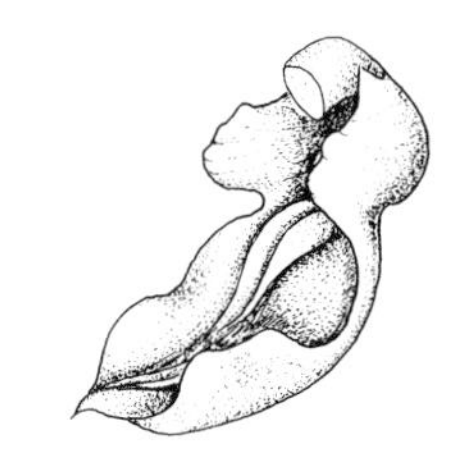

Vanda lamellata – lip.

LIP: shorter than the sepals, the central lobe violet, oblong, notched, with 2–3 conspicuous ridges.
DISTRIBUTION: Philippines.
HARDINESS: G2.
FLOWERING: winter.

9. V. tesselata (Roxburgh) G. Don
SYNONYMS: *Epidendrum tesselatum* Roxburgh; *V. roxburghii* R. Brown.
ILLUSTRATIONS: Botanical Magazine, 2245 (1821); Die Orchidee **22**: 13 (1971); Orchid Digest **43**: 47 (1979); Bechtel et al., Manual of cultivated orchid species, 279 (1981).
Also like *V. tricolor*, except as below.
SEPALS AND PETALS: wavy, tessellated yellow and brown.
LIP: central lobe oblong, slightly waisted, dark violet.
DISTRIBUTION: Tropical Asia from SE India to Malaysia.

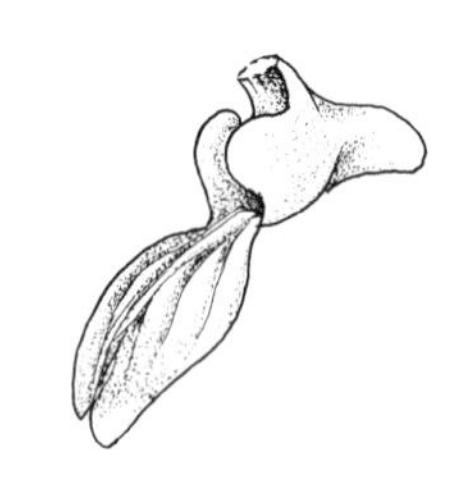

Vanda tesselata – lip.

10. V. denisoniana Benson & Reichenbach
ILLUSTRATIONS: Botanical Magazine, 5811 (1869); Cogniaux & Goossens, Dictionnaire Iconographique des Orchidées, Vanda, t. 8 (1898); Seidenfaden & Smitinand, Orchids of Thailand 4(1): t. 27 (1963).
STEM: short.
LEAVES: to 30 × 2 cm, linear, acutely 2-lobed at the apex.
RACEMES: to 15 cm, with about 6 flowers, arching.
FLOWERS: *c.* 5 cm in diameter.
SEPALS AND PETALS: elliptic with broad claws, greenish to cream-white, the lateral sepals somewhat longer than the rest.
LIP: a little longer than the sepals, white with a yellow patch at the base, lateral lobes small, central lobe oblong at the base, bearing 4 or 5 ridges, expanding towards the apex to a kidney-shaped, notched portion; spur oblong, blunt, laterally compressed.
DISTRIBUTION: Burma, China, Thailand.
HARDINESS: G2.
FLOWERING: spring.

Vanda denisoniana – lip.

11. V. caerulescens Griffith
ILLUSTRATIONS: Botanical Magazine, 5834 (1870); Orchid Digest **43**: 44 (1979).
Like *V. tricolor* except as below.
RACEME: longer, with many flowers.
SEPALS AND PETALS: pale violet.
LIP: darker violet, the spur conical, curved downwards.
DISTRIBUTION: NE India, China (Yunnan), Burma, Thailand.

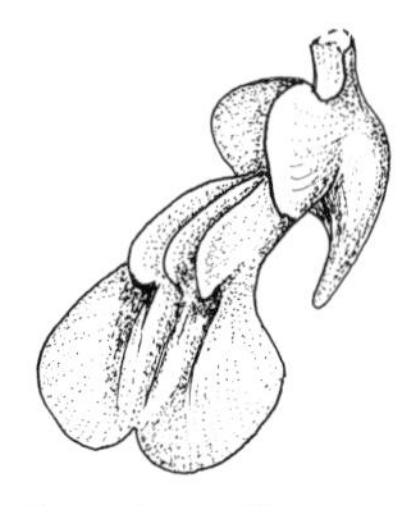

Vanda caerulescens – lip.

12. V. bensonii Bateman
ILLUSTRATIONS: Botanical Magazine, 5611 (1866); Cogniaux

Vanda bensonii – lip.

Vanda insignis – lip.

& Goossens, Dictionnaire Iconographique des Orchidées, Vanda, t. 2 (1898).

Also like *V. triclor* except as below.

SEPALS AND PETALS: olive green with brown spots.

LIP: violet-pink, white at the base, the spur conical.

DISTRIBUTION: Burma, Thailand.

FLOWERING: autumn.

13. **V. insignis** Blume

ILLUSTRATIONS: Botanical Magazine, 5759 (1869); Lindenia **8**: t. 355 (1892); Cogniaux & Goossens, Dictionnaire Iconographique des Orchidées, Vanda, t. 3 (1897).

Also broadly like *V. tricolor* except as below.

FLOWERS: larger in various shades of violet, yellow or white.

LIP: central lobe concave and wavy.

DISTRIBUTION: Indonesia.

14. **V. cristata** (Wallich) Lindley

ILLUSTRATIONS: Botanical Magazine, 4304 (1847); Gartenflora **20**: t. 680 (1871); Orchid Digest **43**: 44 (1979); Bechtel et al., Manual of cultivated orchid species, 278 (1981); American Orchid Society Bulletin **54**: 324 (1985).

STEMS: short, densely leafy.

LEAVES: to 12 × 4 cm, linear, thick and fleshy, flat, 2-lobed at the apex.

RACEMES: short with 2–5 flowers.

SEPALS AND PETALS: to 2.5 cm, curving forwards, oblong, greenish yellow.

LIP: with short, erect lateral lobes and with an oblong central lobe which is 2-lobed at the apex with the lobes diverging, bearing a downwardly pointing horn-like projection between the lobes, the upper surface with 5 tuberculate ridges, all bright yellow to green with brown stripes; spur short, blunt, conical.

DISTRIBUTION: E Himalaya.

HARDINESS: G1.

FLOWERING: winter.

15. **V. alpina** (Lindley) Lindley

ILLUSTRATIONS: Annals of the Royal Botanic Garden Calcutta **8**: t. 289 (1898); Hara, Photo-album of plants of E Himalaya, t. 120 (1968).

Like *V. cristata* except as below.

FLOWERS: hanging.

SEPALS AND PETALS: 1.3–1.8 cm.

LIP: with an oblong, truncate central lobe without a horn-like projection, the base saccate but without a spur.

DISTRIBUTION: Himalaya.

FLOWERING: spring.

16. V. pumila Hooker

ILLUSTRATIONS: Annals of the Royal Botanic Garden Calcutta **5**: t. 68 (1895) & **8**: t. 288 (1898); Botanical Magazine, 7968 (1904).

Also like *V. cristata* except as below.

SEPALS AND PETALS: *c.* 3 cm, greenish or yellowish white.

LIP: with an ovate, concave, red-striped central lobe.

DISTRIBUTION: Himalaya.

FLOWERING: summer.

113. VANDOPSIS Pfitzer

A genus of about 10 species from SE Asia. Cultivation as for *Vanda* (p. 377).

HABIT: large monopodial epiphytes.

LEAVES: mainly at the base of the stem, in 2 ranks, large, flat or v-shaped in section, folded when young.

FLOWERS: in axillary racemes.

SEPALS AND PETALS: similar, fleshy, widely spreading.

LIP: 3-lobed, attached to the column by the base and by the small lateral lobes, which have a flap of fleshy tissue joining them above the ridge of the central lobe; central lobe shorter than the petals, often laterally flattened, saccate at the base and bearing a conspicuous ridge which is interrupted near the base.

COLUMN: short with a projection at its base but without a foot.

POLLINIA: 4 in 2 pairs, borne on a short, broad stipe with reflexed edges, the viscidium broad.

Key to species

1. Central lobe of lip as broad as long, not laterally flattened; leaves to 25 cm **3. parishii**
2. Leaves recurved, flat in section; racemes 25–35 cm **1. gigantea**
3. Leaves straight, V-shaped in section; racemes 1–2 m **2. lissochiloides**

1. V. gigantea (Lindley) Pfitzer

SYNONYMS: *Vanda gigantea* Lindley; *Stauropsis gigantea* (Lindley) Bentham & Hooker.

ILLUSTRATIONS: Botanical Magazine, 5189 (1860); Bateman, Second century of orchidaceous plants, t. 142 (1865); Cogniaux Goossens, Dictionnaire Iconographique des Orchidées, Stauropsis, t. 2 (1898); Orchid Digest **43**: 87 (1979).

STEMS: thick, fleshy.

LEAVES: to 35 × 6 cm, fleshy, recurved, linear-oblong, flat, slightly unequally 2-lobed at the apex.

RACEME: 25–35 cm with up to 15 flowers.

SEPALS AND PETALS: 2.5–3 cm, oblong-obovate, yellow with red-brown blotches.

LIP: longer than broad, mostly yellow, lateral lobes suffused with purple.

DISTRIBUTION: Burma, Thailand, Malaysia.

HARDINESS: G2.

FLOWERING: spring–summer.

2. V. **lissochiloides** (Gaudichaud) Pfitzer

SYNONYMS: *Fieldia lissochiloides* Gaudichaud; *Vanda batemanii* Lindley; *Vanda lissochiloides* (Gaudichaud) Lindley; *Stauropsis lissochiloides* (Gaudichaud) Pfitzer.

ILLUSTRATIONS: Cogniaux & Goossens, Dictionnaire Iconographique des Orchidées, Stauropsis, t. 1 (1898); Orchid Digest 43: 87 (1979); Bechtel et al., Manual of cultivated orchid species, 279 (1981).

HABIT: very large plants.

LEAVES: 30–50 cm, straight, V-shaped in section.

RACEMES: 1–2 m with many flowers.

SEPALS AND PETALS: to 3.5 cm, bright purple or yellowish on the back, yellow with purple spots on the front.

LIP: longer than broad, short, yellow.

DISTRIBUTION: Philippines, Moluccas; ?Thailand.

HARDINESS: G2.

FLOWERING: summer.

3. V. **parishii** (Reichenbach) Schlechter

SYNONYM: *Vanda parishii* Reichenbach; *Hygrochilus parishii* (Reichenbach) Ritter.

ILLUSTRATIONS: Cogniaux & Goossens, Dictionnaire Iconographique des Orchidées, Vanda, t. 11, 11A (1898); Bechtel et al., Manual of cultivated orchid species, 279 (1981).

STEMS: short.

LEAVES: to 25 × 6 cm, oblong-elliptic, unequally 2-lobed at the apex.

RACEME: to 35 cm with 5–6 flowers.

SEPALS AND PETALS: to 2.5 cm, obovate, greenish yellow spotted with brown (rose-pink, whitish at the base in var. **marriottiana** Reichenbach).

LIP: as broad as long, yellowish with red spots, and a conspicuous, vertical ridge.

DISTRIBUTION: Burma, Thailand.

HARDINESS: G2.

FLOWERING: summer.

Both var. **parishii** and var. **marriottiana** have been in cultivation; the latter is apparently more frequent.

114. ESMERALDA Reichenbach

A genus of 2 species from the Himalaya and SE Asia, of which only one is generally grown. Cultivation as for *Vanda* (p. 377).

Literature: Tan, K., Taxonomy of Arachnis, Armodorum, Esmeralda and Dimorphorchis, *Selbyana* **1**: 1–15 (1975), 365–73 (1976).

HABIT: monopodial epiphytes with elongate, scrambling stems.
LEAVES: in 2 ranks, unequally 2-lobed at their apices, folded when young, close, mainly at the bases of the stems.
FLOWERS: in axillary racemes.
SEPALS AND PETALS: similar, spreading.
LIP: freely movable on its attachment to the column, about as long as the petals, 3-lobed with small, erect, oblong lateral lobes and a larger, flat, ovate, auricled central lobe which has a hollow sac concealed within the tissue at the base (perceptible as a hump on the back of the lip), the surface with 3–5 longitudinal ridges.
COLUMN: large, fleshy, without a foot.
POLLINIA: 4 in 2 pairs, stipe rapidly broadening to a broad, disc-like viscidium.

1. E. **cathcartii** (Lindley) Reichenbach
SYNONYMS: *Vanda cathcartii* Lindley; *Arachnanthe cathcartii* (Lindley) Bentham & Hooker; *Arachnis cathcartii* (Lindley) J.J. Smith.
ILLUSTRATIONS: Cogniaux & Goossens, Dictionnaire Iconographique des Orchidées, Arachnanthe, t. 2 (1898); Annals of the Royal Botanic Garden Calcutta **8**: t. 278 (1898); Orchid Digest **43**: 89, 92 (1979).
STEM: to 2 m, thick, hanging.
LEAVES: to 15 × 3–4 cm.
RACEMES: with 3–5 distant flowers.
SEPALS AND PETALS: to 4.5 cm, ovate to broadly elliptic, rounded at their apices, yellow with dense, red-brown, transverse stripes.
LIP: white with red stripes, the margin yellow, irregularly toothed and wavy.
DISTRIBUTION: E Himalaya.
HARDINESS: G2.
FLOWERING: spring–summer.

115. DIMORPHORCHIS Rolfe

A genus from Borneo with 1 or 2 species. The 1 cultivated species is not common in cultivation, despite Holttum's remark (Orchids of Malaya, 619, 1953): 'A large plant with a dozen inflorescences ranks with *Grammatophyllum* in full flower as one of the most remarkable objects in the Orchid world'. Cultivation as for *Vanda* (p. 377).

Literature: Tan, K., Taxonomy of Arachnis, Armodorum, Esmeralda and Dimorphorchis, *Selbyana* **1**: 1–15 (1975), 365–73 (1976); Soon, P.S., Notes on Sabah orchids, part I, *Orchid Digest* **44**: 193–5 (1980).

HABIT: large monopodial epiphytes with elongate stems.

LEAVES: densely set in 2 ranks, folded when young.

FLOWERS: in long, loose, axillary racemes with densely woolly axis and flower-stalks, dimorphic, the first 1–4 flowers of each raceme orange-yellow with fine red spots and shorter and broader, more rounded sepals and petals than the rest, which are larger and have wavy sepals and petals which are greenish with large, confluent, red-brown spots; in each type of flower the sepals and petals are similar, spreading and woolly outside.

LIP: much shorter than the petals, hollowed out at the base and with a thin, high ridge in the centre which ends in a fine, erect point.

COLUMN: short, without a foot.

POLLINIA: 4 in 2 pairs borne almost horizontally, the stipe short and broad, broadening a little to the large, disc-like viscidium.

1. **D. lowii** (Lindley) Rolfe

SYNONYMS: *Vanda lowii* Lindley: *Renanthera lowii* (Lindley) Reichenbach; *Arachnanthe lowii* (Lindley) Bentham & Hooker; *Vandopsis lowii* (Lindley) Schlechter; *Arachnis lowii* (Lindley) Reichenbach; *Renanthera rohaniana* Reichenbach.

ILLUSTRATIONS: Cogniaux & Goossens, Dictionnaire Iconographique des Orchidées, Arachnanthe, t. 4 (1898); Orchid Digest **44**: 194 (1980).

STEMS: to 2 m.

LEAVES: to 70 × 5–6 cm, oblong, obtuse.

RACEMES: to 2.5 m.

SEPALS AND PETALS: fleshy, 3.5–5 cm.

LIP: acute, whitish, violet within.

DISTRIBUTION: Borneo.

HARDINESS: G2.

FLOWERING: autumn

Both var. **lowii** and var. **rohaniana** (Reichenbach) Tan, which has shorter racemes and slightly smaller, paler flowers, have been in cultivation.

116. RENANTHERA Loureiro

A genus of 10 or more species from E Asia. Cultivation as for *Vanda* (p. 377).

Literature: Mahyar, U.W., Observations on some species of the orchid genus Renanthera, *Reinwardtia* **10**: 399–418 (1988).

HABIT: monopodial epiphytes with stout, branched stems.

LEAVES: spirally arranged or in 2 ranks, well spaced, unequally 2-lobed at their apices, folded when young.

FLOWERS: in long, horizontal, axillary panicles.

UPPER SEPAL AND PETALS: similar, spreading upwards.

LATERAL SEPALS: longer and broader, spreading downwards, often close together, all more than 1 cm.

LIP: much smaller than the petals, 3-lobed, with a conical spur and 2 calluses at the junctions of the central and lateral lobes.

COLUMN: short, without a foot.

POLLINIA: 4, borne on a short, linear-oblong, sometimes slightly waisted stipe, and with an oblong viscidium

Key to species

1. Upper sepal at most 1.5 cm **4. pulchella**
2. Central lobe of lip with 5 yellow calluses at the base, in addition to the 2 at the junction of the lobes **3. imschootiana**
3. Lateral sepals red with deep crimson blotches; leaves 20–25 cm **2. storiei**
4. Lateral sepals without blotches; leaves 6–12 cm **1. coccinea**

1. **R. coccinea** Loureiro

ILLUSTRATIONS: Edwards's Botanical Register **14**: t. 1131 (1828); Botanical Magazine, 2997, 2998 (1830).

STEMS: scrambling, to 2 m.

LEAVES: 6–12 × 3–3.5 cm, unequally 2-lobed at the apex.

PANICLE: to 70 cm.

UPPER SEPAL AND PETALS: oblong, obtuse, 2.5–2.8 cm, reddish spotted with scarlet.

LATERAL SEPALS: oblong-spathulate, obtuse, somewhat wavy, 3.5–4.5 cm, clear scarlet.

LIP: with low, yellow, red-striped lateral lobes and a scarlet central lobe which is yellowish towards the base.

DISTRIBUTION: Thailand, Laos, Vietnam, S China, possibly also in the Philippines, Java & Burma.

HARDINESS: G2.

FLOWERING: spring, autumn.

2. **R. storiei** (Storie) Reichenbach

ILLUSTRATIONS: Botanical Magazine, 7537 (1897); Bechtel et al., Manual of cultivated orchid species, 268 (1981).

Like *R. coccinea* except as below.

STEMS: shorter.

LEAVES: 20–25 × 3.5–4 cm.

LATERAL SEPALS: scarlet with crimson blotches.

DISTRIBUTION: Philippines.

FLOWERING: summer.

3. R. imschootiana Rolfe

ILLUSTRATIONS: Botanical Magazine, 7711 (1900); Bechtel et al., Manual of cultivated orchid species, 268 (1981).

STEM: erect, scarcely scrambling.

LEAVES: to 7 × 2 cm, oblong, shortly 2-lobed at their apices.

PANICLE: to 45 cm.

UPPER SEPAL: 2–2.2 cm.

PETALS: 3.7–4 cm, yellow spotted with red.

LATERAL SEPALS: clawed, wavy, elliptic, scarlet, or yellow flushed with red.

LIP: scarlet with 5 yellow calluses at the base of the central lobe.

DISTRIBUTION: NE India, Burma, Laos.

HARDINESS: G2.

FLOWERING: summer.

4. R. pulchella Rolfe

STEM: to 20 cm.

LEAVES: 6.5–8 × 1.5–2 cm, in 2 ranks, narrowly oblong, unequally lobed at their apices.

PANICLE: with few branches.

UPPER SEPAL: 1.3–1.5 cm, lanceolate, obtuse.

LATERAL SEPALS: *c.* 1.7 cm, clawed, oblong-elliptic, obtuse, all yellow.

PETALS: yellow below, red above.

LIP: with red lateral lobes, the central lobe yellow with 4 distinct teeth at the base.

DISTRIBUTION: Burma, perhaps also in Thailand.

HARDINESS: G2.

FLOWERING: summer.

117. ASCOCENTRUM Schlechter

A genus of about 5 species from E Asia (Himalaya to Borneo). Cultivation as for *Gastrochilus*.

HABIT: monopodial epiphytes with short leafy stems.

LEAVES: linear, unequally 2-lobed and toothed at the apex, folded when young.

FLOWERS: in axillary racemes.

SEPALS AND PETALS: similar, spreading.

LIP: small, consisting mainly of a vertical, cylindric, blunt pouch bearing at its apex 2 small, erect lateral lobes and a linear-oblong, acute central lobe which is much smaller than the petals.

COLUMN: short, without a foot.

POLLINIA: 2, spherical, borne on a short linear stipe and with a small, oblong viscidium.

Key to species

1. Sepals and petals pink to violet-purple; leaves 1.7–2 cm broad **1. ampullaceum**

2. Sepals and petals orange or orange-yellow; leaves rarely exceeding 1 cm in breadth **2. miniatum**

1. A. **ampullaceum** (Lindley) Schlechter

SYNONYM: *Saccolabium ampullaceum* Lindley.

ILLUSTRATIONS: American Orchid Society Bulletin **41**: 820 (1972), **55**: 104 (1986); Orchid Digest **42**: 180 (1979).

LEAVES: stiff, 8–12 × 1.7–2 cm.

RACEME: with many flowers, erect, shortly stalked.

SEPALS AND PETALS: 1–1.2 cm, oblong, blunt, pink to violet-purple.

LIP: with pink to violet-purple pouch and central lobe (which is horizontal), the lateral lobes whitish.

DISTRIBUTION: NE India to Thailand.

HARDINESS: G2.

FLOWERING: spring.

2. A. **miniatum** (Lindley) Schlechter

SYNONYM: *Saccolabium miniatum* Lindley.

ILLUSTRATIONS: Holttum, Orchids of Malaya, 730 (1953); Orchid Digest **42**: 180 (1979); Bechtel et al., Manual of cultivated orchid species, 169 (1981).

LEAVES: *c.* 10–20 × 1 cm, fleshy.

RACEME: shortly stalked.

SEPALS AND PETALS: *c.* 6 mm, orange or orange-yellow.

LIP: similar in colour to sepals and petals, the central lobe bent downwards.

DISTRIBUTION: E Himalaya to Borneo & Java.

HARDINESS: G2.

FLOWERING: spring.

118. GASTROCHILUS G. Don

A genus of 15–20 species from the Himalaya and E Asia, including Japan. Only 3 species are commonly grown. Cultivation as for *Aerides* (p. 352) or *Vanda* (p. 377), but vegetative propagation is difficult.

Literature: Herklots, G.A.C., Nepalese and Indian orchids: Gastrochilus, *Orchid Review* **82**: 354–8 (1974); Christensen, E.A., Gastrochilus G. Don, with a synopsis of the genus, *American Orchid Society Bulletin* **54**: 1111–1116 (1985).

HABIT: monopodial epiphytes with short stems.

LEAVES: close together at the stem bases, in 2 ranks, fleshy, folded when young.

FLOWERS: fleshy, in axillary racemes which are sometimes paired at each node.

SEPALS AND PETALS: similar, spreading.

LIP: consisting mainly of a vertical pouch with 2 lateral lobes joined to the column and a forwardly pointing, often hairy or fringed central lobe which is kidney- or crescent-shaped.

COLUMN: short, without a foot.
ANTHER: ovate, projecting forwards on the column.
POLLINIA: 2 on a linear stipe and with a small, oblong viscidium.

Key to species

1. Central lobe of lip with long white hairs on the upper surface **1. bellinus**
2. Leaves acute but not 2-toothed at the apex; stem 10–30 cm **3. acutifolius**
3. Leaves acute and 2-toothed at the apex; stem 1–3 cm **2. dasypogon**

1. **G. bellinus** (Reichenbach) Kunze

ILLUSTRATIONS: Die Orchidee **23**: 8 (1972); Bechtel et al., Manual of cultivated orchid species, 150 (1981); American Orchid Society Bulletin **54**: 1113 (1985).
STEM: usually less than 5 cm, stout.
LEAVES: 15–30 cm, narrowly oblanceolate, curved, deeply notched at the apex.
FLOWERS: fragrant, in a cluster-like raceme of 4–7.
SEPALS AND PETALS: 9–14 mm, slightly incurved, greenish or pale yellow with brown or purple spots.
LIP: *c.* 1.3 cm, central lobe broadly elliptic with a toothed margin, white with purple or red spots, yellow towards the base, upper surface with long white hairs.
DISTRIBUTION: Burma, Laos, W China, Thailand.
HARDINESS: G2.
FLOWERING: winter–spring.

2. **G. dasypogon** (Lindley) Kunze

ILLUSTRATIONS: Sheehan & Sheehan, Orchid genera illustrated, 95 (1979); Bechtel et al., Manual of cultivated orchid species, 150 (1981).
STEM: 1–3 cm.
LEAVES: 10–15 cm, ovate-oblong, acute, notched at the apex.
RACEME: almost umbellate.
SEPALS AND PETALS: to 8 mm, obovate, pale yellow or greenish, sometimes spotted with purple.
LIP: 5–7 mm, central lobe kidney-shaped with a fringed margin, white or yellow with red or purple spots, not hairy.
DISTRIBUTION: Himalaya?; Thailand & Sumatra.
HARDINESS: G2.
FLOWERING: autumn.

3. **G. acutifolius** (Lindley) Kunze

SYNONYM: *Saccolabium acutifolium* Lindley.
ILLUSTRATIONS: Annals of the Royal Botanic Garden Calcutta 8: t. 302 (1898); Orchid Review **82**: 358 (1974); Die Orchidee **27**: 127 (1976).

STEMS: 10–30 cm.
LEAVES: to 12 × 2 cm, oblong, acute, not notched at the apex.
RACEME: with 10–20 flowers, compressed, umbel-like.
SEPALS AND PETALS: *c.* 1 cm, oblong, blunt, greenish yellow, usually with brown-purple spots.
LIP: white, the central lobe broadly triangular, papillose but not hairy on the upper surface, fringed.
DISTRIBUTION: Nepal, NE India.
HARDINESS: G2.
FLOWERING: winter.

119. SCHOENORCHIS Reinwardt

Ten species ranging from the Himalaya to New Guinea. Cultivation as for *Vanda* (p. 377).

Like *Gastrochilus* except as below.
LEAVES: strap-shaped to terete.
RACEME: simple or branched, with many small flowers.
SEPALS AND PETALS: 2–5 mm.
LIP: 3-lobed, spurred but not consisting mainly of a vertical pouch.
POLLINIA: 4, united in pairs attached to a long slender viscidium.
ROSTELLUM: linear, bifid, very conspicuous.

1. S. juncifolia Blume
STEM: to 30 cm, hanging, branched.
LEAVES: to 16 cm, terete, often flushed with purple.
RACEME: to 10 cm, simple, hanging, dense, with many flowers.
SEPALS AND PETALS: oblong, shortly acuminate, bluish violet.
LIP: longer than sepals, with a more or less curved spur, the central lobe *c.* 2 mm, pale violet or white, recurved.
DISTRIBUTION: Sumatra, Java.
HARDINESS: G2.
FLOWERING: summer.

120. CLEISOSTOMA Blume

A genus of about 100 species from Asia, occurring from Nepal to New Guinea. The flowers are more interesting than attractive and few are grown. The genus is commonly found under the name *Sarcanthus* Lindley. Cultivation as for *Vanda* (p. 377).

Literature: Garay, L., On the systematics of the monopodial orchids I, *Botanical Museum Leaflets, Harvard University* **23**: 168–76 (1972); Seidenfaden, G., Orchid genera in Thailand II: Cleisostoma, *Dansk Botanisk Arkiv* **29**(3): 1–80 (1975).

HABIT: monopodial epiphytes with long stems.
LEAVES: borne all around the stem, terete or flat, folded when young.

FLOWERS: numerous in axillary racemes or rarely panicles.

SEPALS AND PETALS: similar or the petals slightly smaller, all spreading, lateral sepals and petals not inserted on the column foot.

LIP: consisting mainly of a short, broad, blunt, vertical pouch, the blade 3-lobed with small lateral lobes and a larger, usually arrowhead-shaped central lobe; the entrance to the pouch is largely blocked by a complex callus borne on the upper surface of its interior, and there is usually a longitudinal cross-wall from the lower part of its interior as well.

COLUMN: short with a short foot.

POLLINIA: 4, united in 2 round masses with the stipe linear and the viscidium disc-like or the stipe broadening rapidly from the apex to a saddle-shaped viscidium.

Key to species

1. Leaves terete

1.A. Leaves 20–25 cm, borne 1-sidedly on the stem **2. filiforme**

1.B. Stipe of pollinia linear, viscidium disc-like **1. appendiculatum**

1.C. Stipe of pollinia narrow at apex but rapidly broadening to a saddle-shaped viscidium **3. simondii**

2. Leave unequally notched at apex; panicles exceeding the subtending leaves **5. racemiferum**

3. Leaves not notched at apex; racemes shorter than the subtending leaves **4. rostratum**

1. C. **appendiculatum** (Lindley) Jackson

SYNONYM: *Sarcanthus appendiculatus* of various authors.

ILLUSTRATIONS: Annals of the Royal Botanic Garden Calcutta **5**: t. 76 (1895); Hooker's Icones Plantarum **22**: t. 2136 (1893).

STEM: to 40 cm, hanging.

LEAVES: 7–10 cm, terete, spirally arranged.

RACEME: with 10–15 flowers, stalked, hanging, longer than the leaves.

FLOWERS: 1.5–2 cm.

SEPALS AND PETALS: oblong, brownish yellow with red-violet veins.

LIP: with small, erect lateral lobes and a broadly triangular central lobe, rose-pink; pouch cylindric, bright brownish yellow with purple veins.

POLLINIA: stipe linear, viscidium disc-like.

DISTRIBUTION: E Himalaya, Burma.

HARDINESS: G2.

FLOWERING: summer.

2. C. **filiforme** (Lindley) Garay

SYNONYM: *Sarcanthus filiformis* Lindley.

ILLUSTRATION: Bechtel et al., Manual of cultivated orchid species, 184 (1981).

Like *C. appendiculatum*: except as below.

STEMS: longer.

LEAVES: 20–25 cm borne 1-sidedly.

FLOWERS: *c.* 1.1 cm.

DISTRIBUTION: E Himalaya to Thailand.

FLOWERING: summer–autumn.

3. **C. simondii** (Gagnepain) Seidenfaden

SYNONYM: *Sarcanthus teretifolius* (Lindley) Lindley.

Like *C. appendiculatum* except as below.

LEAVES: thick, short, often recurved.

POLLINIA: stipe narrow at the apex but rapidly broadened to a saddle-shaped viscidium.

DISTRIBUTION: Himalaya, Laos, Vietnam, Thailand & S China.

4. **C. rostratum** (Lindley) Garay

SYNONYM: *Sarcanthus rostratus* Lindley.

ILLUSTRATIONS: Lindley, Collectanea Botanica, t. 39 (1821); Edwards's Botanical Register **12**: t. 981 (1826); Dansk Botanisk Arkiv **29**(3): 29 (1975).

ATEM: to 25 cm, hanging.

LEAVES: to 10 × 1–1.5 cm, flat.

RACEME: dense, with many flowers, usually shorter than the subtending leaves.

FLOWERS: to 1.3 cm.

SEPALS AND PETALS: yellowish green with purplish veins.

LIP: whitish with short lateral lobes and an acute, upwardly hooked central lobe which is violet-pink; pouch violet-pink.

DISTRIBUTION: S China, Vietnam, Laos & Thailand.

HARDINESS: G2.

FLOWERING: summer–autumn.

5. **C. racemiferum** (Lindley) Garay

SYNONYMS: *Sarcanthus racemifer* (Lindley) Reichenbach; *S. pallidus* Lindley.

ILLUSTRATION: Bechtel et al., Manual of cultivated orchid species, 184 (1981).

Like *C. rostratum* except as below.

LEAVES: unequally 2-lobed at the apex.

FLOWERS: in erect panicles which exceed the subtending leaves.

DISTRIBUTION: Himalaya to Thailand.

121. ANGRAECUM Bory

A genus of over 200 species centred mainly in tropical Africa but extending to Madagascar and Sri Lanka. Garay (reference below) divides it into 18 sections, but as only 6 or 7 species are found in cultivation, these are not described here. Species Nos. 1–3 belong to section *Arachnangraecum* Schlechter; Nos. **4** & **5** to section *Angraecum*; and No. **6** to section *Dolabrifolia* (Pfitzer) Garay.

Cultivation as for *Phalaenopsis* (p. 360), but the smaller species should be suspended near the glass and the thick-leaved species should be given full air and light while avoiding a dry atmosphere. In winter the plants can be allowed to dry out to some extent, but not for long periods.

Literature: Garay, L., Systematics of the genus Angraecum Bory, *Kew Bulletin* **28**: 495–516 (1973); Hillerman, F.E. & Holst, A.W. *An introduction to the cultivated angraecoid orchids of Madagascar*, 45–110 (1986).

HABIT: monopodial epiphytes with long or short stems.

LEAVES: in 2 ranks, variable, overlapping and folded at least at the base, or distant and flat or terete, sometimes unequally 2-lobed at their apices, folded when young.

FLOWERS: variably resupinate or not resupinate, solitary in the leaf axils, or in axillary racemes.

SEPALS AND PETALS: similar, spreading.

LIP: stalkless, entire, with a flat or concave spreading blade extending backwards into a conspicuous spur.

COLUMN: fleshy, short, without a foot, with 2 swellings, between which is the short, inconspicuous, bifid, tooth-like rostellum, beneath the anther.

POLLINIA: 2, spherical, attached either to a short common stipe, or to individual stipes, viscidium single or double.

Synopsis of characters

Leaves. Terete: **3**; overlapping and folded at their bases only: **4**,**5**; overlapping and folded for most of their length: **6**.

Flowers. Stalkless and single in the leaf axils: **6**; single and stalked, or in racemes in the leaf axils: **1**–**5**.

Lip. Narrowly ovate, longer than broad: **5**; broadly ovate, broader than long: **1**–**4**; helmet-like: **6**.

Spur. Less than 1 cm: **6**; between 4 and 15 cm: **1**–**4**; between 20 and 30 cm: **5**. Sharply bent near the middle: **1**,**2**.

Key to species

1. Leaves overlapping and folded, at least at their bases

1.A. Flowers stalkless; sepals and petals less than 1 cm — **6. distichum**

1.B. Lip longer than broad, narrowly ovate; spur 20–30 cm — **5. sesquipedale**

1.C. Lip broader than long, broadly ovate; spur 6–8 cm — **4. eburneum**

2. Leaves terete in section — **3. scottianum**

3. Leaves flat in section
3.A. Spur 12–13 cm, tapering, curved forwards
2. infundibulare
3.B. Spur *c*. 4.5 cm, sharply bent at the middle
1. eichlerianum

1. A. **eichlerianum** Kränzlin
ILLUSTRATIONS: Botanical Magazine, 7813 (1902); Bechtel et al., Manual of cultivated orchid species, 166 (1981); American Orchid Society Bulletin 53: 804 (1984).
STEMS: to 1 m, hanging, somewhat compressed.
LEAVES: 8–12 × 4.5–5 cm, oblong-elliptic, flat, not overlapping.
FLOWERS: solitary or rarely in 2-flowered racemes in the leaf axils, stalk elongate, variably resupinate.
SEPALS AND PETALS: to 4.5 cm, spreading, oblong-lanceolate, acute, yellowish green.
LIP: broadly ovate, concave, broader than long, with a short point, white shading to greenish yellow at the base, with an elongate ridge; spur *c*. 4.5 cm, funnel-shaped, sharply bent at the middle, obtuse.
POLLINIA: with a common stipe.
DISTRIBUTION: W tropical Africa, from Nigeria to Angola.
HARDINESS: G2.
FLOWERING: summer.

2. A. **infundibulare** Lindley
SYNONYM: *Mystacidium infundibulare* (Lindley) Rolfe.
ILLUSTRATIONS: Botanical Magazine, 8153 (1907); Stewart & Campbell, Orchids of tropical Africa, t. 9 (1970); Bechtel et al., Manual of cultivated orchid species, 166 (1981).
Like *A. eichlerianum* except as below.
LEAVES: broader.
LIP: very broad and concave, the spur 12–13 cm, very wide-mouthed and funnel-shaped, yellowish, curved forwards in the lower half.
DISTRIBUTION: Tropical Africa from Zaire to Ethiopia.
FLOWERING: winter.

3. A. **scottianum** Reichenbach
ILLUSTRATIONS: Botanical Magazine, 6273 (1883); Cogniaux & Goossens, Dictionnaire Iconographique des Orchidées, Angraecum, t. 7 (1898); Stewart & Campbell, Orchids of tropical Africa, t. 12 (1970); Bechtel et al., Manual of cultivated orchid species, 167 (1981); Hillerman & Holst, An introduction to the cultivated angraecoid orchids of Madagascar, pl. 6 (1986).
STEMS: *c*. 30 cm, thin, hanging.
LEAVES: 8–10 cm, terete, not overlapping at the base.

FLOWERS: single in the leaf axils, on long stalks, variably resupinate.

SEPALS AND PETALS: 2.5–3 cm, spreading, oblong-linear, acute, greenish white.

LIP: broader than long, broadly ovate or transversely oblong, with a small point, white; spur 8–10 cm, narrow, tapering, yellow.

POLLINIA: with a common stipe.

DISTRIBUTION: Comoro Islands.

HARDINESS: G2.

FLOWERING: summer.

4. A. **eburneum** Bory

ILLUSTRATIONS: Botanical Magazine, 5170 (1854); Lindenia **5**: t. 236 (1889); Cogniaux & Goossens, Dictionnaire Iconographique des Orchidées, Angraecum, t. 13 (1898); Die Orchidee **30**: cxxi (1979); Bechtel et al., Manual of cultivated orchid species, 166 (1981); Hillerman & Holst, An introduction to the cultivated angraecoid orchids of Madagascar, pl. 2 (1986).

STEM: to 1 m, erect.

LEAVES: 40–50 cm, dense, their bases overlapping and folded, the free part oblong-lanceolate, unequally 2-lobed at the apex.

FLOWERS: not resupinate, 8–15 in axillary, 1-sided racemes, stalked, fragrant.

SEPALS AND PETALS: 2.5–3 cm, linear-oblong, acute, spreading, green or whitish green.

LIP: ovate, broader than long, concave, apiculate, white, often greenish at the base; spur 6–8 cm, narrow, tapering, white with a green apex.

POLLINIA: with a common stipe.

DISTRIBUTION: E tropical Africa, islands of the Indian Ocean.

HARDINESS: G2.

FLOWERING: autumn–winter.

A variable species. Subsp. **eburneum**, subsp. **superbum** (Thouars) Perrier de la Bathie (*A. superbum* Thouars), which has somewhat larger flowers, and subsp. **giryamae** (Rendle) Senghas & Cribb (*A. giryamae* Rendle), which has a smaller lip and a funnel-shaped spur which is 4–5 cm, are grown. This last subspecies occurs in Kenya and Tanzania.

5. A. **sesquipedale** Thouars

ILLUSTRATIONS: Botanical Magazine, 5113 (1859); Cogniaux & Goossens, Dictionnaire Iconographique des Orchidées, Angraecum, t. 4 (1898); Stewart & Campbell, Orchids of tropical Africa, t. 13 (1970); Bechtel et al., Manual of cultivated orchid species, 167 (1981); American Orchid Society Bulletin **53**: 731 (1984), **55**: 140 (1986); Hillerman &

Holst, An introduction to the cultivated angraecoid orchids of Madagascar, pl. 1 (1986).

STEM: to 1 m, thick, erect.

LEAVES: dense, their bases overlapping and folded, the free part to 30 × 4–5 cm, oblong, unequally 2-lobed at the apex.

FLOWERS: usually 2–4 in axillary racemes, rarely single in the leaf axils, stalked, fragrant, resupinate.

SEPALS AND PETALS: to 6 cm, creamy white, spreading, lanceolate, acute, the petals somewhat shorter than the sepals.

LIP: narrowly ovate, longer than broad, tapering, its sides somewhat toothed, creamy white; spur 20–30 cm, abruptly tapering from its junction with the lip.

POLLINIA: with separate stipes.

DISTRIBUTION: Madagascar.

HARDINESS: G2.

FLOWERING: winter.

This is the orchid which led Darwin to predict the existence of a moth with a proboscis 20–30 cm long—a species unknown at the time of his prediction but since discovered in Madagascar.

6. A. distichum Lindley

ILLUSTRATIONS: Edwards's Botanical Register **21**: t. 1781 (1835); Botanical Magazine, 4145 (1845); Die Orchidee **21**: 246 (1970); Bechtel et al., Manual of cultivated orchid species, 166 (1981).

STEM: to 12 cm.

LEAVES: 8–12 cm, fleshy, folded and overlapping for most of their length, the free part short, obtuse.

FLOWERS: solitary in the upper leaf axils, the ovary stalkless, not resupinate.

SEPALS AND PETALS: *c.* 5 mm, white, spreading, the petals slightly shorter and narrower than the sepals.

LIP: helmet-like, 3-lobed at the apex, the central lobe recurved; spur *c.* 8 mm, cylindric.

DISTRIBUTION: Tropical Africa from Guinea to Uganda and Angola.

HARDINESS: G2.

FLOWERING: autumn.

122. AERANGIS Reichenbach

A genus of 35 species from tropical Africa and Madagascar. Cultivation as for *Angraecum* (p. 394).

Literature: Stewart, J.A., A revision of the African species of Aerangis, *Kew Bulletin* **34**: 239–319 (1979); Stewart, J., Stars of the islands, a new look at the genus Aerangis in Madagascar and the Comoro Islands, *American Orchid Society Bulletin* **55**: 792–1125 (1986); Hillerman, F.E. & Holst, A.W., *An introduction to the cultivated angraecoid orchids of Madagascar*,

111–137 (1986); Stewart, J. & La Croix, I., Notes on orchids of southern tropical Africa, 3, *Kew Bulletin* **42**: 215–219 (1897)

HABIT: monopodial epiphytes with short, compressed stems.

LEAVES: leathery, unequally 2-lobed at the apex, folded when young.

RACEMES: with 7–many flowers, hanging from the leaf axils, the axis often conspicuously zig-zag; bracts small, brownish.

FLOWERS: usually white, fragrant.

SEPALS AND PETALS: similar in shape and size (rarely the upper sepal smaller), spreading or somewhat reflexed.

LIP: smaller than the petals but often broader, projecting backwards into a short or long, curved or twisted and contorted spur, which has a distinct, narrow mouth.

COLUMN: narrow, without a foot, rostellum large, usually beaked, usually extended at right angles to the column.

POLLINIA: 2, each with a separate stipe or (in all species included here) attached to a single, linear stipe, viscidium oblong, sometimes slightly 2-lobed.

Synopsis of characters

Habit. Plant very small (leaves 3–5 cm, sepals and petals 5–6 mm): **9**.

Sepals. Upper shorter and blunter than laterals: **7,8**. 2 cm or more: **1,2,5,6**.

Spur. Less than 1 cm: **9**; 2–4 cm: **6–8**; 4–10 cm: **3,4**; more than 10 cm: **1,2,5**.

Contorted and twisted: **5**.

Key to species

1. Leaves 3–5 cm; spur 5.5–6 cm, not or only slightly longer than ovary — **9. hyaloides**

2. Spur at most 3 cm

2.A. Upper sepal shorter than laterals — **7. citrata**

2.B. Upper sepal equal in size to the laterals — **8. luteo-alba**

3. Spur 15–27 cm, twisted and contorted, those of the individual flowers often twisted around each other — **5. kotschyana**

4. Sepals and petals white tipped with pink; spur brownish; rostellum touching the lower side of the stigmatic cavity which appears bilobed — **6. biloba**

5. Raceme to 60 cm; leaves to 25 x 6 cm

5.A. Raceme axis markedly zig-zag — **2. articulata**

5.B. Raceme axis straight — **1. ellisii**

6. Sepals and petals 1.5 cm or more; column hairless — **4. stylosa**

7. Sepals and petals to 1.5 cm; column with a few short hairs on the upper side — **3. modesta**

1. A. ellisii (Reichenbach) Schlechter

SYNONYM: *Angraecum ellisii* Reichenbach.

ILLUSTRATIONS: Floral Magazine n.s., t. 191 (1875); Lindenia **2**: t. 92 (1886); American Orchid Society Bulletin **55**: 793 (1986); Hillerman & Holst, An introduction to the cultivated angraecoid orchids of Madagascar, pl. 13 (1986).

STEM: short.

LEAVES: 20–25 × 5 cm, oblong-obovate, dark green, unequally 2-lobed at the apex.

RACEMES: 40–65 cm with 18–25 flowers.

SEPALS AND PETALS: 2.5–3 cm, white, spreading or reflexed, oblong-elliptic, acute.

LIP: similar to the petals but broader and with its margins reflexed; spur 15–17 cm, curved, tapering.

DISTRIBUTION: Madagascar.

HARDINESS: G2.

FLOWERING: summer–autumn.

Introduced, apparently from Madagascar, in the nineteenth century, but not found there since.

2. A. articulata (Reichenbach) Schlechter

SYNONYM: *Angraecum articulatum* Reichenbach.

ILLUSTRATIONS: American Orchid Society Bulletin **55**: 801 (1986); Hillerman & Holst, An introduction to the cultivated angraecoid orchids of Madagascar, pl. 11 (1986).

Like *A. ellisii* except as below.

RACEME AXIS: markedly zig-zag.

DISTRIBUTION: Madagascar.

HARDINESS: G2.

Both of these species are somewhat dubious. Perrier de la Bathie (Flore de Madagascar **49**(2): 108, 116, 1939) suggests that both are related to *A. stylosa* (see **4** below), *A. ellisii* being perhaps a luxuriant variant of it and *A. articulata* a hybrid of it with some other unknown species.

3. A. modesta (Hooker) Schlechter

SYNONYM: *Angraecum modestum* Hooker.

ILLUSTRATIONS: Botanical Magazine, 6693 (1883); Cogniaux & Goossens, Dictionnaire Iconographique des Orchidées, Angraecum, t. 2 (1898); American Orchid Society Bulletin **55**: 797 (1986); Hillerman & Holst, An introduction to the cultivated angraecoid orchids of Madagascar, pl. 13 (1986).

STEM: to 15 cm.

LEAVES: 6–12 × 1.5–3 cm, ovate-oblong, slightly unequally 2-lobed at the apex.

RACEME: to 30 cm at most, with 6–15 flowers which are very distant.

SEPALS AND PETALS: *c.* 1.2 cm, lanceolate, more or less obtuse, the sepals with 5 veins, white.

LIP: broader than the petals, acute; spur 4–7 cm, curved, tapering.

COLUMN: with a few short hairs on the upper side.

DISTRIBUTION: Madagascar & Comoro Islands.

HARDINESS: G2.

FLOWERING: spring.

4. **A. stylosa** (Rolfe) Schlechter

SYNONYM: *Angraecum stylosum* Rolfe.

ILLUSTRATION: Hillerman & Holst, An introduction to the cultivated angraecoid orchids of Madagascar, pl. 14 (1986).

Like *A. modesta* except as below.

SEPALS AND PETALS: 1.5–2 cm, the sepals with 7 veins.

COLUMN: hairless.

DISTRIBUTION: Madagascar & Comoro Islands.

5. **A. kotschyana** (Reichenbach) Schlechter

SYNONYMS: *Angraecum kotschyanum* Reichenbach; *Angraecum kotschyi* Reichenbach; *Aerangis kotschyi* (Reichenbach) Reichenbach.

ILLUSTRATIONS: Botanical Magazine, 7442 (1895); Williamson, Orchids of southern central Africa, f. 108 & t. 175 (1977); Ball, Southern African epiphytic orchids, 33 (1978); Kew Bulletin **34**: 255 (1979).

STEM: very short.

LEAVES: 10–15 × 4.5–5 cm, oblong-elliptic to oblong-obovate, not or slightly 2-lobed at the apex, sometimes spotted with purple above.

RACEME: to 40 cm with 7–10 (rarely more) flowers.

SEPALS AND PETALS: elliptic-lanceolate, acute, to 2.5 cm, white tinged with pink towards their apices.

LIP: reversed-heart-shaped, waisted above; spur 15–27 cm, twisted and contorted, slightly broadened towards the brownish apex, those of the individual flowers often twisted together.

DISTRIBUTION: tropical Africa, from Guinea and Sudan to Zaire and South Africa (Transvaal).

HARDINESS: G2.

FLOWERING: autumn.

6. **A. biloba** (Lindley) Schlechter

SYNONYM: *Angraecum bilobum* Lindley.

ILLUSTRATIONS: Edwards's Botanical Register **27**: t. 35 (1841); Kew Bulletin **34**: 282 (1979); American Orchid Society Bulletin **53**: 936 (1984).

STEM: very short.

LEAVES: to 15 cm, oblong-obovate, deeply and unequally 2-lobed at the apex.

RACEME: to 25 cm with 7–12 flowers.

SEPALS AND PETALS: *c.* **2** cm, lanceolate or narrowly elliptic, acute, white tipped with pink.

LIP: similar but broader; spur *c.* 4 cm, brownish, curved.

ANTHER: conspicuously crested.

ROSTELLUM: extending over the stigmatic cavity, touching its lower side, the cavity thus appearing 2-lobed.

DISTRIBUTION: tropical W Africa, from Senegal to Cameroun.

HARDINESS: G2.

FLOWERING: summer.

7. A. **citrata** (Thouars) Schlechter

SYNONYM: *Angraecum citratum* Thouars.

ILLUSTRATIONS: Botanical Magazine, 5624 (1867): Lindenia **5**: t. 238 (1889); Stewart & Campbell, Orchids of tropical Africa, t. 2 (1970); American Orchid Society Bulletin **53**: 732 (1984), **55**: 904 (1986); Hillerman & Holst, An introduction to the cultivated angraecoid orchids of Madagascar, pl. 11 (1986).

STEM: very short.

LEAVES: 7–15 × 2–3.5 cm, oblong-obovate, very unequally 2-lobed at the apex.

RACEME: 10–30 cm with 15–60 flowers.

SEPALS: upper 5–7 mm, obtuse, shorter than the laterals, which are clawed, acute and 8–10 mm.

PETALS: slightly shorter than the lateral sepals; white with a faint yellow tinge.

LIP: clawed, obovate; spur 2.5–3 cm, curved, tapering.

DISTRIBUTION: Madagascar.

HARDINESS: G2.

FLOWERING: winter–spring.

8. A. **luteo-alba** (Kränzlin) Schlechter var. **rhodosticta** (Kränzlin) Stewart

SYNONYMS: *Angraecum rhodostictum* Kränzlin; *Aerangis rhodosticta* (Kränzlin) Schlechter.

ILLUSTRATIONS: Hunt & Grierson, Orchidaceae, 110–11 (1973); Kew Bulletin **34**: 311 (1979); American Orchid Society Bulletin **55**: 805 (1986).

Like *A. citrata* on account of its short spur, otherwise differing as below.

SEPALS: equal, 1–1.5 cm.

COLUMN: red.

DISTRIBUTION: Tropical Africa from Cameroun to Tanzania.

9. A. **hyaloides** (Reichenbach) Schlechter

SYNONYM: *Angraecum hyaloides* Reichenbach.

ILLUSTRATIONS: Reichenbach, Xenia Orchidacearum **3**: t. 238 (1890); American Orchid Society Bulletin **55**: 908 (1986).

HABIT: plant very small.

LEAVES: 3–5 cm, oblong-obovate, shortly 2-lobed at the apex.
RACEME: 3–4 cm with 3–4 flowers.
SEPALS AND PETALS: 5–6 mm, whitish, translucent.
LIP: somewhat broader than petals; spur 5.5–6 mm, not or scarcely exceeding the ovary, curved.
DISTRIBUTION: Madagascar.
HARDINESS: G2.
FLOWERING: spring.

123. NEOFINETIA Hu

A genus of a single species from Japan, very rarely grown.

Like *Aerangis*, except as below.
LIP: distinctly 3-lobed.

1. N. falcata (Thunberg) Hu

SYNONYMS: *Angraecum falcatum* (Thunberg) Lindley; *Angraecopsis falcata* (Thunberg) Schlechter.
ILLUSTRATIONS: Kitamura et al., Alpine plants of Japan **3**: f. 114 (1978); American Orchid Society Bulletin **54**: 277 (1985).
STEM: very short.
LEAVES: fleshy, to 7 cm.
RACEME: erect, shorter than the leaves, with 3–7 flowers.
FLOWERS: white, *c.* 3 cm in diameter.
LIP: with short lateral lobes and an oblong central lobe; spur *c.* 4 cm.
DISTRIBUTION: Japan.
HARDINESS: G1.
FLOWERING: summer.

124. CYRTORCHIS Schlechter

A genus of 15 species from Africa, of which 2 are occasionally found in cultivation. Cultivation as for *Angraecum* (p. 394).

Literature: Summerhayes, V.S., African orchids XXVII, *Kew Bulletin* **14**: 143–56 (1960).

HABIT: monopodial epiphytes, stems short or long.
LEAVES: in 2 ranks, thick, fleshy and leathery, unequally 2-lobed at their apices, folded when young.
RACEMES: arching or hanging, arising from the axils of the older leaves (or leaf-bases), with several flowers and conspicuous, brownish bracts.
FLOWERS: fragrant, especially at night, white fading to dull orange if not pollinated.
SEPALS AND PETALS: similar in size and shape, usually recurved.
LIP: similar to the petals but prolonged backwards as a conspicuous, tapering spur.
COLUMN: short, without a foot.

ROSTELLUM: elongate, conspicuous, beak-like.

POLLINIA: 2, each with a separate stipe, attached to an oblong or linear viscidium.

Key to species

1. Stipe considerably broadened upwards; viscidium oblong with a hardened upper part and a translucent lower part **1. arcuata**
2. Stipe scarcely broadened upwards; viscidium linear, entirely translucent **2. monteiroae**

1. C. arcuata (Lindley) Schlechter

SYNONYMS: *Angraecum arcuatum* Lindley; *Listrostachys arcuata* (Lindley) Reichenbach.

ILLUSTRATIONS: Stewart & Campbell, Orchids of tropical Africa, t. 25 (1970); American Orchid Society Bulletin **49**: 1230 (1980); Bechtel et al., Manual of cultivated orchid species, 192 (1981).

STEM: 30–50 cm, thick.

LEAVES: 7–15 × 1.5–2.5 cm, oblong.

RACEMES: with 10–20 densely packed flowers.

SEPALS AND PETALS: 2–4 cm, lanceolate, acute, the petals sometimes a little shorter than the sepals.

SPUR: 3–6 cm, curved, tapering.

DISTRIBUTION: S and tropical Africa northwards to Sierra Leone and Kenya.

HARDINESS: G2.

FLOWERING: spring–summer.

A very variable species, divided by Summerhayes into 5 subspecies; it is not known which of these are in cultivation.

2. C. monteiroae (Reichenbach) Schlechter

SYNONYM: *Listrostachys monteiroae* Reichenbach.

ILLUSTRATION: Botanical Magazine, 8026 (1905).

STEM: to 60 cm, leafy.

LEAVES: 15–17 × 4.5–5.5 cm, oblong.

RACEME: with 1–15 rather distant flowers.

SEPALS AND PETALS: 1.5–2 cm, lanceolate.

SPUR: 4–5 cm, curved, brownish yellow towards the apex.

DISTRIBUTION: tropical Africa, from Sierra Leone to Angola and eastwards to Uganda.

HARDINESS: G2.

FLOWERING: spring.

125. DIAPHANANTHE Schlechter

A genus of a few species from Africa; two of them are occasionally grown. Cultivation as for *Angraecum* (p. 394).

Like *Cyrtorchis* except as below.

SEPALS AND PETALS: erect.

LIP: considerably larger than sepals and petals and bearing a conspicuous tooth-like callus just in front of the narrow mouth of the spur.

Key to species

1. Leaves in a close tuft: lip fringed **1. pellucida**
2. Leaves distributed along the stem; lip not fringed **2. bidens**

1. D. **pellucida** (Lindley) Schlechter

SYNONYMS: *Angraecum pellucidum* Lindley; *Listrostachys pellucida* (Lindley) Reichenbach.

ILLUSTRATIONS: Cogniaux & Goossens, Dictionnaire Iconographique des Orchidées, Listrostachys, t. 1 (1906); Bechtel et al., Manual of cultivated orchid species, 204 (1981); American Orchid Society Bulletin **53**: 735, 941 (1984), **55**: 377 (1986).

STEM: compressed.

LEAVES: in a close tuft, each unequally 2-lobed at the apex.

FLOWERS: white or brownish pink.

LIP: triangular, conspicuously fringed.

DISTRIBUTION: tropical Africa from Sierra Leone to Uganda.

HARDINESS: G2.

FLOWERING: winter.

2. D. **bidens** (Reichenbach) Schlechter

SYNONYM: *Listrostachys bidens* Reichenbach.

ILLUSTRATIONS:: Botanical Magazine, 8014 (1905); Bechtel et al., Manual of cultivated orchid species, 204 (1981).

STEM: to 40 cm, leafy along its length.

LEAVES: very conspicuously unequally 2-lobed.

FLOWERS: brownish pink.

LIP: triangular, notched at the apex, with a small point within the notch.

DISTRIBUTION: Guinea.

HARDINESS: G2.

FLOWERING: summer.

126. EURYCHONE Schlechter

A genus of 2 species from tropical Africa, of which 1 is grown. Cultivation as for *Angraecum* (p. 394).

HABIT: monopodial epiphytes with very short stems.

LEAVES: narrowly elliptic to oblanceolate, unequally 2-lobed at their apices, folded when young.

RACEMES: lateral, with 5–10 flowers, arching, shorter than the leaves.

SEPALS AND PETALS: similar, thin, spreading.

LIP: about as long as the sepals but much broader, funnel-shaped, very obscurely 3-lobed, extended backwards as a wide-mouthed spur which is thickened and recurved at its apex.

COLUMN: without a foot.

ROSTELLUM: beaked, conspicuous.

POLLINIA: 2, almost spherical, borne on a narrow, linear stipe, viscidium large, oblong or oval.

1. E. **rothschildiana** (O'Brien) Schlechter

SYNONYM: *Angraecum rothschildianum* O'Brien.

ILLUSTRATIONS: Gardeners' Chronicle 34: 131 (1903); Stewart & Campbell, Orchids of tropical Africa, t. 29 (1970).

STEM: to 7 cm.

LEAVES: 7–14 × 2–2.5 cm.

FLOWERS: fragrant.

SEPALS AND PETALS: 2–2.5 cm, each white with a pale green band.

LIP: to 2.5 cm wide, white with the centre bright green shading to brown or purple at the base, its sides curving upwards towards the column; spur to 2.5 cm.

DISTRIBUTION: tropical Africa, from Sierra Leone to Uganda.

HARDINESS: G2.

FLOWERING: summer.

127. POLYSTACHYA de Jussieu

A genus of about 200 species from the tropics, centred in Africa. Only 1 species is commonly available. Cultivation as for *Eulophia* (p. 409).

HABIT: epiphytic.

PSEUDOBULBS: variable, simple, ovoid, bearing 2 or more leaves.

LEAVES: lanceolate, pleated, dark green, folded when young.

RACEME OR PANICLE: with many flowers, its axis often finely downy, terminal.

FLOWERS: not resupinate (in ours).

SEPALS AND PETALS: similar in size and shape, arching forwards.

LIP: 3-lobed, not or scarcely exceeding the petals, often hairy or papillose.

COLUMN: with a distinct foot, forming a conspicuous mentum.

POLLINIA: 4, in 2 superposed pairs, almost spherical, borne on a short stipe with a circular viscidium.

1. **P. affinis** Lindley

SYNONYM: *P. bracteosa* Lindley.

ILLUSTRATIONS: Botanical Magazine, 4161 (1845); Bechtel et al., Manual of cultivated orchid species, 265 (1981); Fanfani, Macdonald encyclopaedia of orchids, 127 (1989).

PSEUDOBULBS: 3–5 cm, compressed, circular in outline, each bearing 2 leaves.

LEAVES: to 20 cm, thin, lanceolate, acute.

FLOWERS: in a raceme (rarely slightly branched), axis finely downy.

SEPALS AND PETALS: *c.* 1 cm, yellow with brown stripes.

DISTRIBUTION: W Africa.

HARDINESS: G2.

FLOWERING: summer.

A few other species may occur from time to time in specialist collections.

128. GALEANDRA Lindley

A genus of about 8 species from C & S America. They may be grown in epiphytic or terrestrial compost in pots. Cooler conditions are required during the resting season. Plants are very susceptible to attack by thrips and red spider mite. Propagation by division.

Literature: Rolfe, R.H., Galeandras, *Gardeners' Chronicle* 12(2): **430–1** (1892); Teuscher, H., Die Gattung Galeandra, *Die Orchidee* **26**: 1–5 (1975).

HABIT: epiphytic.

PSEUDOBULBS: ovoid or stem-like, compound, bearing 2–7 leaves.

LEAVES: rolled when young.

RACEME: terminal, open, with few to many flowers.

SEPALS AND PETALS: similar, spreading or reflexed.

LIP: somewhat cup-shaped, rolled around and partially concealing the column, with a short, downwardly directed or straight spur.

COLUMN: without a foot.

POLLINIA: 2, laterally ovoid, stipe more or less triangular, short and broad.

Key to species

1. Pseudobulbs long and stem-like; flowers more than 8 cm in diameter **4. devoniana**
2. Sepals and petals reflexed, yellow or greenish flushed with brown; lip violet-red margined with white **1. batemanii**
3. Sepals and petals pale green suffused with brown; lip white with purple spots **2. lacustris**
4. Sepals and petals olive green; lip white with a central violet blotch **3. nivalis**

1. G. batemanii Rolfe

SYNONYM: *G. baueri* misapplied.

ILLUSTRATION: Edwards's Botanical Register **26**: t. 49 (1840).

PSEUDOBULBS: ovoid, each with 5–7 leaves.

LEAVES: 15–20 cm, linear-lanceolate, acute.

RACEMES: arching, with 12–many flowers which are *c.* 5 cm in diameter.

SEPALS AND PETALS: reflexed, yellow or greenish, flushed with brown.

LIP: violet-red margined with white, deeply notched at the apex; spur straight.

DISTRIBUTION: Mexico, Guatemala.

HARDINESS: G2.

FLOWERING: summer.

Genuine *G. baueri* Lindley, from French Guiana, is probably not in cultivation: it has fewer flowers and a yellow, purple-streaked or purple-flushed lip.

2. G. lacustris Barbosa-Rodriguez

ILLUSTRATIONS: Orchid Review **83**: 165 (1975); Die Orchidee **26**: 3 (1975); Bechtel et al., Manual of cultivated orchid species, 214 (1981).

Like *G. batemanii* except as below.

SEPALS AND PETALS: pale green suffused with purple.

LIP: white with purple spots.

DISTRIBUTION: N & W Brazil, Peru.

3. G. nivalis Masters

Like *G. batemanii* except as below.

SEPALS AND PETALS: olive-green.

LIP: white with a central violet blotch.

DISTRIBUTION: N & W Brazil.

4. G. devoniana Schomburgk

ILLUSTRATIONS: Botanical Magazine, 4610 (1851); Dunsterville & Garay, Venezuelan orchids illustrated **2**: 157 (1961); Die Orchidee **26**: 2 (1975); Bechtel et al., Manual of cultivated orchid species, 214 (1981).

PSEUDOBULBS: cylindric, stem-like, each with 2–5 leaves.

LEAVES: to 20 × 1 cm, linear.

RACEME: arching, with few flowers.

FLOWERS: *c.* 10 cm in diameter.

SEPALS AND PETALS: linear, brown with greenish margins.

LIP: white, lined with violet, with 2 low ridges on the surface; spur lightly curved.

DISTRIBUTION: Tropical S America, from Venezuela to N Brazil.

HARDINESS: G2.

FLOWERING: summer.

129. EULOPHIA R. Brown

A genus of about 300 species, mostly from tropical Africa. Several of those cultivated have at one time or other been placed in the genus *Lissochilus* R. Brown. Cultivation can be difficult, as most species are found in seasonally damp places. Abundant moisture is required during the growing season and drier conditions with maximum light while the plants are resting. Propagation by seed or by division.

Literature: Cribb, P.J., The genus Eulophia in Africa, *Proceedings of the 12th World Orchid Congress, Hiroshima Symposium*, 87–103 (1987).

HABIT: large, terrestrial plants.

PSEUDOBULBS: compound, sometimes stem-like or swollen.

LEAVES: 2–several on each pseudobulb, pleated and with prominent veins, rolled when young.

RACEMES: lateral with 5–many flowers.

SEPALS AND PETALS: similar to each other or the petals larger than the sepals, all spreading or the sepals sometimes reflexed.

LIP: pivoting freely on the base of the column, 3-lobed, shortly spurred or saccate at the base.

COLUMN: without a foot.

POLLINIA: 2, ovoid or laterally ovoid, stipe short and broad, viscidium almost circular.

Key to species

1. Petals about as long as, and much narrower than the sepals

1.A. Leaves to 50 cm; racemes with 5–15 flowers **1. guineensis**

1.B. Leaves more than 50 cm (sometimes to 120 cm); racemes with more than 15 flowers **2. alta**

2. Petals yellowish or greenish yellow; flower to 3.5 cm in diameter; bracts lanceolate **5. streptopetala**

3. Sepals green, petals lilac-purple with darker veins; lateral lobes of the lip purple **3. gigantea**

4. Sepals purple; petals white flushed with pink; lateral lobes of lip green striped with purple **4. horsfallii**

1. E. guineensis Lindley

ILLUSTRATION: Botanical Magazine, 2467 (1824).

PSEUDOBULBS: to 5 cm, each with 2–3 leaves.

LEAVES: 30–45 cm, narrowly elliptic, narrowed to the base, acute.

RACEMES: with 5–15 flowers.

FLOWERS: to 6 cm in diameter.

SEPALS AND PETALS: similar in size, narrow, acuminate, either whitish pink or greenish or brownish purple.

LIP: with the central lobe whitish pink to purple, streaked and spotted with darker purple.

DISTRIBUTION: Tropical Africa, from Gambia to Angola and Uganda.

HARDINESS: G2.

FLOWERING: autumn.

2. E. alta (Linnaeus) Fawcett & Rendle

SYNONYMS: *E. longifolia* Humboldt, Bonpland & Kunth: *E. woodfordii* (Lindley) Rolfe.

ILLUSTRATIONS: Hoehne, Flora Brasilica **12**(6): t. 1 (1942); Hawkes, Encyclopaedia of cultivated orchids, 208 (1965); Dunsterville & Garay, Venezuelan orchids illustrated **1**: 150 (1959); Ospina & Dressler, Orquideas de las Americas, f. 141 (1974).

Like *E. guineensis* except as below.

LEAVES: 100–120 cm.

RACEME: with many flowers which are 3–5.5 cm in diameter, greenish or brown.

LIP: central lobe purple.

DISTRIBUTION: tropical Africa from Ghana and Sudan to Angola and Zimbabwe, tropical America from Mexico to Argentina.

HARDINESS: G2.

FLOWERING: autumn.

3. E. gigantea (Welwitsch) R. Brown

SYNONYM: *Lissochilus giganteus* Welwitsch.

ILLUSTRATION: Gardeners' Chronicle 3: 616–7 (1888).

STEM: to 5 m or more.

LEAVES: to 1.2 m × 10 cm, bright green with prominent yellow midribs.

RACEME: with 20–100 flowers, bracts ovate, wrapped around the bases of the flower-stalks.

FLOWERS: 6–7 cm in diameter.

SEPALS: green.

PETALS: larger than sepals, lilac-purple with darker veins.

LIP: mostly purple throughout.

DISTRIBUTION: W tropical Africa from Congo to Angola.

HARDINESS: G2.

FLOWERING: spring–summer.

4. E. horsfallii (Bateman) Summerhayes

SYNONYMS: *Lissochilus horsfallii* Bateman; *L. porphyroglossa* Reichenbach; *E. porphyroglossa* (Reichenbach) Bolus.

ILLUSTRATION: Botanical Magazine, 5486 (1865).

Like *E. gigantea* except as below.

STEMS: to 2 m.

RACEME: with 60–100 flowers which are 4–6 cm in diameter.

SEPALS: purple.

PETALS: white or white flushed with pink.

LIP: lateral lobes green striped with purple.

DISTRIBUTION: Tropical Africa from Sierra Leone to Mozambique.

HARDINESS: G2.

FLOWERING: winter.

Sometimes found under the name *Lissochilus roseus* Lindley – a name of doubtful application.

5. E. streptopetala Lindley

SYNONYM: *Lissochilus krebsii* Reichenbach.

ILLUSTRATIONS: Botanical Magazine, 2931 (1829), 5861 (1870); Bolus, Orchids of South Africa **3**: t. 9–10 (1913); Flowering plants of South Africa **21**: t. 820 (1941); Gibson, Wild flowers of Natal (coastal region), t. 25 (1975).

STEM: to 1.5 m.

LEAVES: to 50 cm × 7 mm, lanceolate, bright green, with a prominent midrib.

RACEME: with up to 50 flowers, bracts lanceolate, not wrapped around the bases of the flower-stalks.

FLOWERS: *c.* 3 cm in diameter.

SEPALS: green striped with purple.

PETALS: yellow or greenish yellow.

LIP: yellow, the lateral lobes purplish or brownish.

DISTRIBUTION: South Africa, Zimbabwe.

HARDINESS: G2.

FLOWERING: winter.

130. OECEOCLADES Lindley

A genus of about 30 species from the tropics, mostly in Africa and Madagascar, formerly known as *Eulophidium* Pfitzer. Cultivation as for *Eulophia* (p. 409).

Literature: Summerhayes, V.S., The genus Eulophidium Pfitzer, *Bulletin du Jardin Botanique de l'Etat, Bruxelles* **27**: 391–403 (1957); Garay, L.A. & Taylor, P., The genus Oeceoclades Lindl., *Botanical Museum Leaflets, Harvard University* **24**: 249–74 (1976).

Like *Eulophia* except as below.

PSEUDOBULBS: simple, each bearing 1–3 leathery leaves.

LEAVES: folded when young.

POLLINIA: 2, ovoid or pear-shaped, stipe very short or absent, viscidium semicircular or oblong.

Key to species

1. Each pseudobulb with 1 non-pleated, irregularly variegated leaf **1. maculata**
2. Each pseudobulb with 2–3 pleated, green leaves **2. saundersiana**

1. O. maculata (Lindley) Lindley

SYNONYMS: *Eulophia maculata* (Lindley) Reichenbach; *Eulophidium maculatum* (Lindley) Pfitzer; *Eulophidium ledienii* N.E. Brown.

ILLUSTRATIONS: Gartenflora **37**: t. 1288 (1888); Ospina & Dressler, Orquideas de las Americas, t. 141 (1974).

PSEUDOBULBS: each bearing 1 leaf.

LEAVES: 15–25 × 2.5–4 cm, not pleated, fleshy, leathery, irregularly variegated.

RACEMES: 7–15 cm with several flowers.

SEPALS AND PETALS: 8–10 mm, pale brownish green.

LIP: 3-lobed with the central lobe itself 2-lobed, brownish green with 2 reddish or purple spots on the sides.

DISTRIBUTION: tropical America, tropical Africa.

HARDINESS: G2.

FLOWERING: summer–autumn.

2. O. saundersiana (Reichenbach) Garay & Taylor

SYNONYM: *Eulophidium saundersianum* (Reichenbach) Summerhayes.

ILLUSTRATION: Reichenbach, Xenia Orchidacearum **2**: t. 173 (1873).

PSEUDOBULBS: each with 2–3 leaves.

LEAVES: deciduous some distance above the pseudobulb, the persistent portion splitting and forming a stiff, fibrous crown; blades 10–18 × 5–7.5 cm, pleated, green.

RACEME: 1–15 cm with many flowers.

SEPALS AND PETALS: 1–1.2 cm, green, the petals marked with black lines.

LIP: 4-lobed, green with a few black lines.

DISTRIBUTION: tropical Africa from Cameroun to Uganda.

HARDINESS: G2.

FLOWERING: autumn?

131. EULOPHIELLA Rolfe

A genus of 4 species from Madagascar, where they occur in seasonally damp places. Cultivation as for *Eulophia* (p. 409).

Literature: Bosser, J. & Morat, P., Contribution à l'étude des Orchidaceae de Madagascar IX: Les Genres Grammangis et Eulophiella, *Adansonia* **9**: 299–309 (1969); Kennedy, G.C., The genus Eulophiella, *Orchid Digest* **36**: 120–2 (1972).

HABIT: rhizomatous epiphytes.

PSEUDOBULBS: large, compound, each bearing 3–6 leaves.

LEAVES: pleated, prominently veined, rolled when young.

RACEME: basal or lateral with 7–25 flowers with prominent bracts.

FLOWERS: large, almost circular in outline.

SEPALS AND PETALS: all similar or the petals a little smaller, lateral sepals attached to the column foot.

LIP: 3-lobed, not spurred, attached directly to the column foot, without an obvious claw.

COLUMN: with a conspicuous foot.

POLLINIA: 2, ovoid or ellipsoid, stipe very short and broad, oval or conical.

Key to species

1. Leaves 100 × 8–10 cm or more; flower-stalk 4.5–5 cm; flowers 8–10 cm in diameter **1. roempleriana**

2. Leaves 60–80 × 3.5–5 cm; flower-stalk 2–3 cm; flowers 4–5 cm in diameter **2. elizabethae**

1. E. roempleriana (Reichenbach) Schlechter

SYNONYMS: *Grammatophyllum roemplerianum* Reichenbach; *E. peetersiana* Kränzlin; *E. hamelinii* Rolfe.

ILLUSTRATIONS: Botanical Magazine, 7612, 7613 (1898); Gardeners' Chronicle **23**: 200 (1898); Orchid Digest **36**: 120 (1972); Hillerman & Holst, An introduction to the cultivated angraecoid orchids of Madagascar, pl. 27 (1986).

PSEUDOBULBS: *c*. 20 cm, spindle-shaped, without fibrous leaf remains.

LEAVES: 100 × 8–10 cm or more.

AXIS OF RACEME: green, bracts brown.

FLOWERS: 8–10 cm in diameter.

SEPALS: pink tinged purplish and with purple tips.

PETALS: pink.

LIP: with large, pink lateral lobes which curve over the white column, the central lobe purple or white margined with purple and with 3–4 orange-yellow ridges.

DISTRIBUTION: Madagascar.

HARDINESS: G2.

FLOWERING: spring–summer.

2. E. elizabethae Linden & Rolfe

ILLUSTRATIONS: Botanical Magazine, 7387 (1894), n.s., 656 (1973); Orchid Digest **36**: 120 (1972); Hillerman & Holst, An introduction to the cultivated angraecoid orchids of Madagascar, pl. 27 (1986).

PSEUDOBULBS: *c*. 15 cm, spindle-shaped, covered with fibrous leaf remains.

LEAVES: 60–80 × 3.5–5 cm.

AXIS OF RACEME AND BRACTS: red.

FLOWERS: 4–5 cm in diameter.

SEPALS: red lined with white outside, white within.

LIP: with small lateral lobes, white, red at the base and with a large yellow blotch on the central lobe.

DISTRIBUTION: Madagascar.

HARDINESS: G2.

FLOWERING: spring.

132. CYRTOPODIUM R. Brown

A genus of about 30 species from C & S America. Cultivation as for *Eulophia* (p. 409).

HABIT: terrestrial or epiphytic.

PSEUDOBULBS: often stem-like, compound, bearing several leaves.

LEAVES: rolled when young and pleated, with conspicuous veins.

RACEME OR PANICLE: with many flowers, lateral.

SEPALS AND PETALS: similar or the petals somewhat broader and shorter, the lateral sepals attached to the column foot.

LIP: 3-lobed, conspicuously clawed, the claw and lateral lobes spreading, the central lobe directed downwards.

COLUMN: with a small foot.

POLLINIA: 2, ovoid or ellipsoid, stipe triangular, short and broad.

Key to species

1. Sepals and petals equal, unspotted; lip yellow — **1. andersonii**
2. Pseudobulbs stem-like, to 1.5 m — **2. punctatum**
3. Pseudobulbs to 10 cm — **3. virescens**

1. C. andersonii (Lambert) R. Brown

SYNONYM: *Cymbidium andersonii* Lambert.

ILLUSTRATIONS: Botanical Magazine, 1800 (1816); Bechtel et al., Manual of cultivated orchid species, 192 (1981).

PSEUDOBULBS: stem-like, to 1.5 m.

LEAVES: to 50 cm, linear and tapering.

PANICLE: with many flowers and ovate, acute, inconspicuous bracts.

FLOWERS: *c.* 5 cm in diameter.

SEPALS AND PETALS: spreading, more or less equal, yellow faintly flushed with green.

LIP: entirely yellow or orange-yellow, the central lobe spoon-shaped, wavy.

DISTRIBUTION: West Indies, Venezuela.

HARDINESS: G2.

FLOWERING: spring–summer.

2. C. punctatum (Linnaeus) Lindley

SYNONYM: *Epidendrum punctatum* Linnaeus.

ILLUSTRATIONS: Hoehne, Iconografia de orchidaceas do Brasil, t. 153 (1949); Ospina & Dressler, Orquideas de las

Americas, f. 139 (1974); Sheehan & Sheehan, Orchid genera illustrated, 77 (1979); Bechtel et al., Manual of cultivated orchid species, 192 (1981).

Like *C. andersonii* except as below.

BRACTS: conspicuous.

SEPALS: greenish yellow with brown stripes or spots, narrower than the yellow, red-spotted petals.

LIP: with dark red lateral lobes, the central lobe edged with red.

DISTRIBUTION: tropical America from Costa Rica to C Brazil.

HARDINESS: G2.

FLOWERING: spring–summer.

3. **C. virescens** Reichenbach & Warming

ILLUSTRATIONS: Botanical Magazine, 7396 (1895); Hoehne, Iconografia de orchidaceas do Brasil, t. 166 (1949).

Like *C. punctatum* except as below.

PSEUDOBULBS: to 10 cm, each with 3–4 leaves.

FLOWERS: *c.* 3 cm in diameter, borne in a raceme.

LIP: fleshy.

DISTRIBUTION: S Brazil and Paraguay.

HARDINESS: G2.

FLOWERING: spring–summer.

133. ANSELLIA Lindley

A genus of 1 species from the southern half of Africa. Plants require careful potting, using terrestrial compost, and establishment, as they suffer from root disturbance. They should not be allowed to dry out completely during the resting season.

Literature: Summerhayes, V.S., African orchids IX, *Kew Bulletin for* 1937: 461–3 (1938).

HABIT: large epiphytes.

PSEUDOBULBS: compound, elongate and stem-like, bearing several leaves.

LEAVES: folded when young.

RACEME OR PANICLE: with 20–many flowers, terminal.

SEPALS AND PETALS: similar or the petals somewhat broader than the sepals, all spreading, the lateral sepals attached to the column foot, forming a small mentum.

LIP: without a spur, 3-lobed, the lateral lobes wrapped around the column, the central lobe with 2–3 ridges.

POLLINIA: 2, each 2-lobed, appearing as 4, ovoid, stipe very short and broad, narrowly triangular.

1. **A. africana** Lindley

SYNONYM: *A. gigantea* Reichenbach var. *nilotica* (N.E. Brown) Summerhayes.

ILLUSTRATIONS: Botanical Magazine, 4965, excluding f. 3

(1854); Sheehan & Sheehan, Orchid genera illustrated, 39 (1979); Bechtel et al., Manual of cultivated orchid species, 168 (1981).

PSEUDOBULBS: to 80 cm.

LEAVES: 4–7, 15–30 × 3–4 cm, linear to narrowly elliptic, prominently veined.

PANICLE: to 75 cm with 20–many flowers.

FLOWERS: 2–2.5 cm in diameter.

SEPALS AND PETALS: petals broader than sepals or as broad as them, all greenish to yellow, usually with brownish or purplish spots.

LIP: 3-lobed, the lateral lobes greenish striped with brown, the central lobe yellow, oblong to circular with 2 bright yellow ridges.

DISTRIBUTION: Tropical and southern Africa.

HARDINESS: G2.

FLOWERING: winter.

134. CYMBIDIUM Swartz

A genus of about 50 species from Asia. Most of the cultivated Cymbidiums are hybrids, many of them complex. Species Nos. **15** & **16** are sometimes separated into the genus *Cyperorchis* Blume.

They should be grown in pots in an epiphytic compost and in cool conditions with a marked drop in temperature at night, which initiates flower-bud formation. They should be shaded from hot sunshine in the summer and in general should be well ventilated. Plants should be repotted after flowering. Propagation by division and commercially by meristem culture.

Literature: Hunt, P.F., Notes on Asiatic orchids V, *Kew Bulletin* **24**: 75–99 (1970); Pradhan, G.M., The Cyperorchis species of northern India, *Orchid Digest* **40**: 115–17 (1976); Koester, A., *The Cymbidium list*, volume 1 (1979); Seidenfaden, G., Orchid Genera in Thailand XI, Cymbidiae Pfitzer, *Opera Botanica* **72**: 65–93 (1983); Dupuy, D. & Cribb, P., *The Genus Cymbidium* (1988).

HABIT: epiphytic or terrestrial.

PSEUDOBULBS: obvious or inconspicuous, compound, with 2-several leaves.

LEAVES: leathery, conspicuously veined, folded when young.

RACEME: lateral with 2–many flowers.

SEPALS AND PETALS: usually similar, spreading or curving forwards.

LIP: free or somewhat united to the base of the column, scarcely to obviously 3-lobed.

COLUMN: narrow, without a foot.

POLLINIA: 2, laterally ovoid, stipe very short and broad, triangular.

Synopsis of characters

Flowers. 8 or more in each inflorescence: **1–11,16**; inflorescence with up to 7 flowers: **12–15**.
Sepals and petals. Directed forwards, not widely spreading: **15,16**; widely spreading, flower up to 5 cm in diameter: **1–4**; widely spreading, flower more than 5 cm in diameter: **5–14**. White or cream, sometimes flushed with pink: **11,12,14,15**; yellow or greenish, rarely brown, often with dark lines or blotched: **1–10,13,16**.
Raceme. Arching or hanging: **1,2,16**; erect or spreading: **3–15**.

Key to species

1. Flowers 8 or more in each well developed inflorescence *Go to Subkey 1*
2. Flowers 1–7 in each well developed inflorescence *Go to Subkey 2*

Subkey 1. Flowers 8 or more in each well developed inflorescence.

1. Flowers at most 5 cm in diameter
1.A. Sepals and petals directed forwards, the flower not opening widely **16. elegans**
1.B. Raceme erect or spreading
1.B.a. Leaves 60–100 × 2–4 cm; lip white or yellowish with irregular red-brown spots **3. ensifolium**
1.B.b. Leaves 30–55 × 1.5–2 cm; lip white to pink dotted with red and tipped with yellow **4. floribundum**
1.C. Leaves distinctly stalked; lip diamond-shaped, scarcely 3-lobed, usually pink to purple **1. devonianum**
1.D. Leaves not stalked; lip clearly 3-lobed, yellowish or reddish **2. aloifolium**
2. Sepals and petals white or white flushed with pink **11. insigne**
3. Sepals and petals green or yellowish, sometimes marked with red, purple or brown lines and dots
3.A. Flower 6–8 cm in diameter; sepals and petals pale green with faint brownish red lines; lip white or pinkish yellow with a central crimson line and crimson blotches **10. erythraeum**
3.B. Flowers 8–10 cm in diameter; sepals and petals green or yellowish with red or brown stripes; lip yellow spotted with reddish brown, hairy **5. iridioides**
3.C. Flowers 9–16 cm in diameter; sepals and petals pale olive green, unstriped; lip yellow blotched with purple **6. hookerianum**
3.D. Flower 9–11 cm in diameter; sepals and petals greenish yellow with faint red-brown stripes; lip yellow, white or orange, usually with a V-shaped scarlet mark **7. lowianum**
3.E. Flowers 10–15 cm in diameter; sepals and petals yellowish

suffused and lined with red-brown; lip cream to yellowish lined with purple or brown **8. tracyanum**

3.F. Flowers *c.* 9 cm in diameter; sepals and petals green lined with reddish brown; lip yellowish white spotted with red-brown on the ridges, lateral lobes lined with red-brown **9. wilsonii**

Subkey 2. Flowers 1–7 in each well-developed inflorescence.

1. Sepals and petals olive green to yellow, faintly spotted and lined with red or purple; leaves to 15 cm **13. tigrinum**
2. Sepals and petals directed forwards, the flower not opening widely **15. mastersii**
3. Column bright crimson; lateral lobes of lip large, more or less folding over column **14. erythrostylum**
4. Column not crimson; lateral lobes of lip not folding over column **12. eburneum**

1. C. **devonianum** Paxton

ILLUSTRATIONS: Botanical Magazine, 9327 (1933); Hawkes, Encyclopaedia of cultivated orchids, 192 (1965); Orchid Digest **41**: 8 (1977); Bechtel et al., Manual of cultivated orchid species, 190 (1981); Opera Botanica **72**: f. 45 (1983); Dupuy & Cribb, The Genus Cymbidium, pl. 12 & t. 82, 83 (1988).

PSEUDOBULBS: with 2–5 leaves.

LEAVES: 30–40 × *c.* 3 cm, distinctly stalked, oblanceolate.

RACEMES: equalling or exceeding the leaves, arching, with many flowers.

FLOWERS: 2.5–4 cm in diameter.

SEPALS AND PETALS: yellowish green or brown with pale purple streaks, spreading.

LIP: diamond-shaped, scarcely 3-lobed, variable in colour, usually pink to purplish, often with darker spots.

DISTRIBUTION: NE India.

HARDINESS: G1.

FLOWERING: spring–summer.

2. C. **aloifolium** (Linnaeus) Swartz

SYNONYM: *C. pendulum* Swartz; *C. simulans* Rolfe.

ILLUSTRATIONS: Orchid Digest **42**: 126 (1978); Bechtel et al., Manual of cultivated orchid species, 189 (1981); Opera Botanica **72**: t. 43 (1983); Dupuy & Cribb, The Genus Cymbidium, pl. 10 & t. 78, 79 (1988).

PSEUDOBULBS: scarcely obvious.

LEAVES: 30–60 × 1.5–3 cm, not stalked, apex usually notched.

RACEMES: arching with many flowers which are 3.5–4.5 cm in diameter.

SEPALS AND PETALS: brownish yellow, each with a purple median stripe, spreading.

LIP: distinctly 3-lobed, yellowish or reddish with red lines and with 2 curved and interrupted ridges, finely downy.
DISTRIBUTION: from E Himalaya to S China and Malaysia.
HARDINESS: G1.
FLOWERING: spring.

3. C. ensifolium (Linnaeus) Swartz
ILLUSTRATIONS: Botanical Magazine, 1751 (1815); Opera Botanica **72**: f. 38 (1983); Orchid Review **92**: 71 (1984); Dupuy & Cribb, The Genus Cymbidium, pl. 24 & t. 120–123 (1988).
PSEUDOBULBS: scarcely obvious.
LEAVES: 60–100 × 2–4 cm, linear to sword-like, acute.
RACEME: erect or spreading, with 8–12 flowers, exceeding the leaves.
FLOWERS: to 4.5 cm in diameter, fragrant.
SEPALS AND PETALS: greenish yellow with reddish brown lines, spreading.
LIP: clearly 3-lobed, white or yellowish with irregular, red-brown spots.
DISTRIBUTION: India, China, Japan and probably elsewhere in SE Asia.
HARDINESS: G1.
FLOWERING: summer.

4. C. floribundum Lindley
SYNONYM: *C. pumilum* Rolfe.
Like *C. ensifolium* except as below.
LEAVES: 30–55 × 1.5–2 cm.
RACEME: shorter than the leaves.
SEPALS AND PETALS: sometimes brown.
LIP: white to pink, dotted with red and tipped with yellow.
DISTRIBUTION: W China.
HARDINESS: G1.
FLOWERING: summer.

5. C. iridioides D. Don
SYNONYM: *C. giganteum* Wallich.
ILLUSTRATIONS: Botanical Magazine, 4844 (1855); Bechtel et al., Manual of cultivated orchid species, 190 (1981); Dupuy & Cribb, The Genus Cymbidium, pl. 14 & t. 86 (1988).
PSEUDOBULBS: ovoid, several-leaved.
LEAVES: linear, tapering to the acute apex, 60–75 × 2–4 cm.
RACEME: erect or spreading, with 8–15 flowers which are 8–10 cm in diameter.
SEPALS AND PETALS: green or yellowish green with red or brown stripes.
LIP: yellow, spotted with reddish brown, hairy.

DISTRIBUTION: Himalaya, W China.
HARDINESS: G1.
FLOWERING: winter.

This species forms part of a complex of variants distributed in the Himalaya, W China and the northern part of SE Asia. Though various species have been described, they are possibly no more than selections from the general variation. The most important of them form Nos. **6–10** below.

6. C. hookerianum Reichenbach

SYNONYM: *C. grandiflorum* Griffith.
ILLUSTRATIONS: Botanical Magazine, 5574 (1866); Dupuy & Cribb, The Genus Cymbidium, pl. 15 & t. 85 (1988).

Like *C. iridioides* except as below.

FLOWERS: 9–13 cm in diameter (occasionally as much as 16 cm?).
SEPALS AND PETALS: olive green, unstriped.
LIP: yellow blotched with purple.
DISTRIBUTION: Himalaya, W China.

7. C. lowianum (Reichenbach) Reichenbach.

ILLUSTRATIONS: Orchid Digest **42**: 124 (1978); Bechtel et al., Manual of cultivated orchid species, 191 (1981); Dupuy & Cribb, The Genus Cymbidium, pl. 16 & t. 92, 93 (1988).

Like *C. iridioides* except as below.

FLOWERS: 9–11.5 cm in diameter.
SEPALS AND PETALS: greenish yellow with faint red-brown stripes.
LIP: variable, yellow, white or orange, usually with a V-shaped scarlet mark.
DISTRIBUTION: Burma.

8. C. tracyanum Rolfe

ILLUSTRATIONS: Botanical Magazine n.s., 56 (1949); Bechtel et al., Manual of cultivated orchid species, 191 (1981); Die Orchidee, **35**: centre page pull-out (1984); Dupuy & Cribb, The Genus Cymbidium, pl. 13 & t. 84 (1988).

Like *C. iridioides* except as below.

FLOWERS: fragrant, 10–15 cm in diameter.
SEPALS AND PETALS: yellowish suffused and lined with red-brown.
LIP: cream to yellow lined with purple or brown.
DISTRIBUTION: Burma, Thailand, Vietnam.

Sometimes considered to be a hybrid between *C. giganteum* and *hookerianum*.

9. C. wilsonii (Cooke) Rolfe

SYNONYM: *C. giganteum* Wallich 'Wilsonii'.
ILLUSTRATIONS: Botanical Magazine n.s., 704 (1976); Dupuy

& Cribb, The Genus Cymbidium, t. 90, 91 (1988).

Like *C. iridioides* except as below.

FLOWERS: *c.* 9 cm in diameter.

SEPALS AND PETALS: green lined with reddish brown.

LIP: yellowish white spotted with red-brown on the ridges and near the apex, lateral lobes lined with red-brown.

DISTRIBUTION: W China.

10. **C. erythraeum** Lindley

SYNONYM: *C. longifolium* misapplied.

Like *C. iridioides* except as below.

LEAVES: to 1.5 cm broad.

FLOWERS: fragrant, 6–8 cm in diameter.

SEPALS AND PETALS: pale green with faint brownish red lines.

LIP: white or pinkish yellow with a central crimson line and crimson blotches.

DISTRIBUTION: Himalaya, W China.

11. **C. insigne** Rolfe

SYNONYM: *C. sanderi* invalid.

ILLUSTRATIONS: Botanical Magazine, 8312 (1910); Opera Botanica **72**: f. 59 (1983); Dupuy & Cribb, The Genus Cymbidium, pl. 17 & t. 12, 13, 96–100 (1988).

PSEUDOBULBS: spherical.

LEAVES: 50–100 × 1.5–2 cm, linear, tapering to the acute apex.

RACEME: erect, exceeding the leaves, with 10–15 flowers.

FLOWERS: 7.5–10 cm in diameter.

SEPALS AND PETALS: white or white flushed with pink, their bases red-spotted.

LIP: white, flushed, lined and spotted with pink and with 2 yellow ridges.

DISTRIBUTION: Vietnam, Laos, N Thailand, S China.

HARDINESS: G1.

FLOWERING: spring–summer.

12. **C. eburneum** Lindley

ILLUSTRATIONS: Botanical Magazine, 5126 (1859); Orchid Digest **42**: 126 (1978); Bechtel et al., Manual of cultivated orchid species, 191 (1981); American Orchid Society Bulletin **54**: 324 (1985); Dupuy & Cribb, The Genus Cymbidium, pl. 19 & t. 102, 103 (1988).

PSEUDOBULBS: scarcely evident.

LEAVES: 45–60 × *c.* 2 cm, linear, tapering to the acute apex.

RACEME: erect with 1–2 flowers.

FLOWERS: usually 7.5–10 cm in diameter, fragrant.

SEPALS AND PETALS: white, spreading.

LIP: usually white with yellow ridges or a yellow centre, occasionally spotted with red or purple, the lateral lobes extending up to the column but not overlapping it, downy.

DISTRIBUTION: Himalaya, Burma, Vietnam.
HARDINESS: G1.
FLOWERING: spring.
Variable in the colour of the lip (see Pradhan, Orchid Digest **40**: 69–71, 1976).

13. C. tigrinum Hooker
ILLUSTRATIONS: Botanical Magazine, 5457 (1864); Orchid Digest **42**: 129 (1978); Bechtel et al., Manual of cultivated orchid species, 191 (1981); Opera Botanica **72**: f. 51 (1983); Dupuy & Cribb, The Genus Cymbidium, pl. 23 & t. 118, 119 (1988).
PSEUDOBULBS: almost spherical, with 3–4 leaves.
LEAVES: to 15 × 2.5 cm, broadly linear, acute.
RACEMES: erect, exceeding the leaves, with 3–6 flowers.
FLOWERS: usually 5–7 cm in diameter (rarely larger).
SEPALS AND PETALS: olive green to yellow, faintly spotted and lined with red or purple, spreading.
LIP: mostly white, lined and spotted with purple, the margins often mostly purple.
DISTRIBUTION: Burma, Thailand.
HARDINESS: G1.
FLOWERING: summer–autumn.

14. C. erythrostylum Rolfe
ILLUSTRATIONS: Botanical Magazine, 8131 (1907); Orchid Digest **42**: 129 (1978); Dupuy & Cribb, The Genus Cymbidium, pl. 21 & t. 108 (1988).
PSEUDOBULBS: scarcely obvious.
LEAVES: 30–45 × 1–1.5 cm, linear, tapering to the acute apex.
RACEME: erect, loose, with 4–6 flowers.
FLOWERS: 7.5–11 cm in diameter.
SEPALS AND PETALS: white.
LIP: white shaded with yellow and with red lines, the lateral lobes large, curved over the column and more or less overlapping.
COLUMN: bright crimson.
DISTRIBUTION: Vietnam.
HARDINESS: G1.
FLOWERING: winter.

15. C. mastersii Lindley
SYNONYM: *Cyperorchis mastersii* (Lindley) Bentham.
ILLUSTRATIONS: Edwards's Botanical Register **31**: t. 50 (1845); Orchid Digest **41**: 116 (1976) & **42**: 127 (1978); Bechtel et al., Manual of cultivated orchid species, 191 (1981); Opera Botanica **72**: f. 52 (1983); Dupuy & Cribb, The Genus Cymbidium, pl. 20 & t. 104, 105 (1988).
PSEUDOBULBS: scarcely obvious.

LEAVES: 45–70 × 1.5 cm, linear, tapering to the acute apex.
RACEMES: erect with 4–6 (rarely 7) flowers.
FLOWERS: fragrant, *c.* 6 cm long, the sepals and petals forwardly directed, white or ivory white, sometimes flushed with pink.
LIP: white, sometimes spotted with pink, with a yellow centre, downy.
DISTRIBUTION: E Himalaya.
HARDINESS: G1.
FLOWERING: winter.

16. C. elegans (Blume) Lindley
SYNONYM: *Cyperorchis elegans* Blume.
ILLUSTRATIONS: Botanical Magazine 7007 (1888); Orchid Digest **42**: 127 (1978); Orchid Review **92**: 71 (1984); Dupuy & Cribb, The Genus Cymbidium, pl. 22 & t. 109–111 (1988).
PSEUDOBULBS: scarcely obvious.
LEAVES: 40–70 × 1–2.5 cm, numerous, linear, tapering to the acute apex.
RACEME: arching, with many flowers.
FLOWERS: 4–5 cm long.
SEPALS AND PETALS: forwardly directed, little spreading, pale to tawny yellow.
LIP: yellow marked with 2 orange lines, sometimes spotted with red, downy and sparsely ciliate.
DISTRIBUTION: Himalaya, W China.
HARDINESS: G1.
FLOWERING: winter.

135. GRAMMANGIS Reichenbach

A small genus, of which only a single species is grown. Cultivation as for *Cymbidium*.

Like *Cymbidium* except as below.
LATERAL SEPALS: fused at the base and attached to the column foot forming a small but distinct mentum.
PETALS: much smaller than the sepals and differently coloured.

1. G. ellisii Reichenbach.
ILLUSTRATIONS: Botanical Magazine, 5179 (1860); Bechtel et al., Manual of cultivated orchid species, 215 (1981); Hillerman & Holst, An introduction to the cultivated angraecoid orchids of Madagascar, pl. 27 (1986).
SEPALS: *c.* 4 × 2 cm, yellow barred and blotched with brown.
PETALS: white with pink tips.
LIP: white with reddish lines, central lobe small.
DISTRIBUTION: Madagascar.
HARDINESS: G2.
FLOWERING: summer.

136. GRAMMATOPHYLLUM Blume

A genus of about 10 species from SE Asia. They are strong-growing plants requiring epiphytic compost and careful watering. *G. scriptum* benefits from cooler conditions during the resting period. Propagation is by division in early spring.

HABIT: large epiphytes.

PSEUDOBULBS: compound, distinct, or elongate and stem-like, with 2–several leaves.

LEAVES: rolled when young.

RACEME: large, lateral, with many flowers; lower flowers sometimes sterile and abortive.

SEPALS AND PETALS: similar, spreading.

LIP: small, clawed, 3-lobed, the lateral lobes large, extending up to the column, the central lobe smaller, bearing several ridges.

COLUMN: without a foot.

POLLINIA: 2, laterally ovoid to almost spherical, stipe short and broad, divided into 2 branches above.

Key to species

1. Pseudobulbs to 14 cm, distinct, with a few leaves at the apex: flowers to 8 cm in diameter **1. scriptum**
2. Pseudobulbs elongate, stem-like, bearing several leaves along their length; flowers to 10 cm in diameter **2. speciosum**

1. G. scriptum Blume

SYNONYMS: *G. fenzlianum* Reichenbach; *G. measuresianum* Weather; *G. multiflorum* Lindley; *G. rumphianum* Miquel.

ILLUSTRATIONS: Botanical Magazine, 7507 (1896); Sheehan & Sheehan, Orchid genera illustrated, 101 (1979).

PSEUDOBULBS: to 14 cm, ridged, bearing 2–3 leaves at the apex.

LEAVES: very narrowly elliptic, 30–35 × *c.* 8 cm.

RACEME: very long, with many flowers which are all fertile, and up to 8 cm in diameter.

SEPALS AND PETALS: pale green or greenish yellow, suffused or spotted with brown or purplish brown.

LIP: small, hairy.

DISTRIBUTION: Borneo, Philippines, Indonesia (Moluccas).

HARDINESS: G2.

FLOWERING: summer.

Variable as regards the colour of the lip, which varies from white to yellow and is striped with brown or purple.

2. G. speciosum Blume

ILLUSTRATIONS: Botanical Magazine, 5157 (1860); Opera Botanica **72**: f. 54 & t. viiib (1983).

PSEUDOBULBS: elongate, stem-like, leafy along their length.

LEAVES: 45–60 × 3 cm, broadly linear, acute.

RACEME: 2 m or more with many flowers, the lower of which are abortive and sterile with 2 sepals, 2 petals, no lip and an abortive column.

FERTILE FLOWERS: *c.* 10 cm in diameter, sepals and petals pale greenish yellow with dull orange-brown spots.

LIP: small, hairy, yellow striped with red-brown.

DISTRIBUTION: Burma, Laos, Thailand, Vietnam, Malaysia, Indonesia, Philippines.

HARDINESS: G2.

FLOWERING: autumn–winter.

137. MORMODES Lindley

A genus of about 50 species from C & S America which is taxonomically very confused, owing to the wide and little-understood variation in flower colour, shape and size. A few species are grown as curiosities. Cultivation as for *Catasetum* (p. 428).

Literature: Pabst, G.F.J., An illustrated key to the species of the genus Mormodes, *Selbyana* 2: 149–55 (1978).

HABIT: epiphytic.

PSEUDOBULBS: conspicuous, compound, broadly spindle-shaped, each bearing 2-several leaves.

LEAVES: strongly veined, rolled when young.

RACEME: lateral, from the central nodes of the pseudobulb, erect or arching, with inconspicuous bracts.

FLOWERS: variously coloured, often fragrant.

SEPALS AND PETALS: more or less equal, the upper 3 often converging to form a hood over the column.

LIP: clawed, not spurred, entire or 3-lobed, often with the margins rolled under.

COLUMN: cylindric, always twisted to 1 side of the flower which is thus asymmetric.

POLLINIA: 2, ovoid, the stipe wineglass-shaped, longer than the pollinia and broader than them at the apex, the viscidium almost circular.

Key to species

1. Lip clearly 3-lobed, though the lateral lobes may be small

1.A. Flowers to 6 cm in diameter, lip *c.* 5 cm

1.A.a. Upper sepal and petals converging to form a hood over the column **3. luxatum**

1.A.b. Upper sepal and petals not converging to form a hood **1. maculata**

1.B. Flower at most 5 cm in diameter, lip *c.* 3 cm **2. lineatum**

2. Lip hairy above; flower usually uniformly dark purple **4. hookeri**

3. Lip flat or with the sides curved upwards; pseudobulbs 8–10 cm **5. aromatica**
4. Lip ovate, tapering to the apex **6. colossus**
5. Lip broader than long to circular, rounded to notched (though with a small point) at the apex
5.A. Lip 2–2.5 cm wide **7. buccinator**
5.B. Lip 3–3.5 cm wide **8. igneum**

1. M. maculata (Klotzsch) Williams
SYNONYMS: *Cyclosia maculata* Klotzsch; *M. pardinum* Bateman.
ILLUSTRATIONS: Botanical Magazine, 3879 (1842), 3900 (1843); Bechtel et al., Manual of cultivated orchid species, 234 (1981).
PSEUDOBULBS: 8–10 cm with 4–5 leaves.
FLOWERING STEM: to 40 cm, raceme with many flowers.
FLOWERS: *c.* 6 cm in diameter, yellow or pale yellow, with or without red-brown spots.
SEPALS AND PETALS: lanceolate, acute.
LIP: *c.* 5 cm with the margins rolled under, 3-lobed from the middle, the lateral lobes conspicuous.
DISTRIBUTION: Mexico.
HARDINESS: G2.
FLOWERING: autumn.

2. M. lineatum Lindley
ILLUSTRATION: Die Orchidee **22**: 45 (1971).
Like *M. maculata* except as below.
FLOWERS: slightly smaller (*c.* 5 cm in diameter).
LIP: *c.* 3 cm, with very small, incurved, tooth-like lateral lobes.
DISTRIBUTION: Guatemala.

3. M. luxatum Lindley
Like *M. maculata* except as below.
HABIT: plant larger.
FLOWERS: *c.* 7 cm in diameter, sometimes white.
UPPER SEPALS AND UPPER PETAL: converging to form a hood over the column.
DISTRIBUTION: Mexico.

4. M. hookeri Lemaire
ILLUSTRATION: Botanical Magazine, 4577 (1851).
PSEUDOBULBS: oblong-ovoid.
FLOWERING STEM: *c.* 30 cm, raceme with 4–5 distant flowers.
SEPALS AND PETALS: reflexed, dark purple, their sides themselves reflexed.
LIP: pointing forwards, entire, its sides rolled under, hairy on the upper surface, purple but paler than sepals and petals.
DISTRIBUTION: Panama.

HARDINESS: G2.
FLOWERING: winter.

5. **M. aromatica** Lindley
ILLUSTRATIONS: Edwards's Botanical Register **29**: t. 56 (1843); Hamer, Las Orquideas de El Salvador **2**: 124–5 & t. 14 (1974).
PSEUDOBULBS: 8–10 cm with 4–5 leaves.
FLOWERING STEM: weak, to 15 cm, with 5–8 rather distant flowers.
FLOWERS: *c.* 3 cm in diameter, yellowish deeply flushed with violet-brown and with darker spots.
SEPALS AND PETALS: similar, elliptic, acute.
LIP: entire, flat or with the sides curved upwards, with a triangular apex.
DISTRIBUTION: Mexico, El Salvador.
HARDINESS: G2.
FLOWERING: autumn.

6. **M. colossus** Reichenbach
ILLUSTRATIONS: Botanical Magazine, 5840 (1870); Bechtel et al., Manual of cultivated orchid species, 233 (1981).
PSEUDOBULBS: to 30 cm.
FLOWERING STEM: to 70 cm, with 6–10 distant flowers which are up to 15 cm in diameter.
SEPALS AND PETALS: lanceolate, acuminate, reflexed, bright yellow, red towards the base.
LIP: ovate, acuminate, entire, the sides rolled under, bright yellow marked with red at the base.
DISTRIBUTION: Costa Rica.
HARDINESS: G2.
FLOWERING: spring.

7. **M. buccinator** Lindley
SYNONYM: *M. lentiginosa* Hooker.
ILLUSTRATIONS: Botanical Magazine, 4455 (1849), 8041 (1905); Couret, Orquideas Venezolanas, 39–42 (1977).
PSEUDOBULBS: to 30 cm.
FLOWERING STEMS: to 25 cm with 7–10 flowers.
FLOWERS: extremely variable in colour, bright yellow to dark red-brown, variably spotted and lined.
SEPALS AND PETALS: narrowly elliptic, acute, spreading or reflexed at first, later curving forwards.
LIP: 2–2.5 cm wide, entire, broader than long, its sides very markedly rolled under, forming a trumpet shape.
DISTRIBUTION: Venezuela.
HARDINESS: G2.
FLOWERING: spring.

8. M. igneum Lindley & Paxton

Like *M. buccinator* except as below.

FLOWERS: variable in colour.

LIP: 3–3.5 cm wide.

DISTRIBUTION: C America and northern S America.

138. CATASETUM Richard

In the broad sense, a genus of about 70 species from C & S America; about a dozen of these are grown for the sake of their bizarre male flowers. The plants require sharp drainage and full light, combined with a humid atmosphere. They should be rested when the pseudobulbs are ripened. Propagation by seed or by division.

Literature: Mansfeld, R., Die Gattung Catasetum, *Feddes Repertorium* **30**: 257–75 (1932), **31**: 99–125 (1933).

The descriptions given here (with the exception of that for Nos. **1** & **2**) refer to the male flowers; the female flowers are less well known and their identification is difficult. Plants bearing female flowers were at one time referred to the genus *Monachanthus* Lindley. The production of male and female flowers appears to be controlled by light and nutrition (see Dodson, *Annals of the Missouri Botanical Garden* **49**: 35–56,1962). The antennae borne by the male flowers are sensitive, and touching of them by an insect causes the pollinia to be forcibly ejected. For more details on pollination mechanisms in the genus see Hills, Williams & Dodson, *Biotropica* **4**: 61–9 (1972). The genus is divided into 3 genera, *Clowesia*, *Dressleria* and *Catasetum*, by Dodson, *Selbyana* **1**: 130–8 (1975). Under this arrangement, species Nos. **1** & **2** covered here belong to *Clowesia*, the rest to *Catasetum* in the strict sense.

HABIT: epiphytes.

PSEUDOBULBS: large, compound, bearing several leaves.

LEAVES: large, pleated, rolled when young.

RACEMES: arising from the bases of the pseudobulbs, erect or arching, with few to many flowers.

FLOWERS: usually not resupinate, monomorphic and bisexual, or, more usually, dimorphic and unisexual, very rarely trimorphic (male, female and bisexual).

SEPALS AND PETALS: similar, free, spreading or incurved.

LIP: variable in size, saccate or not.

COLUMN: large, without a foot usually acute in male and bisexual flowers, with (in all species treated below except Nos. **1** & **2**) **2** processes (antennae) arising from near the apex and directed backwards onto the lip in the male flowers.

POLLINIA: **2**, ovoid or ellipsoid, the stipe elongate, linear or strap-shaped, its sides usually folded upwards, viscidium circular or broadly oblong.

Synopsis of characters

Flowers. Always bisexual: **1,2**; usually unisexual: 3–13.
Lip. Flat, not hollowed or saccate: **7,9**: hollowed or saccate but with a flat blade: **1,2,5,6,8,10–13**; entirely hollowed, with no flat blade: **3,4**. With conspicuous fringed or ciliate margins: **1,2,6,8–12**; with more or less entire margins: **3–5,7,13**.
Antennae. Asymmetric: **3–6**; symmetric: 7–13.

Key to species

1. Flowers all bisexual; column without antennae
1.A. Saccate part of the lip projecting forwards **1. russellianum**
1.B. Saccate part of the lip curved backwards **2. thylaciochilum**
2. Lip not at all saccate or hollowed
2.A. Lip broadly obovate with 3 teeth at the apex, the lateral teeth narrowly triangular, incurved, the central shorter, straight; margins of the rest of the lip entire **7. cernuum**
2.B. Lip triangular with rounded corners, the apex entire, the margin toothed or ciliate **9. trulla**
3. Antennae not symmetrical
3.A. Lip helmet-shaped, completely hollowed out, without a flat blade
3.A.a. Basal angles of the lip projecting inwards, touching each other or almost so, toothed **3. integerrimum**
3.A.b. Basal angles of lip not projecting inwards, entire **4. macrocarpum**
3.B. Lip semicircular, truncate at base, not lobed **5. pileatum**
3.C. Lip oblong, 3-lobed, the central lobe narrow, not truncate at the base **6. saccatum**
4. Lip broadly obovate, fan-shaped, bearing a large, humped callus in front of the mouth of the sac **8. fimbriatum**
5. Lip small, deeper than long, containing a callus composed of hair-like teeth which almost entirely fill the sac **11. microglossum**
6. Lip narrowly ovate, its margin distantly and shortly toothed, green with red spots **13. callosum**
7. Lip margin almost entirely divided into blunt, fleshy, hair-like teeth, its terminal portion narrowly triangular with a few teeth, with a small saccate area in the middle **12. barbatum**
8. Lip margin coarsely and sharply toothed except for the entire, semicircular terminal part; saccation extending from the base for more than half of the lip **10. atratum**

1. **C. russellianum** Hooker
SYNONYM: *Clowesia russelliana* (Hooker) Dodson.
ILLUSTRATIONS: Botanical Magazine, 3777 (1840);

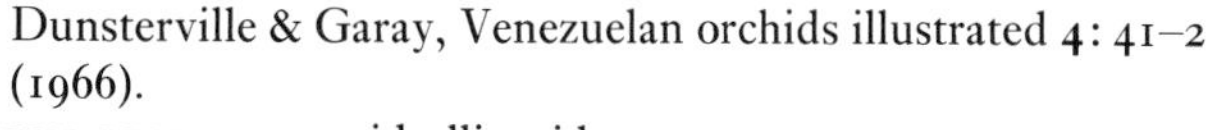

Dunsterville & Garay, Venezuelan orchids illustrated **4**: 41–2 (1966).

Catasetum russellianum – lip from front.

PSEUDOBULBS: ovoid-ellipsoid.

LEAVES: to 36 × 11 cm, broadly lanceolate.

RACEME: arching with many fragrant, non-resupinate flowers.

SEPALS AND PETALS: 3–4 cm, somewhat incurved, ovate-oblong to elliptic, white with green stripes.

LIP: oblong, deeply saccate for about one-third of its length, the sac projecting forwards under the ovate blade; margins lacerate-toothed, the blade bearing an elongate, double ridge with toothed margins; the whole white with green lines.

Catasetum russellianum – lip, oblique.

DISTRIBUTION: C America, Venezuela.

HARDINESS: G2.

FLOWERING: autumn.

2. C. thylaciochilum Lemaire

SYNONYM: *Clowesia thylaciochilum* (Lemaire) Dodson.

Like *C. russellianum* except as below.

LIP: sac curved backwards.

DISTRIBUTION: Mexico.

3. C. integerrimum Hooker

SYNONYM: *C. maculatum* Bateman not Knuth.

ILLUSTRATIONS: Edwards's Botanical Register **26**: t. 42 (1840); Botanical Magazine, 3823 (1840).

PSEUDOBULBS: to 15 cm, spindle-shaped.

LEAVES: to 40 cm, elliptic-lanceolate.

RACEMES: erect, to 35 cm, with 6–12 non-resupinate flowers.

SEPALS: 4.5–4.8 cm, elliptic or oblong-elliptic, acute, green outside, green suffused or spotted with red inside.

PETALS: similar to sepals but slightly smaller.

LIP: helmet-shaped, wholly hollowed, the basal angles projecting inwards and touching each other or almost so, toothed, the whole greenish yellow or yellow, sometimes with red spots outside, greenish with brown spots inside.

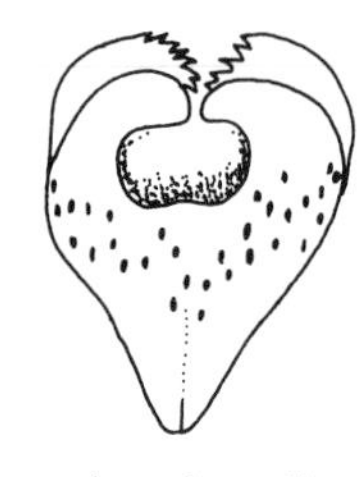

Catasetum integerrimum – lip.

ANTENNAE: asymmetric.

DISTRIBUTION: C Mexico, Guatemala, Honduras & Venezuela.

HARDINESS: G2.

FLOWERING: autumn.

4. C. macrocarpum Knuth

SYNONYM: *C. tridentatum* Hooker.

ILLUSTRATIONS: Edwards's Botanical Register **10**: t. 840 (1824); Botanical Magazine, 2559 (1825), 3329 (1834); Hoehne, Iconografia de Orchidaceas do Brasil, t. 170 (1949); Dunsterville & Garay, Venezuelan orchids illustrated **2**: 60–1 (1961).

Like *C. integerrimum* except as below.

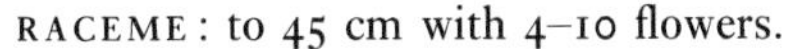

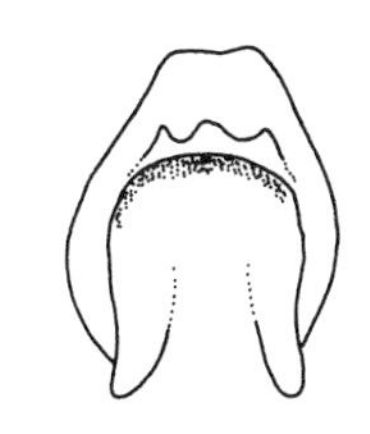

Catasetum macrocarpum – lip.

RACEME: to 45 cm with 4–10 flowers.

LIP: with its basal angles not projecting inwards, and not toothed except for the 3-toothed apex, entirely yellow within and without.

DISTRIBUTION: tropical S America from Trinidad to Brazil.

HARDINESS: G2.

FLOWERING: autumn–winter.

5. C. pileatum Reichenbach

SYNONYM: *C. bungerothii* N.E. Brown.

ILLUSTRATIONS: Botanical Magazine, 6998 (1888); Cogniaux & Goossens, Dictionnaire Iconographique des Orchidées, Catasetum, t. 1, 1A (1897); Hoehne, Iconografia de Orchidaceas do Brasil, t. 163 (1949); Dunsterville & Garay, Venezuelan orchids illustrated 1: 70–1 (1959).

PSEUDOBULBS: 15–25 cm, spindle-shaped.

LEAVES: large, lanceolate.

RACEME: to 30 cm, arching, with 4–10 resupinate (?) flowers.

UPPER SEPAL AND PETALS: to 5 cm, erect.

LATERAL SEPALS: spreading, oblong-lanceolate, acuminate, white, pale yellow or white spotted with red.

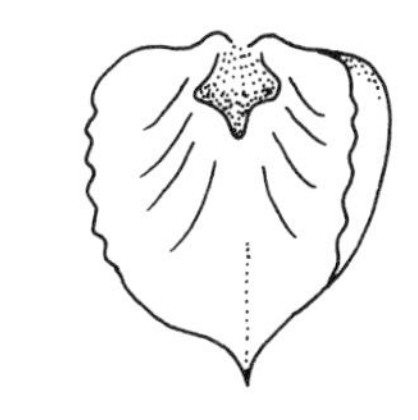

Catasetum pileatum – lip.

LIP: semicircular in outline, truncate at the base, unlobed, saccate with a small sac whose mouth is just below the lip attachment, white, pale yellow or rarely red, the interior of the sac orange-red.

ANTENNAE: asymmetric.

DISTRIBUTION: tropical S America, south to Ecuador and N Brazil.

HARDINESS: G2.

FLOWERING: autumn.

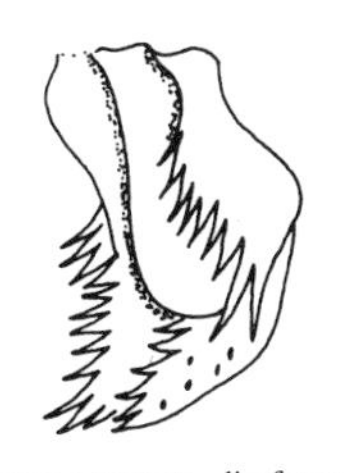

Catasetum saccatum – lip from side.

6. C. saccatum Lindley

SYNONYM: *C. christyanum* Reichenbach.

ILLUSTRATIONS: Botanical Magazine, 8007 (1905); Hoehne, Iconografia de Orchidaceas do Brasil, t. 172 (1949).

PSEUDOBULBS: to 20 cm, fusiform.

LEAVES: to 30 cm or more, lanceolate.

RACEME: arching, with up to 18 non-resupinate flowers.

UPPER SEPAL AND PETALS: close, erect, lateral sepals spreading, all 5–5.5 cm, narrowly lanceolate, acuminate, reddish green with darker red spots.

LIP: oblong, 3-lobed, its margins rolled downwards, lacerate-ciliate, the central lobe oblong, its basal part saccate, all reddish brown with a white callus in front of the mouth of the sac.

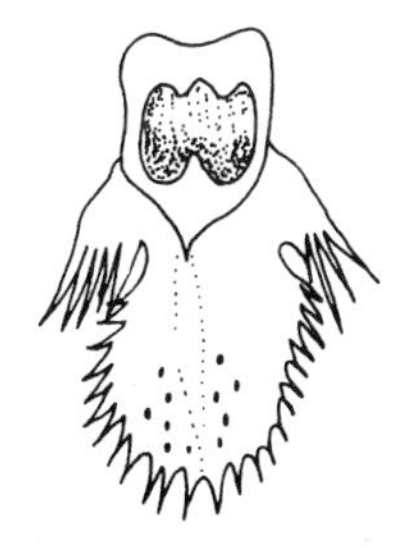

Catasetum saccatum – lip from front.

ANTENNAE: asymmetric.

DISTRIBUTION: Tropical S America from E Peru to Guyana.

HARDINESS: G2.

FLOWERING: winter.

The most commonly cultivated variant is var. *christyanum* (Reichenbach) Mansfeld, which is described above.

7. C. cernuum (Lindley) Reichenbach

SYNONYM: *C. trifidum* Hooker.

PSEUDOBULBS: large.

LEAVES: elliptic.

RACEME: to 30 cm, arching, bearing 1–15 non-resupinate flowers which are themselves downwardly directed.

SEPALS AND PETALS: incurved, up to 3 cm, the petals usually a little shorter than the sepals, all oblong, acute, green with dark brown spots.

LIP: flat, greenish spotted with brown, broadly obovate with 3 teeth at the apex, the laterals elongate-triangular, incurved, the central shorter, straight.

ANTENNAE: symmetric.

DISTRIBUTION: S Brazil.

HARDINESS: G2.

DISTRIBUTION: spring–summer.

Catasetum cernuum – lip.

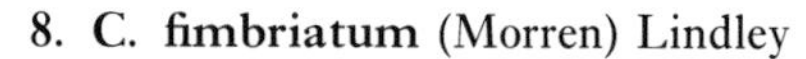

8. C. fimbriatum (Morren) Lindley

ILLUSTRATIONS: Botanical Magazine, 7158 (1891); Dunsterville & Garay, Venezuelan orchids illustrated **4**: 42–3 (1966).

PSEUDOBULBS: to 20 cm, ellipsoid.

LEAVES: 20 cm or more, elliptic.

RACEME: to 40 cm, arching, with 7–15 non-resupinate flowers.

UPPER SEPAL AND PETALS: close, erect.

LATERAL SEPALS: pointing downwards, all to 3.5 cm, lanceolate, acuminate, with margins rolled downwards, yellow or greenish, spotted or striped with purple.

LIP: broadly obovate, fan-shaped, saccate near the base with a large humped callus in front of the mouth of the sac, margins lacerate-ciliate, yellow or greenish spotted with purple towards the base.

ANTENNAE: symmetric.

DISTRIBUTION: tropical and subtropical S America.

HARDINESS: G2.

FLOWERING: spring–summer.

A very variable species.

Catasetum fimbriatum – lip from front.

Catasetum fimbriatum – lip from side.

9. C. trulla Lindley

ILLUSTRATIONS: Edwards's Botanical Register **27**: t. 34 (1841); Hoehne, Flora Brasilica **12**(6): t. 63 (1942).

PSEUDOBULBS: 12–20 cm, variable in shape.

LEAVES: 35–40 cm, oblong-linear.

RACEME: arching, with many non-resupinate flowers.

SEPALS AND PETALS: spreading, the upper sepal and petals

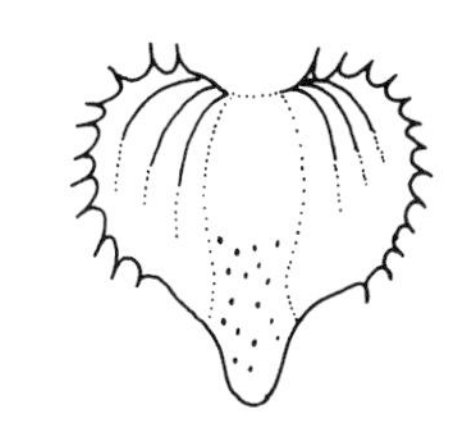

Catasetum trulla – lip.

erect, the lateral sepals outwardly directed, all to 2.5 cm, oblong-elliptic, green.

LIP: flat, triangular with rounded corners, the margin lacerate-ciliate except towards the apex, the whole whitish green tinged with red-brown in the centre, the apical part blunt, brown.

ANTENNAE: symmetric.

DISTRIBUTION: S Brazil.

HARDINESS: G2.

FLOWERING: autumn.

10. C. atratum Lindley

ILLUSTRATIONS: Edwards's Botanical Register **26**: t. 63 (1838); Botanical Magazine, 5202 (1860).

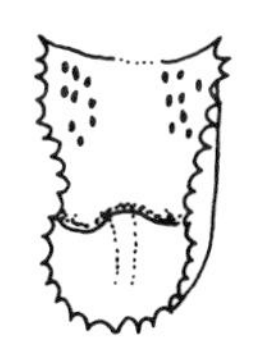

Catasetum atratum – lip from front.

PSEUDOBULBS: 10–13 cm, spindle-shaped.

LEAVES: to 30 cm, narrowly elliptic or lanceolate.

RACEME: to 30 cm, arching, bearing 12–15 resupinate (?) flowers.

SEPALS AND PETALS: to 2.5 cm, all more or less directed downwards, oblong, acute, green, densely marked with small brown transverse stripes.

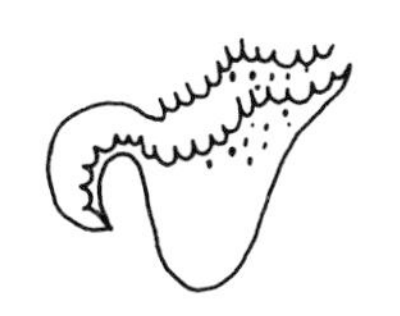

Catasetum atratum – lip from side.

LIP: oblong, saccate for two-thirds of its length from the base, margins lacerate-ciliate or lacerate-toothed, yellowish green with some brown spots and a white apical part.

ANTENNAE: symmetric.

DISTRIBUTION: S & E Brazil.

HARDINESS: G2.

FLOWERING: spring.

11. C. microglossum Rolfe

ILLUSTRATION: Botanical Magazine, 8514 (1913).

PSEUDOBULBS: to 10 cm, spindle-shaped.

LEAVES: to 30 cm, oblong-elliptic to oblanceolate.

RACEME: to 65 cm, arching, with many non-resupinate flowers.

SEPALS AND PETALS: upper sepal and petals erect; lateral sepals directed downwards; all to 2.5 cm, oblong-lanceolate, acute, purplish.

LIP: small, consisting mostly of an obconic sac which is deeper than long, with lacerate margins and containing a callus which terminates in numerous teeth or hair-like processes, yellow.

Catasetum microglossum – lip.

ANTENNAE: symmetric.

DISTRIBUTION: borders of Peru and Colombia.

HARDINESS: G2.

12. C. barbatum (Lindley) Lindley

SYNONYMS: *Myanthum barbatum* Lindley; *M. spinosum* Hooker.

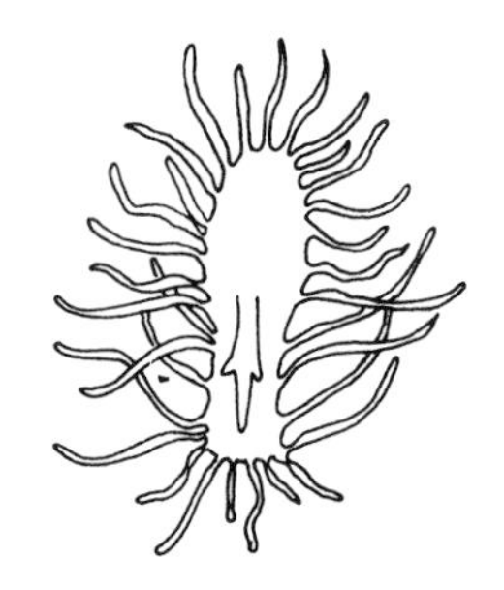

Catasetum barbatum – lip from front.

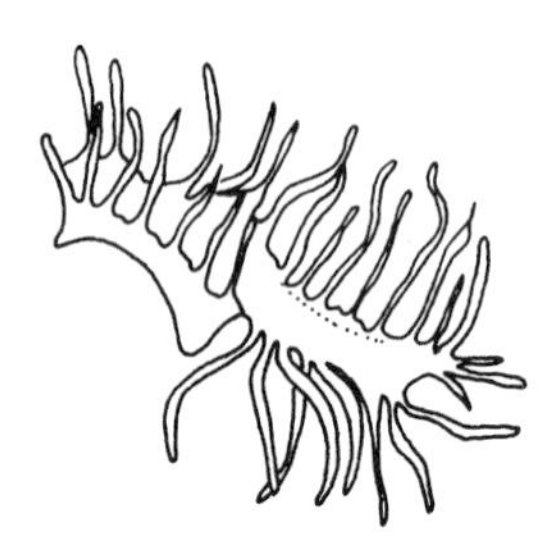

Catasetum barbatum – lip from side.

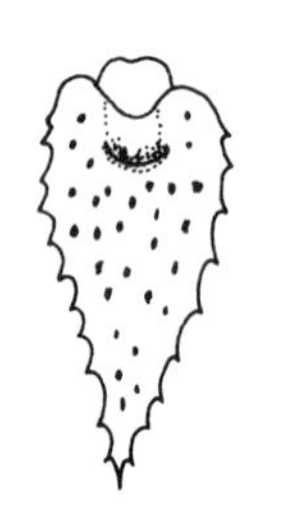

Catasetum callosum – lip from front.

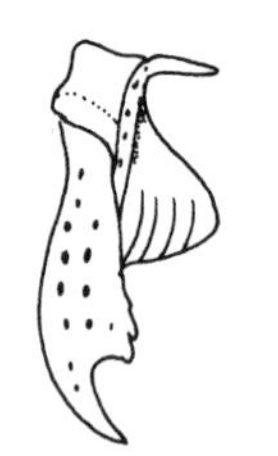

Catasetum callosum – lip from side.

ILLUSTRATIONS: Botanical Magazine, 3802 (1841); Dunsterville & Garay, Venezuelan orchids illustrated **3**: 54–5 (1965).

PSEUDOBULBS: 10–12 cm, narrowly conical to spindle-shaped.

LEAVES: lanceolate, wavy.

RACEME: to 45 cm, arching or erect(?), bearing numerous non-resupinate flowers.

SEPALS AND PETALS: upper sepal and petals erect, lateral sepals spreading, all to 3 cm, narrowly oblong, acute, green with small, dark purple, transverse bars.

LIP: oblong, white or pink, with a small sac at about the middle, the margin deeply divided into fleshy, blunt, hair-like teeth, with a narrowly triangular apical part and a callus divided usually into 3 fleshy teeth situated behind the mouth of the sac.

ANTENNAE: symmetric.

DISTRIBUTION: northern S America.

HARDINESS: G2.

FLOWERING: spring.

The variant most commonly found in cultivation is var. **spinosum** (Hooker) Rolfe, with the callus as described above.

13. C. callosum Lindley

ILLUSTRATION: Botanical Magazine, 6648 (1882).

PSEUDOBULBS: to 8 cm, spindle-shaped.

LEAVES: 15–30 cm, oblanceolate, greyish.

RACEME: to 35 cm, arching or erect, with 10–15 resupinate(?) or non-resupinate flowers.

SEPALS AND PETALS: upper sepal and petals erect, lateral sepals directed downwards, all 4–5 cm, narrowly lanceolate, acute, brownish.

LIP: narrowly ovate, green with red spots, its margin finely and distantly toothed, with a small sac near the base whose mouth is sited just in front of a humped callus.

ANTENNAE: symmetric.

DISTRIBUTION: Venezuela, Guyana, Peru.

HARDINESS: G2.

FLOWERING: winter.

139. CYCNOCHES Lindley

A genus of 7 species from C & S America, whose classification is complicated and based almost entirely on the male flowers, which are more frequently seen than the female. Gregg (*Selbyana* **2**: 217, 1978) reports that 'plants growing in full sunlight with adequate moisture and embedded in a suitable substrate are robust and generally produce female flowers. Less robust plants often growing in shade or lacking adequate nutrition tend to produce male flowers'. Cultivation as for *Catasetum* (p. 428).

Literature: Allen, P.H., The swan orchids: a revision of the genus Cycnoches, *Orchid Journal* 1: 173–84, 225–30, 349–54, 397–403 (1952).

HABIT: epiphytic.

PSEUDOBULBS: cylindric, tapering slightly towards the apex, compound.

LEAVES: pleated, rolled when young.

RACEMES: with several flowers, lateral from the axils of the upper leaves.

FLOWERS: not resupinate, usually unisexual.

SEPALS AND PETALS: similar, smaller in the female flowers than in the male.

LIP: fleshy and undivided (all female flowers and male flowers of No. 1) or variously divided or conspicuously appendaged (male flowers of Nos. 2 and 3), with a basal callus.

COLUMN: short and fleshy in the female flowers, elongate, broadened above and arched in male flowers.

POLLINIA: 2, ovoid, stipe narrowly ovoid, longer and narrower than the pollinia, viscidium oblong, 2-lobed.

Key to species

1. Male inflorescence with up to 9 flowers; lip of male flowers simple and entire

1.A. Sepals and petals yellowish green — 1. **ventricosum**

1.B. Sepals and petals greenish brown — 2. **loddigesii**

2. Lip of male flower linear or lanceolate, flat or wavy, with a forwardly pointing process on the claw and 2 linear, rounded teeth on each side — 5. **pentadactylon**

3. Lip of male flower clawed, ovate to circular or strap-shaped, the margin drawn out into several processes which are club-shaped or forked at their apices

3.A. Lip ovate to circular — 3. **egertonianum**

3.B. Lip strap-shaped — 4. **maculatum**

1. **C. ventricosum** Bateman

MALE FLOWERS: 9–15 cm in diameter.

SEPALS AND PETALS: green or yellowish green, spreading.

LIP: 6–6.5 cm, simple, fleshy, stalkless or with an elongate claw which may be narrowly winged, white marked with dark green at the base.

COLUMN: 2.8–3.2 cm.

DISTRIBUTION: tropical America, from S Mexico to French Guiana.

HARDINESS: G2.

FLOWERING: summer.

A variable species, divided by Allen into several varieties; the most commonly grown is var. **chlorochilon** (Klotzsch) Allen (*C. chlorochilon* Klotzsch) which has large flowers with a stalkless lip (see Hoehne, Flora Brasilica **12**(6): t. 101, 1942; Dunsterville & Garay, Venezuelan orchids illustrated **2**: 81–2, 1961; Couret, Orquideas Venezolanas, 36, 1977; Sheehan & Sheehan, Orchid genera illustrated 71, 1979; Bechtel et al., Manual of cultivated orchid species, 189, 1981).

2. C. **loddigesii** Lindley

ILLUSTRATIONS: Hoehne, Flora Brasilica **12**(6): t. 100 (1942); Couret, Orquideas Venezolanas, 37 (1977).

Like *C. ventricosum* except as below.

FLOWERS: larger.

SEPALS AND PETALS: greenish brown.

LIP: with a conspicuously winged claw.

DISTRIBUTION: Venezuela & NE Brazil.

3. C. **egertonianum** Bateman

ILLUSTRATIONS: Hoehne, Flora Brasilica **12**(6): t. 109 (1942); Bechtel et al., Manual of cultivated orchid species, 189 (1981).

FLOWERS: 5–6 cm in diameter.

SEPALS AND PETALS: greenish or greenish brown, spotted with brown or purple, sometimes yellow or almost white (var. **aureum**), more rarely pink.

LIP: of male flowers stalked, green or white, the apical part concave, ovate or circular, the margin with several elongate, club-shaped teeth which may be forked at their apices.

COLUMN: *c.* 20 mm.

DISTRIBUTION: C & S America.

HARDINESS: G2.

FLOWERING: autumn.

Variable in flower colour; both var. **egertonianum,** with greenish or greenish brown, darker spotted perianth, and var. **aureum** (Lindley) Allen (*C. aureum* Lindley), with yellow to almost white perianth, are grown.

4. C. **maculatum** Lindley

Like *C. egertonianum* except as below.

LIP: apical part drawn out into a strap-shaped lobe, and the marginal teeth not forked at their apices.

DISTRIBUTION: Venezuela.

5. C. **pentadactylon** Lindley

ILLUSTRATIONS: Flora Brasilica **12**(6): t. 104 (1942); Bechtel et al., Manual of cultivated orchid species, 189 (1981).

FLOWERS: 7–10 cm in diameter.

SEPALS AND PETALS: green blotched with dark brown.

LIP: fleshy and rigid, linear-lanceolate, flat or wavy, clawed and with a forwardly pointing process on the upper surface of the claw, the blade narrow, with 2 large, linear, rounded teeth on each side.

DISTRIBUTION: E & C Brazil.

HARDINESS: G2.

FLOWERING: spring.

140. LOCKHARTIA Hooker

A genus of about 20 species from C & S America. Only 1 species is readily available in Europe, though others are grown in the USA (see Teuscher, H., The genus Lockhartia, *American Orchid Society Bulletin* **43**: 399–405, 1974). Cultivation as for *Oncidium* (p. 493).

HABIT: epiphytic, without pseudobulbs.

STEMS: erect, covered with leaves in 2 ranks.

LEAVES: strongly keeled, consisting mainly of overlapping sheaths, the free blades small, folded when young.

FLOWERS: borne laterally, either singly or in open, few-flowered racemes, usually conspicuously bracteate with stem-clasping bracts.

SEPALS AND PETALS: similar, 7–15 mm, spreading or somewhat reflexed, yellow.

LIP: simple or more frequently complex with 2 lateral lobes and a central lobe which is often further divided and warty at its base.

COLUMN: spreading, winged towards the apex, without a foot.

POLLINIA: 2, attached directly to a small, circular viscidium.

Key to species

1. Sepals and petals similar in size — **1. oerstedtii**
2. Petals somewhat smaller than sepals — **2. lunifera**

1. L. oerstedtii Reichenbach

SYNONYMS: *L. verrucosa* Reichenbach; *L. robusta* Schlechter.

ILLUSTRATIONS: Saunders' Refugium Botanicum **2**: t. 76 (1869); Bechtel et al., Manual of cultivated orchid species, 222 (1981).

STEMS: 15–30 cm.

LEAVES: rounded or somewhat acute at the apex.

SEPALS AND PETALS: golden yellow.

LIP: with blunt, downwardly deflected, yellow and often red-spotted lateral lobes and a yellow central lobe which is often red-spotted at the base, oblong or reversed-heart-shaped, deeply notched at the apex and with 2 lateral, somewhat reflexed, lobe-like prolongations near its junction with the lateral lobes.

DISTRIBUTION: C America.

HARDINESS: G2.
FLOWERING: summer.

2. L. lunifera Reichenbach
Like *L. oerstedtii* except as below.
FLOWERS: smaller.
SEPALS: *c.* 7 mm.
PETALS: slightly smaller than the sepals.
LIP: lateral lobes sickle-shaped.
DISTRIBUTION: N & E Brazil.

141. ACINETA Lindley

A genus of 10–15 species from C & S America. They should be grown in suspended baskets or rafts so that the hanging racemes can develop properly. An epiphytic compost should be used and abundant moisture provided when in growth. A lower temperature is required during the resting season. Propagation is by seed or division.

Literature: Schlechter, R., Die Gattung Acineta, *Orchis* **11**: 21–47 (1917).

HABIT: epiphytic.
PSEUDOBULBS: distinct, simple, ovoid or spindle-shaped, grooved, each bearing 2–4 leaves.
LEAVES: large, pleated, rolled when young.
FLOWERS: not resupinate, fleshy, bell-shaped, in lateral, hanging racemes which arise from the bases of the pseudobulbs.
SEPALS AND PETALS: similar in size or the sepals somewhat larger, incurved.
LIP: clawed, 3-lobed, the lateral lobes erect and borne near the middle of the claw, reaching up to the column, the central lobe relatively small, with a prominent callus between the lateral lobes.
COLUMN: finely downy on the back, with a short foot to which the lateral sepals are attached.
POLLINIA: 2, ellipsoid, stipe broadly linear, almost as long as the pollinia, the viscidium with a truncate or rounded basal part.

Key to species

1. Sepals and petals red or pale brown and red-spotted; flowering in spring **3. superba**
2. Flowers 4–5 cm in diameter; pseudobulbs each with 2 leaves; raceme with 12–15 flowers **1. barkeri**
3. Flowers *c.* 6 cm in diameter; pseudobulbs each with 3 or 4 leaves; racemes usually with 15 or more flowers **2. chrysantha**

1. A. **barkeri** Lindley

ILLUSTRATION: Botanical Magazine, 4203 (1846); Orchid Review **93**: 399 (1985).

PSEUDOBULBS: *c.* 10 cm, with 2 leaves, ovoid.

LEAVES: to 50 × 10 cm.

RACEME: with 12–15 flowers which are 4–5 cm in diameter.

SEPALS AND PETALS: ovate, golden yellow.

LIP: yellow with a purple-red spot or spots near the base.

DISTRIBUTION: S Mexico.

HARDINESS: G2.

FLOWERING: autumn.

2. A. **chrysantha** (Morren) Lindley & Paxton

SYNONYM: *A. densa* Lindley & Paxton.

ILLUSTRATIONS: Botanical Magazine, 7143 (1890); Bechtel et al., Manual of cultivated orchid species, 161 (1981).

Like *A. barkeri* except as below.

HABIT: larger.

PSEUDOBULBS: with 3–4 leaves.

RACEMES: usually with more than 15 flowers.

PETALS: red-spotted.

LIP: with very prominent callus.

DISTRIBUTION: tropical America from Mexico to Colombia.

3. A. **superba** (Humboldt, Bonpland & Kunth) Reichenbach

SYNONYM: *A. humboldtii* Lindley.

ILLUSTRATIONS: Dunsterville & Garay, Venezuelan orchids illustrated 1: 40–1 (1959); Bechtel et al., Manual of cultivated orchid species, 161 (1981).

PSEUDOBULBS: with 3 leaves.

RACEMES: loose, with 6–12 flowers which are *c.* 6 cm in diameter.

SEPALS AND PETALS: red, or pale brown and red-spotted.

LIP: yellow, reddish towards the base, and with dark purple spots.

DISTRIBUTION: tropical S America from Venezuela to Ecuador.

HARDINESS: G2.

FLOWERING: spring.

The variant with brownish, red-spotted flowers is referred to as var. **fulva** (Hooker) Schlechter (see Botanical Magazine, 4156, 1845).

142. PERISTERIA Hooker

A genus of about 10 species from C & S America. All can be grown like *Acineta*, but *P. elata*, which has erect racemes, is suitable for pot-culture, and will tolerate a terrestrial compost.

HABIT: epiphytic.

PSEUDOBULBS: simple, ovoid, spindle-shaped, grooved and bearing 3–5 leaves.

LEAVES: stalked, often large, pleated, rolled when young.

RACEMES: arising from the bases of the pseudobulbs, erect or hanging, with few to many flowers.

FLOWERS: spherical to bell-shaped, not widely open, often fragrant, resupinate or not.

SEPALS AND PETALS: similar or the petals somewhat smaller, fleshy, ovate to circular, incurved.

LIP: complex, joined to the base of the column by a basal part which bears 2 lateral lobes which stand upright on either side of the column, and an erect or outwardly curving anterior part which is distinguished from the basal part by a waist which is sharply bent or folded; the anterior part is freely movable on the rest.

COLUMN: thick and fleshy, sometimes with a small wing on either side, the rostellum prominently beaked.

POLLINIA: 2, ellipsoid, attached independently to the triangular viscidium, stipe absent.

Key to species

1. Raceme erect, stiff, with 12–25 flowers **1. elata**
2. Column with a narrow, downwardly pointing wing on each side; basal part of lip with a crested surface **3. pendula**
3. Column without wings; basal part of lip smooth **2. cerina**

1. P. elata Hooker

ILLUSTRATIONS: Botanical Magazine, 3116 (1832); Dunsterville & Garay, Venezuelan orchids illustrated 1: 238–9 (1959); Ospina & Dressler, Orquideas de las Americas, t. 159 (1974); Sheehan & Sheehan, Orchid genera illustrated, 131 (1979).

PSEUDOBULBS: to 13 × 10 cm.

LEAVES: to 95 cm, narrowly elliptic, acute.

RACEME: to 1.5 m, held stiffly erect, with 12–25 flowers.

SEPALS AND PETALS: waxy, white, flower 4.5–5 cm in diameter.

LIP: with anterior part entire, curving outwards.

COLUMN: unwinged.

DISTRIBUTION: Tropical America from Costa Rica to Venezuela and Colombia.

HARDINESS: G2.

FLOWERING: summer.

Known as the dove orchid because of a resemblance between the column and the lateral lobes of the lip and a dove with outstretched wings.

2. P. cerina Lindley

ILLUSTRATIONS: Edwards's Botanical Register 23: t. 1953 (1837); Hoehne, Flora Brasilica 12(6): t. 119 (1942).

PSEUDOBULBS: to 11 × 5 cm.

LEAVES: to 45 cm, narrowly elliptic, acute.

RACEMES: hanging, 10–15 cm, with 7–13 flowers which are *c.* 3 cm in diameter.

SEPALS AND PETALS: yellow.

LIP: with the basal part smooth, the anterior part erect, somewhat hooded, the margins orange-yellow and crisped.

COLUMN: unwinged.

DISTRIBUTION: Trinidad, N Brazil.

HARDINESS: G2.

FLOWERING: summer.

3. P. pendula Hooker

ILLUSTRATION: Botanical Magazine, 3479 (1836).

Vegetatively similar to *P. cerina.*

RACEMES: hanging, with 4–7 flowers which are 4.5–5 cm in diameter.

SEPALS AND PETALS: white or pale yellow, spotted with red.

COLUMN: with a narrow, downwardly directed wing from the apex on either side.

DISTRIBUTION: tropical S America from Trinidad to N Brazil and N Peru.

HARDINESS: G2.

FLOWERING: spring.

143. LACAENA Lindley

A genus of 2 species from C America; both are grown in conditions like those for *Acineta* (p. 438).

Literature: Jenny, R., Die Gattung Lacaena Lindley, *Die Orchidee* **30**: 55–61 (1979).

HABIT: epiphytic.

PSEUDOBULBS: simple, ovoid or spindle-shaped, grooved, each bearing 2 leaves.

LEAVES: stalked, elliptic, pleated, rolled when young.

RACEMES: hanging, arising from the bases of the pseudobulbs, with 8–30 flowers.

FLOWERS: openly bell-shaped, not resupinate.

SEPALS AND PETALS: similar, incurved.

LIP: attached to the column foot, complex in shape, 3-lobed, with a callus between the short, erect, basal, lateral lobes, the central lobe hairless, clawed, ovate or oblong, its anterior part not freely movable on the rest.

POLLINIA: 2, ellipsoid, stipe linear, longer than the pollinia, somewhat expanded above, the viscidium with an acute base.

Key to species

1. Flowers 4.5–5 cm in diameter, lip with a hairy callus and ovate central lobe **1. bicolor**
2. Flowers 2.5–3 cm in diameter, lip with a hairless callus and oblong central lobe **2. spectabilis**

1. L. **bicolor** Lindley

ILLUSTRATIONS: Botanical Magazine n.s., 330 (1959); Ospina & Dressler, Orquideas de las Americas, f. 155 (1974).

PSEUDOBULBS: 6–8 cm, ovate.

LEAVES: 20–25 cm, elliptic.

RACEMES: with 10–30 flowers which are fragrant and 4.5–5 cm in diameter.

SEPALS AND PETALS: greenish white, greenish yellow or pinkish, scurfy outside.

LIP: white with a hairy, purple callus and purple spots on the shortly stalked, ovate central lobe.

DISTRIBUTION: Mexico, Guatemala, Honduras.

HARDINESS: G2.

FLOWERING: spring–summer.

2. L. **spectabilis** (Klotzsch) Reichenbach

ILLUSTRATION: Botanical Magazine, 6516 (1880).

Like *L. bicolor* except as below.

RACEMES: with 8–12 flowers which are 2.5–3 cm in diameter.

SEPALS AND PETALS: pinkish white.

LIP: with hairless callus and an oblong, spade-like, conspicuously stalked, purple central lobe.

DISTRIBUTION: Mexico.

HARDINESS: G1.

FLOWERING: spring.

144. LUEDDEMANNIA Reichenbach

A small genus, not commonly grown. Cultivation as for *Acineta* (p. 438).

Like *Lacaena* except as below.

PSEUDOBULBS: with 3–5 leaves.

SEPALS AND PETALS: somewhat more spreading.

LIP: simpler, 3-lobed, with curving, obliquely triangular lateral lobes and a narrowly triangular, downy central lobe, bearing a humped, downy callus.

1. L. **pescatorei** (Lindley) Linden & Reichenbach

ILLUSTRATION: Botanical Magazine, 7123 (1890).

SEPALS: broader than petals, brown and scaly-hairy outside, yellow striped with red inside.

PETALS: yellow.

LIP: yellow with red spots.

DISTRIBUTION: Colombia.
HARDINESS: G2.
FLOWERING: summer.

145. HOULLETIA Brongniart

A genus of about 10 species from C & S America. Cultivation as for *Acineta* (p. 438) except that Houlletias are best grown in pots, as the large leaves make suspension difficult.

HABIT: usually epiphytic.
PSEUDOBULBS: simple, to 6 cm, each bearing 1 leaf.
LEAVES: stalked, elliptic, acute, pleated, rolled when young.
RACEME: erect or arching, arising from the base of the pseudobulb, with 4–10 non-resupinate flowers which are often fragrant, axis hairless.
SEPALS AND PETALS: widely spreading, similar or the petals slightly smaller.
LIP: complex, attached to the column foot, divided by a narrow waist into a clawed basal part which is narrowly oblong and bears on each side a narrow, pointed lateral lobe which curves backwards and upwards, then forwards; anterior part stalkless or shortly stalked, oblong or ovate, arrowhead-shaped at the base with pointed lobes.
POLLINIA: 2, irregularly ellipsoid, stipe linear, longer than pollinia, slightly expanded above, viscidium elongate, acute; rostellum conspicuously beaked.

Key to species

1. Sepals and petals oblong, of uniform colour, dark red or dark brown, not spotted or blotched **5. odoratissima**
2. Raceme arching; lip whitish with violet spots; petals each with a large tooth on the inner margin **1. lansbergii**
3. Lateral sepals united **4. wallisii**
4. Bases of sepals and petals entirely brown; anterior part of lip yellow with very dark brown spots **3. picta**
5. Bases of petals yellow with brown spots; anterior part of lip purple **2. brocklehurstiana**

1. H. lansbergii Linden & Reichenbach

ILLUSTRATIONS: Botanical Magazine, 7362 (1894); Fieldiana **26**: 525 (1953).
PSEUDOBULBS: *c.* 6 cm, ovate.
LEAVES: to 40 cm, elliptic.
RACEME: arching, with 5–10 flowers, each *c.* 8.5 cm in diameter.
SEPALS AND PETALS: ovate to elliptic, yellow with conspicuous brown spots, each petal with a single tooth on the upper margin.
LIP: whitish with violet spots.
DISTRIBUTION: Costa Rica.

HARDINESS: G2.
FLOWERING: autumn.

2. H. brocklehurstiana Lindley
ILLUSTRATIONS: Botanical Magazine, 4072 (1844); Hoehne, Flora Brasilica **12**(6): t. 136 (1942).
Vegetatively similar to *H. lansbergii*.
RACEME: erect with 5–10 flowers, each *c*. 7 cm in diameter.
SEPALS AND PETALS: ovate to elliptic, yellow with brown spots, petals untoothed.
LIP: basal part yellow with red spots, the anterior part purple.
DISTRIBUTION: E Brazil.
HARDINESS: G2.
FLOWERING: autumn–winter.

3. H. picta Linden & Reichenbach
ILLUSTRATION: Botanical Magazine, 6305 (1877).
Like *H. brocklehurstiana* except as below.
SEPALS AND PETALS: bases entirely brown.
LIP: anterior part yellow with very dark brown spots.
DISTRIBUTION: Colombia.

4. H. wallisii Linden & Reichenbach
ILLUSTRATION: Ospina & Dressler, Orquideas de las Americas, f. 154 (1974).
Like *H brocklehurstiana* except as below.
LATERAL SEPALS: united.
PETALS: rather narrow.
DISTRIBUTION: Colombia.

5. H. odoratissima Lindley & Paxton
ILLUSTRATION: Lindenia **7**: t. 324 (1891).
Vegetatively similar to *H. lansbergii*.
RACEME: erect with 6–9 flowers each *c*. 7 cm in diameter.
SEPALS AND PETALS: oblong, dark brown or dark red, petals entire.
LIP: mostly white.
DISTRIBUTION: Colombia.
HARDINESS: G2.
FLOWERING: autumn.
A variant with red sepals and petals has been called var. **antioquiensis** Anon.

146. POLYCYCNIS Reichenbach

A small genus, rarely grown. Cultivation as for *Acineta* (p. 438).

Like *Houlletia* except as below.
HABIT: plants not so large.

RACEME: axis downy.

LIP: with the basal part somewhat clawed and bearing upcurving, oblong-triangular lateral lobes, the central lobe heart-shaped.

COLUMN: slender, arching, abruptly widened at the apex.

1. **P. barbata** Reichenbach

ILLUSTRATION: Botanical Magazine, 4479 (1849).

RACEME: axis and flower-stalks purple, downy.

SEPALS AND PETALS: 2.5–2.7 cm, yellowish flushed with red and with brown-purple spots.

LIP: similar in colour, prominently but sparsely hairy.

DISTRIBUTION: Costa Rica, Colombia.

HARDINESS: G2.

FLOWERING: winter.

147. **PAPHINIA** Lindley

A genus of about 7 species from C & S America. They require warm, humid conditions when growing and infrequent watering when resting. They are best grown in epiphytic compost in small pans, and given some shade during the summer.

Literature: Jenny, R., Die Gattung Paphinia Lindley, *Die Orchidee* **29**: 207–15 (1978).

HABIT: epiphytic.

PSEUDOBULBS: rather small, simple, each usually bearing 2 leaves.

LEAVES: pleated, rolled when young.

RACEMES: hanging, arising from the bases of the pseudobulbs, each with 1–5 flowers which are resupinate and 8–14 cm in diameter.

SEPALS AND PETALS: widely spreading, the petals somewhat smaller than the sepals.

LIP: complex, 3-lobed, smaller than the petals, the claw and basal part ridged and crested, bearing the erect, forwardly directed lateral lobes; the central lobe movable on the rest, diamond-shaped, bearing a tuft of club-shaped, hair-like processes on the margin towards and at the apex.

COLUMN: narrow with small wings or teeth towards the apex.

POLLINIA: 2, ellipsoid, their surfaces roughened, stipe linear-oblong with the sides curved upwards near the top, viscidium very small; rostellum conspicuously beaked.

Key to species

1. Flowers to 8 cm in diameter; lip red with white crests, ridges and hair-like processes **1. cristata**

2. Flowers 12–14 cm in diameter; lip dark purple, white towards the base, hair-like processes white **2. grandiflora**

1. **P. cristata** Lindley

ILLUSTRATIONS: Fieldiana **26**: 527 (1952); Hoehne, Flora Brasilica **12** (7): t. 3 (1953); Orchid Review **93**: 377 (1985); American Orchid Society Bulletin **56**: 520 (1987).

PSEUDOBULBS: *c.* 4 cm, somewhat compressed.

LEAVES: 10–15 cm, lanceolate, acute.

FLOWERS: to 8 cm in diameter.

SEPALS AND PETALS: translucent white with brown or red stripes or spots.

LIP: red with white ridges, crests and hair-like processes.

DISTRIBUTION: tropical S America, from Colombia and Trinidad to N Brazil.

HARDINESS: G2.

FLOWERING: autumn.

2. **P. grandiflora** Rodriguez

ILLUSTRATION: Hoehne, Flora Brasilica **12** (7): t. 1 (1953).

PSEUDOBULBS: to 5 cm.

LEAVES: 20–25 cm, lanceolate to elliptic.

FLOWERS: 12–14 cm in diameter.

SEPALS AND PETALS: white, striped and spotted with red-purple, especially towards the apex.

LIP: dark purple, white towards the base and with white hair-like processes.

DISTRIBUTION: NW Brazil.

HARDINESS: G2.

FLOWERING: autumn.

The flowers are said to produce a strong, rather unpleasant smell.

148. STANHOPEA Hooker

A genus of from 25 to 50 species whose classification is chaotic. Most of the species are poorly known, variable and difficult to distinguish. About 10 species are reputedly in cultivation; they should be grown in the same manner as *Acineta* (p. 438).

Literature: Dodson, C.H. & Frymire, G.F., Preliminary studies in the genus Stanhopea, *Annals of the Missouri Botanical Garden* **48**: 137–72 (1961); Dodson, C.H., Clarification of some nomenclature in Stanhopea, *Selbyana* **1**: 46–55 (1975).

HABIT: epiphytic.

PSEUDOBULBS: small, simple, 1-leaved.

LEAVES: large, leathery, elliptic, stalked, pleated, rolled when young.

RACEMES: arising from the bases of the pseudobulbs, hanging, with 2–10 flowers which have prominent, coloured bracts.

FLOWERS: usually fragrant, not resupinate.

SEPALS AND PETALS: similar, or petals sometimes a little

smaller, all spreading or reflexed, the lateral sepals joined at the base.

LIP: complex, very fleshy, usually clearly divided into 3 parts, the basal part hollowed or saccate, the middle usually bearing a horn-like lateral lobe on either side, the anterior part simple or 3-toothed or 3-lobed at the apex, jointed to the median part; more rarely the whole lip forming an irregular pouch.

COLUMN: arched, slender, often winged towards the apex.

POLLINIA: 2, ellipsoid, stipe linear, longer than pollinia, somewhat expanded above, its sides turned upwards, viscidium elongate.

Key to species

1. Lip an irregular pouch, not clearly divided into 3 parts **1. ecornuta**
2. Lateral lobes (horns) borne on the basal part of the lip **2. grandiflora**
3. Horns less than 1 cm **3. lewisae**
4. Basal part of lip globular or saccate, its hollow centre linked to the middle part by a short groove; anterior part usually 3-toothed or 3-lobed

4.A. Sepals to 5.5 cm; central lobe of anterior part of lip shorter than the lateral lobes **6. saccata**

4.B. Sepals more than 7.5 cm; column very broadly winged **7. tigrina**

4.C. Basal part of lip purple, the rest yellow, spotted with red **4. insignis**

4.D. Lip whitish spotted with red throughout **5. hernandezii**

5. Basal part of lip oblong when viewed from the side, its sides notched **8. oculata**
6. Basal part of lip oblong when viewed from the side, its sides not notched **10. wardii**
7. Basal part of lip boat-shaped when viewed from the side **9. jenishiana**

1. S. ecornuta Lemaire

ILLUSTRATIONS: Botanical Magazine, 4885 (1855); Fieldiana **26**: 530 (1952); Annals of the Missouri Botanical Garden **48**: 145 (1961); Die Orchidee **27**: 173 (1976); American Orchid Society Bulletin **53**: 362 (1984); Orchid Review **94**: 252 (1986).

LEAVES: 35–40 cm.

RACEMES: with 1–3 flowers.

SEPALS AND PETALS: creamy white, the petals usually spotted with purple towards the base, sepals 4.5–6.5 cm.

LIP: simple, mostly yellow, forming an irregularly margined pouch, without obvious lateral lobes.

DISTRIBUTION: C America, from Guatemala to Costa Rica.

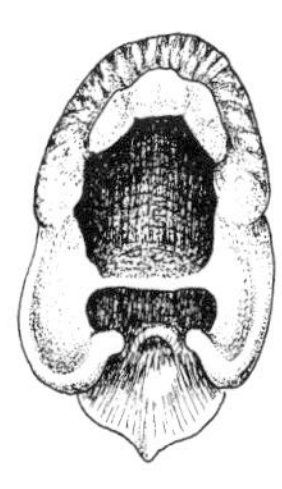

Stanhopea ecornuta – lip.

HARDINESS: G2.

FLOWERING: summer.

2. S. grandiflora (Loddiges) Lindley not Reichenbach

SYNONYM: *S. eburnea* Lindley.

ILLUSTRATIONS: Botanical Magazine, 3359 (1834); Hoehne, Flora Brasilica **12**(6): t. 117 (1942); Annals of the Missouri Botanical Garden 48: 153 (1961); Couret, Orquideas Venezolanas, 28 (1977); American Orchid Society Bulletin **53**: 359, 362 (1984).

LEAVES: to 30 cm.

RACEME: with 2–4 flowers, stalks and ovaries with sparse, short, blackish hairs.

SEPALS: *c.* 8 cm, white.

PETALS: white, sometimes spotted with violet around their margins, violet at the base.

LIP: clearly 3-partite, the basal part purple-spotted, hollowed and bearing small, horn-like lateral lobes, the middle part solid, oblong, whitish spotted and streaked with purple, the anterior part ovate, acute, white.

Stanhopea grandiflora – lip.

DISTRIBUTION: tropical S America from Trinidad to N Brazil.

HARDINESS: G2.

FLOWERING: summer.

3. S. lewisae Ames & Correll

ILLUSTRATIONS: Fieldiana **26**: 533 (1952); Annals of the Missouri Botanical Garden **48**: 155 (1961).

LEAVES: 40–50 cm.

RACEMES: with 3–5 flowers.

SEPALS AND PETALS: creamy white with reddish or purple spots, sepals 5–5.7 cm.

LIP: with a short, hollow, deep yellow, purple-spotted basal part, the middle part bearing horns which are less than 1 cm long on each side, the anterior part broadly ovate, obtuse, white densely spotted with red.

DISTRIBUTION: Guatemala.

HARDINESS: G2.

FLOWERING: summer.

A relatively recently described species whose distribution in cultivation is uncertain. It is possibly a naturally occurring hybrid between *S. ecornuta* and *S. jenishiana* (see Oestereich, Orchidata **7**: 209–10, 1969).

4. S. insignis Hooker

ILLUSTRATIONS: Botanical Magazine, 2948, 2949 (1829); Annals of the Missouri Botanical Garden **48**: 158 (1961).

LEAVES: 20–30 cm.

RACEMES: with 2–5 flowers.

Stanhopea insignis – lip.

SEPALS AND PETALS: yellowish white spotted with red or purple, the sepals *c.* 7 cm.

LIP: basal part spherical, saccate, purple, linked to the yellow, red-spotted middle part by a very short groove; horns 2–3 cm, borne on the middle part; anterior part broadly ovate, weakly 3-lobed at the apex, yellowish white spotted with red.

COLUMN: broadly winged.

DISTRIBUTION: SE Brazil.

HARDINESS: G2.

FLOWERING: autumn

S. insignis is the best known of a complex of species which are very difficult to distinguish among themselves. Three other species of this complex are reputedly in cultivation, and are briefly described below.

5. **S. hernandezii** (Knuth) Schlechter

SYNONYM: *S. devoniensis* Lindley.

SEPALS: 5.5–6.5 cm, yellow or orange-yellow spotted with red.

LIP: white spotted with purple, the anterior part broadly ovate with 3 equal, blunt teeth.

COLUMN: unwinged or very narrowly winged.

DISTRIBUTION: Guatemala.

HARDINESS: G2.

FLOWERING: autumn.

6. **S. saccata** Bateman

ILLUSTRATIONS: Fieldiana **26**: 535 (1952); Annals of the Missouri Botanical Garden **48**: 161 (1961).

SEPALS: 5–5.5 cm, greenish white to cream spotted with red or brown.

LIP: orange-yellow spotted with red or brown, the anterior part 3-lobed at the apex, the central lobe shorter than the laterals.

COLUMN: slightly winged.

DISTRIBUTION: Guatemala, El Salvador.

HARDINESS: G2.

FLOWERING: autumn.

7. **S. tigrina** Lindley

ILLUSTRATIONS: Botanical Magazine, 4197 (1845); Annals of the Missouri Botanical Garden **48**: 159 (1961); Die Orchidee **31**: clxv (1980); Bechtel et al., Manual of cultivated orchid species, 275 (1981); American Orchid Society Bulletin **53**: 364 (1984); Orchid Review **94**: 257 (1986).

SEPALS: *c.* 9 cm, yellowish white with confluent red-purple spots.

LIP: yellow with violet spots, the anterior part broadly ovate with 3 equal, acute teeth.

COLUMN: very broadly winged.

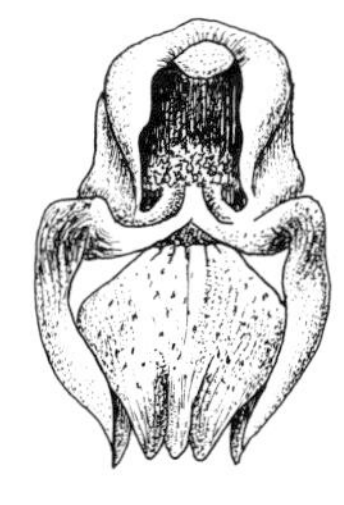

Stanhopea tigrina – lip.

DISTRIBUTION: S Mexico.
HARDINESS: G2.
FLOWERING: autumn.
This species has the largest flowers in the genus.

8. S. oculata (Loddiges) Lindley
ILLUSTRATIONS: Botanical Magazine, 5300 (1862); Hoehne, Flora Brasilica **12** (6); t. 115 (1942); Sheehan & Sheehan, Orchid genera illustrated, 161 (1979); Bechtel et al., Manual of cultivated orchid species, 275 (1981); Orchid review **94**: 253 (1986).
LEAVES: 30–45 cm.
RACEMES: with 4–10 flowers.
SEPALS AND PETALS: sepals 5.5–7 cm, all variable in colour, white, yellow or orange, usually red-spotted.
LIP: basal part oblong when viewed from the side, hollowed only near its base, where the sides are notched, linked to the middle part by an elongate groove; middle part bearing horns which are 2–3 cm long; anterior part broadly ovate, acute or obtuse, not usually 3-lobed.

Stanhopea oculata – lip.

COLUMN: winged.
DISTRIBUTION: tropical America from S Mexico to Venezuela and N Peru.
HARDINESS: G2.
FLOWERING: summer–autumn.
As with *S. insignis*, this is the best known of a complex of species, of which 2 more (and perhaps others) are in cultivation.

9. S. jenishiana Reichenbach
SYNONYMS: *S. bucephalus* misapplied; *S. graveolens* misapplied; *S. grandiflora* Reichenbach, not (Loddiges) Lindley.
ILLUSTRATIONS: Botanical Magazine, 5278 (1861), 8517 (1913); Annals of the Missouri Botanical Garden **48**: 165 (1961); Orchid Review **94**: 257 (1986).
SEPALS AND PETALS: orange-yellow, sparsely purple-spotted, sepals *c.* 7 cm.
LIP: basal part boat-shaped when viewed from the side, its sides not notched.

Stanhopea jenishiana – lip.

COLUMN: winged.
DISTRIBUTION: Ecuador to C Peru.
HARDINESS: G2.
FLOWERING: summer.

10. S. wardii Loddiges
ILLUSTRATIONS: Botanical Magazine, 5289 (1862); Fieldiana **26**: 537 (1952); Annals of the Missouri Botanical Garden **48**: 167 (1961); Couret, Orquideas Venezolanas, 29–31 (1977); Orchid Review **94**: 252 (1986).

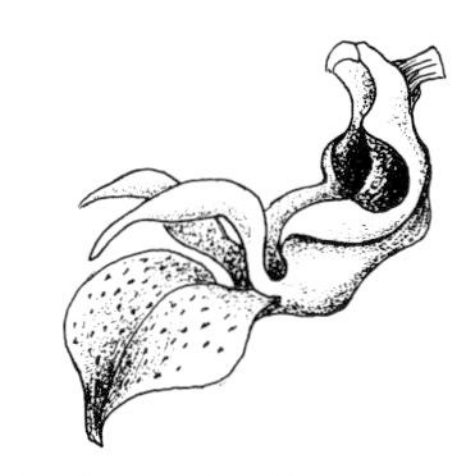
Stanhopea wardii – lip.

SEPALS AND PETALS: orange-yellow with red-purple spots near the base, sepals *c.* 7 cm.

LIP: basal part oblong when viewed from the side, its sides not notched.

COLUMN: very broadly winged.

DISTRIBUTION: tropical America, from S Mexico to Venezuela & Peru.

HARDINESS: G2.

FLOWERING: summer.

149. GONGORA Ruiz & Pavon

A genus of about 25 species from Tropical America. A few of them are cultivated in the same manner as *Acineta* (p. 438), for the sake of their bizarre flowers.

HABIT: epiphytic.

PSEUDOBULBS: simple, ovoid, grooved, each usually with 2 leaves.

LEAVES: elliptic to narrowly elliptic, stalked or not, pleated, rolled when young.

RACEMES: hanging, arising from the bases of the pseudobulbs, with many flowers.

FLOWERS: often fragrant, resupinate.

SEPALS: spreading, the upper attached to the column for about half its length or more.

PETALS: smaller than the sepals, attached to the side of the column for part of their length.

LIP: borne on the column foot, complex, fleshy, strongly keeled, divided into 2 parts, the basal part often bearing narrow lateral lobes (horns), the anterior part forwardly directed, usually acuminate.

ANTHER: at the apex of the column.

POLLINIA: 2, elongate-ovoid, stipe linear, longer than the pollinia, viscidium narrow; rostellum prominently beaked.

Key to species

1. Lip without horns, its apex forming an upcurved hook; apex of petals truncate, the angles obtuse or prolonged into curved points — **1. galeata**
2. Petals very small, less than half of the length of the column — **2. truncata**
3. Sepals yellow with red spots, or mostly red; basal part of the lip with 2 spreading lobes at the base as well as 2 horns at its apex — **3. quinquenervis**
4. Sepals mostly purple, the colour made up of confluent purple spots on a white or greenish ground; basal part of the lip without lobes at the base — **4. bufonia**

1. G. galeata (Lindley) Reichenbach

SYNONYMS: *Maxillaria galeata* Lindley; *Acropera loddigesii* Lindley.

ILLUSTRATIONS: Botanical Magazine, 3653 (1834); Bechtel et al., Manual of cultivated orchid species, 215 (1981).

PSEUDOBULBS: *c.* 6 cm, ovoid.

LEAVES: 25–35 cm.

SEPALS: to 2.2 cm, ovate, concave, brownish yellow.

PETALS: brown, joined to the column only at the base, truncate at the apex, the angles obtuse or prolonged into curved points.

LIP: yellow spotted with red, hornless, the apex prolonged into an upwardly curving, hook-like point.

DISTRIBUTION: Mexico.

HARDINESS: G2.

FLOWERING: summer.

2. G. truncata Lindley

ILLUSTRATIONS: Edwards's Botanical Register **31**: t. 56 (1845); Addisonia **2**: t. 46 (1917).

PSEUDOBULBS: 6 cm or more.

LEAVES: to 30 cm.

SEPALS: to 2.3 cm, ovate, pale yellow spotted with purple-brown.

PETALS: very small, less than half the length of the column, acute.

LIP: yellow, unspotted, the basal part with 2 lobes at its base and 2 upcurving horns at its apex, the anterior part curved, abruptly acute.

DISTRIBUTION: S Mexico.

HARDINESS: G2.

FLOWERING: summer.

3. G. quinquenervis Ruiz & Pavon

SYNONYM: *G. maculata* Lindley.

ILLUSTRATIONS: Botanical Magazine, 3687 (1835); Lindenia **5**: t. 208 (1889); Die Orchidee **30**: cxxiii–cxxiv (1979); American Orchid Society Bulletin **54**: 599 (1985).

PSEUDOBULBS: 5–12 cm.

LEAVES: 20–60 cm.

SEPALS: 1.5–2.4 cm, yellow with red spots or red, occasionally yellow with whitish spots, lanceolate, acuminate.

PETALS: purplish, acute, attached to the column for about half their length.

LIP: mostly yellow with 2 spreading lobes near its base and 2 upcurving horns at the apex of the basal part, the anterior part very strongly keeled, acuminate.

DISTRIBUTION: C & S America.

HARDINESS: G2.
FLOWERING: summer.

4. **G. bufonia** Lindley
ILLUSTRATIONS: Edwards's Botanical Register **26**: t. 2 (1841); Hoehne, Flora Brasilica **12** (6): t. 129 (1942).
PSEUDOBULBS: 6 cm or more.
LEAVES: to 30 cm, pale whitish green.
SEPALS: *c.* 2.3 cm, mostly purple, the colour made up of confluent purple blotches on a white or greenish ground.
PETALS: similar in colour to the sepals, attached to the column for about half their length, acute.
LIP: purple, the basal part with 2 upcurving horns at its apex, the anterior part keeled, acuminate.
DISTRIBUTION: S & E Brazil.
HARDINESS: G2.
FLOWERING: spring–summer.

150. CIRRHAEA Lindley

A small genus. Cultivation as for *Acineta* (p. 438).

Like *Gongora* except as below.
PSEUDOBULBS: each with 1 leaf.
UPPER SEPAL AND PETALS: free from the back of the column.
LIP: 3-lobed, the lateral lobes directed backwards, the central lobe narrowly triangular, directed forwards, the whole with a narrow arrowhead shape.
ANTHER: on the back of the column.

1. **C. dependens** Reichenbach
SYNONYM: *C. tristis* Lindley.
ILLUSTRATION: Hoehne, Iconografia de orchidaceas do Brasil, t. 191 (1949).
SEPALS: 2–2.5 cm, greenish with purple spots or flushed with purple.
PETALS: smaller than sepals, similar in colour.
LIP: violet.
DISTRIBUTION: Brazil.
HARDINESS: G2.
FLOWERING: summer.

151. CORYANTHES Hooker

A genus of about 12 species from C America and the northern part of S America with striking, if bizarre flowers which are difficult to describe; study of the illustrations cited is helpful in understanding the structure of the lip and column and their spatial relations. Nectar drips from the glands on the basal part of the lip and fills up the bucket-like anterior part, from where it overflows via the apical lobes, which form a spout.

Cultivation as for *Acineta* (p. 438); the plants should not be allowed to dry out completely during the resting period.

Literature: Kennedy, G., Some members of the genus Coryanthes, *Orchid Digest* **42**: 31–7 (1978); Gerlach, G., Beiträge zur Gattung Coryanthes, *Die Orchidee* **38**: 292–297 (1987).

HABIT: epiphytic.

PSEUDOBULBS: small, simple, grooved, ovoid or almost spindle-shaped, each bearing 1 or 2 leaves.

LEAVES: lanceolate, pleated, rolled when young.

RACEMES: hanging, arising from the bases of the pseudobulbs, with 2–6 flowers.

FLOWERS: fragrant.

SEPALS: the upper reflexed, smaller than the 2 lateral sepals which are spreading with their apices folded over backwards.

PETALS: erect, smaller than the sepals, twisted or with wavy margins.

LIP: extremely complex, divided into 3 parts. The basal part, which consists of a stalk borne at an acute angle to the ovary, terminating in a hollow, cup-like structure whose axis is at right angles to the stalk, is confluent with the column foot and bears 2 white, peg-like, nectaries near its base. The middle part, which is narrow and keeled, arises from the margin of the cup. The anterior part is bucket-like, hollowed out and with 3–5 lobes towards the apex, which lies close to the apex of the column.

COLUMN: slender, its apex reflexed at right angles to the rest.

POLLINIA: 2, ellipsoid, folded backwards along the surface of the curved stipe but arching away from it, viscidium narrowly triangular, almost as long as the stipe.

Key to species

1. The sides of the middle part of the lip corrugated and ridged; lateral sepals at least 12 cm **1. macrantha**

2. Basal and middle parts of the lip finely downy outside; anterior part of the lip yellowish brown inside and out **2. speciosa**

3. Basal and middle parts of the lip not downy outside; anterior part of the lip yellow marbled with purple, at least within **3. maculata**

1. C. macrantha (Hooker) Hooker

ILLUSTRATIONS: Botanical Magazine, 7692 (1900); Dunsterville & Garay, Venezuelan orchids illustrated 3: 63 (1961); Couret, Orquideas Venezolanas, 35 (1977); Orchid Digest **42**: 35 (1978).

PSEUDOBULBS: to 12 cm, narrow.

LEAVES: to 30 cm.

RACEME: usually with 2–4 flowers.

LATERAL: sepals 12–13 cm, yellow with many elongate red spots.

PETALS: with wavy margins, reddish yellow, red-spotted at the base.

LIP: basal part with a purplish stalk *c.* 2.5 cm, the cup orange-red; middle part yellow with red spots or entirely red, its sides ridged and corrugated; anterior part yellow, densely red-spotted.

DISTRIBUTION: tropical S America, from E Peru to Trinidad.

HARDINESS: G2.

FLOWERING: spring.

2. **C. speciosa** Hooker

ILLUSTRATIONS: Hoehne, Flora Brasilica 12 (6): t. 121, 122 (1942); Fieldiana **26**: 543 (1953); Couret, Orquideas Venezolanas, 34 (1977); Orchid Digest **42**: 35 (1978); Orchid Review **92**: 370 (1984).

PSEUDOBULBS: 7–15 cm.

LEAVES: 35–55 cm.

RACEME: with 2–5 flowers.

SEPALS: yellow, the laterals 6–8 cm.

PETALS: yellow, somewhat twisted and with wavy margins.

LIP: basal part with a brownish stalk 1–1.2 cm, and a brownish cup, downy outside; middle part brown, its sides smooth, downy; anterior part pale yellowish brown.

DISTRIBUTION: tropical America, from Guatemala to Peru & E Brazil.

HARDINESS: G2.

3. **C. maculata** Hooker

ILLUSTRATIONS: Botanical Magazine, 3102 (1831), 3747 (1845); Orchid Digest **42**: 32 (1978); American Orchid Society Bulletin **56**: 911 (1987).

PSEUDOBULBS: to 12 cm.

LEAVES: to 30 cm.

RACEME: with 3–6 flowers.

SEPALS: *c.* 9 cm, yellowish.

PETALS: yellowish, margins wavy.

LIP: basal part with a yellowish or purplish stalk *c.* 1 cm, the cup yellow, hairless outside; middle part yellow or purplish, its sides smooth, hairless; anterior part yellow outside, yellow marbled with purple or rarely entirely purple within.

DISTRIBUTION: Guyana.

HARDINESS: G2.

FLOWERING: spring–summer.

152. TRICHOCENTRUM Poeppig & Endlicher

A genus of about 20 species from C & S America. Cultivation as for Paphinia (p. 445), but good drainage is essential.

HABIT: epiphytic.
PSEUDOBULBS: small, simple, to 1 cm, each with 1 leaf.
LEAVES: thick, narrowly elliptic, acute, folded when young.
RACEMES: spreading or arching, arising from the bases of the pseudobulbs, with 1–2 flowers which are 2.5 cm or more in diameter.
SEPALS AND PETALS: similar in size and colour, spreading.
LIP: spurred, with a narrow basal part which often bears a number of ridges and a spreading, usually broad, notched blade.
COLUMN: short, without a foot, conspicuously winged, the wings variable in shape, sometimes coarsely fringed.
ANTHER: downy.
POLLINIA: 2, ovoid, stipe strap-shaped, longer and broader than the pollinia, viscidium small, circular,

Key to species

1. Lip without ridges but with 5 violet lines **2. orthoplectron**
2. Sepals and petals white spotted with brown **3. pfavii**
3. Spur about one-quarter of the length of the lip; lip with 4 yellow ridges **1. albo-coccineum**
4. Spur much shorter; lip with 3 yellow ridges **4. tigrinum**

1. T. albo-coccineum Linden
SYNONYMS: *T. albo-purpureum* Reichenbach; *T. alboviolaceum* Schlechter.
ILLUSTRATIONS: Botanical Magazine, 5688 (1868); Lindenia **16**: t. 748 (1901); Hoehne, Flora Brasilica **12**(7): t. 165 (1953).
LEAVES: to 8 cm.
FLOWERS: to 5 cm in diameter.
SEPALS AND PETALS: oblong-obovate, olive green outside, brown fading to yellow at the tips inside.
LIP: notched at the apex, mostly white with 2 reddish violet patches at the sides near the base and with 4 yellow ridges; spur about a quarter the length of the lip.
COLUMN: wings erect, acuminate.
DISTRIBUTION: E Peru & Brazil.
HARDINESS: G1.
FLOWERING: summer–autumn.

2. T. orthoplectron Reichenbach
Like *T. albo-coccineum* except as below.
LEAVES: 7–14 cm.

LIP: without ridges but with 5 violet lines.
DISTRIBUTION: Brazil.

3. **T. pfavii** Reichenbach
ILLUSTRATIONS: Gardeners' Chronicle **17**: 117 (1882); Illustration Horticole **33**: t. 587 (1886).
Like *T. albo-coccineum* except as below.
FLOWERS: slightly smaller with white, brown-spotted sepals and petals.
LIP: white with a red spot near the base, the spur very short.
COLUMN: wings oblong, truncate.
DISTRIBUTION: Costa Rica.
FLOWERING: autumn.

4. **T. tigrinum** Linden & Reichenbach
ILLUSTRATIONS: Illustration Horticole **24**: t. 282 (1877); Botanical Magazine, 7380 (1894); Bechtel et al., Manual of cultivated orchid species, 277 (1981); American Orchid Society Bulletin **55**: 1133 (1986).
Like *T. albo-coccineum* except as below.
FLOWERS: *c*. 6 cm in diameter, sepals and petals yellow spotted with brown.
LIP: white with lateral violet spots, 3 yellow ridges and 2 small yellow subulate teeth on the margins near the edges; the spur very short.
COLUMN: wings coarsely fringed.
DISTRIBUTION: C America.
FLOWERING: spring.

153. IONOPSIS Humboldt, Bonpland & Kunth

A genus of 8 species widely distributed in tropical and subtropical America. Plants can be grown successfully on cork bark or blocks of tree fern hung near the glass in a warm house. Frequent spraying is necessary during the summer. Propagation by division.

HABIT: small, rhizomatous epiphytes.
PSEUDOBULBS: simple, to 1.5 cm, each with 1 leaf.
LEAVES: (some leaves borne directly on the rhizomes) oblong-linear to narrowly elliptic, fleshy or not, folded when young.
FLOWERS: in panicles arising from the apices of the pseudobulbs.
SEPALS: upper free, the laterals united at their bases and forming a spur under the flower.
PETALS: similar to the upper sepal.
LIP: with a narrow claw which usually bears 2 ridges, and a reversed-heart-shaped blade.
COLUMN: short, unwinged, without a foot.

POLLINIA: 2, ellipsoid, stipe linear, slightly broadened above, longer than the pollinia, viscidium small, oblong, acute, forming an acute angle with the stipe.

1. **I. utricularioides** (Swartz) Lindley

SYNONYM: *I. tenera* (Steudel) Lindley.

ILLUSTRATIONS: Ames, Orchidaceae **1**: t. 5 (1905); Fieldiana **26**: 598 (1953); Dunsterville & Garay, Venezuelan orchids illustrated **1**: 174–5 (1959); Bechtel et al., Manual of cultivated orchid species, 217 (1981).

PSEUDOBULBS: cylindric.

LEAVES: flat, not very fleshy, narrowly elliptic.

SEPALS AND PETALS: to 8 mm, pinkish white.

LIP: pinkish white with the claw intense purple, the blade deeply notched.

DISTRIBUTION: tropical and subtropical America from USA (Florida) to Peru and Paraguay.

HARDINESS: G2.

FLOWERING: spring & autumn.

I. paniculata Humboldt, Bonpland & Kunth, from Brazil, is sometimes treated as a separate species; it is reputedly larger than *I. utricularioides*, with oblong, fleshy leaves, sepals and petals to 12 mm, and the blade of the lip less deeply notched.

154. **RODRIGUEZIA** Ruiz & Pavon

A genus of about 30 species from C & S America, concentrated in Brazil. Cultivation as for Paphinia (p. 445), but *R. decora* should be grown on a raft or on a piece of bark. All are subject to attack by red spider mite and thrips.

Literature: Bock, I., Die Gattung Rodriguezia, *Die Orchidee* **39**: 145–150 (1988).

HABIT: epiphytes with short or long rhizomes.

PSEUDOBULBS: distant or close, simple, each with 1–3 leaves.

LEAVES: oblong, folded when young.

RACEMES: with few to many flowers, arching or somewhat hanging, arising from the apices of the pseudobulbs.

UPPER SEPAL AND PETALS: similar, incurved.

LATERAL SEPALS: fused at their bases, and forming a short spur.

LIP: directed forwards with a distinct claw and a notched or 2-lobed blade, bearing 2 ridges on the claw and a single appendage projecting back into the sepal spur.

COLUMN: short with 2 large or small wings at the apex, without a foot.

POLLINIA: 2, broadly ovoid, stipe very narrow, longer than pollinia, viscidium small, oblong, borne at an acute angle to the stipe.

Key to species

1. Rhizomes long, pseudobulbs separated; lip with claw longer than the blade **1. decora**
2. Raceme 1-sided; flowers usually pink, occasionally orange or purple **2. secunda**
3. Raceme not 1-sided; flowers white with a yellow blotch on the lip **3. venusta**

1. **R. decora** (Lindley) Reichenbach

ILLUSTRATIONS: Hoehne, Iconografia de Orchidaceas do Brasil, t. 239 (1949); Bechtel et al., Manual of cultivated orchid species, 270 (1981).

PSEUDOBULBS: *c.* 10 cm apart, each compressed and with 1 leaf.

LEAVES: 6–9 cm, oblong.

RACEME: with 6–10 flowers, arching, loose.

FLOWERS: *c.* 5 cm in diameter.

SEPALS AND PETALS: oblong, acute, yellow or white with red-purple spots.

LIP: much longer than sepals, the claw longer than the blade and bearing 2 toothed erect white- and red-spotted or rarely entirely red ridges; blade white, broader than long, deeply notched.

COLUMN: wings elongate, ciliate at their ends.

DISTRIBUTION: S & C Brazil.

HARDINESS: G2.

FLOWERING: autumn.

2. **R. secunda** Humboldt, Bonpland & Kunth.

ILLUSTRATIONS: Dunsterville & Garay, Venezuelan orchids illustrated 1: 376–7 (1959); Couret, Orquideas Venezolanas, 58–9 (1977); Sheehan & Sheehan, Orchid genera illustrated, 153 (1979); Bechtel et al., Manual of cultivated orchid species, 270 (1981); American Orchid Society Bulletin 53: 1270 (1984).

PSEUDOBULBS: close or touching, each with 2–3 leaves.

LEAVES: to 15 cm, oblong.

RACEME: arching, 1-sided, with many closely packed flowers.

SEPALS AND PETALS: ovate, pink, rarely orange or purple.

LIP: scarcely longer than the sepals, pink, with a short claw gradually broadening into the oblong blade which is slightly notched; ridges on the claw in the form of 2 swellings.

COLUMN: wings minute, not ciliate.

DISTRIBUTION: tropical America, from Panama to Ecuador, Trinidad & French Guiana.

HARDINESS: G2.

FLOWERING: spring–summer.

3. R. venusta (Lindley) Reichenbach

SYNONYM: *R. fragrans* Lindley.

ILLUSTRATION: Hoehne, Iconografia de Orchidaceas do Brasil, t. 240 (1949).

Like *R. decora* except as below.

FLOWERS: very fragrant, borne in a raceme which is not 1-sided.

SEPALS AND PETALS: narrower, acute, white.

LIP: white with a yellow centre, its blade almost circular.

DISTRIBUTION: E Brazil.

HARDINESS: G2.

FLOWERING: autumn.

155. COMPARETTIA Poeppig & Endlicher

A genus of about 8 species from tropical America. Cultivation as for *Rodriguezia* (p. 458).

HABIT: rhizomatous epiphytes.

PSEUDOBULBS: small, simple, close or touching, each with 1 leaf.

LEAVES: narrowly elliptic, folded when young.

RACEMES: arising from the bases of the pseudobulbs, arching or somewhat hanging, loose, with 6–10 flowers.

SEPALS AND PETALS: upper sepal and petals similar or the petals slightly broader, erect; lateral sepals united at the base and prolonged into a backwardly pointing spur.

LIP: much larger than the petals, with a short claw and broad blade, and with 2 appendages projecting back into the sepal spur.

COLUMN: unwinged and without a foot.

POLLINIA: 2, ovoid, stipe longer than pollinia, broadened above, viscidium small, circular or heart-shaped.

Key to species

1. Sepals and petals more than 1.5 cm; spur 3 cm or more **3. macroplectrum**
2. Sepals and petals rose-pink to violet **2. falcata**
3. Sepals and petals bright shining scarlet **1. coccinea**

1. C. coccinea Lindley

ILLUSTRATIONS: Edwards's Botanical Register **24**: t. 68 (1838); Hoehne, Iconografia de Orchidaceas do Brasil, t. 238 (1949); Bechtel et al., Manual of cultivated orchid species, 188 (1981).

PSEUDOBULBS: 2–2.5 cm.

LEAVES: to 10 cm, oblong, acute.

FLOWERS: bright shining scarlet.

SEPALS AND PETALS: *c.* 1.3 cm.

LIP: yellow outside; spur to 1.8 cm.

DISTRIBUTION: S & C Brazil.
HARDINESS: G2.
FLOWERING: summer.

2. C. falcata Poeppig & Endlicher
SYNONYM: *C. rosea* Lindley.
ILLUSTRATIONS: Botanical Magazine, 4980 (1857); Fieldiana **26**: 603 (1953); Dunsterville & Garay, Venezuelan orchids illustrated **1**: 86–7 (1959); Bechtel et al., Manual of cultivated orchid species, 188 (1981).

Vegetatively very like *C. coccinea*.

SEPALS AND PETALS: *c.* 1.3 cm, rose-pink to violet.
LIP: deeper violet than the petals inside, pink or violet outside; spur to 2 cm.
DISTRIBUTION: tropical America, from S Mexico & the West Indies to Bolivia.
HARDINESS: G2.
FLOWERING: winter.

3. C. macroplectrum Reichenbach & Triana
ILLUSTRATIONS: Botanical Magazine, 6679 (1883); Lindenia **14**: t. 664 (1899); Sheehan & Sheehan, Orchid genera illustrated, 69 (1979); Bechtel et al., Manual of cultivated orchid species, 188 (1981).

Vegetatively like *C. coccinea*.

FLOWERS: clear violet with purple spots.
SEPALS AND PETALS: 1.5–2 cm.
LIP: violet outside; spur 3–5 cm, very conspicuous.
DISTRIBUTION: tropical S America from Ecuador to N Brazil.
HARDINESS: G2.
FLOWERING: autumn.

156. TRICHOPILIA Lindley

A genus from C and northern S America of an uncertain number (15–30?) of doubtfully distinguished species. Various species, formerly included in the genus, have now been placed elsewhere (see Garay, *Orquideologia* 7: 191–6, 1972). Five species of the reduced genus are found in cultivation.

They are grown in epiphytic compost which must be well drained in pots or tied on to sections of bark. Only occasional watering is required in the winter. Propagation is by division.

HABIT: epiphytic.
PSEUDOBULBS: simple, small or large, each with 1 leaf.
LEAVES: more or less elliptic or elliptic-oblong, acute, folded when young.
RACEMES: arising from the bases of the pseudobulbs, arching or somewhat pendent with 1–5 flowers which are often fragrant.

SEPALS AND PETALS: similar, spreading, narrow, sometimes twisted.

LIP: rolled around the column at its apex, forming a tube, obscurely 3-lobed, margins wavy.

COLUMN: elongate, without a foot, with 3 fringed lobes at the apex.

POLLINIA: 2, ovoid to almost spherical, stipe linear, slightly broadened above, longer than pollinia, viscidium small, forming an acute angle with the stipe.

Key to species

1. Sepals and petals conspicuously twisted **4. tortilis**
2. Pseudobulbs to 4 cm, almost circular in outline; sepals and petals 6 cm or more **1. suavis**
3. Lip red within **5. marginata**
4. Sepals and petals 5.5–6 cm; lip white with a yellow blotch within **2. fragrans**
5. Sepals and petals to 5 cm; lip white with a faint yellow blotch and red-brown spots within **3. galeottiana**

1. T. **suavis** Lindley & Paxton

ILLUSTRATIONS: Botanical Magazine, 4654 (1852); Lindenia 9: t. 423 (1893); Sheehan & Sheehan, Orchid genera illustrated, 165 (1979); Bechtel et al., Manual of cultivated orchid species, 277 (1981).

PSEUDOBULBS: 3–4 cm, almost circular in outline, compressed.

LEAVES: to 20 cm, elliptic, broadened abruptly to almost cordate at the base.

RACEME: with 2–3 flowers.

SEPALS AND PETALS: 6–7 cm, white, sometimes faintly yellowish or greenish, rarely red-spotted.

LIP: white with a yellow blotch inside and red spots on the central lobe, or rarely (var. **alba** Anon.) without spots.

DISTRIBUTION: Costa Rica, Panama & Colombia.

HARDINESS: G2.

FLOWERING: spring.

2. T. **fragrans** (Lindley) Reichenbach

ILLUSTRATIONS: Cogniaux & Goossens, Dictionnaire Iconographique des Orchidées, Trichopilia, t. 3 (1900); Dunsterville & Garay, Venezuelan orchids illustrated 3: 316–17 (1965); Bechtel et al., Manual of cultivated orchid species, 277 (1981).

PSEUDOBULBS: 10–13 cm, elongate in outline, compressed.

LEAVES: to 17 cm, elliptic, tapered to the base.

RACEMES: with 2–5 flowers.

SEPALS AND PETALS: 5.5–6 cm, white or greenish white.

LIP: white with a yellow blotch inside.

DISTRIBUTION: tropical America from the West Indies to Bolivia.

HARDINESS: G2.

FLOWERING: winter.

3. T. galeottiana A. Richard

SYNONYM: *T. picta* Lemaire.

ILLUSTRATIONS: Illustration Horticole 6: t. 225 (1859); Cogniaux & Goossens, Dictionnaire Iconographique des Orchidées, Trichopilia, t. 5 (1900).

Like *T. fragrans* except as below.

RACEMES: with 2–3 flowers.

SEPALS AND PETALS: to 5 cm, greenish, sometimes with brown lines.

LIP: white outside, yellowish or whitish with red-brown spots inside.

DISTRIBUTION: Mexico.

FLOWERING: summer.

4. T. tortilis Lindley

ILLUSTRATIONS: Botanical Magazine, 3739 (1839); Cogniaux & Goossens, Dictionnaire Iconographique des Orchidées, Trichopilia, t. 6 (1900); Bechtel et al., Manual of cultivated orchid species, 278 (1981).

PSEUDOBULBS: 4–8 cm, elongate in outline, compressed.

LEAVES: 10–15 cm, elliptic, tapered to the base.

RACEME: with 1–2 flowers.

SEPALS AND PETALS: to 6 cm, conspicuously twisted, green, striped and spotted with red-brown.

LIP: white with red-brown spots inside.

DISTRIBUTION: C America from S Mexico to Honduras.

HARDINESS: G2.

FLOWERING: winter.

An albino variant, with pure white flowers, has been in cultivation.

5. T. marginata Henfrey

SYNONYMS: *T. coccinea* Lindley; *T. crispa* Lindley.

ILLUSTRATIONS: Paxton's Flower Garden **2**: t. 54 (1851–52); Botanical Magazine, 4857 (1855); Cogniaux & Goossens, Dictionnaire Iconographique des Orchidées, Trichopilia, t. 1, 2, 2A (1900).

PSEUDOBULBS: 5–7 cm, elongate in outline, compressed.

LEAVES: to 17 cm, elliptic, narrowed to the base.

RACEME: with 2–3 flowers.

FLOWERS: rather variable in colour.

SEPALS AND PETALS: to 6 cm, white or greenish white, variously tinged, spotted or lined with red.

LIP: white outside, mostly red within, the margin sometimes white.

DISTRIBUTION: C America, Colombia.

HARDINESS: G2.

FLOWERING: spring.

157. COCHLIODA Lindley

A genus of about 6 species from S America. Cultivation as for *Odontoglossum* (p. 468).

HABIT: epiphytic, with short rhizomes.

PSEUDOBULBS: simple, compressed, each with 1 or 2 leaves.

LEAVES: linear to oblong, usually blunt, folded when young.

FLOWERS: in erect, arching or hanging racemes or rarely panicles, arising laterally.

SEPALS AND PETALS: spreading.

LIP: with an erect claw which is fused to the column base and a spreading and usually 3-lobed blade bearing 2 or 4 swellings near the base, the lateral lobes rounded and often reflexed, the central lobe pointing forwards, often laterally expanded from a narrow base, notched or 2-lobed.

COLUMN: erect, curved, without a foot, with 2 or 3 lobes at the apex; stigmatic surface divided into 2.

POLLINIA: 2, broadly ovoid, stipe linear, somewhat longer than the pollinia, viscidium oblong, almost as long as the stipe and forming an acute angle with it.

Key to species

1. Lateral sepals united; lip without any obvious lateral lobes, tapering to its tip — **5. sanguinea**
2. Bract equalling the ovary and its stalk; flower white or cream — **3. densiflora**
3. Central lobe of lip oblong, not expanded laterally from a narrow base; column 3-lobed at apex — **4. rosea**
4. Central lobe of lip expanded laterally from a narrow base; column finely toothed at apex

4.A. Flowers rosy purple — **1. vulcanica**

4.B. Flower scarlet, lip with yellow blotches — **2. noezliana**

1. C. vulcanica (Reichenbach) Veitch

ILLUSTRATIONS: Botanical Magazine, 6001 (1872); Cogniaux & Goossens, Dictionnaire Iconographique des Orchidées, Cochlioda, t. 1 (1898).

PSEUDOBULBS: 2.5–6 cm, clustered, compressed, each with 2 leaves.

LEAVES: 7.5–15 × 1.5–3.5 cm, elliptic-oblong to oblong-linear.

RACEME: with 6–many flowers, almost erect; bracts shorter than the ovaries and their stalk.

FLOWERS: *c.* 4 cm in diameter, rosy purple.

SEPALS AND PETALS: *c.* 2.2 cm, lanceolate-oblong, acute.
LIP: fused to the lower half of the column, 3-lobed, the central lobe with a narrow base and laterally expanded, 2-lobed, apical part.
COLUMN: red, finely toothed at the apex.
DISTRIBUTION: Ecuador, Peru.
HARDINESS: G1.
FLOWERING: autumn–winter.

2. **C. noezliana** Rolfe
Like *C. vulcanica* except as below.
FLOWERS: scarlet with yellow blotches on the lip.
COLUMN: 3-lobed at the apex.
DISTRIBUTION: Peru, Bolivia.

3. **C. densiflora** Lindley
PSEUDOBULBS: to 5 cm, compressed, each with 1 leaf.
LEAVES: to 14 × 1.8 cm, oblong.
RACEMES: with *c.* 9 flowers, almost erect; bracts as long as the ovaries and their stalks.
FLOWERS: *c.* 3 cm in diameter, white or cream.
SEPALS AND PETALS: *c.* 1.5 cm, the lateral sepals narrower than the rest.
LIP: fused to the lower half of the column, 3-lobed, the central lobe with a narrow base and laterally expanded, 2-lobed apical part.
COLUMN: with 2 auricles at the apex.
DISTRIBUTION: Peru.
HARDINESS: G1.

4. **C. rosea** (Lindley) Bentham
ILLUSTRATION: Cogniaux & Goossens, Dictionnaire Iconographique des Orchidées, Cochlioda, t. 3 (1898); Bechtel et al., Manual of cultivated orchid species, 185 (1981).
PSEUDOBULBS: 3–5 cm, compressed, clustered, each with 1 or 2 leaves.
LEAVES: 5.5–20 × 1–2.5 cm, elliptic-oblong to linear.
RACEME: with up to 20 flowers, erect or arched, rarely branched above; bracts much shorter than the ovaries and their stalks.
FLOWERS: 2–3.5 cm in diameter, rose-red.
SEPALS AND PETALS: *c.* 1.1 cm, narrowly elliptic to lanceolate.
LIP: fused to the lower third of the column, 3-lobed, the central lobe oblong.
COLUMN: 3-lobed at the apex.
DISTRIBUTION: Ecuador, Peru.
HARDINESS: G1.
FLOWERING: spring.
The name *C. stricta* Cogniaux has been applied to plants

resembling *C. rosea*, but thought to originate in Colombia. It is not certain what these plants are, or whether any is in cultivation now.

5. **C. sanguinea** (Reichenbach) Bentham

PSEUDOBULBS: 3–5 cm, ovoid, compressed, each with 2 leaves.

LEAVES: to 25 × 1.5 cm, linear.

RACEME: with many flowers, almost erect, often branched above; bracts much shorter than the ovaries and their stalks.

FLOWERS: *c.* 3 cm in diameter, red, the apex of the column and the base of the lip white.

SEPALS AND PETALS: *c.* 1.5 cm, the lateral sepals united for about half their length.

LIP: small with scarcely perceptible lateral lobes and a tapering central lobe, fused to the lower half of the column.

COLUMN: weakly 3-lobed at apex.

DISTRIBUTION: Ecuador, Peru.

HARDINESS: G2.

FLOWERING: autumn.

Sometimes placed in the genus *Symphyglossum*, where its correct name is *S. sanguinea* (Reichenbach) Schlechter.

158. GOMESA R. Brown

A genus of about 8 species from S America, mostly from Brazil. They are grown in a similar manner to *Odontoglossum* but require more heat in winter.

HABIT: epiphytic, with short rhizomes.

PSEUDOBULBS: simple, clustered, oblong, compressed, each with 2 leaves.

LEAVES: oblong or very narrowly elliptic, folded when young.

RACEMES: arising from the bases of the pseudobulbs, arching, many-flowered.

SEPALS AND PETALS: spreading, greenish yellow, the upper sepal and the petals erect, the lateral sepals fused at the bases to some extent, spreading downwards.

LIP: greenish yellow, shorter than the sepals, with an erect basal part and a downwardly deflected, ovate or oblong-ovate blade which bears 2 large, finely toothed ridges from the base to the middle, which reach up to and slightly clasp the column.

COLUMN: white, reddish around the rostellum, erect, without a foot.

POLLINIA: 2, ovoid, stipe oblong, shorter than pollinia, viscidium small, making an acute or right angle with the stipe.

Key to species

1. Lateral sepals united only at the extreme base; blade of lip oblong-ovate **1. crispa**

2. Lateral sepals united for more than half their length; blade of lip ovate **2. planifolia**

1. G. crispa (Lindley) Klotzsch & Reichenbach

ILLUSTRATIONS: Pabst & Dungs, Orchidaceae Brasilienses **2**: f. 1914 (1977); Sheehan & Sheehan, Orchid genera illustrated, 97 (1979); Bechtel et al., Manual of cultivated orchid species, 215 (1981).

PSEUDOBULBS: 6–10 cm.

RACEME: many-flowered.

FLOWERS: *c.* 2 cm in diameter.

SEPALS AND PETALS: 9–10 mm, wavy, oblong, rounded at the apex, the lateral sepals united only at the base.

LIP: oblong-ovate.

DISTRIBUTION: Brazil.

HARDINESS: G1.

FLOWERING: spring.

2. C. planifolia (Lindley) Klotzsch & Reichenbach

SYNONYM: *G. recurva* Loddiges, not R. Brown.

ILLUSTRATIONS: Loddiges' Botanical Cabinet **7**: t. 660 (1822); Botanical Magazine, 3504 (1836); Pabst & Dungs, Orchidaceae Brasilienses **2**: f. 1922 (1977).

Very like *G. crispa* except as below.

SEPALS AND PETALS: scarcely wavy, rather broad, the lateral sepals united for about two-thirds of their length.

LIP: blade ovate, rounded.

DISTRIBUTION: Brazil, Paraguay, Argentina.

HARDINESS: G1.

FLOWERING: spring–summer.

The flowers of this species are said to be fragrant.

159. HELCIA Lindley

A genus related to *Trichopilia* but keying near *Gomesa*. Cultivation as for *Odontoglossum*.

Like *Gomesa* except as below.

PSEUDOBULBS: 1-leaved.

FLOWERS: *c.* 7 cm in diameter.

LIP: 3-lobed with small lateral lobes reaching up to the column and central lobe is much larger, conspicuously wavy and without ridges extending to the column.

1. H. sanguinolenta Lindley

SYNONYM: *Trichopilia sanguinolenta* (Lindley) Reichenbach.

ILLUSTRATIONS: Botanical Magazine, 7281 (1893); Bechtel et al., Manual of cultivated orchid species, 216 (1981).

SEPALS AND PETALS: yellow with large brown spots.

LIP: white with violet or purple spots.

DISTRIBUTION : Ecuador.
HARDINESS : G2.
FLOWERING : winter.

160. ODONTOGLOSSUM Humboldt, Bonpland & Kunth
A large genus of about 200 species from C & S America, in need of further study. A very large number of species has been in cultivation in the past, but only about 20 are widely available today. Many hybrids have been raised in the past, and many species are variable in flower-colour, which has formed the basis for the recognition of numerous cultivars in some species.

Plants should be grown in an epiphytic compost. They require humid conditions with good ventilation, and may require shading during the hottest months. Watering should be reduced in winter. Propagation is by division, which is sometimes difficult, or by seed.

Literature. The genus has been studied by numerous workers during the last one hundred years, but an acceptable classification has yet to be worked out. Comprehensive revisions are provided by R. Escobar (El genero Odontoglossum, *Orquideologia* **11**: 21–57, 119–160, 257–302, 1976) and F. Halbinger (Odontoglossum and related genera in Mexico and Central America, *Orquidea (Mexico)* **8**: 155–282, 1982 - in Spanish and English). In two later papers (Cymbiglossum, Ticoglossum y Rhynchostele, tres generos derivados de Odontoglossum en Mexico y Centroamerica, *Orquidea (Mexico)* **9**: 1–12, 1983, and Lemboglossum, a new name for the Odontoglossum cervantesii complex, *ibid.*, 351–354, 1984, both in Spanish and English) Halbinger redefined the genus to exclude several species he included earlier in *Odontoglossum*; these he placed in various genera, notably *Lemboglossum* Halbinger (species **11–17** below), *Mesoglossum* Halbinger (species **18**), *Cuitlauzina* Llave & Lexarza (species **19**), *Osmoglossum* Schlechter (species **20**) and *Rossioglossum* (Schlechter) Garay & Kennedy. He also treats species **8** and **9** below as belonging to *Oncidium*. Further detailed information is provided by Bockemühl, L., Ondontoglossum HBK, Studien zu einer natürlichen Gliederung, *Die Orchidee*, 1987 and continuing, and Gay, J., The genus Odontoglossum, *Orchid Review* **96**: 113 (1988) and continuing.

While it is certain that the genus is somewhat heterogeneous, and its distinction from *Oncidium* is not entirely clear-cut, it seems best to maintain a wide circumscription of the genus for the purposes of the present book.

HABIT : rhizomes long or short, with distant or clustered pseudobulbs.
PSEUDOBULBS : compressed, simple, each bearing 1–3 leaves.
LEAVES : folded when young.

INFLORESCENCE: a terminal raceme or panicle; flowers often fragrant.

SEPALS: spreading, free or the lateral sepals united for a short distance at the base.

PETALS: similar to but often shorter than the sepals.

LIP: complex, usually shorter than the petals, with a short claw ascending parallel with the column (sometimes united to it), with or without small lateral lobes and with complex keels and swellings and a variably shaped blade which is bent backwards near its base.

COLUMN: long and slender, not thickened at the base, without a foot, often with auricles near the apex.

POLLINIA: 2, ellipsoid, stipe linear or linear-oblong, somewhat longer than the pollinia, viscidium almost as long as the stipe and forming an acute angle with it, narrowly oblong or triangular.

Key to species

1. Pseudobulbs bearing a single leaf *Go to Subkey 1*
2. Column without auricles *Go to Subkey 2*
3. Blade of lip broader than long, usually notched at the apex
3.A. Sepals *c.* 2.5 cm, petals similar, white or pink, sometimes with a few reddish spot; inflorescence with usually more than 8 flowers **19. pendulum**
3.B. Sepals 5–8.5 cm, petals shorter and broader, all yellow, the sepals barred with brown, the petals brown in the lower half; raceme with 4–8 flowers **21. grande**
4. Blade of lip triangular to heart-shaped, tapering gradually to the apex
4.A. Blade of lip 2.8 x 2.8 cm or more, pink, irregularly veined with white; petals ovate-elliptic **17. uroskinneri**
4.B. Blade of lip to 2 x 2.5 cm, violet, white or pink; petals elliptic-lanceolate to lanceolate **16. bictoniense**
5. Blade of lip oblong, waisted or not, tapering abruptly to the apex *Go to Subkey 3*

Subkey 1. Pseudobulbs each bearing a single leaf.

1. Column with auricles near the apex
1.A. Sepals and petals conspicuously wavy-crisped; rhizome extensive, pseudobulbs distant **7. brevifolium**
1.B. Lip narrowing from just above its base; auricles formed by 2 hair-like processes **3. cirrhosum**
1.C. Blade of lip toothed or ragged, usually waisted in the middle, the base with a keel formed from several tooth-like processes **4. hallii**
1.D. Blade of lip entire, not waisted, the base bearing 2 entire keels **13. cervantesii**
2. Blade of lip narrowly triangular to heart-shaped, acuminate; raceme with more than 4 flowers **11. cordatum**

3. Blade of lip irregularly toothed or ragged **12. stellatum**
4. Blade of lip entire though wavy **14. rossii**

Subkey 2. Pseudobulbs each bearing 2 or 3 leaves; column without auricles.

1. Blade of lip distinctly waisted just below its middle **9. reichenheimii**
2. Blade of lip longer than broad, tapering to the apex, not notched **15. apterum**
3. Blade of lip shining yellow barred with brown at the base, sometimes with additional brown spots on the surface **18. londesboroughianum**
4. Blade of lip mostly white, purple at base **10. cariniferum**

Subkey 3. Pseudobulbs each bearing 2 or 3 leaves; column with auricles.

1. Flowers to 3 cm in diameter, pure white except for the yellow, red-spotted keels on the lip; leaves oblong-linear **20. pulchellum**
2. Blade of the lip distinctly waisted
2.A. Sepals and petals to 3.5 cm, yellow or greenish with brown bars or red spots; panicle with many flowers
2.A.a. Sepals and petals barred with red-brown; basal part of lip red-purple **8. laeve**
2.A.b. Sepals and petals with red spots; basal part of lip red **5. confusum**
2.B. Sepals and petals at least 4 cm, dark brown with yellow veins; raceme with 4–12 flowers **6. harryanum**
3. Sepals and petals white or pink spotted with brown, elliptic, overlapping, the petals finely toothed **1. crispum**
4. Sepals and petals yellow with brown blotches, ovate-lanceolate or oblanceolate, not overlapping, petals wavy, not finely toothed **2. spectatissimum**

1. **O. crispum** Lindley

SYNONYM: *O. alexandrae* Bateman.

ILLUSTRATIONS: Cogniaux & Goossens, Dictionnaire Iconographique des Orchidées, Odontoglossum, t. 1 (1897); Schlechter, Die Orchideen, edn 2, t. 8 (1927); Williams et al., Orchids for everyone, 31 (1980); Bechtel et al., Manual of cultivated orchid species, 236 (1981).

HABIT: rhizome short.

PSEUDOBULBS: clustered, 4–8 cm, each with 2 leaves.

LEAVES: to 40 cm, linear-elliptic acute.

RACEME: arched, with 8–20 flowers.

SEPALS: 3–4 cm, elliptic, acute, margins wavy, overlapping with the petals, white or pink spotted with brown.

PETALS: like the sepals, but toothed.

Odontoglossum crispum – lip.

LIP: with a short claw bearing a fan-shaped, laciniate yellow keel with 2 longer projections in the middle, the blade oblong, wavy, finely toothed, acute, white pink or pale purple with brown spots.

COLUMN: with 2 rounded, entire auricles.

DISTRIBUTION: Colombia.

HARDINESS: G1.

FLOWERING: winter.

This species is very variable and many selected variants (some of them perhaps accidental hybrids) were grown in the past. The species has also been used widely in hybridisation.

2. O. spectatissimum Lindley

SYNONYM: *O. triumphans* Reichenbach.

ILLUSTRATIONS: Cogniaux & Goossens, Dictionnaire iconographique des Orchidées, Odontoglossum, t. 8, 8A, 8B (1897); Dunsterville & Garay, Venezuelan orchids illustrated **5**: 218–9 (1972).

HABIT: rhizome short, pseudobulbs clustered.

PSEUDOBULBS: 4–8 cm, each bearing 2 leaves.

LEAVES: to 40 cm, linear-elliptic, acute.

RACEME: erect, with 10–20 flowers.

SEPALS AND PETALS: to 4 cm, acute, margins wavy-crisped, yellow with numerous brown blotches.

Odontoglossum spectatissimum – lip.

LIP: with a short claw bearing a fan-shaped crest which is divided into processes, of which the 2 inner are the longest, yellow; blade oblong, acute, white or rarely yellow, the basal part pink or reddish, the apex brown, lacerate.

COLUMN: with finely toothed auricles.

DISTRIBUTION: Colombia, Venezuela.

HARDINESS: G1.

FLOWERING: spring.

3. O. cirrhosum Lindley

ILLUSTRATIONS: Botanical Magazine, 6317 (1877); Gartenflora **41**: t. 1383 (1892); Cogniaux & Goossens, Dictionnaire iconographique des Orchidées, Odontoglossum, t. 9 (1897); Bechtel et al., Manual of cultivated orchid species, 236 (1981); Orchid Review **93**: 42 (1985).

HABIT: rhizome short, pseudobulbs clustered.

PSEUDOBULBS: 5–8 cm, each bearing a single leaf.

LEAVES: 10–30 × 2.5–3 cm, oblong-linear.

PANICLE: arching, many-flowered.

SEPALS: to 4 cm × 7 mm, lanceolate, acuminate with a recurved point, white or cream with red-purple or red-brown blotches.

PETALS: similar to sepals but shorter and broader.

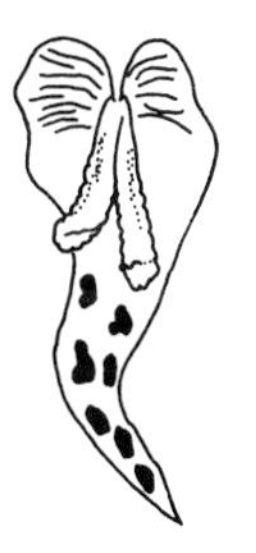

Odontoglossum cirrhosum – lip.

LIP: with a short claw bearing 2 lateral lobes, yellow lined with red and with 2 forwardly projecting keels; blade white spotted with red, narrowly heart-shaped to triangular, long-acuminate.

COLUMN: bearing 2 hair-like auricles near the apex.

DISTRIBUTION: Peru, Ecuador.

HARDINESS: G1.

FLOWERING: spring.

4. **O. hallii** Lindley

ILLUSTRATIONS: Botanical Magazine, 6237 (1876); Lindenia 4: t. 158 (1888); Bechtel et al., Manual of cultivated orchid species, 237 (1981); Orchid Review **93**: 43 (1985).

HABIT: rhizomes short, pseudobulbs clustered.

PSEUDOBULBS: 5–10 cm, each usually bearing a single leaf.

LEAVES: 13.5–30 × 2–4.5 cm, elliptic-oblong.

RACEME: loose, many-flowered, arching.

SEPALS: 3.5–6 cm, ovate-lanceolate, acuminate, yellow with large purple-brown blotches.

PETALS: similar to sepals but shorter and broader.

Odontoglossum hallii – lip.

LIP: claw short, bearing a large crest of several, yellow, tooth-like keels which extend on to the blade; blade white, variously spotted with red or purple, oblong, usually waisted, toothed to lacerate.

COLUMN: bearing auricles which are divided into a number of tooth-like processes.

DISTRIBUTION: Colombia, Ecuador, Peru.

HARDINESS: G1.

FLOWERING: spring.

5. **O. confusum** Garay

SYNONYMS: *O. schroederianum* Reichenbach; *Miltonia schroederiana* (Reichenbach) Veitch.

ILLUSTRATION: Cogniaux & Goossens, Dictionnaire iconographique des Orchidées, Miltonia, t. 3 (1897).

Like *O. hallii* except as below.

SEPALS AND PETALS: green spotted with red.

LIP: red in the lower half.

COLUMN: with a transverse ledge below the stigmatic area.

DISTRIBUTION: Costa Rica.

HARDINESS: G1.

FLOWERING: winter.

6. **O. harryanum** Reichenbach

ILLUSTRATIONS: Lindenia 3: t. 142 (1887); Cogniaux & Goossens, Dictionnaire iconographique des Orchidées, Odontoglossum, t. 11 (1897); Williams et al., Orchids for everyone, 140 (1980).

HABIT: rhizomes short, pseudobulbs clustered.

PSEUDOBULBS: 6–8 cm, each bearing 2 leaves.

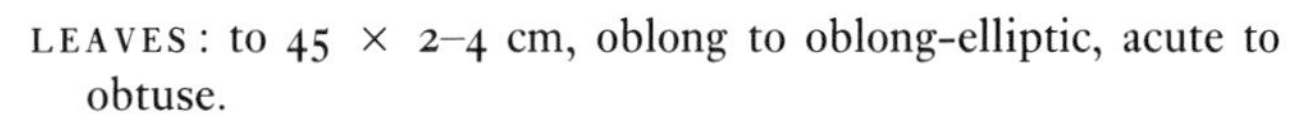

LEAVES: to 45 × 2–4 cm, oblong to oblong-elliptic, acute to obtuse.

RACEME: 4–12-flowered.

SEPALS: 4.5 × 1.3–2.5 cm, elliptic-oblong to elliptic, acute, brown to dark brown, mottled darker, with yellow veins.

PETALS: like the sepals but somewhat smaller, base whitish with purple lines.

LIP: with a short claw with a prominent yellow callus which is lacerate towards the front; blade oblong, waisted below the middle, yellowish at the base, the sides and the centre with purple lines on a white or yellow ground, the apical part white.

COLUMN: with small, finely-toothed auricles.

DISTRIBUTION: Colombia, Peru.

HARDINESS: G1.

FLOWERING: summer–autumn.

Odontoglossum harryanum – lip.

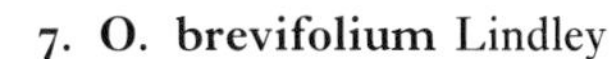

7. O. brevifolium Lindley

SYNONYMS: *O. chiriquense* Reichenbach; *O. coronarium* Lindley; *Otoglossum brevifolium* (Lindley) Garay & Dunsterville.

ILLUSTRATIONS: Cogniaux & Goossens, Dictionnaire iconographique des Orchidéees, Odontoglossum, t. 18 (1897); Botanical Magazine, 7687 (1899), 8725 (1917).

HABIT: rhizome creeping.

PSEUDOBULBS: distant, 4–11 cm, each with 1 leaf.

LEAVES: 10–30 × 6–9 cm, ovate to elliptic.

RACEME: with few to many flowers, erect or arching.

SEPALS AND PETALS: similar, 1.6–3 × 1.6–2 cm, obovate or obovate-oblong, obtuse, brown edged with yellow or yellow blotches with brown, margins conspicuously undulate-crisped.

LIP: with an indistinct claw bearing small lateral lobes and irregular keels, the blade oblong, usually waisted, whitish or yellow with transverse brown bars near the base.

COLUMN: with conspicuous, variably developed auricles which are finely toothed.

DISTRIBUTION: tropical America from Panama to Peru.

HARDINESS: G1.

FLOWERING: irregular.

Odontoglossum brevifolium – lip.

8. O. laeve Lindley

ILLUSTRATIONS: Botanical Magazine, 6265 (1876); Bechtel et al., Manual of cultivated orchid species, 237 (1981).

HABIT: rhizomes short.

PSEUDOBULBS: clustered, 5–12 cm, each with 2 leaves.

LEAVES: 15–45 × 2.5–5.5 cm, oblong to narrowly elliptic, rounded to almost acute.

PANICLE: with many flowers, arching.

SEPALS AND PETALS: 2.5–3.5 cm × 6–9 mm, linear-elliptic to

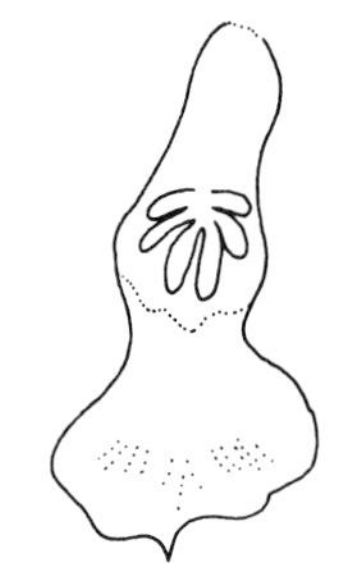

Odontoglossum laeve – lip.

linear-oblanceolate, yellowish blotched and barred with reddish brown.

LIP: with a short, thickened claw bearing 2–5 inconspicuous keels which run on to the blade, blade oblong, waisted, the basal part pale purple, the rest white, apex with a short point.

COLUMN: without a ledge below the stigmatic surface and with finely toothed auricles.

DISTRIBUTION: S Mexico, Guatemala.

HARDINESS: G1.

FLOWERING: summer

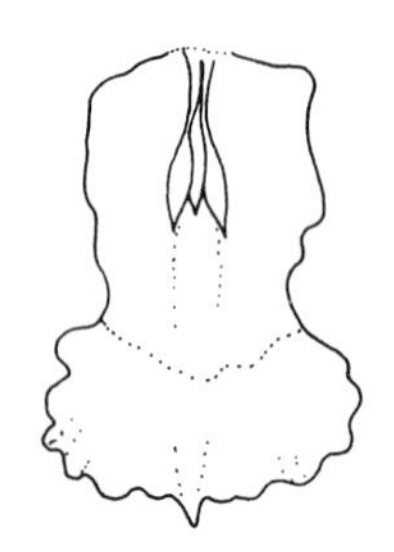

Odontoglossum reichenheimii – lip.

9. **O. reichenheimii** Linden & Reichenbach

ILLUSTRATION: Bechtel et al., Manual of cultivated orchid species, 239 (1981).

Like *O. laeve* except as below.

LIP: with a broader, oblong claw.

COLUMN: entirely without auricles.

DISTRIBUTION: Mexico.

For further information on the complex involving *O. laeve*, *O. reichenheimii* and *O. confusum* see Garay, Die Orchidee **6**: 213–18 (1962), and Garay & Stacy, Bradea 1(40): 393–424 (1974) in which all these species are considered to form part of the genus *Oncidium*.

10. **O. cariniferum** Reichenbach

ILLUSTRATIONS: Williams et al., Orchids for everyone, 138 (1980); Bechtel et al., Manual of cultivated orchid species, 235 (1981).

HABIT: rhizome short.

PSEUDOBULBS: close, 8–10 cm, each with 2 leaves.

LEAVES: to 30 × 3–3.5 cm, oblong-elliptic, acute.

PANICLE: with many flowers, arching.

SEPALS AND PETALS: similar, *c.* 3.5 cm, lanceolate, acute, purplish brown.

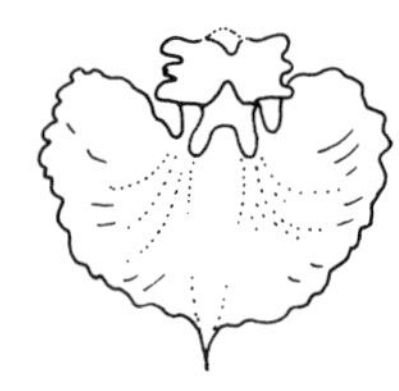

Odontoglossum cariniferum – lip.

LIP: with a short claw bearing 2 small, oblong, purple lateral lobes and 2 divergent, purple keels; blade broader than long, heart-shaped at the base, notched and with a small point within the notch at the apex, white.

COLUMN: without auricles.

DISTRIBUTION: tropical America, from Costa Rica to Venezuela.

HARDINESS: G1.

FLOWERING: spring.

11. **O. cordatum** Lindley

SYNONYMS: *Cymbiglossum cordatum* (Lindley) Halbinger; *Lemboglossum cordatum* (Lindley) Halbinger.

ILLUSTRATIONS: Lindenia 9: t. 430 (1893); Hamer, Orquideas de El Salvador **2**; 153 & t. 17 (1974); Williams et al., Orchids

for everyone, 139 (1980); Bechtel et al., Manual of cultivated orchid species, 236 (1981).
HABIT: rhizome short.
PSEUDOBULBS: clustered, 4.5–7.5 cm, each with 1 leaf.
LEAVES: 9–30 × 3–4.5 cm, narrowly elliptic, acute.
RACEME: erect with few to many flowers.
SEPALS: 3.5–5 cm × 6–12 mm, lanceolate, acuminate.
PETALS: 2.5–4 cm × 7–11 mm, ovate-lanceolate to elliptic-lanceolate, acuminate, greenish, whitish or yellowish blotched with brown.
LIP: with a short claw with complex yellow keels which extend on to the blade; blade narrowly triangular to heart-shaped, acuminate, finely scalloped, white at the base with pink around the keels, the apex yellow spotted with brown.
COLUMN: very shortly downy, without auricles.
DISTRIBUTION: C America, Venezuela.
HARDINESS: G1.
FLOWERING: summer.

Odontoglossum cordatum – lip.

12. O. stellatum Lindley

SYNONYMS: *Cymbiglossum stellatum* (Lindley) Halbinger; *Lemboglossum stellatum* (Lindley) Halbinger.
ILLUSTRATION: Hamer, Orquideas de El Salvador **2**: 159 & t. 18 (1974).
HABIT: rhizome short.
PSEUDOBULBS: clustered, 2–6 cm, each with 1 leaf.
LEAVES: 6.5–15 cm × 8–25 mm, ovate to elliptic or oblanceolate.
RACEME: erect with 1–2 flowers.
SEPALS AND PETALS: similar, 2–2.8 cm × 3–5 mm, yellowish barred with brown, sometimes entirely brown, lanceolate, acute.
LIP: with a narrow claw bearing keels which extend onto the base of the blade as a short bifid plate; blade white or pink tinged with mauve, triangular-ovate, obtuse, margin toothed to lacerate.
COLUMN: papillose, without auricles.
DISTRIBUTION: Mexico, Guatemala, El Salvador.
HARDINESS: G1.
FLOWERING: winter–spring.

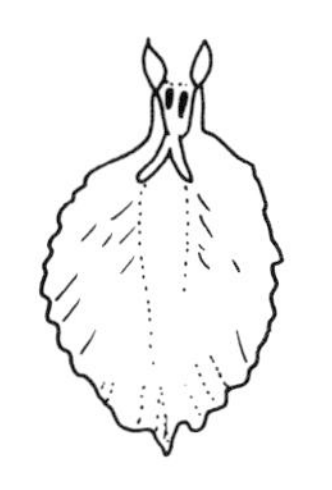
Odontoglossum stellatum – lip.

13. O. cervantesii Llave & Lexarza

SYNONYMS: *Cymbiglossum cervantesii* (Llave & Lexarza) Halbinger; *Lemboglossum cervantesii* (Llave & Lexarza) Halbinger.
ILLUSTRATIONS: Cogniaux & Goossens, Dictionnaire Iconographique des Orchidées, Odontoglossum, t. 16 (1897); Journal of the Orchid Society of Great Britain **20**: 9 (1971); Williams et al., Orchids for everyone, 138 (1980); Bechtel et

al., Manual of cultivated orchid species, 236 (1981); American Orchid Society Bulletin **53**: 269 (1984); Kew Magazine **5**: pl. 97 (1986).

HABIT: rhizome short.

PSEUDOBULBS: clustered, 2–6.5 cm, each with 1 leaf.

LEAVES: 4–30 × 1–3.5 cm, variable, ovate-lanceolate to elliptic-oblong.

RACEME: with 1–6 flowers, erect or arching, loose; flowers fragrant.

SEPALS AND PETALS: sepals lanceolate, petals elliptic, otherwise similar, 1.8–3 cm × 6–20 mm, white to pink, marked in the lower third with concentrically arranged red spots and bars.

Odontoglossum cervantesii – lip.

LIP: with a thick claw, white or pink, purple striped at the sides and with a yellow, papillose callus extending on to the blade as 2 points; blade heart-shaped, irregularly toothed, with a few spots.

COLUMN: papillose with 2 circular auricles.

DISTRIBUTION: S Mexico, Guatemala.

HARDINESS: G1.

FLOWERING: winter–spring.

14. **O. rossii** Lindley

SYNONYMS: *Cymbiglossum rossii* (Lindley) Halbinger; *Lemboglossum rossii* (Lindley) Halbinger.

ILLUSTRATIONS: Edwards's Botanical Register **25**: t. 48 (1839); Cogniaux & Goossens, Dictionnaire Iconographique des Orchidées, Odontoglossum, t. 6, 6A (1897).

HABIT: rhizome short.

PSEUDOBULBS: loosely clumped, 3–6 cm, each with 1 leaf.

LEAVES: 5–20 × 1.5–4 cm, elliptic or elliptic-lanceolate.

RACEME: with 2–4 flowers, erect or arched.

SEPALS AND PETALS: sepals 2.5–4.5 cm × 5–11 mm, oblong-elliptic to oblong-lanceolate, acute or acuminate, petals 2.5–3.8 cm × 8–19 mm, broadly elliptic to oblong-elliptic, obtuse to acute; all yellow, white or pink, the sepals and the bases of the petals spotted with brown.

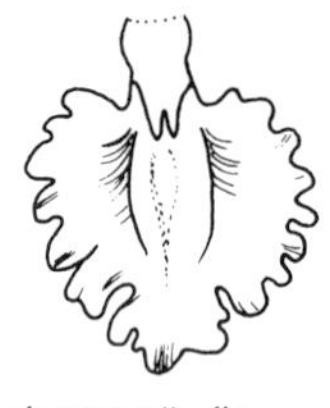

Odontoglossum rossii – lip.

LIP: with a narrow claw bearing a yellow, brown-spotted callus which is notched at the apex with divergent tips; blade circular or nearly so to ovate, white or pink, wavy.

COLUMN: without auricles.

DISTRIBUTION: C America southwards to Nicaragua.

HARDINESS: G1.

FLOWERING: spring.

15. **O. apterum** Llave & Lexarza

SYNONYMS: *O. nebulosum* Lindley; *Cymbiglossum apterum* (Llave & Lexarza) Halbinger; *Lemboglossum apterum* (Llave & Lexarza) Halbinger.

ILLUSTRATION: Lindenia **8**: t. 350 (1892); Orchid Review **93**: 49 (1985).
HABIT: rhizome short.
PSEUDOBULBS: clustered, to 10 cm, each with 2 leaves.
LEAVES: to 20 × 6 cm, elliptic or narrowly elliptic.
RACEME: with 5–7 flowers, loose, arching.
SEPALS AND PETALS: *c.* 3 cm, ovate, sepals acute, petals broader and more rounded at the apex, all white with red or greenish spots in the lower half.
LIP: shortly clawed, the claw bearing 2 yellow keels which project as divergent teeth over the base of the triangular (longer than broad), acute, slightly wavy, white, red- or yellow-spotted blade.
COLUMN: without auricles.
DISTRIBUTION: Mexico.
HARDINESS: G1.
FLOWERING: summer.

Odontoglossum apterum – lip.

16. O. bictoniense (Bateman) Lindley
SYNONYMS: *Cymbiglossum bictoniense* (Lindley) Halbinger; *Lemboglossum bictoniense* (Lindley) Halbinger.
ILLUSTRATIONS: Edwards's Botanical Register **26**: t. 66 (1840); Hamer, Orquideas de El Salvador **2**: 151 & t. 18 (1974); Williams et al., Orchids for everyone, 138 (1980); Bechtel et al., Manual of cultivated orchid species, 235 (1981).
HABIT: rhizome short.
PSEUDOBULBS: clustered, to 10 cm, each with 2 or 3 leaves.
LEAVES: 10–45 × 1.5–5.5 cm, elliptic-oblong to linear.
RACEME: with many flowers, rarely branched, erect.
SEPALS AND PETALS: similar, 1.8–2.7 cm × 5–8 mm, elliptic-lanceolate to lanceolate, acute, pale yellowish green banded or spotted with brown.
LIP: with a short claw bearing a finely downy callus which clasps the column at the base and forms convex lobes at the apex, the blade 2 × 2.5 cm, heart-shaped, tapering to the apex, violet, white or pink, margin crisped.
COLUMN: with 2 flap-like auricles.
DISTRIBUTION: Mexico, Guatemala, El Salvador.
HARDINESS: G1.
FLOWERING: winter–spring.

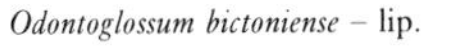
Odontoglossum bictoniense – lip.

17. O. uroskinneri Lindley
SYNONYMS: *Cymbiglossum uroskinneri* (Lindley) Halbinger; *Lemboglossum uroskinneri* (Lindley) Halbinger.
ILLUSTRATIONS: Cogniaux & Goossens, Dictionnaire Iconographique des Orchidées Odontoglossum, t. 3, 3A (1897); Bechtel et al., Manual of cultivated orchid species, 239 (1981).

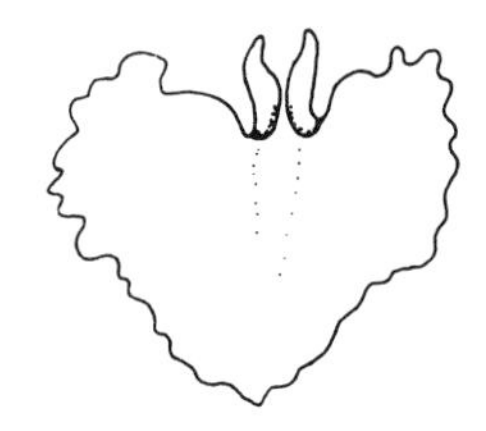
Odontoglossum uroskinneri – lip.

Like *O. bictoniense* except as below.
LEAVES: lanceolate.
SEPALS AND PETALS: 2.5–3 cm × 7–13 mm, dark red to greenish with brown bars and spots.
LIP: blade 2.8 cm or more, pink, veined or spotted with white.
DISTRIBUTION: Guatemala, Honduras.
HARDINESS: G1.
FLOWERING: summer–autumn.

18. O. londesboroughianum Reichenbach
SYNONYM: *Mesoglossum londesboroughianum* (Reichenbach) Halbinger.
ILLUSTRATION: Illustration Horticole **30**: t. 497 (1883).
HABIT: rhizome creeping.
PSEUDOBULBS: distant, 6–8 cm, each bearing 2 (rarely 1) leaves.
LEAVES: to 45 cm, lanceolate.
RACEME: with many flowers, occasionally branched, arching.
SEPALS AND PETALS: similar, 1.5–2 cm, ovate, obtuse, yellow with concentrically arranged red-brown spots towards the base.
LIP: with a short, keeled claw and a blade which is broader than long, heart-shaped at the base, notched at the apex, usually shining yellow, rarely spotted with brown, barred with brown at the base.
COLUMN: without auricles.
DISTRIBUTION: Mexico.
HARDINESS: G1.
FLOWERING: winter.

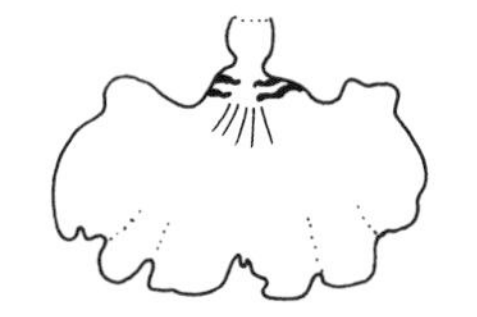
Odontoglossum londesboroughianum – lip.

19. O. pendulum (Llave & Lexarza) Bateman
SYNONYMS: *O. citrosmum* Lindley; *Cuitlauzina pendula* (Llave & Lexarza) Halbinger.
ILLUSTRATIONS: Cogniaux & Goossens, Dictionnaire Iconographique des Orchidées, Odontoglossum, t. 19 (1897); Williams et al., Orchids for everyone, 139 (1980); Bechtel et al., Manual of cultivated orchid species, 238 (1981).
HABIT: rhizome short.
PSEUDOBULBS: clustered, 5–8 cm, each with 2 leaves.
LEAVES: to 30 × 7 cm, oblong-elliptic, obtuse.
RACEME: with many lemon-scented flowers, dense, arching.
SEPALS AND PETALS: similar, *c.* 2.5 cm, ovate, shortly clawed, obtuse, white or pink, sometimes with a few reddish spots.
LIP: with an elongate claw bearing 2 yellow, red-spotted lateral lobes, without distinct keels; blade broader than long, heart-shaped at the base, notched at the apex, purple or pink.
COLUMN: with 2 toothed auricles.
DISTRIBUTION: Mexico.
HARDINESS: G1.
FLOWERING: spring and autumn.

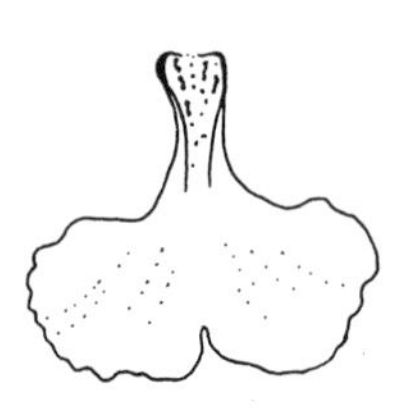
Odontoglossum pendulum – lip.

20. O. pulchellum Lindley

SYNONYM: *Osmoglossum pulchellum* (Lindley) Halbinger.

ILLUSTRATIONS: Botanical Magazine, 4104 (1844); Cogniaux & Goossens, Dictionnaire Iconographique des Orchidées, Odontoglossum, t. 18 (1897); Hamer, Orquideas de El Salvador **2**: 157 & t. 17 (1974); Bechtel et al., Manual of cultivated orchid species, 238 (1981); American Orchid Society Bulletin **55**: 917 (1986).

HABIT: rhizome short.

PSEUDOBULBS: clustered, to 10 cm, each with 2 leaves.

LEAVES: 10–35 × 8–15 mm, oblong-linear, acute.

RACEME: with 6–10 flowers, loose, arching; flowers very fragrant.

SEPALS: 1–2 cm × 6–13 mm, obovate to elliptic, the laterals fused at their bases for *c.* 5 mm, white.

PETALS: 1.3–2 cm × 7–15 mm, obovate, white.

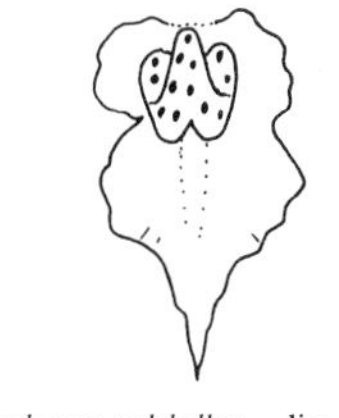

Odontoglossum pulchellum – lip.

LIP: complex, its claw ascending, the rest sharply curved outwards and then under, claw with a large, yellow, red-spotted callus extending as 2 blunt projections over the blade; blade waisted, acute, white, margins crisped.

COLUMN: with 3 coarsely toothed, pinkish auricles.

DISTRIBUTION: Mexico, Guatemala, El Salvador.

HARDINESS: G1.

FLOWERING: winter.

21. O. grande Lindley

SYNONYM: *Rossioglossum grande* (Lindley) Garay & Kennedy.

ILLUSTRATIONS: Schlechter, Die Orchideen, edn 2, t. 9 (1927); Fieldiana **26**: 615 (1953); Die Orchidee **31**: clxviii (1980); Williams et al., Orchids for everyone, 140 (1980).

HABIT: rhizome short.

PSEUDOBULBS: 4–10 cm, each with 2 (rarely 1) leaves.

LEAVES: 10–20 × 3–6.5 cm, elliptic to lanceolate, leathery, acute.

RACEME: with 4–8 flowers, erect.

SEPALS: 5.5–8.5 × 1–2 cm, lanceolate, acuminate and recurved at the apex, yellow, heavily barred with brown.

PETALS: oblanceolate to oblong, rather blunt at the apex, brown in the lower half, yellow above, margins wavy.

LIP: more or less without a claw, with 2 small lateral lobes at the base, bearing a callus with variable horns and projections; blade broader than long, tapered to the base, not or scarcely notched at the apex, white with reddish, brownish or yellowish spots near the base and on the margins.

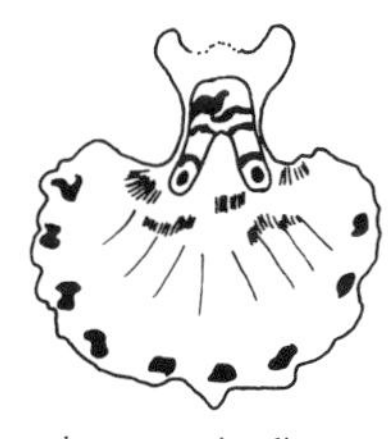

Odontoglossum grande – lip.

COLUMN: very finely downy with 2 oblong or rounded auricles.

DISTRIBUTION: Mexico, Guatemala.

HARDINESS: G1.

FLOWERING: winter.

Rather variable in flower colour. This species and 5 others compose the genus *Rossioglossum* Garay & Kennedy (Orchid

Digest 40: 139, 1976)—see Die Orchidee **31**: clxvi–clxx (1980) and the papers by Halbinger cited at the beginning of this account; the other species are found only in specialist collections.

161. ASPASIA Lindley

A genus of 5 species widely distributed in C and northern S America. Cultivation as for *Odontoglossum* (p. 468).

Literature: Williams, N.H., The taxonomy of Aspasia, *Brittonia* **26**: 333–46 (1974).

HABIT: rhizomatous epiphytes.

PSEUDOBULBS: simple, elliptic to oblong, somewhat compressed, often furrowed, bearing 1–2 leaves.

LEAVES: lanceolate to linear-lanceolate, acuminate, folded when young.

RACEME: lateral with 1–10 flowers.

SEPALS: spreading, pointed, the upper united to the base of the column.

PETALS: usually similar to the sepals but broader, both united to the base of the column.

LIP: united to the column by its lateral margins for about half the length of the column or more, then bent at right angles into a broad, 3-lobed blade, the lateral lobes small, the base bearing 2–4 keels.

COLUMN: cylindric, short, without a foot.

POLLINIA: 2, ellipsoid, stipe slightly longer than pollinia, narrowly club-shaped, viscidium oblong, slightly shorter than the stipe and forming an acute angle with it.

Key to species

1. Free part of column much shorter than the part joined to the lip; apex of lip entire, pointed, not curved upwards **3. lunata**
2. Lip 1.7–2.1 cm wide, its apex curving upwards; anther-cap not beaked **1. epidendroides**
3. Lip 1.3–1.5 cm wide, its apex not curving upwards; anther-cap prominently beaked **2. variegata**

1. A. epidendroides Lindley

ILLUSTRATIONS: Botanical Magazine, 3962 (1842); Fieldiana **26**: 625 (1953); Bechtel et al., Manual of cultivated orchid species, 169 (1981).

PSEUDOBULBS: 5–12 cm, flattened on one side, convex on the other, bearing 2 leaves.

LEAVES: 13–30 cm.

SEPALS AND PETALS: 1.5–2.5 cm, elliptic-obovate, yellow or yellow-green with transverse brown bars.

LIP: 1.7–2.1 cm wide, hairless at the base, white with purple markings towards the apex, yellow towards the base where there is a callus of 2–4 ridges, the apex notched, curving upwards.

COLUMN: lower half united to the lip.

ANTHER CAP: not beaked.

DISTRIBUTION: C America, Colombia.

HARDINESS: G2.

FLOWERING: spring.

2. **A. variegata** Lindley

ILLUSTRATIONS: Edwards's Botanical Register **22**: t. 1907 (1836); Botanical Magazine, 3679 (1838); Pabst & Dungs, Orchidaceae Brasilienses **2**: t. 2082 (1977).

Like *A. epidendroides* except as below.

PSEUDOBULBS AND LEAVES: a little smaller.

LIP: 1.3–1.5 cm wide, downy at the base, its apex not curving upwards.

ANTHER CAP: prominently beaked.

DISTRIBUTION: tropical S America from Colombia to C Brazil.

3. **A. lunata** Lindley

ILLUSTRATIONS: Lindenia **14**: t. 669 (1899); Pabst & Dungs, Orchidaceae Brasilienses **2**: t. 2080 (1977).

PSEUDOBULBS: compressed.

LEAVES: 14–25 cm, thin.

SEPALS: 2.4–4.5 cm, linear-lanceolate, yellow-brown with dark red spots.

PETALS: shorter and broader than the sepals.

LIP: 2–4 × 1.7–4 cm, oblong, conspicuously waisted, the apex entire, with a small point, the margin toothed, white on the margin and on the apical part, violet in the centre.

COLUMN: united to the lip for most of its length, the free part very short.

ANTHER CAP: not beaked.

DISTRIBUTION: C Brazil.

HARDINESS: G2.

FLOWERING: spring.

162. **ADA** Lindley

A genus of 2 very similar species (often considered as 1), both of which are cultivated. Conditions as for *Odontoglossum* (p. 468).

HABIT: epiphytic, with short rhizomes.

PSEUDOBULBS: simple, ovoid, tapered, somewhat compressed, each with 1–3 leaves.

LEAVES: lanceolate to narrowly elliptic, acute, folded when young.

RACEMES: erect, arising from the bases of the pseudobulbs, with 8–13 shortly stalked, rather erect flowers.

SEPALS: clawed, very narrowly elliptic, acuminate, orange.

PETALS: similar to sepals but shorter, sometimes with a dark red line or spots inside; all incurved, spreading only at their tips.

LIP: shorter than the petals, very narrowly elliptic, acuminate, with 1–2 elongate, raised ridges, the lip spreading away from the short column.

COLUMN: without a foot.

POLLINIA: 2, ovoid, stipe slightly longer than the pollinia, expanded above, viscidium shorter than the stipe and forming an acute angle with it.

Key to species

1. Lip orange throughout — **1. aurantiaca**
2. Lip white with an orange ridge — **2. lehmannii**

1. A. aurantiaca Lindley

ILLUSTRATIONS: Botanical Magazine, 5435 (1864); Cogniaux & Goossens, Dictionnaire Iconographique des Orchidées, Ada, t. 1 (1897); Bechtel et al., Manual of cultivated orchid species, 161 (1981); American Orchid Society Bulletin **53**: 284 (1984).

PSEUDOBULBS: 6–10 cm.

LEAVES: 15–25 cm.

SEPALS AND PETALS: sepals 3–3.5 cm, petals 2–2.5 cm.

LIP: orange, bearing 2 elongate, irregularly toothed ridges.

DISTRIBUTION: tropical S America, from Venezuela to Ecuador.

HARDINESS: G1.

FLOWERING: winter.

2. A. lehmannii Rolfe

Like *A. aurantiaca* except as below.

LIP: white, bearing an orange ridge.

DISTRIBUTION: Colombia.

Perhaps not specifically distinct from *A. aurantiaca*.

163. BRASSIA Lindley

A genus of about 50 species from tropical and subtropical America, notable for its spider-like flowers with long, narrow petals and sepals. Six species, all belonging to section *Brassia* are available in Europe and are dealt with below; several other species, belonging to the section *Glumacea* Lindley are grown in the USA (see Teuscher, Memoires du Jardin Botanique de Montreal **55**: 15–21, 1962, reprinted in American Orchid

Society Bulletin **42**: 1089–94, 1973); these have recently been placed in the genus *Ada*.

Plants should be grown in hanging baskets or on rafts or pieces of bark, in epiphytic compost. The atmosphere should be humid, though watering can be reduced during the winter. Propagation is by division or by seed.

Literature: Kooser, R.G. & Kennedy, G.C., The genus Brassia, *Orchid Digest* **43**: 164–72 (1979).

HABIT: epiphytic, with stout, creeping rhizomes which give rise to leaves as well as pseudobulbs.

PSEUDOBULBS: large, simple, ovoid, compressed, each bearing 1–3 leaves.

LEAVES: leathery, variable in shape, folded when young.

RACEMES: arising from the bases of the pseudobulbs, arching, with 6–many flowers.

SEPALS: spreading, linear-lanceolate, acuminate, the laterals often longer than the upper, all 6 or more times longer than broad.

PETALS: similar to the sepals but shorter.

LIP: shorter than the sepals, stalkless at the base of the column, directed away from it at an obtuse angle, simple, bearing a keel which is grooved or composed of 2 ridges.

COLUMN: without a foot.

POLLINIA: 2, ovoid, attached independently to the small, oblong viscidium, stipe absent.

Key to species

1. Blade of lip with green warts or excrescences **6. verrucosa**

2. Blade of lip conspicuously broadened towards the apex so that its maximum breadth is at least twice the breadth at the base

2.A. Blade of lip white or yellowish with purple spots; petals yellowish green with purple-brown spots towards the base **1. maculata**

2.B. Blade of lip yellow with red-brown spots; petals yellow in the upper half, brown in the lower **2. gireoudiana**

3. Ridges of the keel of the lip each ending in a separated tooth **5. caudata**

4. Sepals *c.* 6 cm; petals *c.* 3 cm, yellow with brown spots **3. lanceana**

5. Sepals *c.* 7 cm; petals *c.* 4 cm, yellow, unspotted **4. lawrenceana**

1. **B. maculata** R. Brown

ILLUSTRATIONS: Botanical Magazine, 1691 (1814); Hamer, Orquideas de El Salvador **1**: 81 & t. 5 (1974); Orchid Digest **43**: 167 (1979); American Orchid Society Bulletin **55**: 1093 (1986).

PSEUDOBULBS: 6–8 cm, each with 2 leaves.
LEAVES: to 30 cm, oblong-linear.
RACEME: with 10–15 flowers.
SEPALS: 6–7.5 cm, greenish yellow with a few brown-purple spots near the base.
PETALS: 4–5.5 cm, similar in colour to the sepals.
LIP: white or yellowish, purple-spotted, parallel-sided near the base then widely expanded towards the rounded or obtuse, though mucronate apex; ridges of the lip yellow, grooved, slightly downy.
DISTRIBUTION: C America, West Indies.
HARDINESS: G2.
FLOWERING: autumn and spring.

2. **B. gireoudiana** Reichenbach & Warscewicz
ILLUSTRATIONS: Reichenbach, Xenia Orchidacearum 1: t. 32 (1855); Orchid Digest **43**: 166 (1979); Bechtel et al., Manual of cultivated orchid species, 172 (1981).
Like *B. maculata* except as below.
SEPALS: 10–13 cm, bright yellow with brown spots.
PETALS: *c.* 5 cm, bright yellow in the upper half, brown in the lower.
LIP: yellow with red-brown spots and deep yellow, hairless ridges.
DISTRIBUTION: Costa Rica, Panama.
FLOWERING: summer.

3. **B. lanceana** Lindley
ILLUSTRATIONS: Botanical Magazine, 3577 (1837); Dunsterville & Garay, Venezuelan orchids illustrated **4**: 38–9 (1966); Orchid Digest **43**: 166 (1979).
PSEUDOBULBS: 10–12 cm, each with 2 leaves.
LEAVES: to 30 cm, lanceolate.
RACEMES: loose, with 7–12 flowers.
SEPALS: *c.* 6 cm, yellow with brown spots towards the base.
PETALS: *c.* 3 cm, similar in colour to the sepals.
LIP: oblong, acuminate, pale yellow-green with a few brown spots near the base; ridges yellow, downy.
DISTRIBUTION: tropical S America from Colombia to E Brazil.
HARDINESS: G2.
FLOWERING: autumn.

4. **B. lawrenceana** Lindley
ILLUSTRATIONS: Edwards's Botanical Register **27**: t. 18 (1841); Dunsterville & Garay, Venezuelan orchids illustrated **2**: 48–9 (1961); Orchid Digest 43: 166 (1979).
Like *B. lanceana* except as below.
SEPALS: *c.* 7.5 cm, yellow with red-brown spots.

PETALS: *c.* 4 cm.

LIP: clear yellow or green, unspotted, sometimes a little widened towards the base.

DISTRIBUTION: tropical S America, southwards to N Brazil and NE Peru.

FLOWERING: summer.

5. **B. caudata** (Linnaeus) Lindley

ILLUSTRATIONS: Fieldiana **26**: 625 (1953); Dunsterville & Garay, Venezuelan orchids illustrated **4**: 36–7 (1966); Luer, Native orchids of Florida, t. 76, f. 5,6 (1972); Sheehan & Sheehan, Orchid genera illustrated, 51 (1979); American Orchid Society Bulletin **54**: 983 (1985).

PSEUDOBULBS: 6–15 cm, each with 2 leaves.

LEAVES: to 35 cm, oblong-elliptic, lanceolate or oblong-lanceolate.

RACEME: loose, with 7–12 flowers.

SEPALS: orange-yellow, spotted with red-brown, the upper 3.5–7.5 cm, the laterals 7.5–18 cm.

PETALS: 1.5–3.5 cm, similar in colour to the sepals.

LIP: oblong, broadening slightly towards the apex, yellow or greenish with a few brown blotches towards the base; ridges 2, hairless, each ending in a distinct, separated tooth.

DISTRIBUTION: tropical and subtropical America from USA (Florida) to Bolivia and N Brazil.

HARDINESS: G2.

FLOWERING: autumn & spring.

6. **B. verrucosa** Lindley

SYNONYM: *B. brachiata* Lindley.

ILLUSTRATIONS: Edwards's Botanical Register **33**: t. 29 (1847); Cogniaux & Goossens, Dictionnaire Iconographique des Orchidées, Brassia, t. 1 (1897); Fieldiana **26**: 629 (1953); Orchid Digest **43**: 169 (1979); Bechtel et al., Manual of cultivated orchid species, 173 (1981); American Orchid Society Bulletin **55**: 1095 (1986).

PSEUDOBULBS: 6–10 cm, clustered, each with 2 leaves.

LEAVES: to 45 cm, oblong-elliptic to oblanceolate.

RACEME: with 6–15 flowers, loose.

SEPALS: yellowish green with dark brown spots towards the base, the upper 8–15 cm, the laterals 10–20 cm.

PETALS: 4–8.5 cm, similar in colour to the sepals.

LIP: white, red-spotted, bearing green warts or excrescences, oblong at the base, widened towards the apex; ridges ending in attached teeth.

DISTRIBUTION: tropical America from S Mexico to Venezuela.

HARDINESS: G2.

FLOWERING: spring–summer.

164. MILTONIA Lindley

A genus of about 20 species from C & S America, of which 13 are generally available. Species Nos. 1–5 are sometimes separated off as the genus *Miltoniopsis* Godefroy. Much hybridisation has been done within the genus, and many hybrids are cultivated.

Cultivation as for *Odontoglossum* (p. 468) but the plants require a higher winter temperature.

Literature: Brieger, F.G. & Lückel, E., Der Miltonia-Komplex, eine Neuüberteilung, *Die Orchidee* **34**: 128–133 (1983).

HABIT: epiphytic with rhizomes.

PSEUDOBULBS: clustered or distant, grey-green or yellow-green, simple, usually compressed, each with 1–2 leaves.

LEAVES: oblong, oblong-linear, or oblong-lanceolate, usually acute, folded when young.

RACEMES: arising from the bases of the pseudobulbs, usually erect, with 1–several flowers; bracts sometimes conspicuous.

FLOWERS: often rather flat.

SEPALS AND PETALS: similar.

LIP: spreading away from the column base, usually simple, usually larger than the sepals, with 2 or more longitudinal ridges (or rarely a patch of hairs) at the base.

COLUMN: short, with a narrow wing on each side at the base, without a foot, with auricles at the apex.

POLLINIA: 2, ovoid to ellipsoid, stipe variable, shorter or longer than pollinia, viscidium shorter than the stipe and forming an acute angle with it.

Synopsis of characters

Rhizome. Long, pseudobulbs distant: **1–5**; short, pseudobulbs clustered: **6–13**.

Pseudobulbs. Greyish green, surrounded by the bases of a number of leaves: **1–5**; yellow-green, not surrounded by leaf-bases: **6–13**.

Raceme. With 4 or more flowers: **1–3,8–13**; with fewer than 4 flowers: **4–7**. Axis flattened: 7.

Sepals. White, pink, red or purple: **1–6,8**; green to yellow, often spotted with brown: **7–13**.

Lip. Longer than sepals: **1–8**; shorter than sepals: **9,13**; about as long as sepals: **10–12**. Clearly waisted: **3–5,9–11**; with oblong claw and abruptly widened blade: **7,12**; widening gradually from the base and more or less flat: **1,2,6,8**; widening gradually from the base, the sides curving up to the column: **13**. With 2 backwardly projecting auricles: **1,2**.

Key to species

1. Pseudobulbs and leaves greyish green, the pseudobulbs

surrounded by the bases of a number of leaves or leaf-sheaths; pseudobulbs usually distinct, usually each with 1 leaf *Go to Subkey 1*

2. Pseudobulbs and leaves yellowish green, the pseudobulbs not surrounded by leaf-bases; pseudobulbs usually clustered and usually each with 2 leaves *Go to Subkey 2*

Subkey 1. Pseudobulbs and leaves greyish green, the pseudobulbs surrounded by the bases of a number of leaves or leaf-sheaths; pseudobulbs usually distant, usually each with 1 leaf.

1. Lip with 2 backwardly-pointing projections at the base, the blade not waisted

1.A. Sepals and petals white, acuminate, petals usually with a deep purple blotch at the base; leaves striped with dark green beneath **2. roezlii**

1.B. Sepals and petals entirely pink, rounded to blunt points; leaves not striped beneath **1. vexillaria**

2. Sepals and petals oblong or oblong-obovate, strongly undulate-crisped, purple with white tips **5. warsewiczii**

3. Flower *c.* 7 cm in diameter; lip with 2 small reddish or violet spots near the base, the rest white **3. endressii**

4. Flowers to 5.5 cm in diameter; lip mostly violet, the margins white **4. phalaenopsis**

Subkey 2. Psuedobulbs and leaves yellowish green, pseudobulbs not surrounded by leaf-bases; pseudobulbs usually clustered and usually each with 2 leaves.

1. Sepals and petals of a uniform colour, not spotted or blotched

1.A. Bract much shorter than ovary and flower-stalk together; base of lip with 5–9 ridges **8. regnellii**

1.B. Flowers 6–12 in each raceme; sepals and petals yellow; lip with a patch of hairs at the base **9. flavescens**

1.C. Axis of raceme flattened; lip white with purple lines **7. anceps**

1.D. Axis of raceme not flattened; lip purple or reddish **6. spectabilis**

2. Sides of the lip curving upwards to meet the column **13. candida**

3. Blade of lip narrowly oblong at base, abruptly widened to an apical part which is broader than long **12. cuneata**

4. Sepals and petals little spreading, directed forwards; lip purple or pink with darker spots **10. russelliana**

5. Sepals and petals widely spreading; base of lip purple with a median yellow stripe, apical part white **11. clowesii**

1. **M. vexillaria** (Bentham) Nicholson

SYNONYMS: *Odontoglossum vexillarium* Bentham; *Miltoniopsis vexillaria* (Bentham) Godefroy-Lebeuf.

ILLUSTRATIONS: Cogniaux & Goossens, Dictionnaire Iconographique des Orchidées, Miltonia, t. 1 (1900); Schlechter, Die Orchideen, edn 2, t. 10 (1927); Bechtel et al., Manual of cultivated orchid species, 233 (1981).

PSEUDOBULBS: 5–6 cm, grey-green, distant, compressed, each with 1 leaf, subtended by 3–4 leaf-bases on each side.

LEAVES: 15–25 cm, grey-green, linear-oblong.

RACEME: with 4–6 flowers, loose.

SEPALS AND PETALS: *c.* 4 cm, oblong-elliptic, rounded to blunt points, pink.

LIP: much larger than the sepals, almost circular, with 2 backwardly directed projections and 3 ridges at the base, deeply notched at the apex, very variable in colour.

DISTRIBUTION: Colombia, N Ecuador.

HARDINESS: G2.

FLOWERING: spring–summer.

A very attractive species with the lip varying from pink to white, usually with red lines and a yellow blotch towards the base. Many cultivars exist and are widely grown.

2. **M. roezlii** (Reichenbach) Nicholson

SYNONYMS: *Odontoglossum roezlii* Reichenbach; *Miltoniopsis roezlii* (Reichenbach) Godefroy-Lebeuf.

ILLUSTRATIONS: Lindenia **2**: t. 78 (1886); Cogniaux & Goossens, Dictionnaire Iconographique des Orchidées, Miltonia, t. 6 (1900); Sheehan & Sheehan, Orchid genera illustrated, 121 (1979); Bechtel et al., Manual of cultivated orchid species, 233 (1981); American Orchid Society Bulletin **54**: 145 (1984).

Like *M. vexillaria* except as below.

LEAVES: with dark green longitudinal lines beneath.

SEPALS AND PETALS: acuminate, white, the petals with a purple blotch at the base.

LIP: with 5 ridges at the base.

DISTRIBUTION: Colombia.

FLOWERING: winter.

3. **M. endresii** (Reichenbach) Nicholson

SYNONYM: *M. superba* Schlechter.

ILLUSTRATION: Cogniaux & Goossens, Dictionnaire Iconographique des Orchidées, Miltonia, t. 10 (1900).

PSEUDOBULBS: 5–6 cm, grey-green, subtended by 3–4 leaf-bases on each side, each 1-leaved, compressed, usually distant on the rhizome.

LEAVES: 15–25 cm, grey-green, linear-oblong.

RACEME: with 4–6 flowers, loose.

FLOWERS: *c.* 7 cm in diameter.

SEPALS AND PETALS: 3–3.5 cm, ovate-elliptic, obtuse, white with red or purple stripes at the base.

LIP: larger than the sepals, broadly obovate, slightly waisted near the base, deeply notched and with a small point in the sinus at the apex, white with 2 small red or purple blotches at the base, and with 3 short papillose ridges.

DISTRIBUTION: Costa Rica.

HARDINESS: G2.

FLOWERING: winter.

4. M. phalaenopsis (Linden & Reichenbach) Nicholson

SYNONYMS: *Odontoglossum phalaenopsis* Linden & Reichenbach; *Miltoniopsis phalaenopsis* (Linden & Reichenbach) Garay & Dunsterville.

ILLUSTRATIONS: Lindenia 7: t. 334 (1891): Cogniaux & Goossens, Dictionnaire Iconographique des Orchidées, Miltonia, t. 11 (1900); Bechtel et al., Manual of cultivated orchid species, 233 (1981).

PSEUDOBULBS: 2.5–4 cm, ovoid, grey-green, usually distant on the rhizome, subtended by 3–4 leaf-bases on each side, each with 1 leaf.

LEAVES: 15–22 cm, green, linear.

RACEME: with 2–4 flowers, loose.

FLOWERS: to 5.5 cm in diameter.

SEPALS AND PETALS: ovate-elliptic, white.

LIP: larger than the sepals, broadly oblong, waisted towards the base, notched and with a small point within the sinus at the apex, white, mostly overlaid by red-purple in the form of streaks or lines which join together, covering most of the surface; callus with 3 short teeth at the base.

DISTRIBUTION: Colombia.

HARDINESS: G2.

FLOWERING: autumn.

5. M. warsewiczii Reichenbach

SYNONYMS: *Odontoglossum weltonii* invalid; *Oncidium fuscatum* Reichenbach; *M. vexillaria* var. *warsewiczii* (Reichenbach) Schlechter; *Miltoniopsis warsewiczii* (Reichenbach) Garay & Dunsterville.

ILLUSTRATIONS: Botanical Magazine, 5843 (1870); Lindenia 8: t. 384 (1892); Cogniaux & Goossens, Dictionnaire Iconographique des Orchidées, Miltonia, t. 8 (1900); Bechtel et al., Manual of cultivated orchid species, 253 (1981).

Like *M. phalaenopsis* except as below.

SEPALS AND PETALS: oblong or oblong-obovate, conspicuously undulate-crisped, mostly purple, the tips white.

LIP: mostly purple, the margins white, usually with a yellow blotch or variegation in the centre.

DISTRIBUTION: Peru, Ecuador, Colombia.

FLOWERING: spring.

6. M. spectabilis Lindley

ILLUSTRATIONS: Edwards's Botanical Register **23**: t. 1992 (1837); Cogniaux & Goossens, Dictionnaire Iconographique des Orchidées, Miltonia, t. 2, 2A (1900); Bechtel et al., Manual of cultivated orchid species, 232 (1981).

PSEUDOBULBS: to 7 cm, clustered, ovoid, compressed, each with 2 leaves.

LEAVES: to 15 cm, narrowly linear-oblong, green.

RACEME: with 1 or 2 flowers and a terete axis; bracts exceeding the flower-stalk and ovary together, sheathing.

SEPALS AND PETALS: to 4.5 cm, spreading, oblong, obtuse, uniformly white, red or purple.

LIP: broadly obovate, widening gradually from the base, purple or reddish, with 3 yellow ridges at the base.

DISTRIBUTION: Venezuela, E Brazil.

HARDINESS: G2.

FLOWERING: summer.

7. M. anceps Lindley

SYNONYM: *Odontoglossum anceps* (Lindley) Klotzsch.

ILLUSTRATIONS: Botanical Magazine, 5572 (1866); Pabst & Dungs, Orchidaceae Brasilienses **2**: f. 2050 (1977).

PSEUDOBULBS: 4.5–5.5 cm, each with 2 leaves, yellow-green, oblong, strongly compressed, clustered.

RACEME: with 1 flower, a flattened axis and the bract sheathing, exceeding the ovary and flower-stalk together.

SEPALS AND PETALS: 3.5–3.7 cm, spreading, oblong-lanceolate, obtuse, uniformly greenish yellow.

LIP: oblong, slightly waisted, exceeding the sepals, white with purple lines, and with 3 ridges which are yellow, each with a single red stripe at the base.

DISTRIBUTION: SE Brazil.

HARDINESS: G2.

FLOWERING: spring.

8. M. regnellii Reichenbach

SYNONYM: *Oncidium regnellii* (Reichenbach) Reichenbach.

ILLUSTRATIONS: Botanical Magazine, 5436 (1864); Cogniaux & Goossens, Dictionnaire Iconographique des Orchidées, Miltonia, t. 7, 7A, 7B (1900); Hoehne, Iconografia de Orchidaceas do Brasil, t. 267 (1949); Bechtel et al., Manual of cultivated orchid species, 232 (1981).

PSEUDOBULBS: 5–8 cm, each with 2 leaves, yellow-green, clustered, ovoid.

LEAVES: to 30 cm, linear-lanceolate, acute.

RACEME: with 4–7 flowers, erect, loose.

SEPALS AND PETALS: *c.* 3.5 cm, ovate-oblong, acute, white or yellow.

LIP: broadly obovate, larger than the sepals, obscurely 3-lobed,

blunt at the apex, white or violet with white margins, with 7–9 low ridges at the base.
DISTRIBUTION: S Brazil.
HARDINESS: G2.
FLOWERING: summer.

9. M. flavescens (Lindley) Lindley
SYNONYM: *Cyrtochilum flavescens* Lindley.
ILLUSTRATIONS: Gartenflora **39**: t. 1328 (1890); Hoehne, Iconografia de Orchidaceas do Brasil, t. 263 (1949); Pabst & Dungs, Orchidaceae Brasilienses **2**: f. 2051 (1977).
PSEUDOBULBS: 6–8 cm, each with 2 leaves, clustered, yellow-green, oblong.
LEAVES: to 30 cm, linear-oblong, acute.
RACEME: with 6–12 flowers, erect, loose, bracts conspicuous, sheathing, longer than the ovary and flower-stalk together.
SEPALS AND PETALS: *c.* 3.5 cm, yellow, linear-oblong or linear-lanceolate, acute.
LIP: oblong, waisted, shorter than the sepals, pale yellow with reddish lines.
DISTRIBUTION: S & E Brazil, Paraguay, N Argentina.
HARDINESS: G2.
FLOWERING: summer–autumn.

10. M. russelliana Lindley.
ILLUSTRATIONS: Edwards's Botanical Register **22**: t. 1830 (1836); Pabst & Dungs, Orchidaceae Brasilienses **2**: f. 2058 (1977).
PSEUDOBULBS: clustered, yellow-green.
LEAVES: linear-oblong.
RACEMES: several-flowered, erect, bracts inconspicuous.
SEPALS AND PETALS: 2.5–3 cm, oblong, acute, not spreading but directed forwards, green or greenish yellow heavily marbled or spotted with brown.
LIP: about as long as sepals, oblong, somewhat waisted towards the apex, purple or pink with darker spots, with many ridges at the base.
DISTRIBUTION: S Brazil.
HARDINESS: G2.
FLOWERING: winter–spring.

11. M. clowesii Lindley
SYNONYMS: *Odontoglossum clowesii* (Lindley) Lindley; *Brassia clowesii* (Lindley) Lindley.
ILLUSTRATIONS: Botanical Magazine, 4109 (1844); Cogniaux & Goossens, Dictionnaire Iconographique des Orchidées, Miltonia, t. 4 (1900); Bechtel et al., Manual of cultivated orchid species, 231 (1981).

PSEUDOBULBS: 5–8 cm, clustered, yellow-green, elongate-ovoid, each with 2 leaves.

LEAVES: to 30 cm, linear-lanceolate, acute.

RACEME: with 3–7 flowers, erect, loose, bracts inconspicuous.

SEPALS AND PETALS: 3–3.5 cm, spreading, oblong, acute, greenish yellow with brown spots.

LIP: about as long as sepals, waisted, the anterior part ovate, acute, violet in the lower part, white towards the apex, with several short ridges at the base.

DISTRIBUTION: E Brazil.

HARDINESS: G2.

FLOWERING: autumn.

12. M. **cuneata** Lindley

SYNONYM: *M. speciosa* Klotzsch.

ILLUSTRATIONS: Edwards's Botanical Register 31: t. 8 (1845); Cogniaux & Goossens, Dictionnaire Iconographique des Orchidées, Miltonia, t. 9 (1900); Pabst & Dungs, Orchidaceae Brasilienses 2: f. 2055 (1977); Bechtel et al., Manual of cultivated orchid species, 232 (1981).

PSEUDOBULBS: 5–8 cm, oblong-ovoid, yellow-green, clustered, each with 2 leaves.

LEAVES: to 30 cm, oblong-lanceolate.

RACEME: with 5–8 flowers, erect.

SEPALS AND PETALS: 3–3.5 cm, oblong, acute, the margins somewhat wavy, yellow spotted with brown.

LIP: about as long as the sepals, narrowly oblong at the base, abruptly widened to an obovate, heart-shaped apical part which is broader than long, slightly notched at the apex, white, with 2 ridges at the base.

DISTRIBUTION: E Brazil.

HARDINESS: G2.

FLOWERING: spring.

13. M. **candida** Lindley

SYNONYM: *Oncidium candidum* (Lindley) Reichenbach.

ILLUSTRATIONS: Hoehne, Iconografia de Orchidaceas do Brasil, t. 262 (1949); Pabst & Dungs, Orchidaceae Brasilienses 2: f. 2053 (1977); Bechtel et al., Manual of cultivated orchid species, 231 (1981).

PSEUDOBULBS: 5–8 cm, yellow-green, clustered, oblong-ovoid, each with 2 leaves.

LEAVES: to 30 cm, linear-lanceolate, acute.

RACEME: with 3–7 flowers, erect, loose.

SEPALS AND PETALS: 4–4.5 cm, oblong, acute, greenish yellow with many brown spots.

LIP: shorter than the sepals, white or rarely pink, sometimes purplish in the centre, its sides rolled upwards to meet the column, somewhat wavy, with 3 ridges at the base.

DISTRIBUTION: E Brazil.

HARDINESS: G2.
FLOWERING: autumn.

165. ONCIDIUM Swartz

A large genus (*c.* 450 species?) from subtropical and tropical America, of which over 100 species have been in cultivation. It is sometimes difficult to distinguish from both *Miltonia* and *Odontoglossum*, and it has been suggested that it should be broken up into a number of genera. Garay & Stacy (reference below), however, prefer to maintain it as a single genus. They divide the genus into many sections, some of which are interpolated into the present account with very short descriptions.

The species are very variable as to their requirements in cultivation. They are usually grown in epiphytic compost in containers or on rafts. Some have a definite resting period while others do not, and treatment should vary accordingly. Propagation is by seed, division or the removal of small bulbs from mature plants.

Literature: Kränzlin, F., *Das Pflanzenreich* **80**: 25–290 (1922); Garay, L.A., A reappraisal of the genus Oncidium Sw., *Taxon* **19**: 443–67 (1970); Garay, L.A. & Stacy, J.E., Synopsis of the genus Oncidium, *Bradea* **1**(40): 393–424 (1974).

HABIT: rhizomatous epiphytes.
PSEUDOBULBS: usually present, simple, usually bearing 1–3 leaves and subtended by a number of leaves or sheathing leaf-bases.
LEAVES: variable, fleshy, leathery, sometimes terete, folded when young.
FLOWERS: usually numerous in racemes or panicles.
SEPALS AND PETALS: usually spreading, usually clawed, similar in size or the petals larger, the lateral sepals free or variably united at the base.
LIP: entire or 3-lobed, often waisted, very variable in shape and size, firmly united with the base of the column from which it diverges, with a complex tuberculate or ridged callus.
COLUMN: with a fleshy plate below the stigma but without a foot, and with distinct, variably shaped auricles or wings on either side.
ROSTELLUM: short or beaked.
POLLINIA: 2, ovoid to spherical, stipe linear, longer than pollinia, viscidium small, making an acute angle with the stipe.

Key to species

1. Sepals, petals and lip fleshy *Go to Subkey 1*
2. Pseudobulbs absent or rudimentary and less than 4 cm, larger only when the plant is very large (panicle *c.* 2 m) *Go to Subkey 2*

3. Upper sepal and petals erect, antenna-like
3.A. Axis of raceme flattened above **13. papilio**
3.B. Axis of raceme not flattened **14. kramerianum**
4. Column downy, with auricles in the form of forwardly-pointing, fleshy arms at the sides of the stigma **15. pubes**
5. Sepals, petals and lip spread in 1 plane, the lip little more conspicuous than the rest
5.A. Lip white with red lines at the base, yellow towards the apex, lateral lobes triangular, pointing forwards **16. maculatum**
5.B. Lip mostly pink, lateral lobes oblong or almost square, directed outwards **17. hastatum**
6. Rostellum elongate, proboscis-like
6.A. Lip, sepals and petals mostly pink, though with yellow ridges on the lip **18. ornithorrhynchum**
6.B. Lip, sepals and petals yellow, though with white ridges on the lip **19. cheirophorum**
7. Callus of the lip with 2–4 parallel tubercles or ridges **20. concolor**
8. Lateral sepals longer than lip **21. longipes**
9. Lateral sepals united to some extent at the base *Go to Subkey* 3
10. Lateral sepals entirely free *Go to Subkey* 4

Subkey 1. *Sepals, petals and lip fleshy.*
1. Lip oblong, tapering slightly to the apex **3. superbiens**
2. Sepals and petals more than 4 cm; petals yellow **1. macranthum**
3. Sepals and petals less than 2 cm; petals reddish brown **2. microchilum**

Subkey 2. *Sepals, petals and lip not fleshy; pseudobulbs absent or rudimentary and less than* 4 *cm, larger only when the plant is very large (panicle c.* 2 m).
1. Sheaths enveloping the pseudobulbs all leaf-bearing **4. onustum**
2. Leaves terete **5. cebolleta**
3. Leaves fleshy; callus of lip swollen, downy
3.A. Flowers to 1.8 cm in diameter; leaves grey-green, 7.5–15 cm **11. harrisonianum**
3.B. Flowers more than 2 cm in diameter; leaves green, 20–30 cm **12. divaricatum**
4. Pseudobulbs absent, or at most 1.5 cm
4.A. Sepals and petals unspotted, yellow or greenish yellow bordered with yellow or crimson; flowers in racemes **10. bicallosum**
4.B. Petals to 9 mm; lip to 11 mm wide at its widest **8. nanum**

4.C. Petals more than 1.2 cm; lip more than 1.5 cm wide **9. cavendishianum**

5. Panicle *c.* 2 m, with many branches and many flowers; sepals and petals to 2 cm **6. altissimum**

6. Panicle much smaller, with relatively few branches and 9–20 flowers; sepals and petals 2.5–3 cm **7. splendidum**

Subkey 3. *Sepals, petals and lip not fleshy; pseudobulbs relatively large and conspicuous for the size of the plant; lateral sepals united to some extent at the base.*

1. Petals large, much more conspicuous than sepals

1.A. Wings of column sharply and conspicuously toothed **22. crispum**

1.B. Sepals, petals and lip chestnut brown with yellow marbling; flowers usually in racemes **23. forbesii**

1.C. Lip 3 × 4.5 cm; lateral sepals united for half their length **24. gardneri**

1.D. Lip *c.* 4 x 5 cm; lateral sepals united for a third of their length **25. marshallianum**

2. Blade of lip 2.5–4 × 3.5–5.5 cm, with 3 notches towards the apex, producing 4 small lobes **26. varicosum**

3. Blade of lip 1.5–2 × 1.5–1.8 cm, with 1 deep notch at the apex **27. flexuosum**

Subkey 4. *Sepals, petals and lip not fleshy; pseudobulbs relatively large and conspicuous for the size of plant; lateral sepals free to the base.*

1. Petals wider and more conspicuous than the sepals **28. excavatum**

2. Bracts small, scale-like, inconspicuous

2.A. Lip white; sepals and petals to 1.7 cm **31. leucochilum**

2.B. Pseudobulbs each with 1 leaf **29. ampliatum**

2.C. Pseudobulbs each with 2 leaves **30. tigrinum**

3. Leaves 35–60 cm; lip wider across the central lobe than across the lateral lobes **32. sphacelatum**

4. Leaves 20–30 cm; lip wider across the lateral lobes than across the central lobe **33. wentworthianum**

Section **Cyrtochilum** (Humboldt, Bonpland & Kunth) Lindley. Flowers conspicuous, sepals, petals and lip all fleshy; column with fleshy auricles.

1. **O. macranthum** Lindley

SYNONYM: *O. hastiferum* Reichenbach & Warscewicz.

ILLUSTRATIONS: Botanical Magazine, 5743 (1868); The Garden **24**: t. 412 (1883); Cogniaux & Goossens, Dictionnaire Iconographique des Orchidées, Oncidium, t. 14 (1898); Bechtel et al., Manual of cultivated orchid species, 246 (1981).

PSEUDOBULBS: 7–15 cm, oblong-conical, each with 2 leaves.
LEAVES: 25–50 × 2.5–5 cm, oblanceolate to oblong, acute.
PANICLE: to 3 m, each branch with 2–5 flowers.
SEPALS: 2.5–3.5 cm, clawed, the laterals somewhat longer than the upper, dull yellow-brown.
PETALS: circular to ovate, very shortly clawed, crisped-undulate, a little shorter than the sepals, golden yellow.
LIP: smaller than the petals, broadly triangular, abruptly narrowed to the recurved, acuminate apex, red or brownish purple towards the base, yellowish with red lines towards the apex, with a large white callus with 6–7 erect teeth.
DISTRIBUTION: Colombia, Ecuador, Peru.
HARDINESS: G1.
FLOWERING: summer.

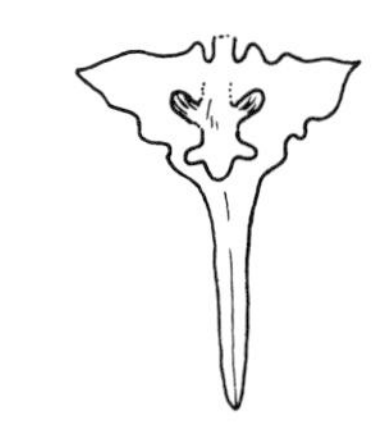

Oncidium macranthum – lip.

2. O. microchilum Lindley

ILLUSTRATIONS: Fieldiana **26**: 655 (1953); Bechtel et al., Manual of cultivated orchid species, 246 (1981).
PSEUDOBULBS: 2.5–3.5 cm, spherical, compressed, each with 1 leaf.
LEAVES: 12–25 × 3–6.5 cm, fleshy, elliptic, margins irregularly scalloped.
PANICLE: to 1.5 m with many flowers.
SEPALS: 1.2–1.4 cm, shortly clawed, circular to elliptic, reddish brown, the upper very concave.
PETALS: almost clawless, oblong-elliptic, otherwise similar to sepals.
LIP: complex, very broadly triangular, much shorter than the petals, white spotted with red; callus dark red, covering most of the centre, with 5 or more blunt tubercles.
COLUMN: with 2 slender, sickle-shaped wings.
DISTRIBUTION: Mexico, Guatemala.
HARDINESS: G2.
FLOWERING: summer.

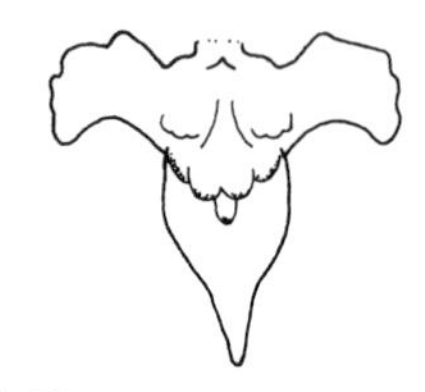

Oncidium microchilum – lip.

3. O. superbiens Riechenbach

ILLUSTRATIONS: Botanical Magazine, 5980 (1872); Cogniaux & Goossens, Dictionnaire Iconographique des Orchidées, Oncidium, t. 31 (1904); Bechtel et al., Manual of cultivated orchid species, 249 (1981).
PSEUDOBULBS: 6–10 cm, clustered, each usually with 2 leaves.
LEAVES: to 70 cm, linear-oblanceolate, acute.
PANICLE: to 6 m, twining, with many flowers.
SEPALS: clawed, ovate to almost circular, the upper 1.6–2 cm, the laterals somewhat longer, all greenish brown or brown, sometimes with yellow margins.
PETALS: smaller than the sepals, oblong-ovate to almost circular, shortly clawed, margins crisped-undulate, yellow or white with reddish or brown bars in the lower half.

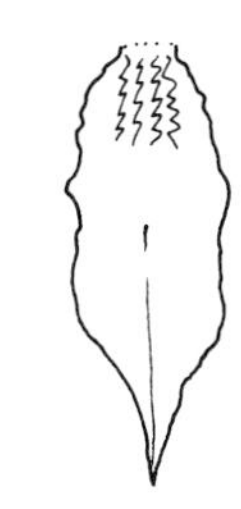

Oncidium superbiens – lip.

LIP: shorter than the petals, oblong, the apex recurved, purple, with a complex, toothed callus.
COLUMN: with sickle-shaped wings.
DISTRIBUTION: Peru, Colombia, Ecuador & Venezuela.
HARDINESS: G1.
FLOWERING: winter.

Section **Onusta** Garay & Stacy. Pseudobulbs well developed, bearing leaves, or leafless with a terminal appendage; sheaths surrounding the pseudobulbs all leaf-bearing; leaves leathery or fleshy, finely toothed towards the apex.

4. **O. onustum** Lindley
ILLUSTRATION: Lindenia **11**: t. 498 (1895).
PSEUDOBULBS: 2–4 cm, subtended by several leaves with blades, green mottled with brown, bearing 1 or 2 leaves or a terminal appendage.
LEAVES: 5–12.5 × 1–1.8 cm, oblong or linear-oblong.
RACEME: (rarely branched) to 40 cm, often 1-sided, with many flowers.
SEPALS: 7–8 mm, ovate or ovate-elliptic, yellow, the upper hooded.
PETALS: *c.* twice as large as the sepals, almost circular, wavy, yellow.

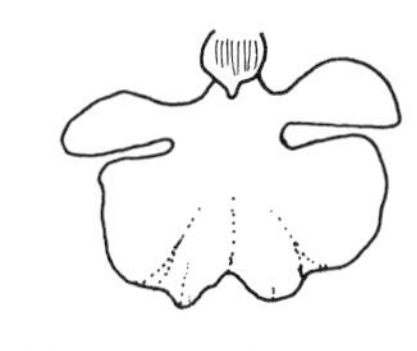

Oncidium onustum – lip.

LIP: larger than the petals, 3-lobed, the lateral lobes small, the central lobe almost circular, bearing a 3-lobed callus.
COLUMN: with large, crescent-shaped auricles.
DISTRIBUTION: Panama, Colombia, Ecuador & Peru.
HARDINESS: G2.
FLOWERING: spring.

Section **Cebolleta** Lindley. Pseudobulbs small, each bearing 1 leaf; leaves terete in section.

5. **O. cebolleta** (Jacquin) Swartz
ILLUSTRATIONS: Botanical Magazine, 3568 (1837); Dunsterville & Garay, Venezuelan orchids illustrated **2**: 246–7 (1961); Pabst & Dungs, Orchidaceae Brasilienses **2**: t. 1945 (1977); Bechtel et al., Manual of cultivated orchid species, 242 (1981).
PSEUDOBULBS: 1.5–2 cm, conical to almost spherical, each with 1 leaf.
LEAVES: 7–40 cm, fleshy, terete, slightly grooved, erect, often tinged or spotted with purple.
PANICLE: to 1.2 m with many flowers, its stalk spotted with purple.
SEPALS: 6–10 mm, obovate, greenish yellow with red-brown spots.
PETALS: similar, margins wavy.

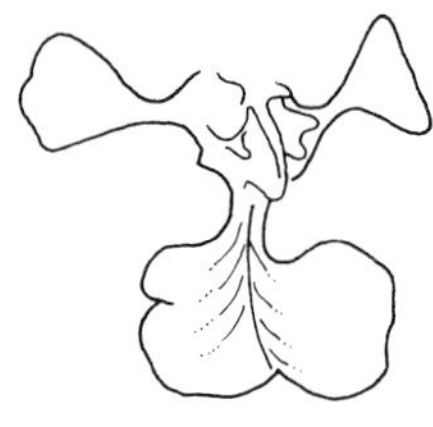
Oncidium cebolleta – lip.

LIP: 3-lobed, with lateral lobes rather large, oblong or obovate, the central lobe larger than the sepals, shortly clawed, kidney-shaped, deeply notched at the apex, yellow, callus consisting of a sharp projecting ridge surrounded by tubercles.

COLUMN: with conspicuous, oblong, sometimes 2-lobed auricles.

DISTRIBUTION: tropical America, from Mexico and the West Indies to N Argentina.

HARDINESS: G2.

FLOWERING: spring.

Section **Pluritubercülata** Lindley. Pseudobulbs absent or very small for the size of the plant; leaves leathery; callus of lip with many tubercles.

6. **O. altissimum** (Jacquin) Swartz

ILLUSTRATION: Edwards's Botanical Register **19**: t. 1651 (1834).

PSEUDOBULBS: 5–7 cm, ovoid, compressed, each with 2 leaves.

LEAVES: 25–30 cm, oblong or oblong-lanceolate, acute.

PANICLE: to 2 m or more, with many flowers.

SEPALS AND PETALS: *c.* 2 cm, narrowly lanceolate or oblanceolate, somewhat wavy, greenish yellow with red or red-brown spots.

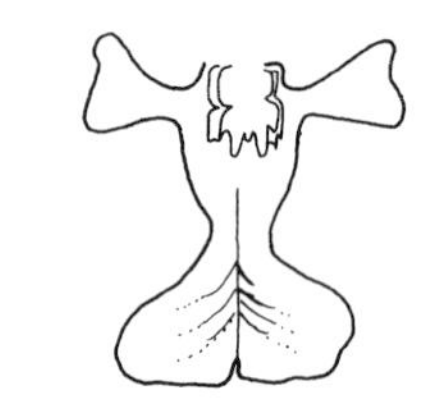
Oncidium altissimum – lip.

LIP: 3-lobed with small, oblong lateral lobes and a clawed, kidney-shaped central lobe which is slightly notched at the apex and bears a complex callus with about 10 tubercles, yellowish with a reddish band across the narrowest part.

COLUMN: with small, triangular, entire auricles.

DISTRIBUTION: West Indies.

HARDINESS: G2.

FLOWERING: spring–summer.

For a discussion of the use of the name *O. altissimum* see Garay & Stacy, Bradea 1(40): 395–6 (1974).

7. **O. splendidum** Duchartre

ILLUSTRATIONS: Cogniaux & Goossens, Dictionnaire Iconographique des Orchidées, Oncidium, t. 7 (1897); Bechtel et al., Manual of cultivated orchid species, 248 (1981).

PSEUDOBULBS: 4–5 cm, clustered, almost spherical, each bearing 1 leaf.

LEAVES: to 100 × 3.5 cm, oblong-elliptic, usually purple-tinged.

PANICLE: to 1.5 m with many flowers.

SEPALS: *c.* 2.5 cm, shortly clawed, linear-elliptic to oblanceolate, wavy, yellow-green with broad, red-brown bands.

PETALS: similar to sepals but to 3 cm.

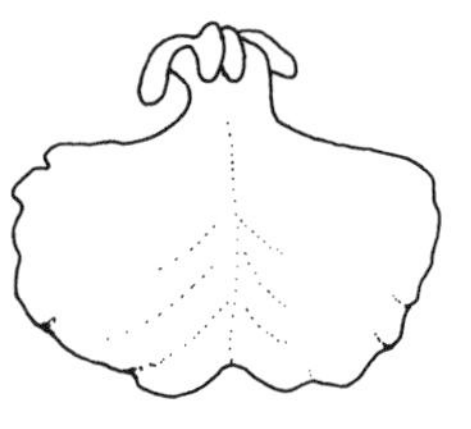
Oncidium splendidum – lip.

LIP: much larger than the sepals, 3-lobed, the lateral lobes very small, the central up to 4 cm wide, almost stalkless, kidney-shaped, deeply notched, somewhat wavy, yellow, with a callus composed of 3 conspicuous, bright yellow ridges.

COLUMN: with small wings.

DISTRIBUTION: Guatemala.

HARDINESS: G2.

FLOWERING: winter–spring.

8. O. nanum Lindley

ILLUSTRATIONS: Dunsterville & Garay, Venezuelan orchids illustrated **1**: 272–3 (1959); Pabst & Dungs, Orchidaceae Brasilienses **2**: t. 1948 (1977); Bechtel et al., Manual of cultivated orchid species, 246 (1981).

PSEUDOBULBS: to 7 mm, cylindric, each with 1 leaf.

LEAVES: 7–24 × 2–7 cm, elliptic or oblong, acute.

PANICLE: to 25 cm with many flowers.

SEPALS: 7–9 mm, obovate, yellow spotted with orange, purple or brown.

PETALS: similar but a little smaller.

Oncidium nanum – lip.

LIP: slightly longer than the sepals, 3-lobed, the lateral lobes small, the central lobe transversely oblong to kidney-shaped, notched at the apex and with a complex, tuberculate callus at the base.

COLUMN: with oblique, slightly toothed auricles.

DISTRIBUTION: tropical S America from Colombia to Peru and C Brazil.

HARDINESS: G2.

FLOWERING: spring.

9. O. cavendishianum Bateman

ILLUSTRATIONS: Cogniaux & Goossens, Dictionnaire Iconographique des Orchidées, Oncidium, t.11 (1898); Bechtel et al., Manual of cultivated orchid species, 242 (1981).

PSEUDOBULBS: absent or very small.

LEAVES: 15–45 × 5–13 cm, elliptic to broadly lanceolate, acute or subobtuse.

PANICLE: 60–150 cm, with many flowers.

FLOWERS: very fragrant.

SEPALS: 1.2–1.7 cm, obovate, obtuse, margins undulate-crisped, yellow spotted with red.

PETALS: similar to sepals, but clawed.

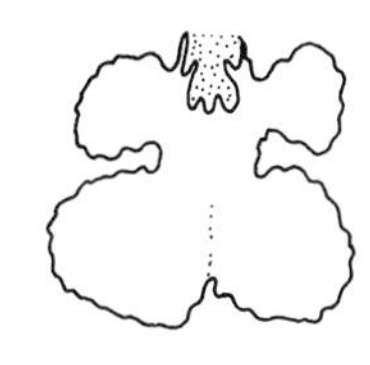

Oncidium cavendishianum – lip.

LIP: larger than the sepals, deeply 3-lobed, the lateral lobes conspicuous, obovate, the central lobe large, transversely oblong to kidney-shaped, notched at the apex, yellow with red spots at the base; callus consisting of 5 tubercles.

COLUMN: with sickle-shaped, red-spotted auricles.

DISTRIBUTION: S Mexico, Guatemala, Honduras.

HARDINESS: G2.

FLOWERING: spring.

10. O. bicallosum Lindley

ILLUSTRATIONS: Edwards's Botanical Register 29: t. 12 (1843); Botanical Magazine, 4148 (1845); Bechtel et al., Manual of cultivated orchid species, 241 (1981).

PSEUDOBULBS: very small or absent.

LEAVES: 15–35 × 4–8.5 cm, oblong-elliptic, obtuse.

RACEME: (rarely branched) 20–65 cm with many flowers.

FLOWERS: fragrant.

SEPALS: 1.5–2 cm, obovate to spathulate, obtuse, greenish yellow to deep yellow, rarely margined with red.

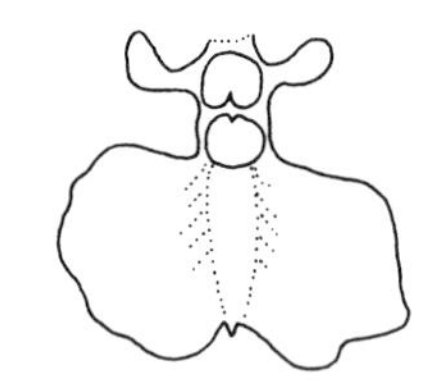

Oncidium bicallosum – lip.

PETALS: similar but with crisped margins.

LIP: larger than the sepals, deeply 3-lobed, the lateral lobes rather small, obovate, the central lobe transversely oblong to kidney-shaped, yellow, with a callus of 2 swellings and a variable number of tubercles, which are white with red spots.

COLUMN: wings fleshy, sickle-shaped.

DISTRIBUTION: S Mexico, Guatemala, El Salvador.

HARDINESS: G2.

FLOWERING: autumn.

Section **Pulvinata** Lindley. Pseudobulbs small, subtended by sheaths without blades; callus of the lip downy.

11. O. harrisonianum Lindley

ILLUSTRATIONS: Edwards's Botanical Register **19**: t. 1569 (1833); Pabst & Dungs, Orchidaceae Brasilienses **2**: t. 1955 (1977); Bechtel et al., Manual of cultivated orchid species, 244 (1981).

PSEUDOBULBS: to 2.5 cm, almost spherical, each bearing 1 leaf.

LEAVES: 7.5–15 cm, greyish green.

PANICLE: with 15–20 flowers, little branched, erect, loose, exceeding the leaves.

SEPALS AND PETALS: 6–9 mm, yellow with red longitudinal lines.

Oncidium harrisonianum – lip.

LIP: longer than the sepals, 3-lobed, the lateral lobes small, the central lobe clawed, transversely oblong or kidney-shaped, notched at the apex, yellow with a callus formed of 5 concave lobes.

COLUMN: wings sickle-shaped.

DISTRIBUTION: E Brazil.

HARDINESS: G2.

FLOWERING: autumn–winter.

12. O. divaricatum Lindley

ILLUSTRATIONS: Pabst & Dungs, Orchidaceae Brasilienses **2**: t. 1951 (1977); Bechtel et al., Manual of cultivated orchid species, 243 (1981).

PSEUDOBULBS: 2.5–4 cm, spherical, each bearing 1 leaf.

LEAVES: 20–30 cm, oblong, obtuse.

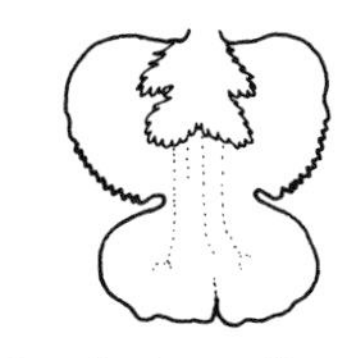

Oncidium divaricatum – lip.

PANICLE: to 1.2 m, very branched, with many flowers.

SEPALS AND PETALS: 1–1.3 cm, clawed, obovate, brownish with yellow apices.

LIP: larger than the sepals, 3-lobed, with large lateral lobes and a rather small central lobe which is notched at the apex, all pale yellow, reddish at the base, with a callus consisting of 4 swellings.

COLUMN: wings sickle-shaped, deflexed.

DISTRIBUTION: E Brazil.

HARDINESS: G2.

FLOWERING: autumn.

Section **Glanduligera** Lindley. Pseudobulbs conspicuous; upper sepal and petals linear, erect, antenna-like, lateral sepals curving downwards, much broader than the upper.

13. **O. papilio** Lindley

ILLUSTRATIONS: Edwards's Botanical Register **11**: t. 90 (1825); Botanical Magazine, 3733 (1840); Cogniaux & Goossens, Dictionnaire Iconographique des Orchidées, Oncidium, t. 3 (1896); Dunsterville & Garay, Venezuelan orchids illustrated **2**: 254–5 (1961); Bechtel et al., Manual of cultivated orchid species, 247 (1981).

PSEUDOBULBS: 3–5 cm, clustered, spherical, each bearing 1 leaf.

LEAVES: 12–25 × 4–7 cm, ovate to elliptic, acute, green mottled with brown.

RACEME OR PANICLE: to 1.2 m, erect, few-flowered, with only 1 or 2 flowers open at any one time, its axis conspicuously flattened towards the apex.

UPPER SEPAL AND PETALS: 8–13 cm, erect, linear, slightly expanded towards the apex, acute, with slightly wavy margins, purplish brown slightly mottled with greenish yellow.

LATERAL SEPALS: oblong-lanceolate, crisped-undulate, brown with yellow transverse bands.

Oncidium papilio – lip.

LIP: shorter than the lateral sepals, 3-lobed, the lateral lobes rather small, the central lobe clawed, almost circular to kidney-shaped, notched at the apex, mottled yellow and brown with a yellow centre, and with an obscurely 3-lobed callus.

COLUMN: wings oblong, lacerate, ending at the top in capitate, fleshy teeth.

DISTRIBUTION: tropical S America from Trinidad to Peru.

HARDINESS: G2.

FLOWERING: throughout the year.

14. **O. kramerianum** Reichenbach

ILLUSTRATIONS: Lindenia **6**: t. 246 (1890); Cogniaux &

Goossens, Dictionnaire Iconographique des Orchidées, Oncidium, t. 26 (1900); Bechtel et al., Manual of cultivated orchid species, 245 (1981).

Like *O. papilio*, except as below.

HABIT: smaller.

PANICLE: axis not flattened above.

LIP: central lobe rather shortly clawed.

DISTRIBUTION: tropical America from Costa Rica to Ecuador.

HARDINESS: G2.

FLOWERING: autumn.

Oncidium kramerianum – lip.

Section **Waluewa** (Regel) Schlechter. Pseudobulbs conspicuous; lateral sepals united for most of their length; column downy with curved, forwardly projecting fleshy arms.

15. **O. pubes** Lindley

ILLUSTRATIONS: Edwards's Botanical Register **12**: t. 1007 (1826); Pabst & Dungs, Orchidaceae Brasilienses **2**: t. 1967 (1977); Bechtel et al., Manual of cultivated orchid species, 247 (1981).

PSEUDOBULBS: 5–7 cm, cylindric, narrowed above, each bearing 1 leaf.

LEAVES: 12–16 cm, lanceolate to elliptic, acute.

PANICLE: 30–50 cm, erect, with many flowers.

SEPALS AND PETALS: 1–1.3 cm, obovate-oblong, brown with yellow transverse bars, the lateral sepals united for the whole of their length.

LIP: slightly shorter than the sepals, 3-lobed with small, spreading lateral lobes and a large, almost circular central lobe, yellow spotted with red-brown and bearing a large, tuberculate, downy callus.

DISTRIBUTION: Brazil.

HARDINESS: G2.

FLOWERING: summer.

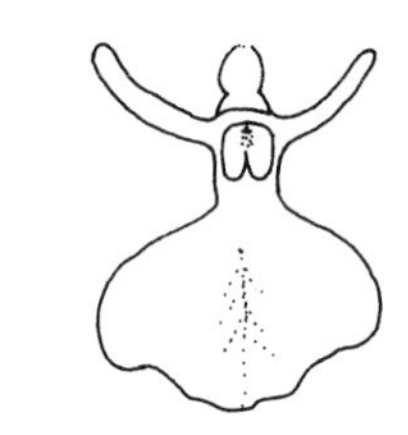

Oncidium pubes – lip.

Section **Stellata** Kränzlin. Pseudobulbs conspicuous; sepals and petals and lip in 1 plane, the lip not very different from the sepals and petals, giving the flower the appearance of a 6-pointed star.

16. **O. maculatum** (Lindley) Lindley

SYNONYM: *Cyrtochilum maculatum* Lindley.

ILLUSTRATION: Cogniaux & Goossens, Dictionnaire Iconographique des Orchidées, Oncidium, t. 19 (1898).

PSEUDOBULBS: 7–10 cm, ovoid, compressed, bearing 2 leaves.

LEAVES: 18–25 × 2.5–5 cm, linear to oblong or oblong-elliptic, obtuse or acute.

RACEME OR PANICLE: erect, up to 1 m.

FLOWERS: fragrant.

Oncidium maculatum – lip.

SEPALS AND PETALS: 1.5–3 cm, oblong-elliptic to oblong-lanceolate, pale yellow or yellow-green blotched with brown.

LIP: oblong, 3-lobed at about the middle, the lateral lobes directed forwards, triangular, the central lobe oblong, notched or rounded at the apex, white with red stripes at the base, yellow at the apex, with a callus of 4 united ridges which end in upright teeth.

COLUMN: wings small and inconspicuous.

DISTRIBUTION: Mexico, Guatemala, Honduras.

HARDINESS: G2.

FLOWERING: summer.

17. **O. hastatum** (Bateman) Lindley

SYNONYM: *Odontoglossum hastatum* Bateman.

ILLUSTRATION: Bechtel et al., Manual of cultivated orchid species, 244 (1981).

Like *O. maculatum* except as below.

LIP: mostly pink with oblong or almost square, outwardly directed lateral lobes, the central lobe ovate, acute.

DISTRIBUTION: S Mexico.

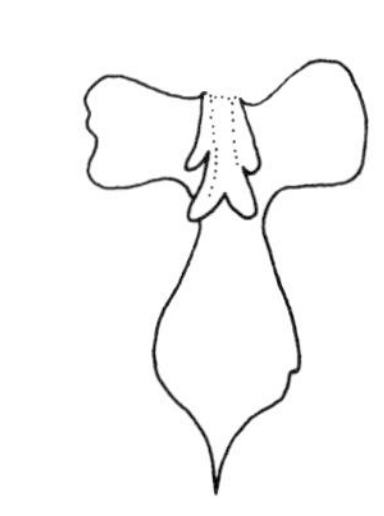

Oncidium hastatum – lip.

Section **Rostrata** Rolfe. Pseudobulbs conspicuous; rostellum elongate, proboscis-like.

18. **O. ornithorrhynchum** Humboldt, Bonpland & Kunth

ILLUSTRATIONS: Edwards's Botanical Register **26**: t. 10 (1840); Botanical Magazine, 3912 (1841); Cogniaux & Goossens, Dictionnaire Iconographique des Orchidées, Oncidium, t. 22 (1899); Bechtel et al., Manual of cultivated orchid species, 247 (1981).

PSEUDOBULBS: 2.5–9 cm, ovoid or ellipsoid, each bearing 2–3 leaves.

LEAVES: 10–40 × 1–3 cm, linear-lanceolate to linear-elliptic, acute.

PANICLE: up to 50 cm, erect.

SEPALS AND PETALS: 7–11 mm, shortly clawed, oblong-elliptic to oblanceolate-elliptic, pink or pinkish purple, obtuse.

LIP: 3-lobed, longer than the sepals, the lateral lobes small, rolled downwards and clasping the bases of the lateral sepals, the central lobe clawed, obovate, deeply notched at the apex, pink or pinkish purple with a small, reddish brown-spotted callus.

COLUMN: wings broadly triangular from a narrow base, finely toothed.

DISTRIBUTION: C America, from S Mexico to Costa Rica.

HARDINESS: G2.

FLOWERING: autumn.

Oncidium ornithorrynchum – lip.

19. O. cheirophorum Reichenbach

ILLUSTRATIONS: Botanical Magazine, 6278 (1877); Lindenia 3: t. 126 (1887); Bechtel et al., Manual of cultivated orchid species, 242 (1981).

PSEUDOBULBS: 3–4 cm, ovoid, each with 1 leaf.

LEAVES: to 15 cm, linear-oblong, acute.

PANICLE: to 25 cm, erect or arching, with many fragrant flowers.

SEPALS AND PETALS: 6–13 mm, ovate, rounded at the apex, yellow.

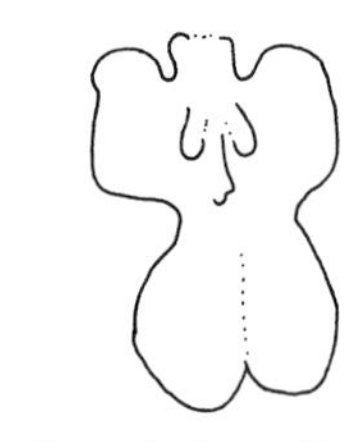

Oncidium cheirophorum – lip.

LIP: 3-lobed, longer than the sepals, the lateral lobes spreading, rounded, the central lobe stalkless, oblong to almost circular, notched at the apex, yellow with a conspicuous white callus with 5 teeth.

COLUMN: wings obovate, entire.

DISTRIBUTION: Costa Rica, Panama, Colombia.

HARDINESS: G2.

FLOWERING: winter.

Section **Concoloria** Kränzlin. Pseudobulbs conspicuous; lateral sepals united; lip entire and bearing a callus made up of 2 or 4 ridges.

20. O. concolor Hooker

ILLUSTRATIONS: Botanical Magazine, 3752 (1840); Cogniaux & Goossens, Dictionnaire Iconographique des Orchidées, Oncidium, t. 32 (1904); Pabst & Dungs, Orchidaceae Brasilienses 2: t. 1989 (1977); Bechtel et al., Manual of cultivated orchid species, 242 (1981).

PSEUDOBULBS: to 5 cm, ovoid, grooved, each bearing 2 leaves.

LEAVES: to 15 cm, linear-lanceolate.

RACEME: with 6–12 flowers, arching.

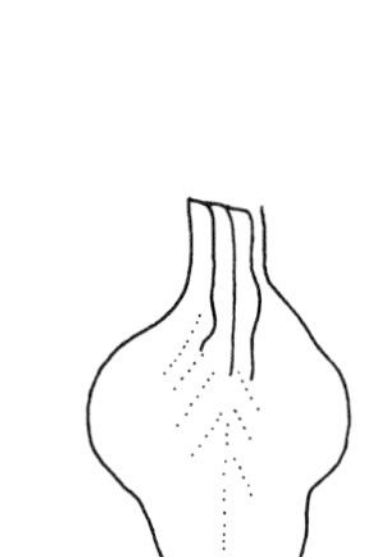

Oncidium concolor – lip.

SEPALS AND PETALS: 2.5–3 cm, oblong-elliptic, acute, yellow, the lateral sepals united for about half their length.

LIP: larger than the sepals, shortly clawed, diamond-shaped, slightly notched at the apex, yellow, bearing 2 red or orange ridges.

COLUMN: wings oblong, entire, directed forwards.

DISTRIBUTION: SE Brazil, N Argentina.

HARDINESS: G2.

FLOWERING: spring.

Section **Barbata** Pfitzer. Pseudobulbs conspicuous; lateral sepals variously united, longer than the lip; callus of lip compound, composed of an uneven number of plates or tubercles.

21. O. longipes Lindley

ILLUSTRATIONS: Botanical Magazine, 5193 (1860); Bechtel et al., Manual of cultivated orchid species, 246 (1981).

PSEUDOBULBS: 2–3 cm, elongate-ovoid, each with 2 leaves.
LEAVES: 1–15 cm, oblong, acute.
RACEME: with 2–6 flowers, loose, scarcely exceeding the leaves.
SEPALS AND PETALS: 1.4–1.8 cm, oblong, rounded and finally pointed at the apex, yellow spotted with red-brown below or dark purple-brown throughout.

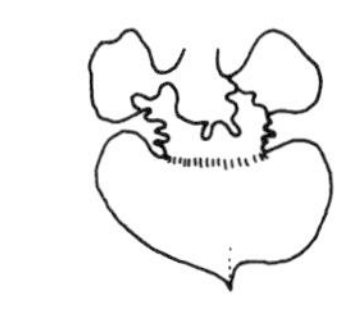

Oncidium longipes – lip.

LIP: 3-lobed, the lateral lobes broadly obovate, attached to the broadly kidney-shaped central lobe by a broad, toothed claw, yellow with a central brown spot, bearing a large, yellow-spotted ridge with 3 teeth at its apex, the central tooth larger and blunter than the other 2.
COLUMN: wings small, toothed.
DISTRIBUTION: SE Brazil.
HARDINESS: G2.
FLOWERING: spring.

Section **Crispa** Pfitzer. Pseudobulbs conspicuous; sepals and petals clawed, crisped, the lateral sepals variously united, the petals more conspicuous than the sepals; lip with a callus formed from an uneven number of tubercles.

22. **O. crispum** Loddiges
ILLUSTRATIONS: Botanical Magazine, 3499 (1836); Edwards's Botanical Register **23**: t. 1920 (1837); Cogniaux & Goossens, Dictionnaire Iconographique des Orchidées, Oncidium, t. 6, 6A (1897); Bechtel et al., Manual of cultivated orchid species, 242 (1981).
PSEUDOBULBS: to 10 cm, ovoid, grooved, each bearing 2 leaves.
LEAVES: to 20 cm, narrowly lanceolate acute.
PANICLE: with many flowers, loose, arching, axis bluish green.
SEPALS: 2.5–3 cm, oblong, crisped, brown with yellow spots towards the base.
PETALS: similar but broader.

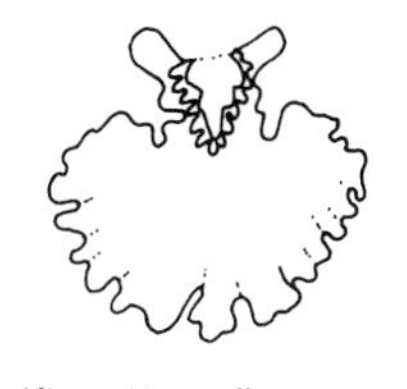

Oncidium crispum – lip.

LIP: a little longer than the sepals, obscurely 3-lobed with very small lateral lobes and a large, kidney-shaped, strongly crisped central lobe which is brown with a yellow centre, bearing a prominent callus with an uneven number of teeth.
COLUMN: wings oblong, finely toothed.
DISTRIBUTION: E Brazil.
HARDINESS: G2.
FLOWERING: autumn–spring.
Autumn-flowering material with rather small flowers may prove to be *O. enderianum* Masters (see Fowlie, Orchid Digest **43**: 190–45, 1979).

23. **O. forbesii** Hooker
ILLUSTRATIONS: Botanical Magazine, 3705 (1839); Cogniaux & Goossens, Dictionnaire Iconographique des Orchidées,

Oncidium forbesii – lip.

Oncidium, t. 1 (1896); Pabst & Dungs, Orchidaceae Brasilienses **2**: t. 2024 (1977); Bechtel et al., Manual of cultivated orchid species, 244 (1981).
Like *O. crispum* except as below.
RACEME: unbranched.
SEPALS AND PETALS: chestnut brown with yellow marbling around the margins.
COLUMN: with entire, red-spotted wings.
DISTRIBUTION: E Brazil
FLOWERING: autumn.

24. **O. gardneri** Lindley
ILLUSTRATIONS: Pabst & Dungs, Orchidaceae Brasilienses **2**: t. 2013 (1977); Bechtel et al., Manual of cultivated orchid species, 244 (1981).
Like *O. crispum* except as below.
SEPALS AND PETALS: oblanceolate, brown with yellow stripes on the margin.
LIP: with very small lateral lobes, the central lobe *c*. 3 × 4.5 cm, mostly yellow with a zone of red-brown spots around the margin, brown at the base.
COLUMN: with entire wings.
DISTRIBUTION: E Brazil.
FLOWERING: summer.

Oncidium gardneri – lip.

25. **O. marshallianum** Reichenbach
ILLUSTRATIONS: Botanical Magazine, 5725 (1868); Cogniaux & Goossens, Dictionnaire Iconographique des Orchidées, Oncidium, t. 8 (1897); Pabst & Dungs, Orchidaceae Brasilienses **2**: t. 2008 (1977).
PSEUDOBULBS: 7–10 cm, ovoid, each with 2 leaves.
LEAVES: 20–30 cm, oblong-lanceolate, acute.
PANICLE: to 1.5 m with many flowers, erect.
SEPALS: 2.3–2.5 cm oblong-obovate, greenish yellow with brown spots or transverse stripes.
PETALS: to 3 cm, notched at the apex, golden yellow with dark brown spots in the centre.
LIP: 3-lobed with small, yellow and red-spotted lateral lobes and a large, broadly kidney-shaped, notched, yellow central lobe bearing a large, many-toothed callus.
COLUMN: wings oblong, entire.
DISTRIBUTION: E Brazil.
HARDINESS: G2.
FLOWERING: spring–summer.

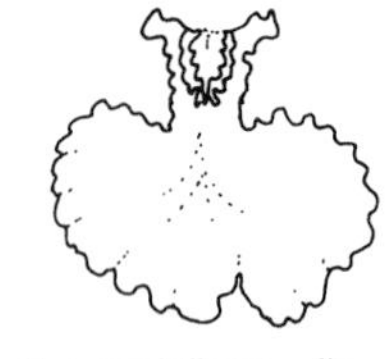
Oncidium marshallianum – lip.

Section **Synsepala** Pfitzer. Pseudobulbs conspicuous; sepals and petals similar, inconspicuous, the lateral sepals united for part of their length; lip with a callus made up of an uneven number of tubercles.

26. O. varicosum Lindley

ILLUSTRATIONS: Cogniaux & Goossens, Dictionnaire Iconographique des Orchidées, Oncidium, t. 18, 18A (1898); Schlechter, Die Orchideen, edn 2, t. 11 (1927); Pabst & Dungs, Orchidaceae Brasilienses **2**: t. 2029 (1977).

PSEUDOBULBS: 8–10 cm, elongate-ovoid, grooved, each bearing 2 leaves.

LEAVES: 15–25 cm, oblong, acute.

PANICLE: 80–150 cm, with many flowers, loosely branched.

FLOWERS: very variable in size.

SEPALS AND PETALS: 5–7 mm, oblong, lateral sepals united for about half their length, yellowish green spotted with brown.

LIP: 3-lobed, much larger than the sepals, the lateral lobes small, rounded, the central lobe 2.5–4 × 3.5–5.5 cm, 3-notched towards the apex to produce 4 small lobes, golden yellow, bearing a callus with several teeth.

COLUMN: with entire wings.

DISTRIBUTION: E & C Brazil.

HARDINESS: G2.

FLOWERING: winter.

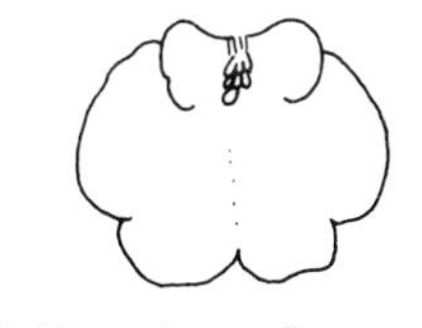

Oncidium varicosum – lip.

27. O. flexuosum Loddiges

ILLUSTRATIONS: Botanical Magazine, 2203 (1821); Reichenbach, Flora Exotica **2**: t. 94 (1834); Pabst & Dungs, Orchidaceae Brasilienses **2**: t. 2024 (1977); Bechtel et al., Manual of cultivated orchid species, 243 (1981).

PSEUDOBULBS: 3–4 cm, compressed, rather distant, each bearing 1 or 2 leaves.

LEAVES: 10–20 cm, oblong, somewhat acute.

PANICLE: to 80 cm, with many flowers.

SEPALS AND PETALS: oblong, acute, yellow with red-brown spots at the base, the lateral sepals united at the base.

LIP: 3-lobed, much longer than the sepals, the lateral lobes small, curved erect, the central lobe 1.5–2 × 1.5–1.8 cm, clawed, kidney-shaped, notched at the apex, golden yellow, bearing a many-toothed callus.

COLUMN: wings small, entire.

DISTRIBUTION: SE Brazil, Paraguay & Argentina.

HARDINESS: G2.

FLOWERING: autumn–winter.

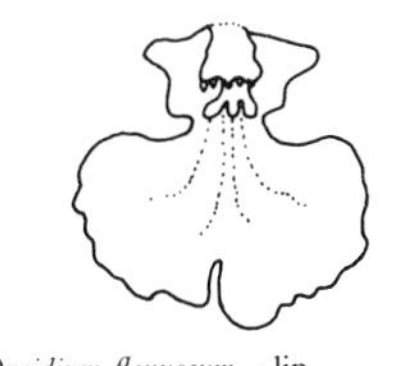

Oncidium flexuosum – lip.

Section **Excavata** Kränzlin. Pseudobulbs conspicuous; sepals and petals clawed, the lateral sepals free, the petals broader and more conspicuous than the sepals; callus of lip with an uneven number of tubercles.

28. O. excavatum Lindley

ILLUSTRATIONS: Botanical Magazine, 5293 (1862); Cogniaux & Goossens, Dictionnaire Iconographique des Orchidées,

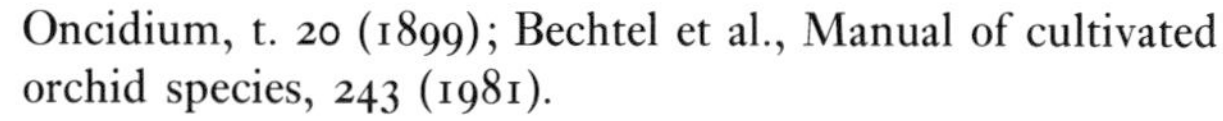

Oncidium, t. 20 (1899); Bechtel et al., Manual of cultivated orchid species, 243 (1981).

PSEUDOBULBS: 7–12 cm, oblong-ovoid, somewhat compressed, each with 1 or 2 leaves.

LEAVES: 30–50 × 2.5–4 cm, linear-oblong, acute.

PANICLE: to 1.5 m, erect, with many flowers.

SEPALS: 1.2–1.6 cm, oblanceolate-oblong, rounded and abruptly acute, the laterals narrower than the upper, all yellow spotted with red towards the base.

PETALS: similar to sepals, but slightly longer and much broader.

LIP: longer than the sepals, 3-lobed, the lateral lobes small, the central lobe 1.2–2.2 × 2 cm, clawed, yellow, red towards the base with a complex toothed and grooved callus.

COLUMN: wings oblong, slightly notched.

DISTRIBUTION: Ecuador, Peru.

HARDINESS: G2.

FLOWERING: spring.

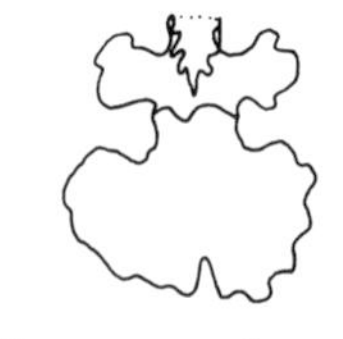

Oncidium excavatum – lip.

Section **Oblongata** Kränzlin. Pseudobulbs conspicuous; lateral sepals free; bracts small and inconspicuous, scale-like.

29. O. ampliatum Lindley

ILLUSTRATIONS: Edwards's Botanical Register 20: t. 1699 (1835); Fieldiana **30**: 859 (1961); Dunsterville & Garay, Venezuelan orchids illustrated **2**: 242–3 (1961); Bechtel et al., Manual of cultivated orchid species, 240 (1981).

PSEUDOBULBS: 3–7.5 cm, spherical or pear-shaped, compressed, each bearing 1 leaf.

LEAVES: 8–50 × 2.5–10 cm, variable in shape, obtuse or broadly rounded.

PANICLE: to 80 cm with many flowers.

SEPALS: upper 7–9 mm, obovate, hooded, the lateral sepals longer and narrower, all yellow with small brown spots.

PETALS: distinctly clawed, yellow with small brown spots.

LIP: longer than the sepals, 3-lobed, the lateral lobes small, the central shortly clawed, kidney-shaped, notched at the apex, yellow, with a small callus which has 2 broad lateral lobes and 3 smaller lobes in front.

COLUMN: wings toothed.

DISTRIBUTION: tropical America from Guatemala to Peru.

HARDINESS: G2.

FLOWERING: spring.

Oncidium ampliatum – lip.

30. O. tigrinum Llave & Lexarza

ILLUSTRATIONS: Illustration Horticole **22**: t. 221 (1875); Cogniaux & Goossens, Dictionnaire Iconographique des Orchidées, Oncidium, t. 4, 4A, 4B (1897). Bechtel et al., Manual of cultivated orchid species, 249 (1981).

PSEUDOBULBS: 7–9 cm, spherical, with 2 leaves.
LEAVES: 20–30 cm, narrowly oblong, acute, thinly leathery.
PANICLE: to 80 cm, little branched, with many fragrant flowers.
SEPALS AND PETALS: similar, 2.5–3.5 cm, spreading, oblong to obovate, yellow with heavy brown spotting and marbling.
LIP: much larger than the sepals, 3-lobed, the lateral lobes small, the central lobe clawed, spherical to kidney-shaped, golden yellow, with a large, 3-lobed callus.
COLUMN: wings oblong, entire.
DISTRIBUTION: Mexico.
HARDINESS: G2.
FLOWERING: winter.

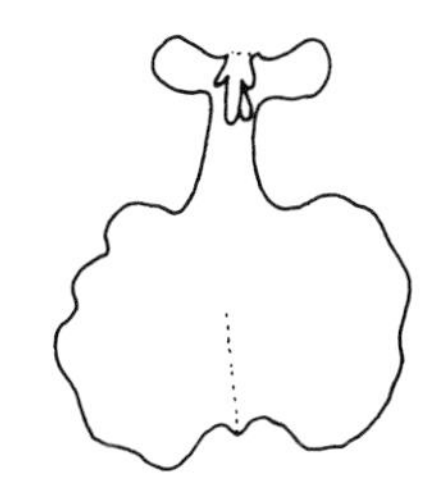
Oncidium tigrinum – lip.

31. **O. leucochilum** Bateman
ILLUSTRATION: Cogniaux & Goossens, Dictionnaire Iconographique des Orchidées, Oncidium, t. 24 (1899).
PSEUDOBULBS: 5–13 cm, ovoid, compressed, each with 1 or 2 leaves.
LEAVES: 10–60 × 1.5–4.5 cm, oblong, obtuse to acute.
PANICLE: to 3 m.
SEPALS AND PETALS: 1.3–2.3 cm, similar, oblong-elliptic to oblanceolate, bright green or greenish white blotched with red.
LIP: larger than the sepals, 3-lobed, the lateral lobes small, oblong to ovate, the central clawed, transversely oblong to kidney-shaped, notched at the apex, white, sometimes with red lines, with a purplish callus which ends in 5–9 slender teeth, some of them upcurved.
COLUMN: wings oblong, irregularly scalloped.
DISTRIBUTION: Mexico, Guatemala, Honduras.
HARDINESS: G2.
FLOWERING: winter–spring.

Oncidium leucochilum – lip.

Section **Planifolia** Bentham & Hooker. Pseudobulbs conspicuous; bracts conspicuous, as long as or little shorter than the stalked ovaries; lip with a callus composed of an uneven number of tubercles.

32. **O. sphacelatum** Lindley
ILLUSTRATIONS: Edwards's Botanical Register **28**: t. 30 (1842); Dunsterville & Garay, Venezuelan orchids illustrated **4**: 196–7 (1966); Bechtel et al., Manual of cultivated orchid species, 248 (1981).
PSEUDOBULBS: to 20 cm, ovoid-ellipsoid, tapering, each with 2 leaves.
LEAVES: to 100 × 3.5 cm, linear-oblong or linear-lanceolate, acute.
PANICLE: to 1.5 m, with many flowers, erect.

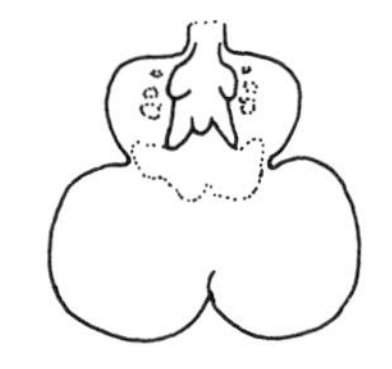

Oncidium sphacelatum – lip.

SEPALS: 10–20 × 3–6.5 mm, clawed, elliptic to elliptic-obovate, yellow with reddish brown blotches.

PETALS: similar to sepals but not clawed.

LIP: about as long as the sepals, 3-lobed, wider across the central lobe than across the lateral lobes, central lobe large, semicircular, notched at the apex and often at the sides as well, yellow, with a downy callus of 5–7 tubercles, some of which are divergent.

COLUMN: with narrow, brown-bordered, slightly scalloped wings.

DISTRIBUTION: C America, Venezuela.

HARDINESS: G2.

FLOWERING: spring.

33. O. wentworthianum Lindley

ILLUSTRATIONS: Paxton's Flower Garden, f. 127 (1851–52); Die Orchidee 31 (5): clxxxv–clxxxvi (1980); Bechtel et al., Manual of cultivated orchid species, 249 (1981).

PSEUDOBULBS: 7–10 cm, ovoid-ellipsoid, compressed, often mottled with brown, each bearing 2 leaves.

LEAVES: 13–35 × 1.5–3 cm, linear-oblong to lanceolate.

PANICLE: up to 1 m, with many flowers, arching.

SEPALS AND PETALS: 1.4–2.2 cm, spreading or somewhat reflexed, elliptic or elliptic-obovate, deep yellow irregularly blotched with brown.

Oncidium wentworthinanum – lip.

LIP: 3-lobed, about as long as the sepals, wider across the lateral lobes than across the central lobe, the lateral lobes small, curved forwards, scalloped on their margins, the central lobe reversed-heart-shaped, deeply notched at the apex, yellow, with a fleshy reddish brown callus consisting of 3 teeth flanked by ridges.

COLUMN: wings triangular, finely scalloped, bordered with red-brown spots.

DISTRIBUTION: Mexico, Guatemala.

HARDINESS: G2.

FLOWERING: summer.

166. SIGMATOSTALIX Reichenbach

A genus of about 12 species from C & S America; only 1 is generally grown. Cultivation as for *Odontoglossum* (p. 468).

Literature: Kränzlin, F., *Das Pflanzenreich* **80**: 301–12 (1922).

HABIT: small, rhizomatous epiphytes.

PSEUDOBULBS: simple, clustered or distant, each bearing 1 or 2 leaves.

LEAVES: very narrow, grass-like, folded when young.

RACEME: erect, arising from the bases of the pseudobulbs (rarely branched).

SEPALS AND PETALS: similar, widely spreading to reflexed.

LIP: conspicuously clawed, usually simple, with a callus near the base.

COLUMN: very slender, arched, widened above, without a foot.

POLLINIA: 2, ovoid, stipe oblong, expanded above with its sides upturned, longer than the pollinia and the small, almost circular viscidium.

1. **S. graminea** (Poeppig & Endlicher) Reichenbach

ILLUSTRATION: Reichenbach, Xenia Orchidacearum 1: t. 8 (1854).

PSEUDOBULBS: 7–12 mm, clustered, oblong-cylindric to cylindric-ellipsoid.

LEAVES: 3–5 cm × 2 mm, narrowly linear, acute or obtuse, obliquely notched at the apex.

RACEME: erect or arching.

SEPALS: 2–2.5 mm, oblong or elliptic-lanceolate, acute.

PETALS: slightly broader than sepals, both yellow with purple spots.

LIP: shortly and broadly clawed, blade ovate to almost square, notched, with rounded angles at the base.

DISTRIBUTION: Peru.

HARDINESS: G2.

INDEX